HEAVY METAL POLLUTION

SAMUEL E. BROWN
AND
WILLIAM C. WELTON
EDITORS

Nova Science Publishers, Inc.
New York

For permission to use material from this book please contact us:
Telephone 631-231-7269; Fax 631-231-8175
Web Site: http://www.novapublishers.com

LIBRARY OF CONGRESS CATALOGING-IN-PUBLICATION DATA

Heavy metal pollution / Samuel E. Brown and William C. Welton, editors.
p. cm.
Includes bibliographical references and index.
ISBN 978-1-60456-899-8 (hardcover)
1. Heavy metals--Environmental aspects. I. Brown, Samuel E. II. Welton, William C.
TD196.M4H286 2008
571.9'543--dc22
2008024682

Published by Nova Science Publishers, Inc. ✢ New York

CONTENTS

PREFACE

Chapter 1 - In this review, the authors analyse the main and best studied aspects of the ciliate-heavy metal interactions. Ciliates are eukaryotic microorganisms with several advantages to be used in toxicological bioassays, being a good alternative to animal tests. They discuss the advantages and disadvantages of the different types of bioassays using ciliates, the diversity of experimental conditions and methodological approaches. The main ecotoxicological conclusions from heavy metal-ciliate interactions are, likewise, discussed. Several heavy metal cellular effects on ciliated protozoa are similar to those found in stress differentiation processes like encystment. Bioaccumulation seems to be the main heavy metal resistant mechanism present in ciliates, although other possibilities are also reported. This bioaccumulation process involves the presence of metallothioneins, which have been studied in several *Tetrahymena* species. These ciliate proteins present unique features with regard classic metallothioneins. The promoters of some ciliate metallothionein genes can be excellent molecular tools to be used in specific gene constructions to design whole-cell biosensors. Reactive oxygen species formation and some aspect of glutathione metabolism in ciliates under heavy metal stress are also analysed. Finally, the authors report that heavy metals induce a ciliate apoptotic process very similar to the higher eukaryotic apoptosis, and the possibility to use ciliates as whole-cell biosensors to assess heavy metal polluted samples is also discussed.

Chapter 2 - Cellular membranes are basically impermeable to ions and have developed specific pathways (transporters, channels or pumps) to facilitate metal translocation. These physiological carriers are not ideally selective and their specificity spectrum may include xenobiotic species, such as toxic metals whose availability in the environment has increased enormously with the onset of the industrial era. Competition between divalent endogenous and noxious metal ions at specific sites on membrane transport proteins and enzymes is referred to as "ionic mimicry".

In this chapter the author will present some studies on the permeation mechanisms through mammalian cell plasma membranes of lead (Pb) and cadmium (Cd), two metals whose toxicity has been linked to their putative ability to mimic calcium (Ca) and zinc (Zn) at specific binding sites. Both metals can permeate through mammalian cell membranes taking advantage of different Ca channels, but, while Cd appears to take advantage mainly of the same pathways as Ca, Pb is also rapidly taken up in different cell types by passive transport systems that are distinct from Ca channels and independent of specific stimuli.

To further elucidate the role of voltage-dependent Ca channels (VDCC) in Cd uptake, the author and colleagues compared the effect of this metal in two Chinese hamster ovary

(CHO) cell lines, a wild type and modified cell line, which was permanently transfected with an L-type VDCC. Both cultures were subjected to brief (30-60 min) exposure to 50-100 μM Cd in serum-free culture medium. Cell death was evident after 18-24 h with comparable features in both cell lines. Although VDCC represent a pathway of Cd entry and participate in Cd-induced toxicity, as demonstrated by the effect of DHP modifiers, expression of L-type Ca channels is not sufficient to modify Cd accumulation and sensitivity to a toxicologically significant extent. This study confirmed that both Cd and Pb can take advantage of VDCC to permeate the membrane, but these transport proteins are not the only, and frequently not the most important, pathways of permeation.

Chapter 3 - Heavy metals are natural constituents of the earth crust; consequently they are present practically in all environments (air, soils, sediments, waters) as a result of erosion and weathering of parent rocks (Sillanpaa, 1972; Bohn et al., 1985). Nevertheless, the metal introduction in ecosystems occurs not only through natural sources, but also through anthropogenic sources, such as smelters, mining, power station industry and the application of metal-containing pesticides, fertilizers and sewage sludge (Sposito et al, 1982; Brimblecomebe, 1994; McGrath et al., 1995; Chuan et al., 1996; Cambier et al., 1997; Akhter and Al-jowder, 1997; Dijkstra, 1998; Ranjbar, 1998; et al., 2000; Pichtel et al., 2000; Odukoya et al., 2000; Robinson et al., 2001; Abdul-Wahab and Yaghi, 2003).

Anthropogenic atmospheric inputs contribute considerably in soil and sediment contamination (Allowey, 1990). Moreover, long-range atmospheric transport is one of the main factors for dispersion and deposition of main pollutants, including Pb, Zn, Cu, Cd and As (Steiness et al., 1997; Loska et al., 2003) and can lead to pollutant deposition even in supposedly pristine areas (De Vries et al., 2002).

At trace levels some heavy metals as Zn, Cu, Ni, and Cr play an important role in organisms and human life (Mokhtar et al., 1991; Juvanovic et al., 1995; Lapitajs et al., 1995) whereas others, such as Cd, Hg and Pb have no known biological and/or physiological function. However, all metals above reported present toxicity at relative low concentrations (Lisk, 1972; Gadd, 1992; Abdul Rida and Bouchè, 1995).

Heavy metals, unlike other toxic pollutants, are not biodegradable in the environment. As consequence, they are able to accumulate in soils, sediments and organisms and can be considered as a major group of compounds involved in soil and sediment pollution (Yaron et al., 1996). Heavy metal concentration levels higher than natural values in soil and sediment are sign of a direct potential toxicity to biota and indirect threats to human health from groundwater contamination and the possible entry into the food chain.

At a stretch the authors report two monitoring studies concerning heavy metal presence in two different environmental compartments (soil and sediment).

Chapter 4 - The presence of heavy metals in fossil fuels or other materials, such as certain types of biofuels and wastes that present enough energy content for its use as fuels, is a problematic issue. There are several national and international programs to control its emissions, namely that of Hg, Cd and Pb, although much more metals are emitted every day from industrial sources. Heavy metals are not destroyed at high temperatures; they suffer complex transformations and different partitioning among the several streams of thermal processes, depending on the operational conditions employed and materials composition. Metals may volatilize at high temperatures and be carried out as vapors or condense as fine particles contributing to atmospheric pollution or may remain attached to the ashes. Atmospheric pollution may be controlled with efficient air pollution control devices or dedicated systems in the case of Hg, but ashes may present high quantities of metals that require appropriate management or reutilization. Management costs depend on their hazardous nature, which are evaluated with appropriate methodologies. Apart from the common use of fossil fuels and Municipal Solid Wastes, two cases are of enormous interest nowadays and will be discussed as case studies: one is the recovery of energy from sewage sludge, either through combustion or gasification, and the other is the use of automobile shredder wastes (ASR) from end life vehicles dismantling. The study of the combustion of ASR in a fluidized bed pilot showed that its energy recover is highly attractive, however, as the residue contains high quantities of heavy metals and chlorine, emissions may be significant, especially that of Hg, thus requiring efficient flue gas treatment. Ash produced retained most of the metals but evaluation of leachability showed that ashes behave as hazardous materials, essentially because of the release of hexavalent Cr, chlorides and sulfates. In the case of sewage sludge, which also contains significant quantities of mineral matter, including mercury, combustion in a fluidized bed pilot led to lower metal emissions because retention in ashes was higher. Evaluation of leachability revealed that metals did not leach, but the release of sulfate and chloride was high in fly ashes. When sewage sludge was gasified to produce syngas, instead of being directly burned, the reducing environment increased the volatilization of some metals, thus the combustible gas produced requires further removal of metals before being used. Metals that were strongly retained in the gasification residues did not leach out and release of sulfates was minimized because gasification process does not favor sulfur oxidation.

Chapter 5 - This study reports field and laboratory experiments in which the impact of cadmium on the growth characteristics of several diatom taxa from multispecific natural biofilms was assessed under non-limiting nutrient availability. The diatom community structures were determined for each time and each treatment, and coinertia analysis with the physicochemical characteristics of the water revealed the strong influence of many "uncontrolled" factors (differentiating field assemblages and microcosms), but also of metal contamination and duration of exposure. Under non-contaminated conditions, the inverse relationship between specific growth rates (μmax) and cell sizes was in accordance with the allometry studies reported by several authors. Metal contamination influenced the growth responses of the diatoms but, under these toxic conditions, the growth rates were not necessarily correlated with biovolumes.

Microcosm and *in situ* results showed marked differences; examination of the standardized variations [ratio of (treated μmax) - (control μmax) over control μmax] revealed that most of the species considered were saprophilous and metal-tolerant, and the various taxa had slightly increased growth rates under cadmium contamination. These trends, similar in both laboratory and *in situ* experiments, underline the fact that microcosm experiments may be valuable alternatives to field studies, when considered with caution.

Chapter 6- Heavy metal pollution is a growing environmental problem in marine coastal waters; for this reason the monitoring of water pollution needs the availability of test-species able to furnish reliable and cheap results. From this point of view, copepods are important test organisms owing to their wide distribution and for their key-position within the food web. As *Tigriopus fulvus* seems to be a promising target-species in ecotoxicology, in this chapter the toxicity of different concentrations of four heavy metals (chrome, zinc, cadmium, mercury) was evaluated. Treatments were performed after 24, 48 and 72 hours on nauplii and after 24, 48, 72 and 96 hours on adult *Tigriopus fulvus*; the percent survival and LC_{50} were evaluated and a method to obtain same-aged nauplii from detached egg sacs from ovigerous females is described. The results showed that mercury is the most dangerous metal for *T. fulvus* nauplii (LC_{50} 72hrs=0.39±0.02) and adults (LC_{50} 96hrs=0.34±0.13). Zinc and cadmium exhibited similar toxicity and chrome was the least toxic metal. All metals affected seriously egg sac and nauplii production of females with a 30% reduction of nauplii produced per egg sac; cadmium and mercury highly affected survival, reproduction and hatching of nauplii. On the whole, the data were repeatable and *Tigriopus fulvus* showed a satisfactory sensitivity to tested metals. Therefore, for its sensitivity and its biological and ecological characteristics, as well as for its adaptability to laboratory conditions and for its easy and cheap maintaining *Tigriopus fulvus* can be eligible as a test-species highly suitable in aquatic ecotoxicology.

Chapter 7 - The goal of this study was to synthesize a new set of Lauryl poly-glucoside as an economic surfactant, which is produced exclusively with renewable sources having good dermatological properties and absence of eco-toxic properties. The authors have synthesized alkyl glucoside surfactant by direct glucosidation reaction by reacting glucose and n-dodecanol. The suitable conditions for the reaction were obtained, i.e., reaction temperature (120□C), mole ratio of catalyst (glucose to n-dodecanol 0.05:1:5) and reaction time (5 h). The yield of the product was found 54%.

Chapter 8 - Heavy metal pollution is one of the current most troublesome environmental problems due to the widespread use of metals for industrial and agricultural purposes. Heavy metals discharged often from the industrial operations and sewage sludge, cause a phenomenal increase in the extent of heavy metals and thus become a major global threat to the soil environment. Moreover, the introduction of pollutants to soil environment can often overwhelm the self cleaning capacity of recipient ecosystem and thus, result in their accumulation to even harmful level and consequently makes the soil unsuitable for sustainable agriculture. Legumes are used mainly for human consumption as well as for animal feeds and serve as a rich source of protein and contribute a substantial amount of N to the agronomic soils. They are grown in tropical and semi arid tropical countries. Once grown in the contaminated soils, the metals are taken up by plant organs to an undesired level, which in turn limits the growth and physiological processes, like the synthesis of macromolecules including protein and consequently affect the legume yields severely. However, the toxicity of heavy metals to legumes could be reduced/prevented through the application of phytoremediation technology that involves the use of plants to remove, destroy or sequester hazardous substances from the contaminated environment. Also, many plants possess efficient enzymatic and non-enzymatic antioxidant defense system that allows the scavenging of reactive oxygen species and provides protection to plants from oxidative damage. The antioxidant enzymes have been reported to be produced by plants under metal stress that in turn, reduces the toxicity of metal to plants. With these considerations, this chapter focuses mainly on the two aspects of metal in relation to legumes- (i) response of legumes to metal stressed environment and (ii) different strategies adopted by plants to protect itself from metal toxicity, which may help in assessing the ecological significance of metal stress.

Chapter 9 - Heavy metals are significant contaminants in aqueous systems. All heavy metals can produce toxicity when ingested in sufficient quantities, but there are several important ones such as lead, mercury, copper, cadmium, arsenic, nickel and silver. These heavy metals are so pervasive and can produce toxicity at very low concentrations. Moreover, they may build up in biological systems and become a significant health hazard. Humic substances are macromolecules that naturally occur in all environments in which vegetation matters are present. Humic acid (HA) plays a fundamental role in many ecosystems since it interacts with toxic metal ions present in the system, resulting in a decrease in the bioavailability of such ions. Moreover, the availability of humic acid in water can react with other chemical compounds, such as chlorine to form trihalomethanes (including chloroform) and causes an increasing risk of cancer and may be linked to heart, lung, kidney, liver, and central nervous system damage.

Adsorption is approved as an effective and simple method for water and wastewater treatment. Many adsorbents then are developed for use in adsorption process such as montmorillonite, peat, activated carbon, zeolite, etc. In this research, humic acid and some heavy metals (Cu and Pb) were mainly selected for adsorption study. In the sorption experiment, several adsorbents such as synthesised MCM-22 zeolite (SZ), natural zeolite (NZ), powdered activated carbon (PAC), and fly ash (FA), were selected to examine the application for HA and heavy metal removal both in individual and simultaneous adsorption processes. The characteristics and interactions of the adsorbents with HA and heavy metals were systematically studied by batch laboratory experiments. The results show that the four adsorbents exhibit varying adsorption behaviour depending on the adsorbent characteristics. The synthesised zeolite MCM-22 presents higher adsorption for heavy metals and humic acid. The presence of HA will affect heavy metal adsorption on solids. Complexation of HA with metals will result in an increase of metal ions on adsorbent. However, the competitive adsorption can also occur which results in a decrease of HA and metal adsorption on solids.

Chapter 10 - The binding of heavy metal cations in bacterial EPS is of significant ecological relevance: it may cause an undesired accumulation of heavy metals in naturally occurring biofilms, but may as well be used to remove heavy metals from polluted waste water in a biological approach. In an experimental study, the interaction between charged polysaccharides from bacterial EPS and bivalent cations was investigated by conductometry, viscosimetry, and nuclear magnetic resonance spectroscopy. It is shown that bacterial alginate, xanthan and hyaluronic acid bind various bivalent cations (Ca^{2+}, Mg^{2+}, Mn^{2+}, Fe^{2+}, Cu^{2+}, Zn^{2+}, Cd^{2+} and Pb^{2+}) in aqueous solutions. Molar relations between the bivalent cations and the disaccharide units in the resulting complexes were determined as well as their relative stabilities. Regarding the structure of the dissolved chain molecules, the addition of the cations has opposite effects on the polysaccharides: in case of alginate, the bivalent cations preferably form intermolecular cross-links, leading to the formation of a growing molecular network accompanied by the increase of the solution viscosity. In case of xanthan and hyaluronic acid, an increasing cation concentration is connected to a growing number of intramolecular links, reduced hydrodynamic radii and decreasing solution viscosities. Based on these data and results from NMR studies using a paramagnetic center, possible structures of the polysaccharide-cation complexes are proposed.

Chapter 11 - Phytoremediation is a promising method for the treatment of media contaminated with toxic heavy metals. The extraction efficiencies of other technologies are often ineffective due to the low metal concentrations present. Certain plants, such as *Thlaspi caerulescens*, *Brassica juncea*, and *Helianthus annuus* are classified as hyperaccumulators since they are able to uptake much higher than average levels of metal ions (by a factor of 10-100) without exhibiting symptoms of toxicity. However, how such ability might be acquired is poorly understood. One possibility is that when plants are exposed to low levels of toxic metal, over time they adapt to increase their capacity to sequester metals in their leaves or other tissues by inducing metal binding proteins. This chapter will review phytoremediation efficiency of hydroponic and soil based plants and the effect of certain metal chelators. Efficiency will be based on the total metal uptake, selectivity, and sequestration location of the heavy metals. Preliminary data on the isolation and characterization of proteins used to bind and transport the specific metals will also be discussed. The identification of the proteins and possible enzymes involved in the root-to-shoot transport or regulation of metal sequestration should provide researchers with novel targets for engineering metal accumulation.

Chapter 12 - Biosorption of Pb(II), Cd(II) and Zn(II) from binary metal solutions onto algal derived biomass was investigated in a batch system at pH 5.3 and 20°C. Discrete and continuous equilibrium models were able to predict equilibrium results at different equilibrium concentrations and pH values. The following sequence was obtained for the maximum uptake capacity and equilibrium affinity constants of metal ions to the active sites: Pb(II) > Cd(II) > Zn(II). The "iso-concentrations" cuts from the three dimensional sorption isotherms, show that the uptake capacity of Cd(II) and Zn(II) decreased drastically in the presence of Pb(II). On the other hand, for the Cd(II)/Zn(II) system it was concluded that Zn(II) uptake capacity is more sensitive to the presence of Cd(II) than the opposite. From discrete model the Langmuir affinity parameter, K_L, can be determined as a function of pH and co-cation concentration. The K_L values for Cd(II) and Zn(II) decreased drastically in the presence of Pb(II), showing a higher affinity of the binding sites to Pb(II).

Chapter 13 - Motor neuron disorders are progressive neurodegenerative diseases involving the motor neurons of at least one of the following areas: cerebral cortex, brainstem and spinal cord. Amyotrophic lateral sclerosis is the most common motor neuron disorder of adult onset. Even in cases of a familial inheritance of the pathologic trait that suggests a genetic pattern, the pathogenesis of this affection often remains unclear. Based on epidemiological surveys and similitudes with toxic disorders, environmental factors should be considered.

Heavy metals have been suspected owing to their proved neurotoxicity. Some heavy metals are key constituents of metalloproteins and their accumulation leads to systemic disease with neurologic symptoms. Such heavy metals might enter the nervous system by different pathways such as inhalation, digestion, skin absorption, and retrograde axonal transport.

ALS cases related to heavy metals intoxication and professional exposure have been reported. Some epidemiological surveys showed that occupational exposure to heavy metals such as mercury or lead could be considered as a risk factor for the development of motor neuron disorder. This relationship between heavy metals intoxication and motor neuron disease was also underscored through neuropathological examination that showed accumulation of metals within cortical and spinal motor neurons. Experimental studies based on autometallography analysis demonstrated such accumulation in mice or rats after heavy metal exposure. Animal studies or experiments also demonstrated the pathological role of metals in the motor neuron homeostasis.

Chapter 14 - Bismuth electrodes are very promising in the field of electrochemical analysis. Bismuth film electrodes (BiFEs) can be formed on the same substrates as mercury, which is used most frequently for simultaneous determination of heavy metals (such as Cu(II), Pb(II), Cd(II), Zn(II)) by anodic stripping voltammetry (ASV), and have a potential to replace toxic mercury. The most important results with bismuth film and mercury hemispherical microelectrodes that the authors reached in this field on different substrates are summarized in this work. High conductive (0.008 - 0.024 Ωcm) p- or n-type silicon substrates have been investigated as a support for mercury plating. The influence of many parameters on mercury plating has been studied. The stripping current response of trace heavy metals (Cu(II), Pb(II) and Cd(II) in concentration 3.3×10^{-8} mol/L of each) has been demonstrated by square wave (SW) ASV. Mercury-plated as well as bismuth-plated nitrogen-doped diamond-like carbon (DLC) microelectrode arrays were investigated for Cu(II), Pb(II) and Cd(II) determination by differential pulse (DP) ASV or SW ASV. The current responses of all metals were linear with concentration in the range from 1.6×10^{-8} to 8.3×10^{-8} mol/L (in the case of mercury) and 2×10^{-8} to 1.2×10^{-7} mol/L (in the case of bismuth). The stripping current responses of the bare nitrogen-doped DLC microelectrode array reached under the same conditions were much lower in comparison with the mercury or bismuth plated arrays. Interesting results have been obtained on a single pencil-lead graphite electrode with a bismuth film. Detection limits 2.4×10^{-9} mol/L for Pb(II), 2.9×10^{-9} mol/L for Cd(II) and 1.2×10^{-8} mol/L for Zn(II) were estimated in this case. Concerning the design of electrodes, the authors used a single disc graphite electrode (0.5 mm in diameter) or microelectrode arrays consisting of 50 625 microdiscs with 2 μm in diameter and interelectrode distances of 20 μm.

In: Heavy Metal Pollution
Editors: S. E. Brown and W. C. Welton

ISBN: 978-1-60456-899-8

Chapter 1

CILIATES AS CELLULAR TOOLS TO STUDY THE EUKARYOTIC CELL: HEAVY METAL INTERACTIONS

Juan C. Gutiérrez**, A. Martín-González, S. Díaz, F. Amaro, R. Ortega, A. Gallego and M.P. de Lucas

Dpto. Microbiología-III. Facultad de Biología, C/. José Antonio Novais
2 Universidad Complutense (UCM), 28040 Madrid, Spain.

ABSTRACT

In this review, we analyse the main and best studied aspects of the ciliate-heavy metal interactions. Ciliates are eukaryotic microorganisms with several advantages to be used in toxicological bioassays, being a good alternative to animal tests. We discuss the advantages and disadvantages of the different types of bioassays using ciliates, the diversity of experimental conditions and methodological approaches. The main ecotoxicological conclusions from heavy metal-ciliate interactions are, likewise, discussed. Several heavy metal cellular effects on ciliated protozoa are similar to those found in stress differentiation processes like encystment. Bioaccumulation seems to be the main heavy metal resistant mechanism present in ciliates, although other possibilities are also reported. This bioaccumulation process involves the presence of metallothioneins, which have been studied in several *Tetrahymena* species. These ciliate proteins present unique features with regard classic metallothioneins. The promoters of some ciliate metallothionein genes can be excellent molecular tools to be used in specific gene constructions to design whole-cell biosensors. Reactive oxygen species formation and some aspect of glutathione metabolism in ciliates under heavy metal stress are also analysed. Finally, we report that heavy metals induce a ciliate apoptotic process very similar to the higher eukaryotic apoptosis, and the possibility to use ciliates as whole-cell biosensors to assess heavy metal polluted samples is also discussed.

---

* E-mail: juancar@bio.ucm.es

INTRODUCTION

The importance of metals in microbial life is well known. Indeed, some of them are key components in many proteins necessary for cellular metabolism and growth (Lovley, 2000). However, certain heavy metals are among the most abundant and toxic inorganic environmental pollutants. Anthropogenic sources, mainly mining and industrial activities, have contributed to substantially increasing heavy metal content in air and in many terrestrial and aquatic ecosystems (Nriagu and Pacyna, 1988; Peñuelas and Filella, 2002). In fact, heavy metals are one of the most persistent pollutants in environment. Unlike complex organic pollutants, they cannot be degraded by microorganisms but they can be accumulated by organisms and also undergo a bioamplification process throughout the food chain (Kong et al., 1995; Martín-González et al., 1999). Almost all heavy metals, especially those which are not included in the composition of biological systems, may produce toxic effects on organisms and ecological disturbances. So, metals can produce directly or indirectly, by the generation of reactive oxygen species (ROS), remarkable alterations on proteins, DNA and cellular lipids (Chang, 1996; Leonard et al., 2004; Valko et al., 2005) that can originate cell death by necrosis or apoptosis (Kang, 1997; Pulido and Parrish, 2003). Furthermore, certain heavy metals are involved in human carcinogenesis (Valko et al., 2006). Due to their ecological, sanitary and even economic consequences, some heavy metals have been considered priority environmental pollutants (Chang, 1996; Keith and Telliard, 1979).

The interest on the interactions of heavy metals with microorganisms has increased in the last few years. These studies have focused on elucidating the different metal resistance mechanisms, especially those used by bacteria and fungi, as well as on which of these processes can be used for bioremediation of polluted habitats (Gadd, 1993; Gadd and White, 1993; Lovley, 2000; Poole and Gadd, 1989; Silver, 1996; Silver and Phung, 1996). Microorganism-heavy metal interactions involve a general cellular stress response united to an specific metal resistance mechanism, which protects the cell against the heavy metal toxicity. The origin of a given metal resistance mechanism is primarily dependent upon the interactions of the metal with the cell (Rouch et al., 1995). Both, essential (like Zn or Cu) and non-essential (Cd) metal ions enter into the cell through specific and non-specific transporters (non-essential metals enter into the cell "parasiting" the non-specific transporters used by essential metals). These non-specific transporters bring in metal ions even during metal excess, which provides the basis as to why the heavy metals become toxic. Several heavy metal resistance mechanisms in microorganisms have been described, but the main microbial strategies to protect themselves against metal toxicity might be classified in two types; avoidance and sequestration. From this classification, we can distinguish at least six basic heavy metal resistance mechanisms present in both prokaryotic or eukaryotic microorganisms and higher eukaryotic organisms (a schematic representation is shown in figure 1); 1- metal exclusion by a permeability barrier (Gowri et al., 1996), blocking the habitual "gate" for this metal (e.g., resistance to Cr in *Pseudomonas aeruginosa* or Co in *Salmonella thyphimurium*); 2- exclusion by active export of the metal from the cell (Mago and Srivastava, 1994), when the metal intracellular concentration is increasing, it is eliminated from the cell using an ATP-dependent membrane efflux pump (e.g., P-type ATPases involved in the Cd or Zn resistance of *Staphylococcus aureus*) (Nies, 2003; Resing at el., 1999) or a cation/proton antiporter effluxing cations from the cell (e.g., the *Eschericha coli* R733 *ars* operon conferring arsenite resistance) ; 3- intracellular sequestration (bioaccumulation) of the metal by binding proteins or peptides (metallothioneins, phytochelatins) or other molecules (glutathione, polyphosphate granules) to prevent the metal-sensitive cellular target interactions (Klaassen et al., 1999; Vasak and Haslerm, 2000); 4- extracellular sequestration or biosorption (Lovley, 2000; Nies,

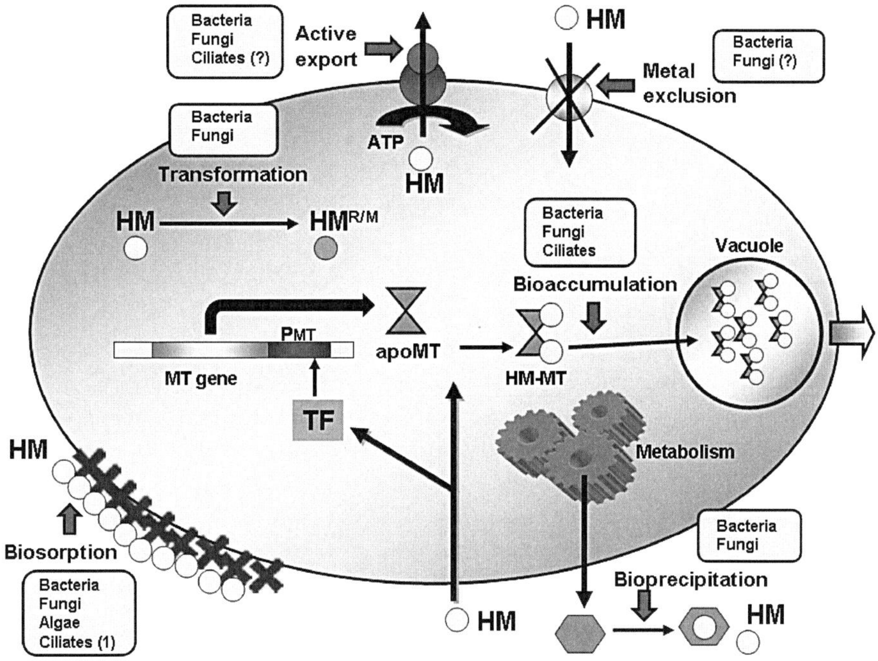

Figure 1. Schematic representation of the main heavy metal resistance mechanisms present in prokaryotic and eukaryotic organisms. Into yellow boxes we indicate the microorganisms presenting a specific resistant mechanism. (1): Ciliates with encystment capacity may have a metal biosorption mechanism, based on the presence of a resting cyst wall. HM: heavy metal. $HM^{R/M}$: methilated or reduced HM. MTgene: metallothionein gene. P_{MT}: metallothionein promoter. apoMT: apo-metallothionein (MT without HM), HM-MT: metallothionein-heavy metal complex. TF: transcription factor.

1999), in which metal cations are passively adsorbed by extracellular polymers, cell walls or capsules. This mechanism is present mainly in bacteria (Brierley, 1990) and algae (Greene et al., 1987), 5- transformation and detoxification (generally re-dox chemistry or methilation to convert more toxic to less toxic metal ion species), it is present in bacteria and fungi (Gadd, 2000; Wakatsuki et al., 1991; Yannai et al., 1991), and 6- bioprecipitation, in this process excreted cell metabolism products from the cell can immobilize toxic metal cations present in the environment converting to less toxic metal ions (e.g., excreted organic acids from the fermentative metabolism of heterotrophic fungi) (Gadd, 1993). There is little available information on the mechanisms involved in the heavy metal resistance in ciliated protozoa, being the bioaccumulation (Martín-González et al., 2006) the resistance mechanism best known in these eukaryotic microorganisms, but we must not discard the potential existence of other mechanisms (in figure 1, we have pointed out some other possibilities).

In this comparative review, we analyse our actual knowledge on ciliate-heavy metal interactions to be used as a microbial model to study the eukaryotic cell-heavy metal interactions. We will analyse here the potentialities and advantages to use ciliates with regard other microorganisms to study these interactions, the types of toxicological bioassays using ciliates, their advantages and disadvantages, the heavy metal toxic effects on ciliates, the ciliate stress response (resistance mechanisms) to heavy metals (mainly ciliates metallothioneins and glutathione related metabolism), ciliate apoptosis induced by heavy metals and, finally, ciliates as whole-cell biosensors and/or as a potential source of molecular biosensors to detect heavy metals from polluted habitats.

Ciliates as Eukaryotic Microbian Models

In general, microorganisms show a rapid growth rate, reaching large cell populations in a short time, which is important in some ecotoxicological assays to obtain an amplified response for detecting the pollutant-induced change, when microorganisms are used as whole-cell biomarkers or biosensors. Besides, it is possible to use genetically engineered microorganisms, which are very useful in "turn on" assays, and to make up appropriate genetic constructions to detect a specific pollutant (e.g. heavy metal-specific biosensors). Likewise, specific mutants or microorganisms with several gene reporters might be used. Genotoxicity studies are simpler and more practicable using microorganisms because they can more easily be subjected to genetic dissection. Ciliates have all these advantages previously indicated, but, furthermore, they present, at least, two additional advantages: a)- In contrast to bacteria and yeasts, ciliates are unicellular organisms without a cell wall in the vegetative stage (figure 2A). A major limitation to the use of bacteria or yeasts as whole-cell biosensors is the uncertainty concerning diffusion of substrates and products through the cell wall, which may results in a slower or less effective response as compared to enzyme-based sensors. To obviate this problem such cells must be permeabilised using physico-chemical or enzymatic methods to allow the free diffusion of small molecules across the cell membrane, but the cells are no longer normal viable organisms. The use of ciliates, instead of bacteria or yeasts, might avoid this serious problem. The absence of a cell wall in these eukaryotic microorganisms might offer a higher sensitivity to environmental pollutants, and therefore, a faster cellular response (Gutiérrez et al., 2003a; Martín-González et al., 1999), and b)- They are eukaryotic microorganisms with some metabolic traits that are more similar to those of human cells than bacteria or yeasts. After completing the genome projects in two model ciliates, often used in ecotoxicological studies, such as *Tetrahymena thermophila* and *Paramecium tetraurelia* (Aury et al., 2006; Dessen et al., 2001; Eisen et al., 2006) have shown that they share a higher degree of functional conservation with human genes than do other microbial eukaryotic models. It is shown by better matches of relevant ciliate coding sequences to those in humans, as compared with non-ciliate microbial models (e.g., humans and *T. thermophila* share more orthologs with each other (2,280) than are shared between humans and the yeast *Saccharomyces cerevisiae* (2,097) (Eisen et al., 2006). Therefore, it seems to be more reasonable to use these eukaryotic cells in ecotoxicological studies as models for humans, and they also represent an alternative to animal tests (Gutiérrez et al., 2003a).

From an ecological point of view, ciliated protozoa are cosmopolitan eukaryotic microorganisms adapted for life in soils and aquatic ecosystems (Finlay and Fenchel, 1996). They are believed to be important grazers of bacteria and other microorganisms and this activity appears to stimulate the rates of carbon and nitrogen cycling in soils (Finlay et al., 2000). Moreover, in some artificial ecosystems such as activated sludge wastewater treatment plants, ciliates significantly improve effluent quality (Nicolau et al., 2001). Furthermore, the community structure of ciliate species is an effective biological indicator of sewage plant functional conditions (Madoni, 1994). Many ciliate species present a facultative cell differentiation process named encystment-excystment cycle (figure 2B), which is composed of two antagonist steps; encystment (cryptobiosis), that involves the formation of resting cysts (figure 2B) from the vegetative cells, and excystment which involves the vegetative cells emergence from the mature resting cysts (Gutiérrez et al., 1990; 1998; 2001). The encystment is induced by different environmental stress conditions (e.g., starvation, temperature changes, different toxic compounds, etc.) that make difficult the normal progression of the cellular growth-division cycle (figure 2A), so it constitutes a protective system against general stress conditions. The encystment process involves a drastic cell differentiation with the formation of a complex resting cyst wall, composed by several biochemically different cyst layers (Gutiérrez et al., 2003b). Also, the resting forms contribute to a passive transport of these species (by air, animals, etc.), through different biogeographical regions or ecosystems.

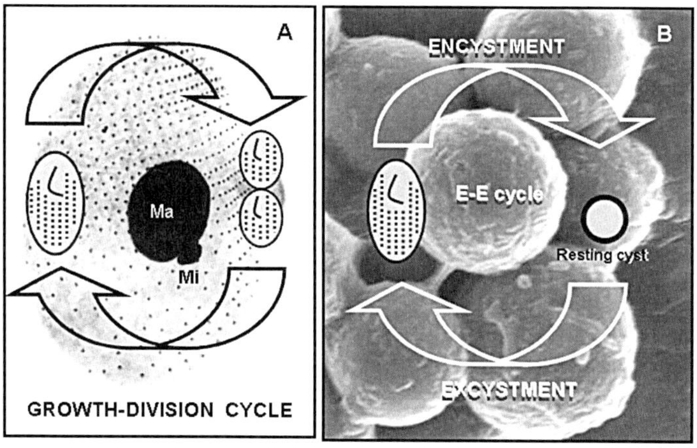

Figure 2. (A): Schematic representation of a ciliate Growth-Division cycle on a micrograph showing a vegetative form of the ciliate *Colpoda cuculus* (Ma: macronucleus, Mi: micronucleus). (B): schematic representation of an ciliate Encystment-Excystment cycle on an scanning electron micrograph of resting cysts from *Colpoda inflata*.

Two main model ciliates have risen through the ciliatology history; *Tetrahymena* and *Paramecium*, due to their "domestication" or adaptation to the microbial laboratory work; they grow rapidly obtaining a high cell density (about 10^6 cells/ml) in a variety of media and culture conditions (pure culture and define medium), they are relative big microorganisms (40-50 μm) which do more easy their micro-manipulation, they present a typical eukaryotic living cycle and sexual interaction (conjugation), doing possible the genetic analysis. In both ciliates, several good genetic analysis tools (e.g. nullisomic and knockouts strains in

Tetrahymena) and molecular genetic tools have been developed. All these advantages have promoted the extensive use of these ciliates in several basic biological areas, including ecotoxicology (Martín-González et al., 1999), contributing to fundamental biological discoveries such as; lysosomes and peroxisomes, cilia components, the discovery of the first eukaryotic retro-transcriptase (telomerase), the auto-catalytic RNA capacity (ribozyme) and the function of histone acetylation in gene expression regulation. And, finally, the recent completion of the macronuclear genome sequencing in both ciliates (Aury et al., 2006; Eisen et al., 2006), will be excellent tools for functional genomic studies to address biological and biotechnological questions of main scientific interest.

CILIATE HEAVY METAL ECOTOXICOLOGY

After reviewing published data on heavy metal ecotoxicological analysis or bioassays using ciliates, we might conclude that the diversity of experimental conditions and methodological approaches is generally too great to make comparisons among results from different authors (including those using the same ciliate species). The following points illustrate this generalized problem, which might be also extended to other bioassays using non-ciliate microorganisms:

a) Heavy metal concentration parameters. Metal concentrations are expressed in diverse forms, generally as weight/volume (ppm or mg/L) or molarity (μM or mM). When authors report experimental data properly, the conversions among these are easily possible, but problems arise if concentrations are expressed in less conventional ways, as a fraction of total proteins or of dry weight, for instance.
b) Culture medium composition. The differences in culture medium composition can represent a significant problem in some circumstances, because the organic matter concentration and dissolved salts or the presence of sediments or clay are not usually reported. These factors, as well as pH changes, might reduce or increase (acidity) the heavy metal availability and, therefore, decrease the toxic impact. For instance, the cell growth of *Tetrahymena pyriformis* is not affected up to a dose of 8 mg/L Cd (in 2% proteose-peptone and 0.1% yeast extract medium) (Sauvant et al., 1999). By contrast, an EC_{50} (effective concentration for 50% growth inhibition) value, at 48h, of 0.78 mg/L Cd (in a medium composed by 1% proteose-peptone and 1% yeast extract) has been reported (Schaefer et al., 1994) for the same ciliate. A protective effect of organic matter has been also observed in other ciliates (Martín-González et al., 1999).
c) Assessment of toxic effect. In general, ecotoxicological assays using ciliates have been conducted to assess sub-lethal effects, such as growth inhibition, variation in feeding rate, respiration inhibition, etc.
d) Endpoints. The LD_{50}/LC_{50} (median lethal dose/median lethal concentration) is the most frequently used toxicological endpoint in ciliates, but other measures of physiological effects are also used, such as IG_{50} (median growth inhibition concentration), also named EC_{50} (effective concentration for 50% growth inhibition) or IC_{50} (median inhibitory concentration), and other more unusual endpoints, such as

LOED/LOEC (lowest observed effective dose or concentration) and TLm (median tolerance limit).

e) Duration of heavy metal exposure. In ciliate toxicological studies the duration of heavy metal exposure is also very diverse. The majority of authors measure the toxic effect at 24h (Coppellotti, 1994; Díaz et al., 2006; Madoni et al., 1992; 1994; Martín-González et al., 2006; among others), but others use very different exposure times, such as; 1h (Nilsson, 1989), 48 and 72h (Schaefer et al., 1994), and so on.

f) Cell concentration. This is another very diverse parameter among ciliate ecotoxicological assays. The range of cell concentrations is really large; some authors use concentrated cell cultures (e.g. about 10^3 cells/ml, Martín-González et al., 2006) and others very few cells (e.g. about 12 cells, Madoni et al., 1994). Obviously, LC_{50} is lower in cultures with low cell number, because the pollutant amount per cell is higher than that in cultures with elevate cell number. So, Stauber et al. (2002), found that as the initial algal cell density of the fresh-water algae *Chlorella* sp. and *Selenastrum capricornutum* was increased, the toxicity of copper decreased. The initial cell density has been proved to be an important factor which modulates results in bioassays. To use a low cell number might be more similar to natural habitat conditions, but, on the other hand, to use very few cells in laboratory experimental conditions may introduce some errors in endpoint values. In assays using few cells, the cell viability might be damaged before cells are exposed to a pollutant, because individual cells must be transferred (by using a micro-pipette) to the medium containing the pollutant and this micro-manipulation may alter the cellular integrity and accelerate the cell death, independently of the real toxic effect of the pollutant exposure.

g) Microbial physiological stage. Another important point to be considered in ecotoxicological studies is the physiological stage of the reporter cell. It was pointed out by Nilsson (1989) that *Tetrahymena* cells in stationary growth phase are more sensitive to Cd than those in exponential growth phase at the same metal concentration, and it is probably due to the increased acidity of the culture, which increases the Cd solubility and releases more Cd toxic form.

h) Bio-geographical parameter or the "strain effect". Strains of the same ciliate species isolated from different geographical origin or habitat can show different heavy metal resistance levels. We name to this parameter as the "strain effect" (Díaz et al., 2006). A study on the evaluation of heavy metal toxicity and bioaccumulation in colpodid ciliates using different strains or isolates of the same colpodid species *Colpoda steinii*, which is the most representative and common ciliate of soils and semi-terrestrial biotopes (Foissner, 1993), have shown this "strain effect" (Díaz et al., 2006). In table-1, we show Cd, Zn or Cu toxicity, expressed as MRD (maximum resistance dose) and LC_{50} (adapted from Díaz et al., 2006), for three different strains of *C. steinii* isolated from different habitats (AZ1 strain from a heavy metal polluted superficial soil sample from Aznalcóllar (Seville, Spain) after a pyrite mine spill (Aguilar et al., 2004) (with a high heavy metal concentration; 1 ppm Cd, 100 ppm Cu, 112 ppm Zn), FM1 strain was isolated from a compost sample of a Spanish enterprise, and 010A strain was isolated from a garden soil sample of Rome, Italy). Other authors have obtained similar results using the same *Colpoda* species (Campbell et al., 1997; Forge et al., 1993) or with other *Colpoda* species, so Xu et al.

(1997) found that six soil isolates of the species *C. inflata* from several biogeographic realms showed different toxic responses to Cd and Zn, but analogous sensitivities to Cu. Authors attributed these toxicological variations to genetic differences and to the distinct geographic location. It is not clear if this "strain effect" is related with the metal soil content, at least in *C. steinii* AZ1 strain (table 1) with the highest resistance values to Cd and Zn might be related with the Cd or Zn soil content, but not the resistance to Cu (although the soil was very rich in Cu). However, other *Colpoda* species (*C. inflata* AZ2 strain), isolated from the same polluted soil, was rather sensitive to Cd (MRD: 5 mg/L; CL_{50}: 1.8 mg/L). The occurrence of differences in tolerance to single toxic pollutants among strains or even individuals and clones within a species is a well-known phenomena (Forbes, 1998), and it illustrates the importance of genotype-environmental interactions in responses to toxicants.

Table 1. Cd, Zn or Cu toxicity, expressed as MRD and LC_{50} for three strains of the same ciliate species

Ciliate strains	MRD (mg/L) Cd Zn Cu	LC_{50} (mg/L) Cd Zn Cu
C. steinii 010A	10 85 15	4.2 33.9 8.1
C. steinii AZ1	12 250 13	5 147.4 5.4
C. steinii FM1	1 150 12.5	0.5 78.7 5.5

After considering all these potential sources of variability present in ciliate heavy metal bioassays, we would like propose here that people working on ciliate ecotoxicology or carrying out metal toxicological bioassays using ciliates, should standardise and validate ciliate-specific bioassays, as it has been achieved in other organisms (microalgae, *Daphnia*, etc).

Types of Bioassays Using Protists

Two main types of bioassays using protists (including ciliates) can be distinguish; single-species or multi-species toxicity tests, which of both is better or more representative of the ecosystem health is still under discussion (Cairns et al., 1996; Wells et al., 1998). We think that both types of methodologies are complementary and the use of one or another depend on the type of research that we wish carry out (Martín-González et al., 1999).

Single-species toxicity tests are formed by pure cultures of one species. This selected species should be sensitive to the toxicant or environmental stress and representative (bioindicator or biomarker). The single-species assays are very useful to study the toxic effects of one or several mixed pollutants. These assays provide valuable information about the physiological / genetic response of the organism to toxicants, including information on the resistance mechanism. In the practice, at least using protists, the species used in this type of bioassays do not seem to be the most representative or frequent in the type of polluted ecosystem (aquatic, edaphic, etc) to be tested, but rather those species better adapted to the laboratory conditions. Among ciliates, members from the genera *Tetrahymena*, *Paramecium, Colpidium* and *Euplotes*, are the most used in these tests, and among photosynthetic microorganisms; the genera *Chlamydomonas, Dunalliella, Chlorella, Scenedesmus* and *Selenastrum*.

In multi-species toxicity tests, which are clearly supported by some expert ecotoxicologist (Cairns et al., 1985; 1996), many species of different trophic levels are tested simultaneously. According to Cairns et al. (1996), only the multi-species or community toxicity tests are suitable to predict effects at ecosystem level. The single-species test underestimates toxicant effects in the field because do not take into account interespecific relationships, the complex interactions and chemical transformations among toxicants that take place in natural ecosystems, producing toxicity changes. The multi-species test has a high degree of environmental realism; it can be developed in microcosm and mesocosm, and many species from different trophic levels are tested. However, in this type of tests more than one hundred species must be identified and the effects on biovolume, cell growth, total biomass must be measured in the endpoints as soon as possible. The determination of these parameters, specially species identification, need expert personnel doing difficult its routine use.

General Heavy Metal Ecotoxicological Conclusions Using Ciliates

In this section, we summarise the main general conclusions on the ciliate heavy metal ecotoxicology data from diverse and recent published works.

a)- Bioassays with single metals. The heavy metals most used in ciliate bioassays are Cd, Cu or Zn, which are usually found in mining effluents and electroplating industries (Dayed et al., 2005). Cadmium is a non-essential metal with the highest biotoxicity, able to produce genotoxic and mutagenic events. It can substitute to other metal ions (mainly Zn^{2+}, Cu^{2+} and Ca^{2+}) in metalloenzymes and shows a very strong affinity to biological structures containing –SH groups (Waalkes, 2003). Copper is an essential trace element that plays a vital role as a catalytic co-factor for a variety of enzymes. At the same time, Cu is toxic to both prokaryotic and eukaryotic due to the generation of reactive oxygen species (ROS), via the so-called Fenton reaction (Mehta et al., 2006). Finally, Zn is considered the most important trace intracellular element and plays a crucial role in a wide range of cellular processes, including cell proliferation, reproduction, immune function and protection against free radicals, but at high concentration it can also produce cellular damage (Ho, 2004).

For the majority of ciliated protozoa the toxicity order rank for these three heavy metals is Cd > Cu >>Zn, being the Zn remarkably less toxic than Cd or Cu. But, however, some exceptions have been reported, such as; in *Colpoda steinii* AZ1 (Díaz et al., 2006) Cd and Cu cytotoxicities are rather similar after using both endpoints (LC_{50} and MRD), and in other ciliate species have been reported (Madoni and Rom, 2006; Martín-González et al., 2006) that Cu is a more biotoxic metal than Cd. Therefore, although ciliates are not phototrophic eukaryotic microorganisms, like microalgae (Peterson and Nyholm, 1993), they present two different toxicity orders; Cd > Cu >> Zn or Cu > Cd >> Zn, depending on considered species

(Dias and Lima, 2002; Díaz et al., 2006; Gallego et al., 2007; Janssen et al., 1995; Martín-González et al., 2006; Nicolau et al., 1999; Pauli and Berger, 1997).

A comparative analysis about Cd, Cu and Zn cytotoxicity from other protists and metazoa lead us to achieve at least two different conclusions; in first place, a high intra- and inter-specific variability on the biological responses and biotoxicity data to heavy metals exits. This variability is also increased by the great experimental conditions diversity used in the bioassays, including the considered endpoints and even, the methodology applied to assess a determinate biological endpoint, as for instance, the percentage of cell death (Martín-González et al., 1999). In a second place, nevertheless the high diversity of biological responses and experimental conditions, unicellular microorganisms are, in general, more sensitive to these metals than vertebrate cells lines or pluricellular metazoa, independently to the level of organization and taxonomical group.

Also, differences among soil and aquatic ciliates, with regard to the heavy metal sensitivity, have been reported (Díaz et al., 2006). Colpodid ciliates (the most usual soil ciliates) (Foissner, 1993) are more resistant to heavy metal (Cd, Cu, Zn) pollution than ciliates from aquatic habitats, probably because they are adapted to a habitat (soil) in which heavy metals are more persistent after a metallic discharge. Soil protozoa occupy a critical niche in soil food, as they are direct predators of bacteria and fungi and food sources for nematodes (Ekelund et al., 2002). Due to their selective grazing, there is an increasing evidence that they produce changes in the rhizobacteria composition (Bonkowski et al., 2004). Besides, the presence of protozoa in the rhizosphere enhances the nitrogen uptake and increase biomass of plants (Clarholm, 2005). These last effects are not only due to the selective feeding, but soil protozoa can also affect the plant root systems (Bonkowski, 2004). Furthermore, it seems to be that the presence of microorganisms, particularly rhizobacteria, enhances both the growth and the removal of toxic heavy metals by plants (phytoextraction) (Glick, 2003). These complex interactions between microorganisms and plants present in heavy metal polluted soils should be elucidated, in order to increase the metal phytoremediation yield.

Recently, for first time in ciliates, flow cytometry to assess the Cd, Cu or Zn cytotoxicity, using cell populations of the model ciliate *T. thermophila,* has been applied (Gallego et al., 2007). Flow cytometry methods are considered suitable and accurate methodologies to evaluate the cytotoxicity of chemicals, and they present the following advantages: (1) short time is required for bioassays, (2) it is possible to investigate several cytotoxical parameters simultaneously, (3) mixed populations and environmental samples can be analyzed, and it is possible to discriminate or even to sort among different species, and (4) this technology can be applied to samples with few cells (environmental samples) or alternatively to abundant populations (laboratory cultures) (Tuschl and Schwarb, 2004). In fact, it has been used in ecotoxicological studies with microorganisms, including bacteria and microalgae (Stauber et al., 2002; Chen and Li, 2005) as well as with mammalian cells (Tuschl and Schwarb, 2005).

b)- Bioassays with heavy metal mixtures. It is known that metal pollution has mainly an anthropogenic origin and these pollutants discharges from mining and industrial practices usually contain more than one heavy metal. Besides, it is also well known that when several metals are exhibiting toxicity simultaneously in the same environmental, metal interactions can be originated, which could lead to enhance or reduce the biotoxicity of each individual metal. These interactions must be considered for a predicted ecological risk assessment and to establish both, an effective monitoring program and the adequate legislations for environmental pollution control. Despite experimental evidences support strongly the above affirmations, insufficient studies about the toxicity of metal mixtures, especially if we compare with those about single metal biotoxicity, have been carried out. In fact, in the excellent review of metal mixture interaction analyses and their effects on aquatic biota by Norwood et al. (2003), only 156 tests with binary mixtures have been reported. From this total number, 21 and 19 tests have been carried out to detect the interactions between Zn and Cu or Cd, respectively. The interaction between these last two metals (Cd-Cu) has been only studied in eight bioassays (Norwood et al., 2003). Likewise, at least nine different methods have been reported for calculating and evaluating the interactions between metals. These ecotoxicological studies have been carried out using mainly microalgae, invertebrates and plants as biological systems. As Norwood et al. (2003) reported, the responses to the metal mixtures are very variable, even for the same binary metal combinations. A substantial part of this variability can be explained by the different used methodological procedures, but it is clear that the different species, and probably strains, present distinct responses to the same bimetallic treatment.

To determine the interaction type between the two metals of the mixture, some authors (Gallego et al., 2007) have considered that, there is not interaction or the interaction is additive between metals when the quantitative toxicity results are strictly additive, so the mixture toxicity is identical to the expected toxicity resulted from the sum of individual toxicities of two metals at the same concentrations. The interaction is synergistic or "more than additive" if the observed mortality rate is greater than the predicted, and, finally, there is antagonism or an interaction "less than additive", if the observed percentage of mortality in the mixture is less than that predicted, considering the individual effects produced by each metal at the same concentration.

Cd and Zn are two heavy metals that usually co-exist in the same ores and have similar atomic and intrinsic features. About Cd-Zn interactions, in colpodid ciliates (Díaz et al., 2006) the Cd biological toxic effect decreases in the presence of low and moderated Zn concentrations. Similar results have been described in the ciliate *Tetrahymena pyriformis* (Dunlop and Chapman, 1981), some microalgae (Sunda and Huntsman, 1996), the invertebrate *Daphnia* (Barata et al., 2002) and other invertebrates (Fargasová, 2001). Likewise, in several ciliate species isolated from urban wastewater treatment plants (Martín-González et al., 2006), the presence of Zn produced a remarkable increase in Cd MRD values in all species used. This might indicate that Zn caused an increase in ciliated protozoan resistance to Cd or, on the contrary, that the cytotoxic effect of Cd decreased under these environmental conditions. In *Euglena gracilis* (Watanabe and Suzuki, 2001), it has been reported that Zn supplementation significantly suppressed the incidence of Cd-induced abnormalities in the cells. Therefore, from all these published data, we might conclude that an antagonism between Cd and Zn exists. However, it is not ever so clear, because in other previously reported ecotoxicological studies, the type of interaction for Cd-Zn pairs is really diverse; in 9 tests it has been evaluated as less than additive (or antagonistic), in 5 tests as strictly additive and in other 5 tests as more than additive (or synergistic) (Norwood et al., 2003).

A most detailed study using the model ciliate *T. thermophila*, flow cytometry and applying the Concentration Addition model based in the Toxic Unit approach to know the type of interaction between metals and to analyse the effects of binary metal mixtures on the cell viability has been carried out (Gallego et al., 2007). Results from this study have showed that not only the metal concentrations but also the ratio between the two metallic components in the mixture are relevant factors for the biotoxicity, at least in the case of bimetallic samples or discharges. For instance; the combined effects of Cd-Zn should be considered as antagonistic in *T. thermophila*, but under high environmental concentrations (1 and 1.25 TUs) the interaction changes to additive or synergistic if the ratio Cd/Zn is 3:1 and 4:1, respectively, so Cd is the main responsible for biotoxicity (Gallego et al., 2007). In other organisms, as the marine microalgae *Phaseosactylum trinutum* (Wang et al., 1995), the shrimp *Callinassa australensis* (Negilski et al., 1981) and two species of plants (Sharma et al., 1999; Otitoloju, 2002) more than one type of interaction has been assessed, depending on the metal ratio used. It is possible that this phenomenon occurs in other biological systems but it has remained undetected because only one or two different ratios of metallic concentrations have been tested in many of these studies. Two main mechanisms have been proposed to explain antagonistic metal interactions; competition for the same uptake pathway or bioaccumulation by metal binding SH-rich proteins (like metallothioneins) (Wang et al., 1995). So, the detected Cd-Zn antagonism in *T. thermophila* (Gallego et al., 2007), might be due to metal bioaccumulation by metallothioneins, which already exist in this ciliate (Díaz et al., 2007). In relation with the cell protection by Zn to Cd toxicity, it has been reported that, in mammalian cells, low or moderate Zn concentrations can prevent both, apoptosis and ROS production induced by Cd. Whereas at high Zn concentrations, it acts as pro-apoptotic and oxidant metal (Watjen et al., 2002; Valko et al., 2005).

In the ciliate *T. thermophila* (Gallego et al., 2007), toxic combined effects of Cu-Zn might be considered as antagonistic and synergistic depending on the toxicity of treatments (TUs) and specially of the Cu:Zn ratio. With regard to these heavy metals, literature data show results very diverse, depending on the type of organism used in bioassays (Norwood et al., 2003). Like in *T. thermophila*, a change from antagonism to synergism depending on metal ratio and concentration, with or without one intermediate additive phase, has also been detected in bacteria (Ince et al., 1999), microalgae (Wang et al., 1995), crustaceans (Negilski et al., 1981), insects (Fargasová, 2001) and plants (Otitoloju, 2002). The reasons that might explain Cu-Zn antagonism are still obscure, it might be that low Zn concentrations act as antioxidants, neutralising the ROS formation originated by Cu.

The Cd-Cu interaction is the less studied interaction among all considered (using Cd, Cu or Zn) metal mixtures. In the scarce previous papers (only eight studies), it is usually considered as a synergistic interaction (four cases) (Norwood et al., 2003). By the contrary, and like in the ciliate *T. thermophila* (Gallego et al., 2007) and some pluricellular organisms (An et al., 2004; Fargasová, 2001), the three types of interactions have been detected, depending on the metal ratio and concentrations. In plants (An et al., 2004), it has been suggested that both, changes in the usual bioaccumulation patterns and uptake of each single metal take place in binary (Cd-Cu) metal combinations.

In table 2, we summarise the main general heavy metal ecotoxicological results from ciliate bioassays.

Table 2. General heavy metal ecotoxicological results from ciliates

Characteristic	Result
Toxicity order ranks	Cd > Cu >> Zn (majority of species)
	Cu > Cd >> Zn (some species)
Resistance / Habitat	soil ciliates > aquatic ciliates
Cd-Zn interactions	Antagonism
	Additive or Synergism (1)
Cu-Zn interactions	Antagonism or Synergism
Cd-Cu interactions	3 types of interactions (1)
Bioaccumulation	Zn >> Cd > Cu

(1): It is depending on metal ratio and concentrations.

CILIATE CYTOLOGICAL ALTERATIONS UNDER HEAVY METAL EXPOSURE

Little is known about the morphological changes induced in ciliates by exposure to sub-lethal heavy metal concentrations, existing very few studies on this topic that includes the following ciliates; *Tetrahymena* (Dunlop and Chapman, 1981; Pyne et al., 1983; Piccinni et al., 1987; Nilsson, 1996; 1997), a strain of *Uronema marinum* isolated from Antarctica (Coppellotti, 1994), two edaphic ciliates (colpodids) isolated from a composting plant and *Drepanomonas revoluta* isolated from an activated sludge wastewater treatment plant (Martín-González et al., 2005).

A remarkable effect of heavy metal treatment is the induction of an abnormal ciliate encystment process. Ciliate encystment is a differentiation process induced by environmental stress, mainly starvation (Gutiérrez et al., 2001; Gutiérrez and Martín-González, 2002), in which vegetative cells are transformed in resting cysts (cryptobiotic stage). This differentiated resting form presents a complex resting cyst wall composed by several biochemically and ultrastructurally distint layers (Gutiérrez et al., 2003b). The abnormal immature resting cysts formed under heavy metal exposure presumably are non-viable, because they show signs of cellular degeneration such as; numerous vacuoles, mitochondrial alterations and an immature cyst wall (Martín-González et al., 2005).

Mitochondrial degeneration is characteristic of Zn and, especially, Cd treatments in ciliates. Disintegration of cristae and/or mitochondrial degeneration has been reported in ciliates exposed to Cd, such as *Tetrahymena pyriformis* (Krawczynska et al., 1989; Pyne et al., 1983) and *Uronema marinum* (Coppellotti, 1994), and exposed to Zn, such as; *Drepanomonas revoluta* and *Cyrtolophosis elongata* (Martín-González et al., 2005). This is not surprising, because certain heavy metals induce ROS formation (Ercal et al., 2001), and according to some authors (Bagchi et al., 2000; Pinot et al., 2000), mitochondria are the cellular organelles most effected by ROS.

Another important change detected in ciliates after heavy metal treatments is the formation of diverse vacuoles/vesicles in the cytoplasm. The appearance of cytoplasmic vacuoles seems to be a very general response to metallic stress. It has been observed in microalgae (Nassiri et al., 1997; Smith, 1983) and the ciliates *Tetrahymena* sp., *U. marinum*, *D. revoluta* and *C. elongata* after exposure to various heavy metals (Coppellotti, 1994;

Dunlop and Chapman, 1981; Martín-González et al., 2005; Nilsson, 1997). According to Smith (1983), vacuolisation is due to the osmotic disorganisation that results from membrane alterations induced by heavy metals. This mechanism is especially pertinent to organisms without cell wall, such as ciliates.

Large amounts of membranous debris scattered throughout the cytoplasm has been detected in *D. revoluta* after Cd or Zn treatment (Martín-González et al., 2005), and in *T. pyriformis* after Bi or Au exposure (Nilsson, 1996, 1997). Similar membranous structures have been observed in some species of microalgae treated with Pb, Cd or Cu (Nassiri et al., 1997a, b; Visviki and Rachlin, 1994). This material generally consists of irregular concentric or juxtaposed masses of membranes and it can be formed into vacuoles. In some cases, these vacuoles containing membranous material show very electron-dense irregular deposits, which might be metallic accumulations. These structures might therefore indicate a detoxification mechanism; the toxic free cations could be neutralised or immobilised by the membranous structures (metal cations could be non-specifically bound by the negative charges of the membrane phospholipids). According to Nilsson (1997), this mechanism is responsible for the compartmentalisation of metal deposits in *Tetrahymena*. The cytoplasmic dense granules (probably, heavy metal-metallothionein complexes) fuse with digestive vacuoles, and the metallic material can be eliminated by defecation. So, in both types of microorganisms (algae and ciliates), cytoplasmic membranous structures seem to be related to metal detoxification. On the other hand, similar membranous structures have been observed in the cytoplasm of early precystic cells in some ciliate species (Gutiérrez et al., 1990). Therefore, at least in ciliates, these membranous structures might be involved in the general response to environmental stress.

Another representative change induced by toxic metal treatment, observed in all studied ciliates, is the formation of numerous cytoplasmic autophagosomes. A high autophagic activity, after Cd treatment, has been reported in *T. pyriformis* (Krawczynska, 1989), *Euplotes vannus* (Iftode et al., 1985) and *C. steinii* (Martín-González et al., 2005). However, this phenomenon was not detected when *T. pyriformis* was treated with gold or bismuth (Nilsson, 1996; 1997). Traditionally, the formation of autophagosomes (macroautophagy) has been considered a specific mechanism induced by starvation or nutritional stress, at least in ciliates (Gutiérrez et al., 2001; Gutiérrez and Martín-González, 2002) and yeasts (Abeliovich and Klionsky, 2001). Autophagy is the major route of protein degradation under starvation conditions (Klionsky and Ohsumi, 1999), and, likewise, it is related to programmed cell death (Bursch, 2001), in fact it is considered as the programmed cell death type-II.

Ciliates also undergo nuclear changes after exposure to Cd or Zn, which involve both, micro- and macronucleus. The ciliate macronucleus usually contain a composite nucleolus made up of more than six nucleolar masses (Raikov, 1982). However, when ciliates are treated with Cd or Zn the nucleolar masses fuse together forming only 1-3 masses. It has been reported in both, *C. steinii* and *C. elongata*, after Zn treatment (Martín-González et al., 2005). Such nucleolar alterations have been also detected in early precystic cells of *D. revoluta*, after encystment induction by Cd or Zn, in Cd-treated cells of *Tetrahymena* (Krawczynska et al., 1989; Pyne et al., 1983) and *U. marinum* (Coppellotti, 1994). Ultrastructural modifications of macronuclear nucleoli, including fusion, have been reported during normal ciliate encystment induced by starvation (Gutiérrez et al., 1998; Gutiérrez and Martín-González, 2002). In ciliates without encystment capacity, such as *Paramecium* sp. and *Tetrahymena* sp., similar nucleolar changes occur during starvation, stationary phase, actinomycin D exposure or ultraviolet irradiation (Raikov, 1982). Nucleolar condensation has been found in animals and plants exposed to heavy metals, especially Cd (Marcano et al., 2002). Metal treatments also induce macronuclear chromatin condensation, being it another typical alteration in ciliates after metallic treatment or starvation (ciliate encystment inducer) (Coppellotti, 1994; Gutiérrez et al., 1998; Iftode et al., 1985). A macronuclear fragmentation has been only described in Cd-treated *U.marinum* (Coppellotti, 1994), but not in other ciliates. In general, no significant ultrastructural changes in the ciliate germinative nucleus (micronucleus) have been reported, but, however in Zn-treated cells of *C. steinii* the micronuclear membrane showed some disorganisation with digitiform projections (Martín-González et al., 2005).

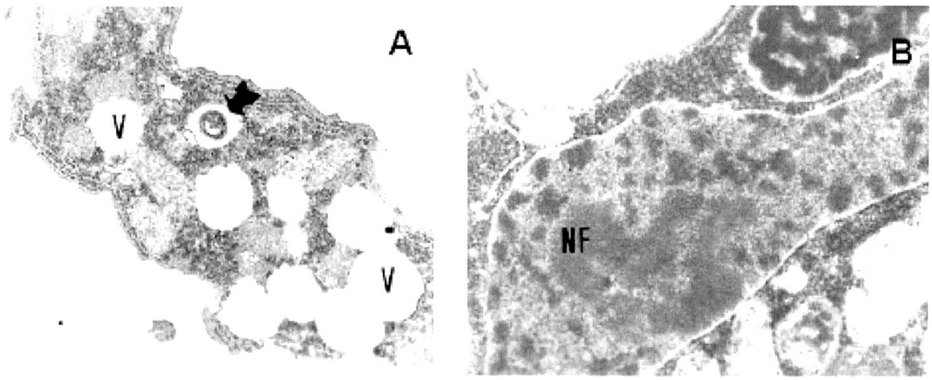

Figure 3. Electron micrographs of cytoplasmic ultrastructural changes caused by Zn (75 mg/L) exposure in the ciliate *Cyrtolophosis elongata* (strain FM2). (A): Cytoplasmic vacuolisation, membranous debris and mitochondrial alterations (arrow indicates a putative metallic accumulation (x 32,000). (B): Macronucleus showing chromatin condensation and nucleolar fusion (x 48,000). V: vacuole, NF: nucleolar fusion.

In summary (figure 3), heavy metal exposure in ciliates induces very diverse changes, such as inhibition of cell motility, increase of autophagic activity, appearance of vesicles/vacuoles in the cytoplasm, mitochondrial degeneration, macronuclear chromatin condensation and nucleolar fusion. Most of these alterations are common with those reported under starvation stress (without encystment) or during the ciliate encystment process. Also, curiously, most of these ultrastructural alterations have been noted after staurosporine treatment in the ciliate *T. thermophila* (Christensen et al., 1998). Staurosporine is a potent inhibitor of both protein kinases and cellular proliferation, and it also induces apoptosis. Likewise, it has been shown that both cytosolic and mitochondrial ROS generation are involved in staurosporine-induced apoptosis (Gil et al., 2003). Therefore, some of the ultrastructural changes induced by heavy metals in ciliates might be an evidence of ROS generation and/or the presence of an apoptotic route.

Ciliate Heavy Metal Resistance Mechanisms

At present, little is known on the potential metal resistance mechanisms present in ciliated protozoa. Among the six general microbial well-known possible heavy metal resistance mechanisms, at least three of them are present in ciliates; bioaccumulation (the best studied), active export (very probable) and biosorption (present in resting cyst forms). Active export is based on the existence of ATP-dependent membrane efflux pumps, which export the metal from inside to outside of the cell. We think that this metal resistant mechanism might be present in ciliates by two reasons; a) after the macronuclear genome sequencing projects from two model ciliates (*T. thermophila* and *P. tetraurelia*) become public and preliminary analysed (Aury et al., 2006; Eisen et al., 2006), numerous membrane transporters families have been detected in both ciliate genomes. For instance, in *T. thermophila* macronuclear genome a total of 485 putative genes encoding membrane transporters for inorganic cations have been reported (Eisen et al., 2006), from which about 91 are P-ATPase. It is difficult think that from all these pumps any of them becomes dedicated to metal efflux pumps, and b) from several expression gene libraries obtained from *T. thermophila* after Cd, Pb or As treatments (unpublished data from our research group), we have identified about 4.6% (from the total identified cDNAs) of genes encoding membrane transporters in the Cd(3h treatment) cDNA population, and about 2.6% in As(1h treatment) cDNA population. These data are showing the relative importance of these genes in Cd or As cellular detoxification process of this ciliate.

A biosorption-like process has been studied in our research group (unpublished data), using mature (> 3 months old) resting cysts from the ciliate *Colpoda inflate*, and *T. thermophila* as a reporter-ciliate to value the biotoxicity of the sample. Mature resting cysts of *C. inflata* (with a complete and mature resting cyst wall) were exposed to Cd (0.5 mg/ml, during 30 min, 1 or 2h) in Tris-HCl buffer, then resting cysts were removed by centrifugation, and the obtained supernatant was exposed (24h) to a vegetative growing population of *T. thermophila*. The mortality rates were calculated with regard a control (the same Cd concentration in Tris buffer without a previous resting cyst exposition). Results (table 3) show a clear detoxification effect by the presence of resting cysts in the Cd toxic solution, because the mortality of *Tetrahymena* is drastically reduced. Our assumption is that the ciliate resting cyst wall (composed mainly by glycoproteins) (Benítez et al., 1991; Izquierdo et al., 2000; Gutiérrez et al., 2003b) might adsorb the Cd ions by different components of the extracellular polymer (cyst wall), removing the toxic ions and reducing their bioavailability from the medium. It is the first time that an experimental evidence of a potential biosorption phenomenon is described in ciliates.

Table 3. Detoxification effect of Cd biosorption by mature resting cysts of *Colpoda inflata*

	Control-1	+RC(30min)	Control-2	+RC(1h)	Control-3	+RC(2h)
Mortality (%)	86	0	77	0	78	0

(+RC): heavy metal solution previously exposed (30 min, 1h or 2h) to resting cysts. For more details see the text.

Ciliate Heavy Metal Bioaccumulation

This heavy metal resistance mechanism is the best known in ciliates. It involves the intracellular sequestration of the metal by proteins (metallothioneins) or peptides (phytochelatins), forming metal-protein complexes which are accumulated into vacuoles and, finally, released from the cell as non-toxic metallic complexes.

Like cytotoxicological data, metal bioaccumulation seems to be a specific feature for each ciliate strain, for instance; *C. steinii* AZ1 (isolated from Cd polluted soil) presented an average value of Cd bioaccumulation of 40% when using 1 mg/L as initial metal concentration, while other strain of the same ciliate species *C. steinii* FM1 (isolated from a compost sample) shown a Cd bioaccumulation value of 62.5%, using the same initial Cd concentration (Díaz et al., 2006). The highest, at present, registered ciliate Cd bioaccumulation value has been for *D. revoluta* BQ1 strain with a 79% of Cd bioaccumulation at 1 mg/L of initial Cd concentration (Martín-González et al., 2006). Likewise, the strain *C. steinii* AZ1 is that presenting the highest Zn bioaccumulation value (96%), at present reported (Díaz et al., 2006). And, with regard Cu, the strain BQ2 of *Uronema nigricans* presents the highest value (95%) for Cu bioaccumulation (Martín-González et al., 2006).

In general, the amount of metal bioaccumulated increases after increasing the metal concentration in the medium, even when ciliates were exposed to metal concentrations similar or higher than their LC_{50} for the corresponding metal, but several exceptions exist. Also, in general, the bioaccumulation order rank for Cd, Zn and Cu in ciliates is Zn >> Cd > Cu (table 2), it is not surprising because Zn is an essential metal and the lowest toxic among them. Cadmium is also well-bioaccumulated metal, probably due to the existence of ciliate Cd/Zn-metallothioneins, as it has been demonstrated in different *Tetrahymena* species (Díaz et al., 2007). Copper bioaccumulation levels are rather low among ciliates, excepting in some ciliate strains, although Cu-metallothioneins exist as it has been reported in *Tetrahymena* species (Boldrin et al., 2003; Díaz et al., 2007), and probably they also exist in other ciliates. Another resistance mechanism such as ATPase efflux pumps have been usually involved in the Cu resistance in both uni- or pluricellular organisms (Mercer et al., 2003; Nies, 2003). Both mechanisms might coexist in ciliates, and therefore the Cu bioaccumulation levels could be lower than those for other metals.

In bimetallic mixtures, Cd bioaccumulation decreases in the presence of Zn in most of studied ciliate strains (Díaz et al., 2006; Martín-González et al., 2006), excepting for the strains more resistant to this metal. So, in general, ciliates have a similar behaviour that some organisms, such as fishes (Glynn, 2001) and crustaceans (Rainbow et al., 2000). Besides, Zn bioaccumulation decreases in treatments with bimetallic mixtures (Cd+Zn) and this decrease is higher than that for Cd, being consistent with the higher Cd cytotoxicity level with regard to Zn. Furthermore, in ciliates Cd is a better inducer of Cd/Zn-metallotioneins than Zn (Boldrin et al., 2002; Díaz et al., 2007), in the majority of Cd-metallothioneins. On the other hand, as it has been previously mentioned, ciliates present a higher Cd resistance when they are treated with Zn simultaneously. It might be due to the induction of a Cd/Zn efflux, in the presence of Zn, as an additional mechanism to metal resistance. In some organisms, the influx and efflux of Cd and Zn occur through common pathways (Gaither and Eide, 2001; Nies, 2003). During ciliate heavy metal bioaccumulation, Cd can displace Zn when both are present in the medium. Therefore, ciliates respond in a similar way than pluricellular organisms and mammalian tissues, in which the presence of Cd may cause a cellular Zn deficiency (Chang, 1996). This competition between Cd and Zn has been detected in several ciliates, such as; *Tetrahymena* sp. (Nilsson, 1989), *Uronema marinum* (Coppellotti, 1994), *Drepanomonas revolute, Euplotes* sp. and *Uronema nigricans* (Díaz et al., 2006).

Ciliate heavy metal bioaccumulation has been detected by transmission electron microscopy (Dunlop and Chapman, 1981; Iftode et al., 1985; Martín-González et al., 2005; Nilsson, 1996). These metallic inclusions appear as very electron-dense granules into the cytoplasm, which seem to be formed inside electrolucent small vacuoles (figure 4). No similar granules appear in untreated vegetative cells, precystic cells or resting cysts (Martín-González et al., 2005). They have been observed in all heavy metal exposed ciliates, independently of the type of used metal; Cd (Coppellotti, 1994; Krawczynska et al., 1989; Martín-González et al., 2005), Au (Nilsson, 1997), Bi (Nilsson, 1996) or Pb (Nilsson, 1989). In addition, they have also been observed in certain microalgae after treatment with Cd, such as *Dunaliella bioculata* (Heuillet at al.,1986), *Skeletonema costatum* (Nassiri et al., 1997a) and *Tetraselmis suecica* (Nassiri et al., 1997b). It is likely that these cytoplasmic granules represent complexes formed by both metallic cations (Cd^{2+}, Zn^{2+}) and metallothioneins. At this stage, heavy metals are not bioavailable and therefore they are not toxic toward cells.

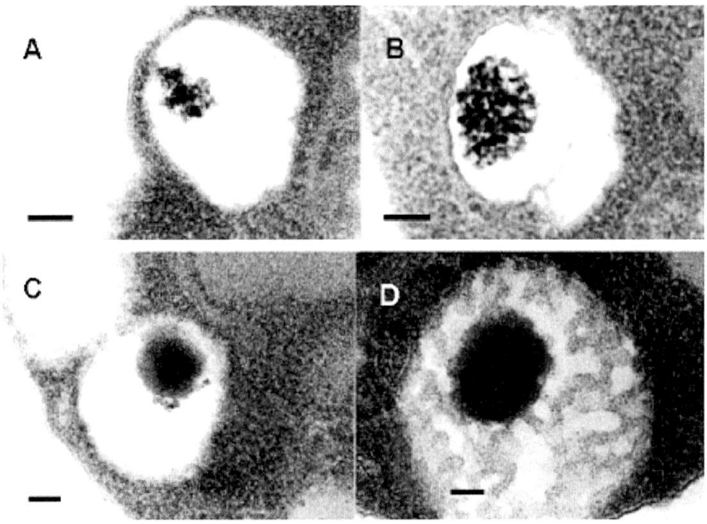

Figure 4. Transmission electron micrographs showing different formation phases (A-D) of putative Zn deposits in treated cells (25 mg Zn/L) of the ciliate *Drepanomonas revoluta* (strain BQ1). Bar: 50 nm.

Such metallic bioaccumulation can be also revealed by fluorescence microscopy. In 2006, we report (Díaz et al., 2006; Martín-González et al., 2006), for the first time in ciliates, the use of specific heavy metal fluorophores to detect metal bioaccumulation. Results from fluorescence microscopy suggest that this methodology is only useful for locating cytoplasmic metallic deposits when cells are exposed to rather high heavy metal concentrations and metallic accumulations inside ciliates are rather large. Furthermore, some fluorophores (e.g., TSQ = [N-(6-metaoxy-8quinolyl-p-toluenesulfonamide)], which is selective for Zn^{2+} in the presence of physiological concentrations of Ca^{2+} and Mg^{2+} ions) seem to be more specific than others. For these reasons, in our opinion, this method for detecting metallic deposits inside cells should be only used after obtaining experimental evidence that cells remove a particular heavy metal from the medium. An additional adventage is that cellular fluorescence can be analyzed by flow cytometry, thereby obtaining quantification of metal bioaccumulation at the cell population level. So, it might be a complementary method that should be joined to a transmission electron microscopy analysis. With regard to Zn bioaccumulation in ciliates, it must be pointed out that each genus shows a specific pattern of cytoplasmic granules, independently of the metal resistance level (figure 5).

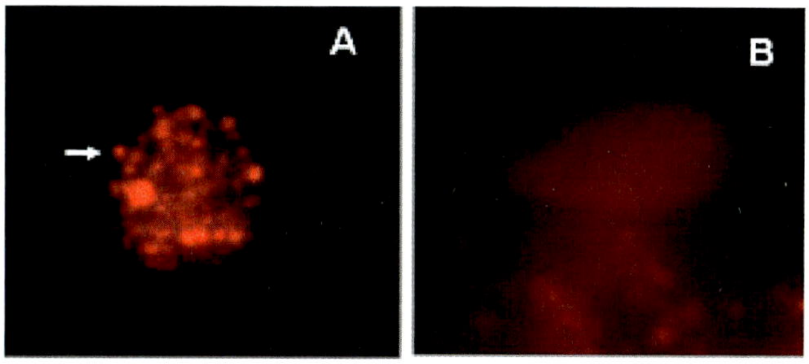

Figure 5. Fluorescent images of putative Zn deposits (arrow) made evident by using the TSQ fluorophore in vegetative cells of the ciliate *Colpoda steinii* (strain 010A). (A): treated cell with Zn (50 mg/L) during 24h. (B): untreated control (400 x).

Ciliate Heavy Metal Chelating Molecules

There are two types of peptide molecules with heavy metal binding capacity, and generally involved in cellular metal detoxification; 1- enzymatically biosynthesized molecules (such as; glutathione, phytochelatins), and 2- proteins encoding by genes (metallothioneins), all of them present numerous free -SH groups (from cysteine residues) in their molecules because they constitute the reactive groups for chelating heavy metals. In ciliated protozoa, metallothioneins (MTs) have been, still now, only detected in *Tetrahymena* species (Díaz et al., 2007), but probably they exist in other ciliates, while phytochelatins have not been detected, at present, in ciliates, but genes encoding putative phytochelatin synthases have been reported in both, *T. thermophila* and *P. tetraurelia* (Amaro et al., 2006).

Phytochelatins (PCs) are synthesized from glutathione (GSH: the tripeptide [γGlu-Cys]-Gly) by phytochelatin synthase (PCS) (γ-glutamylcysteinyltranspeptidase), which present a general structure $(\gamma\text{-Glu-Cys})_n\text{-Gly}$ (n=2-11), chelate heavy metals, especially Cd^{2+}, with a high affinity and facilitate the vacuolar sequestration of these inorganic pollutants. The PCS enzyme catalyzes the transpeptidation of the γ-Glu-Cys moiety of GSH onto a second GSH molecule forming PC_2 or onto a preformed PC_n molecule to form the corresponding PC_{n+1}. PCs were firstly discovered in the yeast *Schizosaccharomyces pombe* (Kondo et al., 1984), and subsequently found in all plants examined to date (Grill et al., 1988), in several fungi (Grill et al., 1986), some microalgae (Grill et al., 1989) and, more recently, in the worm *Caenorhabditis elegans* (Vatamaniuk et al., 2001). Likewise, genes encoding PCSs have been isolated and cloned from plants (Ha et al., 1999; Vatamaniuk et al., 1999), fungi (Clemens et al., 1999) and nematodes (Vatamaniuk et al., 2001). A number of putative PCS sequences have been retrieved recently from EST and genome sequence databases from representatives of all eukaryotic kingdoms (Cobbet and Goldsbrough, 2002), including a group of bacterial PCS-like proteins (Harada et al., 2004; Tsuji et al., 2004), whose main activity is the conversion of GSH to γ-glutamylcysteine rather than PC biosynthesis (Harada et al., 2004). For the first time in ciliates, we have reported (Amaro et al., 2006) a cDNA encoding a putative PCS from the ciliate *T. thermophila*, the full-lenght TtΨPCS cDNA was isolated by RT-PCR from RNA of a Cd-treated (24h) culture of this ciliate. Likewise, a similar (40% alignment score) sequence (PtPCS) has been identified in the macronuclear genome database of *P. tetraurelia* (Amaro et al., 2006). The N-terminal region of TtΨPCS includes the three residues, C72, H198 and D216 (equivalent to C56, H162 and D180 in AtPCS1 from *Arabidopsis thaliana*), that form the catalytic triad of the enzyme (Rea et al., 2004). The most conserved feature of the C-terminal region of known PCSs is the occurrence of multiple (6-14) cysteine residues often arranged as vicinal pairs as in metallothioneins. In contrast, only 3 (TtΨPCS) and 2 (PtPCS) Cys residues are present in the C-terminal region of the putative ciliate enzymes. The total amount of Cys residues is higher for PCSs from higher eukaryotes (3-4%) than for the enzymes from ciliates (~ 2%) and bacteria (< 1%). In addition, similar to the bacterial enzymes, but in contrast with biosynthetically active PCSs, only one Cys residue of TtΨPCS (C72) matches the Cys residue (C56 in AtPCS1) previously shown to be essential for PC synthesis (Tsuji et al., 2005).

By real-time quantitative RT-PCR the relative expression levels of the TtΨPCS gene have been reported (Amaro et al., 2006) under exposure to diverse heavy metals, and the only significant induction (about 8-fold) value was obtained after 1h Cd treatment. Previous studies (Cobbett, 2000) have shown that PCS enzyme levels are generally unaffected by heavy metal exposure, however a few exceptions (Cd-dependent upregulations) have been reported (Lee et al., 2003; Vatamaniuk et al., 1999). And an even closer similarity with the above described TtΨPCS response has been reported recently (Ramos et al., 2007). Various analytical methods (HPLC, capillary electrophoresis and MS-HPLC) have been used to probe *in vivo* PC synthesis in Cd-treated cultures of *T. thermophila* (Amaro et al., 2006), but any of them showed a positive result. On the other hand, these HPLC results have shown that GSH hydrolysis, rather than transpeptidation and PC synthesis, is the main GSH-related activity in Cd-treated *T. thermophila* cells. GSH hydrolysis and γ-Glu-Cys formation is also the only activity of the PCS-like protein (NsPCS) from the cyanobacterium *Nostoc* (Harada et al., 2004). According to Blum et al. (2007), this hydrolytic activity is the most ancient function of PCS and the C-terminal domain was acquired later in evolution to provide regulatory capacity and to contribute to heavy metal tolerance. Such domain is not present in NsPCS, and it has an about 3-fold reduced Cys content in the case of TtΨPCS. This observation, along with the lack of any detectable PC or PC synthetic activity in Cd-treated *T. thermophila* extracts, strongly suggests that, similar to NsPCS, TtΨPCS is a PCS biosynthetically inactive enzyme mainly devoted to GSH hydrolysis and, perhaps, γEC conjugate formation. The detection of Cys-Gly in extracts from Cd-treated cells suggests that TtΨPCS might be endowed with γ-glutamyltransferase (γGT) activity. Since no γGT homolog is present in the sequenced macronuclear genomes of both, *T. thermophila* and *P. tetraurelia*, it is tempting to speculate that in ciliates a surrogate γGT-like activity might be provide by PCS-like enzymes such as TtΨPCS. In conclusion, ciliate pseudo-PCSs are the closest relatives of the bacterial PCS-like enzymes and may represent the connecting link between the prokaryotic and the eukaryotic members of the PCS superfamiliy.

At present, metallothioneins (MTs) seem to be the main molecules involved in ciliate heavy metal bioaccumulation and detoxification. Classic MTs constitute a superfamily of ubiquitous, cytosolic, small (25-82 amino acids, <10 KDa), cysteine-rich proteins (7-21 conserved Cys residues), absence of aromatic amino acids and histidine and capacity to bind heavy metal ions (mainly Cd, Zn and Cu) via metal-thiolate clusters (Coyle et al.,2002; Miles et al., 2000). They are multifunctional proteins whose synthesis is enhanced by heavy metals and other environmental stressors (Coyle et al., 2002; Díaz et al., 2007; Miles et al., 2000). MTs are found, not only throughout the animal kingdom, but also in other eukaryotes (protists, yeasts and higher plants) and in prokaryotic microorganisms (cyanobacteria) (Robinson et al., 2001). At present. in ciliated protozoa (exclusively in different *Tetrahymena* species), two subfamilies of MTs have been reported (Díaz et al., 2007); subfamily 7a composed by 7 CdMTs (cadmium-binding MTs) including; two *T. pyriformis* MTs (Piccinni et al., 1999; Fu and Miao 2006), three *T. thermophila* MTs (Díaz et al., 2007; Dondero et al., 2004; Shang et al., 2002), one *T. tropicalis* MT (Shuja and Shakoori, 2007) and one from *T. rostrata* (Amaro et al., 2007), and subfamily 7b constituted by 4 CuMTs (copper-binding MTs) which includes two *T. thermophila* MTs (Díaz et al., 2007; Boldrin et al., 2008), one *T. pigmentosa* MT (Boldrin et al., 2002) and one from *T. rostrata* (Amaro et al., 2007). Both subfamilies differ in the induction pattern by heavy metals (Cd or Cu) (Boldrin et al., 2002; Díaz et al., 2007; Dondero et al., 2004; Shang et al., 2002) and the pattern of Cys residues clustering (Díaz et al., 2007). Ciliate MTs present exclusive features with regard classic MTs; they are longer proteins (96-181 amino acids, 10-19 KDa), have a higher content in Cys residues (22-54) and aromatic amino acids and histidine are present in several of them (Díaz et al., 2007). In table-4, we compare the main features of ciliate MTs with regard classic MTs (present in the majority of organisms).

Table 4. Comparison between ciliate and classic MTs

Characteristic	Classic MTs	Ciliate MTs(1)
Size (MW / amino acids)	2.5-8 KD (25-82 aa)	10-19 KD (96-181aa)
Cys residues	7-21	22-54
aromatic aa / His	absent	present
Cys cluster "CCC" type	rare	CdMTs
CdMT metal binding capacity	7 Cd atoms/molecule	8-19 Cd atoms/molecule
CuMT metal binding capacity	12 Cu atoms/molecule	13-19 Cu atoms/molcule
K / R ratio	1-13 / 0-6	11-23 / 0-1

(1): Exclusively *Tetrahymena* MTs

An almost exclusive feature of *Tetrahymena* CdMTs is the presence of Cys clusters "CCC" type, which do not exist in *Tetrahymena* CuMTs and they are rare in classic MTs. Considering that in classic MTs all Cys residues are involved in heavy metal binding and, therefore, the stoichiometry is $Cd_7(Cys)_{20}$ for CdMTs or $Cu_{12}(Cys)_{20}$ for CuMTs (Klaassen et al., 1999), we might calculate the theoretical binding capacity of *Tetrahymena* MTs, for instance; TtherMTT1 might bind about 17 Cd atoms, TtherMTT3 (15 Cd) y TtherMTT5 (8 Cd), all from *T. thermophila* (Díaz et al., 2007), TrosMTT1, from *T. rostrata* (Amaro et al.,

2007), 12 Cd atoms, TtropMTT1 (from *T. tropicalis*) (Shuja and Shakoori, 2007) 16 Cd, TpyrMT-2 (from *T. pyriformis*) (Fu and Miao 2006) 19 Cd and TpyrMT-1 (Piccinni et al., 1999) 11 Cd, this last inferred data has been experimentally corroborated by two different methodological approaches; 12 Cd^{2+} / mol of protein (Piccinni et al., 1987) and 11 Cd ions / polypeptide (Domènech et al., 2008). With regard *Tetrahymena* CuMTs; TtherMTT2 and TtherMTT4 (Díaz et al., 2007) might bind about 19 Cu atoms, TpigMT-2 (from *T. pigmentosa*) (Boldrin et al., 2002) 16 Cu and TrosMTT2 (Amaro et al., 2007) 13 Cu. Therefore, the binding capacity of these MTs is considerable higher than classic MTs.

The mRNA levels of *TpyrMT-1* (Santovito et al., 2000), *TpyrMT-2* (Fu and Miao, 2006), *TtherMTT1* (Díaz et al., 2007; Dondero et al., 2004; Shang et al., 2002), *TtherMTT5* (Díaz et al., 2007) and *TrosMTT1* (Amaro et al., 2007) increase significantly with Cd concentration in the medium, while *TtherMTT3* gene is mainly induced by Zn (Díaz et al., 2007). Other heavy metals and different stressors, such as; Pb, Cu, Ni, As, Hg, hydrogen peroxide, heat and cold-shock, interleukin-6, paraquat, acid or basic pH conditions, starvation, can induce (at different levels) several of these MT genes (Díaz et al., 2007; Dondero et al., 2004). Thus *Tetrahymena* CdMTs, and probably CuMTs too (*TrosMTT2* gene is mainly induced by Cu, but it is also induced by other metals, like Zn, Cd and Pb) (Amaro et al., 2007), are not strictly heavy metal-responsive proteins, but they are induced under more general stress conditions (so, we name them as multi-stress-inducible MTs).

An interesting feature of ciliate subfamily 7a (CdMTs) is their remarkably regular and hierarchical modular organization (Díaz et al., 2007). So, TpyrMT-1 (107 aa), TtherMTT5 (99 aa) and TrosMTT1 (113 aa) have bi-modular structures, TtherMTT1, TherMTT3 (both with 162 aa) and TtropMTT1 (156 aa) have tri-modular structures and the biggest CdMT TpyrMT-2 (181 aa) presents a penta-modular structure. These modules can be divided in two types of sub-modules (type-1 and type-2) and united by linkers of 3-5 amino acids (Díaz et al., 2007). These modules appear to have evolved through the combinatorial accretion of repeat units of increasing length and complexity, built from two elementary types of motifs; C_2X_6 and CKCXXCKCCK (consensus) (where "X" represents any amino acid). With only minor modifications, these motifs have been successively combined into submodules, modules and finally completed and functional MTs, so an evolutionary history based on intra-gene duplications to originate these molecules might be possible (Díaz et al., 2007). *Tetrahymena* CuMTs do not present a so clear modular structure like CdMTs. Both, Cd and CuMTs from different *Tetrahymena* species, present 3D structures (inferred from their amino acid sequences and with different estimated precision values) similar to those from classic MTs, which possess a dumb-bell-like shape with almost spherical C-terminal (α-) and N-terminal (β-domains) (Klaassen et al., 1999; Vasák, 2005). In figure 6, two examples are shown.

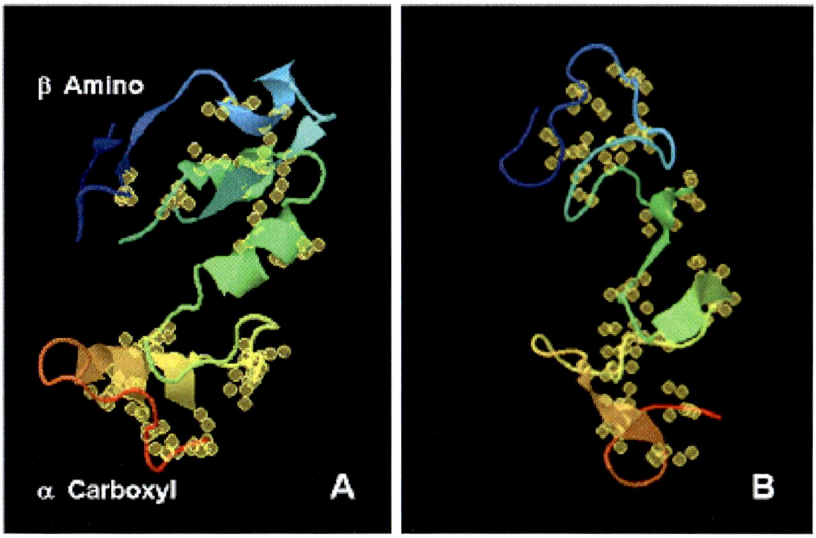

Figure 6. 3D-structures (inferred from the amino acid sequences) of two *Tetrahymena* MTs. Yellow balls: atoms of Cys radicals. (A): *T.thermophila* MTT3 (CdMT) (3D-estimated precision (EP) to standard MT-family: 90%, E-value: 0.11). (B): *T. thermophila* MTT2 (CuMT) (3D-EP: 95%, E-value: 0.073).

Some *Tetrahymena* CdMT promoters (like those from *TherMTT1*, *TherMTT5* or *TrosMTT1* genes) can be considered as "strong-promoters", because their gene expression levels are considerable, in a short time (≤ 1h), after metal stress induction. Indeed, *TherMTT1* (Shang et al., 2002), *TherMTT2/MTT4* (Boldrin et al., 2007) and *TherMTT5* (unpublished data from our research group) promoters have been used in gene constructions to express heterologous proteins. After a rigorous *in silico* analysis about the 5′-UTR sequences (putative promoter regions) of the three *T. thermophila* CdMT (*MTT1*, *MTT3* and *MTT5*) genes, we have reported (Díaz et al., 2007) the existence of a conserved motif (5′TGANTCATGA3′ or 5′TGANNNNTGA3′, where "N" means any nucleotide), which is present 11 times in *MTT5*, 6 times in *MTT1* (with two inverted copies contributing sequence to a perfect palindrome) and only 2 times en *MTT3* gene (the number of these motifs are correlated with the average expression levels for these genes (*MTT5* >> *MTT1* > *MTT3*). Likewise, this motif is present (5 times) in *TpyrMT-1* gene. This motif includes the sequence "TGANTCA" similar to that binding the eukaryotic AP-1 transcription factor (a member of the basic-leucine zipper (bZIP) superfamily) (Shaulian and Karin, 2002). Thus transcription factors with AP-1 like binding specificity may have a regulatory function in the expression of the subfamily 7a (CdMTs) of *T. thermophila*. A comparative analysis of potential transcription-related elements among *Tetrahymena* CdMTs has been carried out (Díaz et al., 2007), and, although it has been reported in the putative promoter regions of *TtherMTT1* or *TtherMTT2* (Boldrin et al., 2003; Boldrin et al., 2008), no canonical MRE (metal responsive elements) motifs have been identified after our computational search (Díaz et al., 2007). A metal responsive, zinc finger transcription factor (MTF-1) binds to promoter proximal MREs to induce transcription (Haq et al., 2003). While the *T. thermophila* genome contains numerous Zn finger proteins, none appears to be a true ortholog of MTF-1, therefore it seem to be that transcriptions of *Tetrahymena* MT genes are not regulated by MRE-like elements, similar to that occurring in the yeast *Saccharomyces cerevisiae*.

REACTIVE OXYGEN SPECIES FORMATION AND GLUTATHIONE METABOLISM IN CILIATES UNDER HEAVY METAL STRESS

Heavy metals may induce (directly or indirectly) oxidative stress in both prokaryotic and eukaryotic cells. Heavy metals with redox activity (Cu, among others) can directly give rise to reactive oxygen species (ROS) by Fenton/Haber-Weiss reactions or auto-oxidation. Besides, heavy metals without redox activity (like Cd or Zn) can indirectly also generate oxidative stress (superoxide, hydroxyl radicals, hydrogen peroxide) (Valko et al., 2005) by blocking or decreasing cellular antioxidant defenses. These defenses may be enzymatic (antioxidant enzymes, such as; glutathione peroxidase, catalase or superoxide dismutase) or non-enzymatic (gluthatione or metallothioneins, which have a protective effect against oxidative stress).

Studies on oxidative stress and reactive oxygen species generation induced by heavy metal exposure has been almost exclusively focused to mammalian cells. In protists the metal-induced ROS formation has been only studied in both photosynthetic and non-photosynthetic strains of *Euglena gracilis* exposed to Cd (Watanabe and Suzuki, 2002; Watanabe et al., 2003), and in ciliates, the only, at present, reported study has been in *T. thermophila* using specific fluorophores to detect mitochondrial peroxides formation and flow cytometry (Gallego et al., 2007). In this ciliate, Cd-treated cells generated H_2O_2 and probably also hydroxyl and peroxyl radicals in mitochondria, as it has been make evident by dihydrorhodamine (Rh123) derived fluorescence (Gallego et al., 2007) (figure 7). This fluorescence increased with increasing Cd concentration, like it was also reported in *E. gracilis* using the same methodology (Watanabe et al., 2003), reaching its highest level under a Cd concentration near of its LC_{50}. These findings support the affirmation that mitochondria are the main target organelles for the Cd toxic action (Valko et al., 2005).

Copper is an essential metal whose redox properties and coordination chemistry are useful for a number of catalytic and transport functions in living cells and their organelles, particularly mitochondria (Mehta et al., 2006). An excess of Cu causes cell toxicity which is usually explained by the assumption that Cu ions are prone to participate in the ROS formation. In *T. thermophila*, Cu is a less toxic metal than Cd but it originates quite similar fluorescence intensity than Cd, indicating a similar mitochondrial peroxides formation (Gallego et al., 2007).

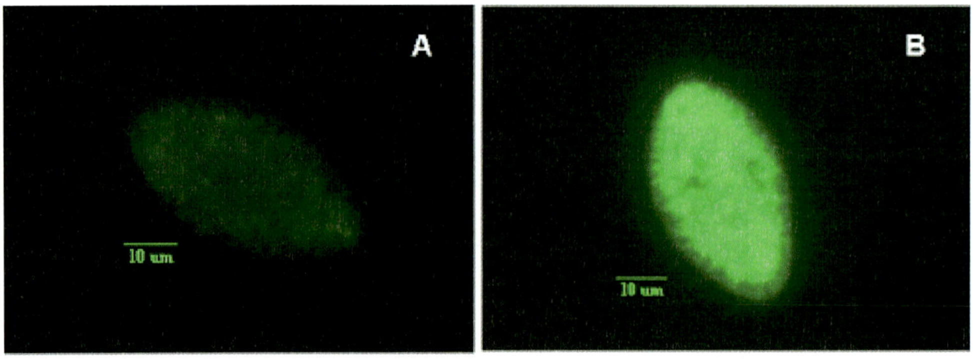

Figure 7. Fluorescence photomicrographs of *Tetrahymena* sp. (strain RT2) incubated with Rh123. (A): control. (B): Cd (0.5 mg/L) treated cells. Bar: 10 μm.

Zn is a redox inert metal that has been usually proposed as an antioxidant (Valko et al., 2005). However, increasing data have been published that support the ROS generation by an excess of Zn in mammals (Chimienti et al., 2001), invertebrates (Geret and Bebianno, 2004) and plants (Prasad et al., 1999). In *T. thermophila*, mitochondrial peroxides generation induced by Zn is quite similar to that generated by Cd (Gallego et al., 2007), despite the toxic effects of these two metals on this ciliate are very different. Another important point is the drastic increase of fluorescence intensity when cells were exposed to Zn concentrations higher to its LC_{50}. In author opinion, ROS generation in mitochondria might be one of the factors involved in the toxicity under an excess of Zn. Under these circumstances, cells become dead by apoptosis or necrosis.

About ROS generation using binary mixtures of heavy metals, when *T. thermophila* was exposed to Cd-Zn mixtures (at concentrations in which antagonism exists) a decreasing of rhodamine fluorescence was obtained (Gallego et al., 2007). According the authors, the presence of these two metals at low concentrations, produce an increase in the synthesis of Cd/ZnMTs, and, therefore, the new biosynthesized MTs are able to neutralize most of radicalts generated and, likewise, to produce an important decline in bioavailable metallic cations. *Tetrahymena* species possess MTs whose synthesis can be co-induced by Cd and Zn (Díaz et al., 2007). When metallic levels raised to 1 TU, the ratio between Cd/Zn affected to both fluorescence emission and biotoxicity, but under higher metal levels (1.25 TUs), only mortality was affected by the metal ratio, whereas the ROS generation was maintained quite similar in all Cd:Zn ratios. Results of ROS generation under Cu-Zn mixtures were quite similar to those from Cd-Zn treatments (Gallego et al., 2007). It is clear that at low concentrations, Zn acts as an antioxidant. Unlike other reported metal binary mixtures, in *T. thermophila* the metallic ratio is an important factor for both cytotoxicity and mitochondrial peroxides formation in Cd-Cu interaction. In mammals, it has been proposed that Cd can replace Cu in various cytoplasmic and membrane proteins, thus increasing the amount of unbound free Cu ions participating in oxidative stress via Fenton reactions (Valko et al., 2005). This Cd-induced displacement, may explain the enhanced toxicity that has been detected in *Tetrahymena* at high metal concentrations (1.25 TUs), as well as the peroxides generation increase that is detected under certain Cd/Cu ratios (Gallego et al., 2007).

Results from both cytotoxicity and mitochondrial ROS generation, by single and binary metal mixtures on the ciliate *T. thermophila*, support again the crucial role that metal concentration ratio plays on the bimetallic effect and therefore, on the type of metal interaction. The types of metal interactions, the factors controlling them and the influence of these interactions on cytotoxicity in unicellular and pluricellular organisms should be considered in environmental regulations and water quality guidelines.

Ciliate Glutathione Metabolism Enzymes and Heavy Metal Stress

Glutathione (GSH) is a tripeptide (L-γ-glutamyl-Lcysteinylglycine), which represents the most abundant cellular non-protein thiol in almost all eukaryotic groups (excepting parasitic protozoa lacking mitochondria) and in many Gram-negative and some Gram-positive bacteria. This molecule has multiple physiological roles in biological systems: scavenger of reactive oxygen species, detoxification of toxic agents (like heavy metals) (Anderson, 1998), amino acid import, protein and DNA synthesis and as cellular redox buffer (Meister and

Anderson, 1983). This tripeptide is synthesized by two dependent sequential reactions (figure 8); ligation of L-glutamate and L-cysteine by the catalytic action of the glutamylcysteine ligase (GCL) (or glutamylcysteine synthetase), and then the addition of glycine by the glutathione synthetase (GS). The first step of glutathione biosynthesis is crucial because it is the primary rate-limiting of the pathway, and the dipeptide can assume, in the absence of GSH, some roles of this molecule, e.g. as antioxidant (Grant et al., 1997). During the last years, we have studied several genes encoding diverse enzymes involved in the GSH metabolism using the ciliate *T. thermophila*. In figure 8, we represent the main steps of the GSH metabolism, indicating the enzymes whose gene expression levels we have analyzed (majority unpublished data) under different environmental stressors.

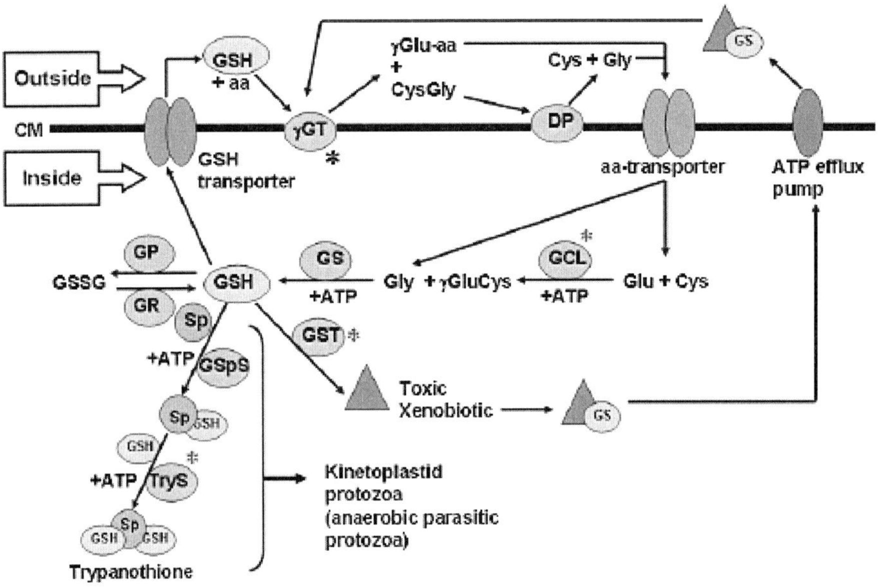

Figure 8. Main enzymes involve in the glutathione (GSH) metabolism. CM: cell membrane, aa: amino acid, GCL: glutamylcysteine ligase, GS: glutathione synthetase, GP: GSH peroxidase, GR: GSH reductase, GSSG: glutathione disulfide, γGT: γ-glutamyl transpeptidase (* the function of this enzyme might be substituted by the TtΨPCS), DP: dipeptidase, GST: glutathione-S-transferase, Sp: spermidine, GSpS: glutamylspermidine synthetase, TryS: trypanothione synthetase. Red-starts indicate genes encoding these enzymes whose expressions have been studied under different stress conditions. For details see the text.

Treatment with BSO (buthionine sulfoximine), which is an inhibitor of GCL, leads to cellular GSH levels decreasing, and its application may provide an useful experimental model of GSH deficiency (Anderson, 1998). Using the fluorophore monobromobimane (mBBr) and flow cytometry, we have analyzed (unpublished data) intracellular molecules with thiol groups (GSH, MTs and others) on cell cultures of the ciliate *Tetrahymena* sp. (strain RT1, isolated from a sample of Tinto river, Spain) treated or not with BSO and heavy metals (Cu or Cd). Results revealed that BSO treated cells showed a (Cu)LC_{50} value lower than control, indicating an increase of Cu toxicity after GSH depletion. Under the presence of Cu or Cd an induction of molecules with thiol groups was detected, while after BSO treatment the amount of these thiol molecules was lower. It might indicates that GSH formation is induced under metal stress, and its absence or lower intracellular amount affect the cellular heavy metal sensitivity level.

By BLAST searching in the already published macronuclear sequenced genome of the ciliate *T. thermophila*, we have found only one gene encoding a putative GCL, while *P. tetraurelia* has two genes (one of them is truncated). This GCL gene encodes a predicted protein of 405 amino acids (Ortega et al., 2007a), and, therefore, it seems to be that the GCL of *T. thermophila* looks like a monomeric enzyme as it occurs in bacteria, parasitic protozoa, fungi and plants (plastids), but unlike invertebrate and vertebrate animals, where GCLs are heterodimers, composed of both catalytic and regulatory subunits (Copley and Dhillon, 2002). *Tetrahymena* GCL sequence is really short in comparison with those from other organisms and microorganisms, including fungi (661-728 aa), parasitic protozoa (795-1161 aa), invertebrates (601-717 aa), mammals (601-637 aa) and even certain γ-proteobacteria or α-proteobacteria. The putative complete GCL sequence from *P. tetraurelia* has 677 aa (into the average length range of the majority of GCLs) (Ortega et al., 2007a). Therefore, the *Tetrahymena* GCL amino acid sequence length (405 aa) is unique among eukaryotic microorganisms, although its highest similarity is to those from other free-living protozoa, such as *P. tetraurelia* (30.2%) and *Dictyostelium discoideum* (21.7%). Then, the higher similarity (about 21%) is to mammals (including humans). A real time quantitative RT-PCR analysis (Ortega et al., 2007a) of this *Tetrahymena* GCL gene has showed that the highest gene expression inductions are detected under menadione (vitamin K3) treatment (an average value of 123 folds induction), acid pH (120 folds induction) and As (89 folds). Menadione is an oxidant or electrophile that leads to ROS formation, such as superoxide, peroxide and highly toxic hydroxyl radicals (Chung et al., 1999). The role of GSH in acid stress is less known, however in diverse bacteria it has been reported that GSH is involved in acid tolerance (Zang et al., 2007). Paraquat (PQ) is a quaternary nitrogen herbicide that is worldwide used in both terrestrial and aquatic environments. However, it is a highly toxic compound for humans and animals due to its capacity to generate superoxide anions, which can lead the hydrogen peroxide and hydroxyl radical formation. Like it occurs in yeasts (Valko et al., 2006), PQ (24h treatment) also induces certain over-expression (6 folds induction) of the *Tetrahymena* GCL gene (Ortega et al., 2007a). As it might be expected, non essential metals (As and Cd) cause a relative high induction of GCL gene expression (Ortega et al., 2007a). In summary, the GCL gene expression in *T. thermophila* is mainly increased as a response to environmental agents inducing oxidative stress, but it is also induced, at a lower level, by other environmental pollutants.

Another important enzymes involved in the GSH metabolism are glutathione-S-transferases (GST), which represent a major group of detoxification enzymes in all eukaryotic (and some prokaryotic) species. One of the main functions of these enzymes is the conjugation of foreign (mainly xenobiotics) compounds with GSH, almost always leading to formation of less reactive (toxic) products that are readily excreted outside the cell by ATP-depending efflux pumps (figure 8). They constitute a super-family composed by three main GST sub-families; soluble or cytosolic, membrane-bound (microsomal) and bacterial GST isoenzymes. Cytosolic enzymes have been classified, based on their amino acid sequence similarities, substrate specificity, 3D-structure and immunological properties into, at least, 12 distinct classes; 3 main-stream classes (Zeta, Theta and Omega), 4 animal-specific (Sigma, Alpha, Mu and Pi), 2 insect-specific (Delta and Epsilon) and 3 plant-specific (Phi, Tau and Lambda) (Hayes et al., 2005). Multigene families arise from single or multiple genes by DNA duplication. Functional redundancy and overlapping specificities are often characteristic of large multigene families. All cytosolic GST have two different subunits (25-27 KDa each), encoded by different genes, and they form homodimers or heterodimers. The gene expression level of GST is a crucial factor in determining the sensitivity of cells to a broad spectrum of toxic chemicals.

Table 5. GST class distribution among selected organisms

Class	Vertebrates		Insects	Plants		Fungi		Ciliates		Prokaryotes		
	A	B	C	D	E	F	G	H	I	J	K	L
Beta										5	1	16
Theta	2	3	4	3	2			5	5			
Zeta	1	1	2		3			2	5			
Omega												
Sigma	2	2	4			3	3	7	7			
	1	1	1									
Alpha	6	6										
Mu	5	6						47	8			
Pi	1	2										
Tau				28	39							
Phi				13	16				2			
Lambda				2	1							
Delta			10									
Epsilon			10						2			
Undetermined								2	2			
Total	18	21	31	46	61	3	3	63	31	5	1	16

A: Homo sapiens, B: Mus musculus, C: Drosophila melanogaster, D: Arabidopsis thaliana, E: Oriza sativa, F: Saccharomyces cerevisiae, G: Aspergillus fumigatus, H: Tetrahymena thermophila, I: Paramecium tetraurelia. J: Escherichia coli, K: Bacillus cereus, L: Pseudomonas aeruginosa.

From the complete, already published, macronuclear genome sequence of the ciliate *T. thermophila*, we have detected, at least 63 genes encoding putative GST (Ortega et al., 2007b). These GST have been assigned to the following classes; 7 to the class Omega (91% identity among them), 5 to the class Theta (86% identity among them), 2 to the class Zeta (65% identity among them), 47 to the class Mu (90% identity among them) and 2 have not been assigned to any known class. Theta and Omega classes are present in all eukaryotic organisms, while the Mu class is vertebrate-specific. In the table-5, we compare the GST class distribution among some selected organisms in comparison with 2 ciliates (*T. thermophila* and *P. tetraurelia*), which are the only, at present reported, eukaryotic microorganisms with Mu GST (57-90% identity for *T. thermophila* to human Mu GST). At present, it constitutes the highest number of Mu GST genes present in an organism, although several of them might be pseudo-genes without a real metabolic relevance. *Tetrahymena* Mu

GST might represent a primitive version of vertebrate Mu class GST. This new knowledge originates the necessity of an intense revision of both the GST classification and the evolution story of these enzymes. Ciliate GST can be classified into 4 main classes (Mu, Omega, Theta and Zeta) (table-5).

Another ciliate GST has been reported in *Blepharisma japonicum* (Takada et al., 2004), but authors consider that it constitutes a new class of GST because the similarity level with regard known GST is really low (9.6-23%). All cytosolic GST act as dimers (Armstrong, 1997), and monomers must be from the same class to form functional dimers. The high number of different subunits from each class make possible a great number of combinations, creating a lot of different isoenzymes from a limited number of genes. In *T. thermophila* (Overbaugh et al., 1988), a cytosolic GST was isolated and purified (but not sequenced), whose functional form is monomeric, being the first case where a functional monomeric GST is reported.

The majority of *Tetrahymena* GST family genes (51%) are clustered in the macronuclear-genome of this ciliate, and due to the high homology among their DNA sequences probably these genes were originated by gene duplication. Information on the gene expression of these genes has been obtained by both, quantitative RT-PCR analysis and screening of *Tetrahymena* EST databases (starvation, conjugation) (Ortega et al., 2007b), from which we can conclude the following points; 1- The majority of expressed GST genes, from the same class, are un-clustered in the macronuclear-genome of *Tetrahymena* (expressed un-clustered: 29%; expressed clustered: 13%), 2- about a 28% of Mu GST genes are expressed under diverse stress conditions, and 3- several GST genes are differentially over-expressed under a specific stress, for instance; the TtGSTM9 (class Mu member) is about 5-folds induced under Zn treatment and 64-folds induced after CDNB (1-chloro-2,4-dinitrobenzene), however this gene is down-expressed under Cd, Cu or PQ treatments (Ortega et al., 2007b). Other microbial GST are also induced by (or related to) heavy metals, such as; GSTIII of *Schizosaccharomyces pombe* (Shin et al., 2002), or *Aspergillus nidulans* (Fraser et al., 2002), where strains with a deletion in the gene *gstA* (encoding a GST) are more sensitive to certain heavy metals (Pb, Se or Ni) than wild-type. Our actual knowledge on the interaction GST-heavy metals is very scanty, although some experimental data, carried out in the yeast *Saccharomyces cerevisiae* (Adamis et al., 2004), have proposed that the formation of "Cd-GSH" complexes is depending on GST activity, therefore a more intensive experimentation is necessary to elucidate these interactions.

Kinetoplastid protozoa (*Trypanosoma, Crithidia, Leishmania*), which are anaerobe parasites, have the biosynthetic capacity of glutathione conjugates formation, such as; N^1-monoglutathionylspermidine (GSp) (initially detected in *Escherichia coli* and other bacteria) and N^1-N^8-bis(glutathionyl)spermidine or trypanothione (Try) (absent in *E. coli*). It is named the trypanothione system which is a metabolic pathway unique to trypanosomatids (Müller et al., 2003). This system replaces the glutathione peroxidase (GP) / glutathione reductase (GR) system (figure 8) present in the majority of aerobe organisms, and it protects the parasites from oxidant damage and toxic heavy metals. Trypanothione is synthesized by two ATP-dependent reactions, from GSH and the polyamine spermidine (Sp), by the enzymes glutathionylspermidine synthetase (GSpS) and trypanothione synthetase (TryS) (figure 8). Both enzymes (GSpS/A and TryS/A) are dimeric, and each monomer has two domains [N-terminal amidase (A) and C-terminal synthetase (S)] with opposed catalytic activities; amide bond-forming and cleaving reactions (Wang and Ballatori, 1998).

After exploring the sequenced macronuclear genomes of the ciliates *T. thermophila* and *P.tetraurelia*, we have detected diverse gene isoforms encoding homologous products to both enzymes (GSpS/A or TryS/A) (Ortega et al., 2007c). *T. thermophila* has 4 isoforms which only present the synthetase domain, although one of them is incomplete without the essential amino acids for catalytic activity, so probably it is not a functional enzyme. The rest of them have an identity range of 62-65% to kinetoplastid TryS and a 64% to bacterial GSpS. The isoform TtTryS4 has a N-terminal domain different to amidase domain, named "von Willebrand" domain. *P. tetraurelia* has 8 isoforms, and contrary to TtTryS they only have the N-terminal amidase domain without synthetase domain. They present an identity range of 63-69% to kinetoplastid TryS and a 63-67% to bacterial GSpS. The homology among all putative TtTryS genes is really high (about 87%), which make difficult to design probes or primers to study their gene expressions. However, a preliminary expression analysis has shown certain expression level after PQ or Cd treatments for some genes (Ortega et al., 2007c). Besides, *T. thermophila* has two putative trypanothione reductase genes, but any of them seem to be expressed under diverse stress conditions. We do not know yet if trypanothione is really biosynthesized by, at least, *T. thermophila*, or what is the real function of these putative TtTryS genes. From several *P. tetraurelia* gene expression libraries (growth-division cycle, conjugation and heat shock stress), at least three cDNA clones encoding putative PtTryA have been isolated, indicating the expression of some of these genes. In a next future will be very necessary to study with much more dedication the real function of these genes (and their products) present in these free-living aerobe protozoa.

Ciliate Apoptosis and Heavy Metals

Apoptosis is a form of physiological programmed cell death present in eubacteria, fungi and multicellular organisms. However, recently, this process has been also detected in both, parasite and free-living protozoa (Fuertes et al., 2008). Several programmed cell death (including apoptosis) have been describe in kinetoplastid protozoa (*Trypanosoma, Leishmania, Crithidia*) (Sen et al., 2008; Soeiro and Souza, 2008), dinoflagellates (Segovia, 2008) and in the free-living ciliate *T. thermophila* (Martín-González et al., 2008). An apoptotic-like phenomenon (programmed nuclear death) exits in ciliate conjugation (e.g., *Tetrahymena* conjugation) and during ciliate encystment (Gutiérrez et al., 2001), and it consists in a controlled macronuclear DNA degradation occurring in both differentiation processes. These processes are still poorly studied at molecular level, and, at present, we do not know the genetic elements involved in them.

Recently, we have started (Gallego et al., 2007) the study of apoptosis induced by heavy metals in the ciliate *T. thermophila*, and the initial results may be summarized in the following points: 1- heavy metals (Cd, Cu or Zn in single form or binary mixtures) induce apoptosis type-I, according to the analysis of some morphological features. 2- all used heavy metals induce the formation of apoptotic macronuclear bodies (AMB) (macronuclear DNA extrusion bodies that are easily detected by DAPI staining) (figure 9). The average AMB number (1-3 /per cell) increases with the heavy metal concentration, being Cd > Zn > Cu the metal order rank inducing cells with AMB. 3- in metal binary mixtures, Cd-Zn or Cu-Zn synergisms and Cd-Cu additive interaction are the metal mixtures inducing the higher numbers of cells with amb. 4- heavy metals induce DNA macronuclear degradation, which can be detected (ladders of bands) by standard electrophoresis. Likewise, inducers of eukaryotic apoptosis (Doxorubicin or Camptothecin) present similar effects on the macronuclear DNA of *T. thermophila*. 5- annexin V-FITC assays (detecting externalization of membrane phosphatidylserine) and flow cytometry quantification (with or without the vital dye propidium iodide (PI) to elucidate between apoptosis and necrosis) during early (1h) and late (24h) apoptosis progression induced by heavy metals, have shown that; a)- the cell number undergoing both processes (apoptosis or necrosis) increase with the metal concentration. b)- at early time (1h) Cd originates more cellular necrosis than apoptosis with regard Cu or Zn (that induce the contrary effect). c)- at late time (24h) the number of cells undergoing necrosis is higher than that for apoptosis, in all heavy metal treatments.

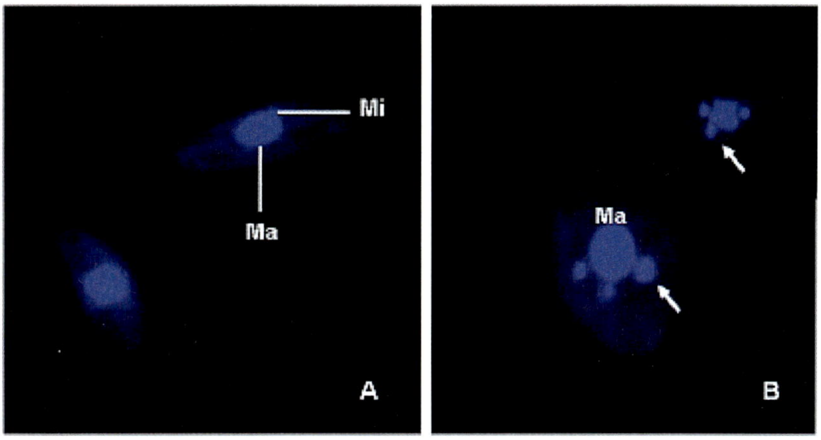

Figure 9. DAPI stain micrographs of *Tetrahymena thermophila*. (A): Control (Ma: macronucleus, Mi: micronucleus). (B): Cd (0.19 mg/L) treated cells (arrows indicate apoptotic macronuclear bodies) (400x).

From these provisional results we might conclude that heavy metals induce in ciliates an apoptotic process very similar to that induced by eukaryotic apoptosis inducers, but showing some specific features.

Ciliates as Whole-Cell Biosensors

In environmental biomonitoring, global parameters such as bioavailability, toxicity and genotoxicity can not be tested using molecular recognition or chemical analysis, but can only be assayed using whole cells. Obviously, in these bioassays the question to be resolved is not "What toxicants does the sample contain?, but rather "How toxic is the sample?. Recently, the concept of whole-cell biosensor has been introduced by several authors (Belkin, 2003; D´Souza, 2001), as a very useful alternative to classical biosensors. A whole-cell biosensors uses the whole prokaryotic or eukaryotic cell as a single reporter incorporating both bioreceptor and transducer elements. In general, living systems to be used as whole-cell biosensors are experimentally modified cells to incorporate the transducer capacity.

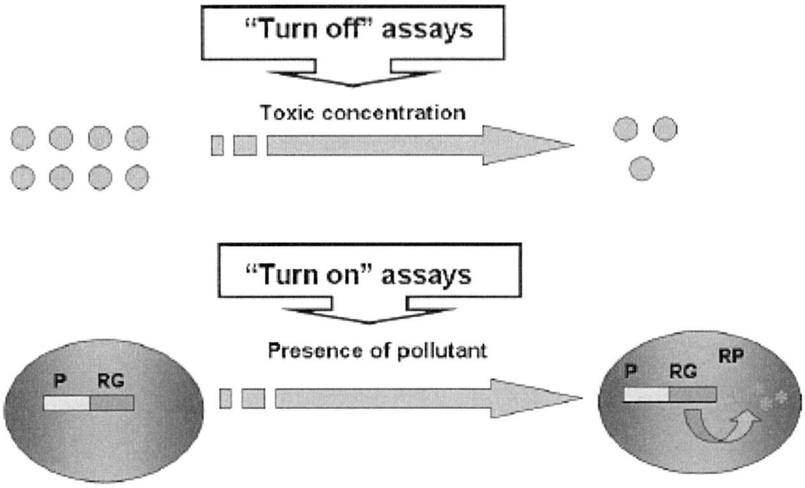

Figure 10. Schematic representation of "Turn off" and "Turn on" bioassays using whole-cell biosensors. P: promoter; RG: reporter gene; RP: reporter protein.

Two types of bioassays using whole-cell biosensors may be considered; "turn off" and "turn on" assays (Belkin, 2003). In "turn off" assays (which are similar to general microbial toxicological bioassays) (figure 10), the sample toxicity is estimated from the degree of inhibition of a cellular activity (e.g. inhibition of growth, respiration, motility or the biosynthesis of a specific molecule), and it is based on the measurement of a decrease in growth rate, light (fluorescence/bioluminiscence) emission, colour-less cell population, motility, etc., as a function of sample toxic concentration. An example of a "turn off" assay using ciliates, may be the bioassay to detect mycotoxins using a melanin precursor overproducer mutant of *T. thermophila* (Martín-González et al., 1997). On the other hand, in "turn on" assays (figure 10) a quantifiable molecular reporter is fused to a specific gene promoter, known to be activated by the target chemical or environmental pollutant (such as heavy metals). Ciliates, as it has been previously discussed in this review, present a lot of very good qualities to be used as whole-cell biosensors to assess environmental pollutants (including heavy metals) (Gutiérrez et al., 2003a; 2004). One of the best candidates to be used as gene promoter to make up a ciliate whole-cell biosensor for "turn on" assays are *Tetrahymena* metallothionein gene promoters (like promoters from *MTT1* or *MTT5* genes of *T. thermophila*), which respond rapidly and strongly to heavy metal exposure (Díaz et al., 2007; Shang et al., 2002). Indeed, in our research group and in collaboration with Dr. A. Turkewitz (University of Chicago, USA), we have developed several types of whole-cell biosensors in *T. thermophila* using *MTT1* or *MTT5* promoters to construct promoter-reporter gene fusions with the system pBTU1/BTU2 (Gaertig et al.,1994) (using the luciferase as reporter gene), and so obtaining stable transformed cells, emitting a quantifiable luminescence response mainly (but not exclusively) under the presence of heavy metals. Actually, we are carrying out the validation of these potential whole-cell biosensors using both artificial and natural heavy metal polluted samples.

Conclusion

From all data reported in this review, we should conclude that ciliates (and in particular the ciliate model *Tetrahymena thermophila*) are eukaryotic microorganism models very useful to study the heavy metal-eukaryotic cell interactions. Their, in general, higher sensitivity to heavy metals is a good feature to develop bioassays for these toxic environmental pollutants. Their fast growth and easy laboratory manipulation make that these microorganisms become an excellent living material for fast bioassays with regard more complex multicellular organisms. Their animal biology similar to human cell biology, as it has been probated after macronuclear genome sequencing of two ciliate models (*Tetrahymena* and *Paramecium*), make more reasonable to use them in ecotoxicological studies as eukaryotic models for humans, and, likewise, it represents a good alternative to animal tests. And, finally, their metallothioneins, involved in their main resistant heavy metal mechanism (bioaccumulation), which present unique and different features with respect to classic metallothioneins, constitute fundamental elements to design classic and /or whole-cell biosensors. Ciliate-heavy metal interaction studies are now starting a new phase, and probably from these studies new surprises and exciting ideas will raise during the next years.

Acknowledgements

The research work on ciliate-heavy metal interactions, carried out during the last five years, has been supported by the following grants; BOS2002-01067, 07M/0029/2002 (CAM) and CGL2005-00548/BOS. Ph.D. scholarships from University Complutense (UCM) are awarded to F.A. and R.O. and from Spanish government (FPI) to A.G.

References

Abeliovich, H. and Klionsky, D.J. (2001) Autophagy in yeasts: Mechanistic insights and physiological function. *Microbiology and Molecular Biology Reviews 65,* 463-479.

Adamis, P.D., Gomes, D.S., Pinto, L.M., Panek, A.D. and Eleutherio, E.C. 2004. The role of glutathione transferases in cadmium stress. *Toxicology Letters 154,* 81-88.

Aguilar, J., Dorronsoro, C., Fernández, E., Fernández, J., García, I., Martín, F. et al. Soil pollution by a pyrite mine spill in Spain: evolution in time. *Environmental Pollution 132,* 395-401.

Amaro, F., Ruotolo, R., Martín-González, A., Faccini, A., Ottonello, S. and Gutiérrez, J.C. (2006). Ciliate phytochelatins? Isolation, cloning, gene characterization and expression analysis of a macronuclear gene encoding a putative phytochelatin synthase from *Tetrahymena thermophila. 2nd FEMS Congress of European Microbiologists.* Madrid (Spain).

Amaro, F., Martín-González, A. and Gutiérrez, J.C. (2007). Two new members of the *Tetrahymena* multi-stress-inducible metallothionein family: *T. rostrata* Cd/Cu metallothioneins. *II International Conference on Environmental, Industrial and Applied Microbiology (BioMicroWorld2007).* Seville (Spain).

An, Y-J., Kim, Y-M., Kwon, T-I. and Jeong, S.-W. (2004). Combined effect of copper, cadmium, and lead upon *Cucumis sativus* growth and bioaccumulation. *Science of the Total Environment 326,* 85-93.

Anderson, M.E. (1998). Glutathione: an overview of biosynthesis and modulation. *Chemico-Biological Interactions 111-112,* 1-14.

Armstrong, R.N. (1997). Mechanicistic imperatives for the evolution of glutathione transferases. *Current Opinion in Chemical Biology 2,* 618-623.

Aury, J.-M., Jaillon, O., Duret, L., Noel, B., Jubin, C. et al. (2006). Global trends of the whole-genome duplications revealed by the ciliate *Paramecium tetraurelia. Nature 444,* 171-178.

Bagchi, D., Joshi, S.S., Bagchi, M., Balmoori, J., Benner, E.J., Kuszynski, C.A. et al. (2000). Cadmium- and chromium-induced oxidative stress, DNA damage, and apoptosis cell death in cultured human myelogenous leukemic K562 cells, promyelocytic leukemic HL-60 cells, and normal peripheral blood mononuclear cells. *Journal of Biochemical and Molecular Toxicology 14,* 33-41.

Barata, C, Markich, S.J., Braid, D.J., Taylor, G. and Soares, A.M.V.M. (2002). Genetic variability in sublethal tolerance to mixtures of cadmium and zinc clones of *Daphnia magna* Straus. *Aquatic Toxicology 60,* 85-99.

Belkin, S. (2003) Microbial whole-cell sensing systems of environmental pollutants. *Current Opinion in Microbiology 6,* 206-212.

Benítez, L., Martín-González, A. and Gutiérrez, J.C. (1991). Protein glycosylation has an important role in the encystment process of the ciliate *Colpoda inflata. Cell Biology International 15,* 221-228.

Blum, R., Back, A., Korte, A., Stengel, A., Letzel, T., Lendzian, K. and Grill, E. (2007). Function of phytochelatin synthase in catabolism of gluthatione-conjugates. *Plant Journal 49,* 740-749.

Boldrin, F., Santovito, G., Irato, P. and Piccinni, E. (2002). Metal interaction and regulation of *Tetrahymena pigmentosa* metallothionein genes. *Protist 153,* 283-291.

Boldrin, F., Santovito, G., Negrisolo, E. and Piccinni, E. (2003). Cloning and sequencing of four new metallothionein genes from *Tetrahymena thermophila* and *Tetrahymena pigmentosa*: evolutionary relationships in *Tetrahymena* MT family. *Protist 154,* 431-442.

Boldrin, F., Santovito, G., Formigari, A., Bisharyan, Y., Cassidy-Hanley, D., Clark, T.G. and Piccinni, E. (2008). MTT2, a copper-inducible metallothionein gene from *Tetrahymena thermophila. Comparative Biochemical Physiology C 147,* 232-240.

Boldrin, F., Santovito, G., Gaerting, J., Wloga, D., Cassidy-Hanley, D., Clark, T.G., and Piccinni, E. (2006). Metallothionein gene from *Tetrahymena thermophila* with a copper-inducible-represible promoter. *Eukaryotic Cell 5,* 422-425.

Bonkowski, M. (2004). Protozoa and plant growth: the microbial loop in soil revisited. *New Phytologist 162,* 617-631.

Brierley, C.L. (1990). Bioremediation of metal contaminated surface and ground water. *Geo-microbiology Journal 8,* 201-233.

Bursch, W. (2001) The autophagosomal-lysosomal compartment in programmed cell death. *Cell Death and Differentiation 8,* 569-581.

Cairns, Jr J. (1986). *Multispecies toxicity testing.* New York: Pergamon Press.

Cairns, J. Jr, Bidwell, J.R. and Amegard, M.E. (1996).Toxicity testing communities: Microcosms, mesocosms and whole-system manipulations. *Reviews in Environmental Contamination and Toxicology 147,* 45-69.

Campbell, C.D., Warren, A., Cameron, C.M. and Hope, S. J. (1997). Direct toxicology assessment of two soils amended with sewage sludge contaminated with heavy metals using a protozoan (*Colpoda steinii*) bioassay. *Chemosphere 34,* 501-514.

Chang, L.W. (1996). *Toxicology of Metals.* Boca Raton, Lewis Publishers.

Chimienti, F., Jourdan, E., Favier, A., Seve, M. (2001). Zinc resistance impairs sensitivity to oxidative stress in HeLa cells : protection through metallothionein expression. *Free Radicals in Biology and Medicine 31,* 1179-1190.

Chen, P.S. and Li, C.S. (2005). Reat-time quantitative PCR with gene probe fluorochrome and floe cytometry analysis. *Journal of Environmental Monitoring 7,* 257-262.

Christensen, S.T., Chemnitz, J., Straarup, E., Kristiansen, K., Wheatley, D.N. and Rasmussen, L (1998). Staurosporine-induced cell death in *Tetrahymena thermophila* shows mixed characteristics to both apoptotic and autophagic degeneration. *Cell Biology International 22,* 581-598.

Chung, S.H., Chung, S.M., Lee, J.Y., Kim, S.R. et al. (1999). The biological significance of non-enzymatic reaction of menadione with plasma thiols: enhancement of menadione-induced cytotoxicity to platelets. *FEBS Letters 449,* 235-240.

Clarholm, M. (2005). Soil protozoa: an under-researched microbial group gaining momentum. *Soil Biology and Biochemistry 37,* 811-917.

Clemens, S., Kim, E.J., Neumann, D. and Schroeder, J.I. (1999). Tolerance to toxic metals by a gene family of phytochelatin synthases from plants and yeast. *EMBO Journal 18,* 3325-3333.

Cobbett,C. (2000). Phytochelatin biosynthesis and function in heavy-metal detoxification. *Current Opinion in Plant Biology 3,* 211-216.

Cobbett, C. and Goldsbrough, P. (2002). Phytochelatins and metallothioneins: roles in heavy metal detoxification and homeostasis. *Annual Reviews in Plant Biology 53,* 159-182.

Copley, S.D. and Dhillon, K.K. (2002) Lateral gene transfer and parallel evolution in the history of glutathione biosynthesis. *Genome Biology 3,* (5): research0025.1-0025.16.

Coppellotti, O. (1994). Effects of cadmium on *Uronema marinum* (Ciliophora, Scuticociliatida) from Antartica. *Acta Protozoologica 33,* 159-167.

Coyle,P., Philcox, J.C., Carey, L.C. and Rofe, A.M (2002). Metallothionein: the multipurpose protein. *Cellular and Molecular Life Sciences 59,* 627-647.

Dayed, V.R., Lynn, D.H. and Bols, N.C. (2005). Cytotoxicity of metals common in mining effluent to rainbow trout cell lines and the ciliated protozoan, *Tetrahymena thermophila. Toxicology in Vitro 19,* 399-410.

Dias, N. and Lima, N. (2002). A comparative study using a fluorescence-based and a direct count assay to determine cytotoxicity in *Tetrahymena pyriformis. Reseach in Microbiology 153,* 313-322.

Díaz, S., Martín-González A. and Gutiérrez, J.C. (2006). Evaluation of heavy metal acute toxicity and bioaccumulation in soil ciliated protozoa . *Environment International 32,* 711-717.

Díaz, S., Amaro, P., Rico, D., Campos, V., Benítez, L., Martín-González, A., Hamilton, E., Orias, E. and Gutiérrez, J.J. (2007). *Tetrahymena* metallothioneins fall into two discrete subfamilies. *PLoS ONE 2,* e291.

Domènech, J., Bofill, R., Tinti, A., Torreggiani, A., Atrian, S. and Capdevilla, M. (2008). Comparative insights into the Zn(II)-, Cd(II)- and Cu(I)-binding features on the protozoan *Tetrahymena pyriformis* MT1 metallothionein. *Biochimica and Biophysica Acta, Proteins and Proteomics* doi: 10.1016/j.bbapap.2008.01.008.

Dondero, F., Cavaletto, M., Ghezzi, A.R., La Terza, A., Banni, M. and Viarengo, A. (2004). Biochemical characterization and quantitative gene expression analysis of the multi-stress inducible metallothionein of *Tetrahymena thermophila. Protist 155,* 157-168.

Dressen; P., Zaguiski, M., Gromadka, R., Plattner, H., Kissmehl, R., Meyer, E., Bétemier, M., Schultz, J.E. et al. (2001). *Paramecium* genome survey: A pilot project. *Treds in Genetics 17,* 306-308.

D'Souza, S.F. (2003). Microbial biosensors. *Biosensors and Bioelectronic 16,* 337-353.

Dunlop, S. and Chapman, G. (1981). Detoxication of zinc and cadmium by the freshwater protozoan *Tetrahymena pyriformis*. II. Growth experiments and ultrastructural studies on sequestration of heavy metals. *Environment Research 24,* 264-274.

Eisen, J.A., Coyne, R.S., Wu, M., Wu, D. and Thiaragarajan M., et al. (2006) Macronuclear genome sequence of the ciliate *Tetrahymena thermophila*, a model eukaryote. *PloS Biology 4,* c286.

Ekelund, F., Frederiksen H.B. and Ronn, R. (2002). Population dynamics of active and total ciliate populations in arable soil amended with wheat. *Applied and Environmental Microbiology 68,* 1096-1001.

Ercal, N., Gurer-Orhan, H. and Aykin-Bums N. (2001).Toxic metals and oxidative stress part I: mechanisms involved metal-induced oxidative damage. *Current Topics in Medicine and Chemistry 1,* 529-535 .

Fargasová, A. (2001). Winter third and fourth-intar larvae of *Chironomus plumosus* as bioassay tools for assessment the acute toxicity of metals and their binary combinations. *Ecotoxicology and Environmental Safety 48,* 1-5.

Finlay, B.J. and Fenchel T. (1996) Role of ciliates in natural environment. In K. Haussmann and P.C. Bradbury (Eds), *Ciliates. Cells as organisms* (pp. 417-440). Stuttgart: Gustav Fischer.

Finlay, B.J., Black H.I.J., Brown, S., Clark, K.J., Esteban G.F., Hinde R.M., Olmo J.L., Rollett, A. and Vickerman, K. (2000). Estimating the growth potential of the soil protozoan community. *Protist 151,* 69-80.

Foissner, W. (1993). *Colpodea (Ciliophora). Protozoenfauna, vol. 1/4.* Stuttgart: Gustav Fischer Verlag.

Forbes, V.E. (1998). *Genetics of ecotoxicology.* Ann Arbor: Taylor and Francis.

Forge, T.A., Berrow, M.L., Darbyshire, J.F. and Warren, A. (1993). Protozoan bioassays of soil ciliate *Colpoda steinii. Biology and Fertility of Soils 16,* 282-286.

Fu, C. and Miao, W. (2006). Cloning and characterization of a new multi-stress inducible metallothionein gene in *Tetrahymena pyriformis. Protist 157,* 193-203.

Fuertes, M.A., Nguewa, P.A., Castilla, J., Alonso, C. and Pérez Martín, J.M. (2008). Programmed cell death in Protozoa: An evolutionary point of view. The example of Kinetoplastid parasites. In J.M. Pérez Martín (Ed). *Programmed Cell Death in Protozoa* (pp. 1-6). Austin, New York: Landes Bioscience,Springer Science.

Fraser, J.A., Davis, M.A. and Hynes, M.J. (2002). A gene from *Aspergillus nidulans* with similarity to URE2 of *Saccharomyces cerevisiae* encodes a glutathione S-transferase which contributes to heavy metal and xenobiotic resistance. *Applied and Environmental Microbiology 68,* 2802-2808.

Gadd, G.M. (2000). Heavy metal pollutants: environmental and biotechnological aspects. In J. Lederberg (Ed). *Encyclopedia of Microbiology* (2 nd edition. Pp. 607-617). New York: Academic Press.

Gadd, G.M. (1993). Interactions of fungi with toxic metals. *New Phytologist 124,* 25-60.

Gadd, G.M. and White, C. (1993). Microbial treatment of metal pollution- a working biotechnology?. *Trends in Biotechnology 11,* 353-359.

Gaertig, J., Gu, L. and Gorovsky, M.A. (1994). High frequency vector-mediated transformation and gene replacement in *Tetrahymena. Nucleic Acid Research 24,* 5391-5398.

Gaither, L.A and Eide, D.J. (2001). Eukaryotic zinc transporters and their regulation. *Biometals 14,* 251-270.

Gallego, A., Martín-González, A., Ortega, R. and Gutiérrez, J.C. (2007). Flow cytometry assessment of cytotoxicity and reactive oxygen species generation by single and binary mixtures of cadmium, zinc and copper on populations of the ciliated protozoan *Tetrahymena thermophila. Chemosphere 64,* 647-661.

Geret, F., Bebianno, M.J. (2004). Does zinc produce reactive oxygen species in *Ruditapes decussatus* ?. *Ecotoxicology and Environmental Safety 57,* 399-409.

Gil, J., Oliveira, C.R. and Rago, A.C. (2003). Cytosolic and mitochondrial ROS in staurosporine-induced retinal cell apoptosis. *Free Radicals in Biology and Medicine 35,* 1500-1514.

Glick, B. R. (2003). Phytoremediation: synergistic use of plants and bacteria to clean up the environment. *Biotechnological Advances 21,* 383-393.

Glynn, A.W. (2001). The influence of zinc on apical uptake of cadmium in the gills and cadmium influx in the circulatory system in zebrafish (*Danio rerio*). *Comparative Biochemistry and Physiology Part C 128,* 165-172.

Gowri, P.M. and Srivastava, S. (1996). Reduced uptake based zinc resistance in *Azospirillum brasilense* sp7. *Current Science 71,* 139-142.

Grant, C.M., MacIver, F.H. and Dawes, I.W. (1997). Glutathione synthetase is dispensable for growth under both normal and oxidative stress conditions in the yeast *Saccharomyces cerevisiae* due to the accumulation of the dipeptide gamma-glutamylcysteine. *Molecular Biology of the Cell 8,* 1699-1707.

Grill, E., Winnaker, E.L. and Zenk, M.H. (1986). Synthesis of seven different homologous phytochelatins in metal-exposed *Schizosaccharomyces pombe* cells. *FEBS Letters 197,* 115-120.

Grill, E., Löeffer, S., Winnacker, E.L. and Zenk, M.H. (1989). Phytochelatins, the heavy-metal-binding peptides from plants, are synthetized from glutathione by a specific gamma-glutamylcysteine dipeptidyl transpeptidase (phytochelatin synthase). *Proceedings of the Natural Academy of Sciences USA 86,* 6838-6842.

Grill, E., Thumann, J, Winnacker, E.L. and Zenk, M.H. (1988). Induction of heavy metal phytochelatins by inoculation of cell cultures in standard media. *Plant Cell Reports 7,* 375-378.

Greene, B., McPherson R, and Darnall, D. (1987). Algal sorbents for selective metal ion recovery. In J.W. Patterson and R. Pasino (eds). *Metal speciation, separation and recovery* (pp. 315-338). Chelsea: Lewis.

Gutiérrez, J.C., Martín-González, A. and Matsusaka, T. (1990). Towards a generalized model of encystment (cryptobiosis) in ciliates: a review and a hypothesis. Biosystems 24, 17-24.

Gutiérrez, J.C., Martín-González, A. and Callejas, S. (1998). Nuclear changes, macronuclear chromatin reorganization and DNA modifications during ciliate encystment. *European Journal of Protistology 34,* 97-103.

Gutiérrez, J.C., Callejas, S., Borniquel, S., Benítez, L. and Martín-González, A. (2001). Ciliate cryptobiosis: a microbial strategy against environmental starvation. *International Microbiology 4,* 151-157.

Gutiérrez, J.C. and Martín-González, A. (2002). Ciliate encysment-excystment cycle: a response to environmental stress. In J. C. Gutiérrez (Ed.), *Microbial development under environmental stress* (pp. 29-49). Kerala: Research Singpost.

Gutiérrez, J.C., Martín-González, A., Díaz, S. and Ortega, R. (2003a). Ciliates as potencial source of cellular and molecular biomarkers/biosensors for heavy metal pollution. *European Journal of Protistology 39,* 461-467.

Gutiérrez, J.C., Díaz,S., Ortega,R. and Martín-González, A. (2003b). Ciliate resting cyst walls: a comparative review. *Recent Research Developments in Microbiology 7,* 361-379.

Gutiérrez, J.C., Martín-González, A., Díaz, S., Ortega, R., Amaro, P., Gallego, A., Campos, V. and Plana, M.P. (2004). Heavy metal bioaccumulation and metallothioneins in ciliates: implications for their use as whole-cell and molecular biosensors. *Current Trends in Microbiology 1,* 51-58.

Haq, F., Mahoney, M. and Koropatnick, J. (2003). Signalling events for metallothionein induction. *Mutation Research 533,* 211-226.

Harada, E., von Roepenack-Lahaye, E. and Clemens, S. (2004). A cyanobacterial protein with similarity to phytochelatin synthases catalyzes the conversión of gluthatione to gamma-glutamylcysteine and lacks phytochelatin synthase activity. *Phytochemistry 65,* 3179-318.

Hayes, J.D., Flanagan, J.V. and Jowsey, I.R. (2005). Glutathione transferases. *Annual Reviews in Pharmacology and Toxicology 45,* 51-88.

Ho, E. (2004). Zinc deficiency, DNA damage and cancer risk. *Journal of Nutritional Biochemistry 15,* 572-578.

Iftode, F., Curgy, J., Fleury, A. and Fryd-Versavel, G. (1985) Action of a heavy ion, Cd2+, and the antagonistic effect of Ca2+, on two ciliates *Tetrahymena pyriformis* and *Euplotes vannus. Acta Protozoologica 24,* 273-279.

Ince, N.H., Dirilgen, N., Apikyan, I.G., Tezcanli, G., Ûstûn, B., 1999. Assessment of toxic interactions of heavy metals in binary mixtures: A statistical approach. Archives of Environmental Contamination and Toxicology 36, 365-372.

Izquierdo, A., Martín-González, A. and Gutiérrez, J.C. (2000). Resting cyst wall glycoproteins isolated from two colpodids ciliates are glycine-rich proteins. *Cell Biology International 24,* 115-119.

Janssen, M.P.M., Oosterhoff, C., Heijmans, G.J.S.M. and Van der Voet, H. (1995). The toxicity of metal salts on the population growth of the ciliate *Colpoda cucullus. Bulletin Environmental Contamination and Toxicology 54,* 597-605.

Kang, Y.J. (1997). Cellular and Molecular mechanisms of the metal toxicities. In E.J. Massaro (Ed). *Handbook of human toxicology* (pp. 256-284). Boca Raton, FL: CRC Press.

Krawczynska, W., Pivovarova, N.N. and Sobota A. (1989). Effects of cadmium on growth, ultrastructure and content and chemical elements in *Tetrahymena pyriformis* and *Acanthamoeba castellanii. Acta Protozoologica 28,* 245-252.

Keith, L.H. and Telliard, W.A. (1979). Priority pollutants: a perspective view. *Environmental Science Technology 13,* 416-423.

Klaassen, C.D., Liu, J. and Choudhuri, S. (1999). Metallothionein: an intracellular protein to protect against cadmium toxicity. *Annual Reviews in Pharmacology and Toxicology 39,* 267-294.

Klionsky, D.J. and Ohsumi, Y. (1999). Vacuolar input of proteins and organelles from the cytoplasm. *Annual Review Cellular and Developmental Biology,* 15: 1-32.

Kondo, N., Imai, K., Isobe, M. and Goto, T. (1984). Cadystin A and B major unit peptides comprising cadmium binding peptides induced in a fission yeast – Separation, revision of structures and synthesis. *Tetrahedron Letters 25,* 3869-3872.

Kong, I.C., Bitton, G., Koopman, B. and Jung, K.-H. (1995). Heavy metal toxicity testing in environmental samples. *Review Environmental and Contamination Toxicology 142,* 119-149.

Lee, S., Moon, J.S., Ko, T.S., Petros, D., Goldsbrough, P.B. and Korban, S.S. (2003). Overexpression of *Arabidopsis* phytochelatin synthase paradoxically leads to hypersensitivity to cadmium stress. *Plant Physiology 131,* 656-663.

Leonard, S.S., Harris, G.K. and Shi, X. (2004). Metal-induced oxidative stress and signal transduction. *Free Radical Biology and Medicine 12,* 1921-1942.

Lovley, D.R. (2000). *Environmental Microbe-Metal Interactions.* Washington, DC : ASM Press.

Madoni, P. (1994). A sludge biotic index (SBI) for the evaluation of the biological performance of activated sludge plants based on microfauna analysis. *Water Research 28,* 67-75.

Madoni, P. and Romeo, M.G. (2006). Acute toxicity of heavy metals towards freswater ciliated protists. *Environmental Pollution 141,* 1-7.

Madoni, P., Davoli, D. and Gorbi, G. (1994). Acute toxicity of lead, chromium, and other heavy metals to ciliates from activated sludge plants. *Bulletin of Environmental Contamination and Toxicology 53,* 420-425.

Madoni, P., Esteban, G. and Gorbi, G. (1992). Acute toxicity of cadmium, copper, mercury, and zinc to ciliates from activated sludge plants. *Bulletin of Environmental Contamination and Toxicology 49,* 900-905.

Mago, R. and Srivastava, S. (1994). Uptake of Zinc in *Pseudomonas* sp. Strain UDG26. *Applied and Environmental Microbiology 60,* 2367-2370.

Marcano, L., Carruyo, L., Del Campo, A. and Montiel, X. (2002) Effect of cadmium on the nucleoli of meristematic cells of onion *Allium cepa* L: An ultrastructural study. *Environmental Research 88,* 30-35.

Martín_González, A., Díaz, S., Gallego, A. and Gutiérrez, J.C. (2008). Programmed nuclear death and other apoptotic-like phenomena in ciliated protozooa. In J. M. Pérez Martín (Ed). *Programmed Cell Death in Protozoa* (pp. 143-160). Austin, New York: Landes Bioscience, Springer Science.

Martín-González, A., Díaz, S., Jareño, C. and Gutiérrez, J.C. (1999). The use of protists in ecotoxicology. *Recent Research in Development and Microbiology 3,* 93-111.

Martín-González, A., Díaz, S., Borniquel, S., Gallego, A. and Gutiérrez, J.C. (2006). Cytotoxicity and bioaccumulation of heavy metals by ciliated protozoa isolated from urban wastewater treatment plants. *Research in Microbiology 157,* 108-118.

Martín-González, A., Borniquel, S., Díaz, S., Ortega, R. and Gutiérrez, J.C. (2005) Ultrastructural alterations in ciliated protozoa under heavy metal exposure. *Cell Biology International 29,* 119-126.

Martín-González, A., Benítez, L., Soto, T., Rodríguez de Lecea, J. and Gutiérrez, J.C. (1997). A rapid bioassay to detect mycotoxins using melatonin precursor overproducer mutant of the ciliate *Tetrahymena thermophila. Cell Biology International 21,* 213-216.

Mehta, R., Templeton, D.M., O'Brien, P.J., 2006. Mitochondrial involvement in genetically determined transition metal toxicity II. Copper toxicity. *Chemico-Biological Interactions 163,* 77-85.

Meister, A. and Anderson, M.E. (1983) Glutathione. *Annual Reviews in Biochemistry 52,* 711-760.

Mercer, J.F. and Llanos, R.M. (2003). Molecular and cellular aspects of copper transport in developing mammals. *Journal of Nutrition 133,* 1481S-1484S.

Miles, A.T., Hawksworth, G.M., Beattie, J.H. and Rodilla, V. (2000). Induction, regulation, degradation and biological significance of mammalian metallothioneins. *Critical Reviews in Biochemistry and Molecular Biology 35,* 35-70.

Müller, S., Liebau, E., Walter, R.D. and Krauth-Siegel, R.L. (2003). Thiol-based redox metabolism of protozoan parasites. *Trends in Parasitology 19,* 320-328.

Nassiri, Y., Ginsburger-Vogel, T., Mansot, J.L and Wery, J. (1997a) Effects of heavy metals on *Tetraselmis suecica*: ultrastructural and energy-disperse X-ray spectroscopic studies. *Biology of the Cell 86,* 151-160.

Nassiri. Y., Mansot, J.L., Wery, J., Ginsburger-Vogel, T. and Amiard, J.C. (1997b). Ultrastructural and electron energy loss spectroscopy studies of sequestration mechanisms of Cd and Cu in the marine diatom *Skeletonema costatum. Archives in Environmental Contamination and Toxicology 33,* 147-155.

Negilski, D.S., Ahsanullah, M. and Mobley, M.C. (1981). Toxicity of zinc, cadmium and copper of the shrimp *Callianassa australiensis.* III. Accumulation of metals. *Marine Biology 64,* 305-309.

Nicolau, A., Mota, M. and Lima, N. (1999). Physiological responses of *Tetrahymena pyriformis* to copper, zinc, cycloheximide and Triton X-100. *FEMS Microbiology Ecology 30,* 209-261.

Nicolau, A., Dias, N., Mota, M. and Lima, N. (2001). Trends in the use of protozoa in the assessment of wastewater. *Research in Microbiology 152,* 621-630.

Nies, D.H. (2003). Effux-mediated heavy metal resistance in prokaryotes. *FEMS Microbiology Reviews 27,* 313-339.

Nies, D.H. (1999). Microbial heavy-metal resistance. *Applied and Environmental Microbiology 51,* 730-750.

Nilsson, J.R (1996). Effects of a bismuth salt on cell proliferation, endocytosis and fine structure of *Tetrahymena. European Journal of Protistology 32,* 283-292.

Nilsson, J.R. (1997). *Tetrahymena* recovering from a heavy accumulation of gold salt. *Acta Protozoologica 36,* 111-119.

Nilsson , J.R. (1989). *Tetrahymena* in cytotoxicity: with special reference to effects of heavy metals and selected drugs. *European Journal of Protistology 25,* 2-25.

Norwood, W.P., Borgmann, U., Dixon D.G. and Wallance, A. (2003). Effects of Metal Mixtures on Aquatic Biota: A review of Observations and Methods. *Human and Ecological Risk Assessment 9,* 795-811.

Nriagu, J.O. and Pacyna, J.M.(1988). Quantitative assessment of worldwide contamination of air, water and soils by trace metals. *Nature 333,* 134-139.

Ortega, R., Martín-González, A. and Gutiérrez, J.C. (2007a). Glutamylcysteine ligase gene of the ciliated protozoan *Tetrahymena thermophila*: A potential tool for pollution monitoring. *II International Congress on Environmental, Industrial and Applied Microbiology (MicroBioWorld2007).* Seville (Spain).

Ortega, R., Martín-González, A. and Gutiérrez, J.C. (2007b). Gluthatione-S-transferase genes of the ciliate *Tetrahymena thermophila*: a large family of key defense enzymes against different stresses. *2nd World Conference of Stress*. Budapest (Hungary).

Ortega, R., Martín-González, A. and Gutiérrez, J.C. (2007c). Caracterización y análisis de genes homólogos de glutation-espermidina sintetasa y trypanotion sintetasa en ciliados aerobios no parásitos (*Tetrahymena thermophila* y *Paramecium tetraurelia*). XII National Congress of Microbiology (SEM07). Seville (Spain).

Otitoloju, A.A. (2002). Evaluation of the joint-action of binary mixtures of the heavy metals against the mangrove periwinkle *Tympanotomus fuscatus* var *radula* (L). *Ecotoxicology and Environmental Safety 53,* 404-415.

Overbaugh, J.M., Lau, E.P., Marino, V.A. and Fall, R. (1988). Purification and preliminary characterization of a monomeric glutathione S-transterase from *Tetrahymena thermophila. Archives of Biochemistry and Biophysic 261,* 227-234.

Pauli, W. and Berger, S. (1997). Toxicological comparisons of *Tetrahymena* species, end ponts and growth media: Supplementary investigations of the pilot ring test. *Chemosphere 35,* 1043-1052.

Peñuelas, J. and Filella, I. (2002). Metal pollution in Spanish terrestrial ecosystems during the twentieth century. *Chemosphere 46,* 501-505.

Peterson, H.G. and Nyholm, N. (1993). Algal bioassays of metal toxicity identification. *Water Pollution Research Journal of Canada 28,* 129-153.

Piccinni, E., Irato, P., Coppellotti, O. and Guidolin, L. (1987). Biochemical and ultrastructural data on *Tetrahymena pyriformis* treated with copper and cadmium. *Journal of Cell Science 88,* 283-293.

Piccinni, E., Bertaggia, D., Santovito, G., Miceli, C. Kraen, A. (1999). Cadmium metallothionein gene of *Tetrahymena pyroformis. Gene 234,* 51-59.

Pinot, F., Kreps, S.E., Bachelet, M., Hainaut, P., Bakonyi, M, and Polla, B.S. (2000). Cadmium in the environment: sources, mechanisms of biotoxicity, and biomarkers. *Reviews in Environmental Health 15,* 299-323.

Poole, R.K and Gadd,G.M (1989). Metal-Microbe Interactions. In *Special Publications of the Society for General Microbiology*, London: Oxford University Press.

Prasad, K.V.S.K., Saradhi, P.P., Sharmila, P. (1999). Concerted action of antioxidant enzymes and curailed growth under zinc toxicity in *Brassica juncea. Environmental Experimental Botany 42,* 1-10.

Pulido M.D. and Parrish, A.R. (2003). Metal-induced apoptosis: mechanisms. *Mutation Research 533,* 227-241.

Pyne, C.K., Iftode, F. and Curgi, J.J. (1983)The effects of cadmium on the growth pattern and ultrastructure of the ciliate *Tetrahymena pyriformis,* and antagonistic effect of calcium. *Biology of the Cell 48,* 121-132.

Raikov, L.B. (1982). *The protozoan nucleus. Morphology and evolution* (pp. 221-242) Cell Biology Monographs 9. New York: Springer.

Rainbow, P.S., Amiard-Triquet, C., Amiard, J.C., Smith, B.D. and Langston, W.J. (2000). Observations on the interaction of zinc and cadmium uptake rates in crustaceans (amphipods and crabs) from the coastal sites in UK and France differently enriched with trace metals. *Aquatic Toxicology 50,* 189-204.

Ramos, J., Clemente, M.R., Naya, L., Loscos, J., Pérez-Rontome, C., Sato, S., Tabata, S., Becana, M. (2007). Phyochelatin synthases of the model legume *Lotus japonicus*. A small multigene family with differential response to cadmium and alternative spliced variants. *Plant Physiology 143,* 1110-1118.

Rea, P.A., Vatamaniuk, O.K. and Ridgen, D.J. (2004). Weeds, Works, and more. Papain's long-lost cousin, phytochelatin synthase. *Plant Physiology 136,* 2463-2474.

Resing, C., Ghosh, M. and Rosen, B.P. (1999). Families of soft-metal-ion-transporting ATPases. *Journal of Bacteriology 181,* 5891-5897.

Robinson, N.J., Whitehall, S.K. and Cavet, J.S. (2001). Microbial metallothioneins. *Advances in Microbial Physiology 44,* 183-213.

Rouch, D.A., Lee, B.T. and Morby, A.P. (1995). Understanding cellular responses to toxic agents: a model for mechanism-choice in bacterial resistance. *Journal of Industrial Microbiology 14,* 132-141.

Santovito, G., Irato, P. and Piccinni, E. (2000). Regulation of metallothionein in *Tetrahymena*: induction of MT-mRNA and protein by cadmium exposure. *European Journal of Protistology 36,* 417-422.

Sauvant, M.P., Pepin, D. and Piccinni, E. (1999). *Tetrahymena pyriformis*: a tool for toxicological studies. *Chemosphere 38,* 1631-1639.

Segovia, M. (2008). Programmed cell death in Dinoflagellates. In J.M. Pérez Martín (Ed). *Programmed Cell Death in Protozoa* (pp. 126-142). Austin, New York: Landes Bioscience, Springer Science.

Sen, N., Banerjee, B. and Majumder, H.K. (2008). Molecular analysis of programmed cell death by DNA topoisomerase inhibitors in kinetoplastid parasite *Leishmania*. In J. M. Pérez Martín (Ed). *Programmed Cell Death in Protozoa* (pp. 49-58). Austin, New York: Landes Bioscience, Springer Science.

Schaefer,H., Hettler, H., Fritsche, U., Pitzen, G., Roederer, G. and Wentzel, A. 1994. Biotests using unicellullar algae and ciliates for predicting long-term effects of toxicants. *Ecotoxicology and Environmental Safety 27,* 64-81.

Shang, Y., Song, X., Bowen, J., Corstanje, R. and Gao, Y. (2002). A robust inducible-repressible promoter greatly facilitates gene knockouts, conditional expression, and over expression of homologous and heterologous genes in *Tetrahymena thermophila*. *Proceedings of the Natural Academy of Sciences USA 97,* 3734-3739.

Sharma, S.S., Schat, H., Vooijs, R. and Van Heerwaarden, L.M. (1999). Combination toxicology of copper, zinc, and cadmium in binary mixtures: concentration-dependent antagonistic, non additive, and synergistic effects on root growth in *Silene vulgaris*. *Environmental Toxicology and Chemistry 18,* 348-355.

Shaulian, E. and Karin, M. (2002). AP-1 as a regulator of cell life and death. *Nature Cell Biology 4E,* 131-136.

Shin, Y.H., Park, E.H., Fuchs, J.A. and Lim, C.J. (2002). Characterization, expression and regulation of a third gene encoding glutathione S-transferase from the fission yeast. *Biochimica and Biophysica Acta 1577,* 164-170.

Shuja, R.N. and Shakoori, A.R. 2007. Identification, cloning and sequencing of a novel stress inducible metallothionein gene from locally isolated *Tetrahymena tropicalis. Gene 405,* 19-26.

Silver, S. (1996). Bacterial metal resistance-a review. *Gene 179,* 9-19.

Silver, S. and Phung, L.T (1996). Bacterial heavy metal resistance: New surprises. *Annual Review of Microbiology 50,* 753-789.

Smith, M.A. (1983). The effect of heavy metals on the cytoplasmic fine structure of *Skeletonema costatum* (Bacillariophyta). Protoplasma 116, 24-33.

Soleiro, M.N.C. and Souza de, E.M. (2008). Programmed cell death in Trypanosomatids: A brief review. In J.M. Pérez Martín (Ed). *Programmed Cell Death in Protozoa* (pp. 24-38). Austin, New York: Landes Bioscience, Springer Science.

Stauber, J.L., Franklin N.M. and Adams, M.S. (2002). Applications of the flow cytometry to ecotoxicology testing using microalgae. *Trends in Biotechnology 20,* 141-143.

Sunda, W.G. and Huntsman, S.A. (1996). Antagonisms between cadmium and zinc toxicity and manganese limitation in a coastal diatom. *Limnology and Oceanography 41,* 373-385.

Takada, Y., Uda, K., Kawamura, K. and Matsuoka,T. (2004). Molecular cloning and characterization of a novel glutathione S-transferase gene induced by light stimulation in the protozoan *Blapharisma japonicum. FEMS Microbiology Letters 231,* 185-189.

Tsuji, N., Nishikori, S., Iwabe, O., Shikari, K., Miyasaka, H., Takagi, M., Hirata, K. and Miyamoto, K. (2004). Characterization of phytochelatin synthase-like protein encoded by alr0975 from a prokaryote, *Nostoc sp. PCC 7120. Biochemistry and Biophysical Research Communications 315,* 751-755.

Tsuji, N., Nishikori, S., Iwabe, O., Matsumoto, S., Shiraki, K., Miyasaka, H., Tagaki, M., Miyamoto, H. and Hirata, K. (2005). Comparative analysis of the two-step reaction catalyzed by prokaryotic and eukaryotic phytochelatin synthase by an ion-pair liquid chromatography assay. *Planta 222,* 181-191.

Tuschl, H. and Schwarb C.E. (2005). The use of flow cytometric methods in acute and long-term in vitro testing. *Toxicology in Vitro 19,* 845-852.

Tuschl, H. and Schwarb, C.E. (2004). Flow cytometric methods used as screening tests for basal toxicity of chemicals. *Toxicology in Vitro 18,* 483-491.

Valko, M., Rhodes, C.J., Moncol, J., Izakovic, M. and Mazur, M. (2006). Free radicals, metals and antioxidants in oxidative stress-induced cancer. *Chemico-Biological Interactions 160,* 1-40.

Valko, M., Morris, H. and Cronin, M.T. (2005). Metals, toxicity and oxidative stress. *Current Medicinal Chemistry 12,* 1161-1208.

Vasák, M. (2005). Advances in metallothionein structure and functions. *Journal of Trace Elements in Medicine and Biology 19,* 13-17.

Vasák, M. and Haslerm, D.W. (2000). Metallothioneins: new functional and structural insights. *Current Opoinion in Chemical Biology 4,* 177-183.

Vatamaniuk, O.K., Bucher, E.A., Ward, J.T. and Rea, P.A. (2001). A new pathway for heavy metal detoxification in animals. Phytochelatin sinthase is required for cadmium tolerance in *Caenorhabditis elegans. Journal of Biological Chemistry 276,* 20817-20820.

Vatamaniuk, O.K., Mari, S., Lu, Y.P. and Rea, P.A. (1999). AtPCS1, a phytochelatin synthase from *Arabidopsis*: isolation and in vitro reconstitution. *Preceedings of the Natural Academy of Sciences USA 96,* 7110-7115.

Visviki, I, Rachlin, J.W. (1994) Acute and chronic exposure of *Dunalliela salina* and *Chlamydomonas bullosa* to copper and cadmium: Effects on ultrastructure. *Archives in Environmental Contamination and Toxicology 26,* 154-162.

Waalkes, M.P. (2003). Cadmium carcinogenesis. *Mutation Research/Fundamentals and Molecular Mechanisms of Mutagenesis 533,* 107-120.

Wang, W. and Ballatori, N. (1998). Endogenous gluthatione conjugates: Occurrence and biological functions. *Pharmacology Reviews 50,* 335-356.

Wang, J., Zhang, M., Xu, J. and Wang, Y. (1995). Reciprocal effect of Cu, Cd, Zn on a kind of marine alga. *Water Reseach 29,* 209-214.

Watanabe, M., Suzuki, T.(2002). Involvement of reactive oxygen species in cadmium-induced cellular damage in *Euglena gracilis. Comparative Biochemistry and Physiology Part C 131,* 491-500.

Watanabe, M. and Suzuki, T.(2001). Cadmium-induced abnormality in strains of *Euglena gracilis*: morphological alteration and its prevention by zinc and cyanocobalamine. *Comparative Biochemistry and Physiology Part C 130,* 29-39.

Watanabe, M., Henmi, K., Ogawa, K., Suzuki,T. (2003). Cadmium-dependent generation of reactive oxygen species and mitochondrial DNA breacks in photosynthetic and non-photosynthetic strains of *Euglena gracilis. Comparative Biochemistry and Physiology Part C 134,* 227-234.

Watjen, W., Haase H., Biagioli, M. and Beyersman, D. (2002). Induction of apoptosis in mammalian cells by cadmium and zinc. *Environmental Health Perspectives 110*, 865-867.

Wakatsuki, T., Hayakara, S., Hatayama, T., Kitamura, T. and Imahara, H. (1991). Solubilization and properties of copper reducing enzyme systems from the yeast cell surface in *Debaryomyces hansenii. Journal of Fermentation and Bioengineering 72,* 79-86.

Wells, P.G, Lee, K. and Blaise, C. (1998). *Microescale testing in Aquatic Toxicology.* Boca Raton, FL: CRC Press.

Xu, Z., Bowers, N. and Pratt, J.R. (1997). Variation in morphology, ecology and toxicological responses of *Colpoda inflata* (Stokes) collected from five biogeographic realms. *European Journal of Protistology 33,* 136-144.

Yannai, S., Berdicevsky, I. and Duek, L. (1991). Transformations of inorganic mercury by *Candida albicans* and *Saccharomyces cerevisiae. Applied and Environmental Microbiology 57,* 245-247.

Zang, J., Fu, R.-Y., Hugenholtz, J., Li, Y. and Chen, Y. (2007). Glutathione protects *Lactococcus lactis* against acid stress. *Applied and Environmental Microbiology 73,* 5268-5275.

In: Heavy Metal Pollution
Editors: S. E. Brown and W. C. Welton
ISBN: 978-1-60456-899-8

Chapter 2

MEMBRANE TRANSPORT OF TOXIC METALS BY IONIC MIMICRY IN MAMMALIAN CELLS: WHERE DO CALCIUM CHANNELS FIT IN?

Carla Marchetti*
Istituto di Biofisica, Consiglio Nazionale delle Ricerche, 16149 Genova, Italy, via De Marini, 6

ABBREVIATIONS

VDCC	voltage-dependent Ca channels
CHO	Chinese hamster ovary
CHOCα	CHOCα9β3α2/δ4
DHP	1,4-dihydropyridine
agatoxin	ω-agatoxin fraction IVA from *Agenelopsis aperta* venom

ABSTRACT

Cellular membranes are basically impermeable to ions and have developed specific pathways (transporters, channels or pumps) to facilitate metal translocation. These physiological carriers are not ideally selective and their specificity spectrum may include xenobiotic species, such as toxic metals whose availability in the environment has increased enormously with the onset of the industrial era. Competition between divalent endogenous and noxious metal ions at specific sites on membrane transport proteins and enzymes is referred to as "ionic mimicry".

In this chapter I will present some studies on the permeation mechanisms through mammalian cell plasma membranes of lead (Pb) and cadmium (Cd), two metals whose toxicity has been linked to their putative ability to mimic calcium (Ca) and zinc (Zn) at specific binding sites. Both metals can permeate through mammalian cell membranes taking advantage of different Ca channels, but, while Cd appears to take advantage mainly of the same pathways as Ca, Pb is also rapidly taken up in different cell types by

* Phone: 39010-6475578; Fax: 39010-6475500; Email marchetti@ge.ibf.cnr.it

passive transport systems that are distinct from Ca channels and independent of specific stimuli.

To further elucidate the role of voltage-dependent Ca channels (VDCC) in Cd uptake, we compared the effect of this metal in two Chinese hamster ovary (CHO) cell lines, a wild type and modified cell line, which was permanently transfected with an L-type VDCC. Both cultures were subjected to brief (30-60 min) exposure to 50-100 μM Cd in serum-free culture medium. Cell death was evident after 18-24 h with comparable features in both cell lines. Although VDCC represent a pathway of Cd entry and participate in Cd-induced toxicity, as demonstrated by the effect of DHP modifiers, expression of L-type Ca channels is not sufficient to modify Cd accumulation and sensitivity to a toxicologically significant extent. This study confirmed that both Cd and Pb can take advantage of VDCC to permeate the membrane, but these transport proteins are not the only, and frequently not the most important, pathways of permeation.

INTRODUCTION

Some metals, namely sodium (Na), potassium (K), calcium (Ca) and magnesium (Mg) are among the essential nutrients of living cells, but several metals can be fatal to mammalian cells even in tiny amounts. A common classification tends to sort these potentially toxic metals into two classes: those required by living organisms as essential micronutrients, such as cobalt, (Co), copper (Cu), iron (Fe), manganese (Mn), molybdenum (Mo) and zinc (Zn) and those devoid of any biological function and thus potentially toxic even at very low concentrations, such as cadmium (Cd), chromium (Cr), mercury (Hg) and lead (Pb). These elements are often referred to as "heavy metals", because of their large atomic weight and/or density, and their noxious effects as "heavy metal toxicity". However, this terminology is somewhat abused and recently Duffus (2002) has pointed out that it lacks a precise meaning and can be misleading. These substances will be referred to simply as "toxic metals" in this chapter.

The difference between toxic and non-toxic metal is hard to define. Even micronutrients can be detrimental to living organisms, when present in excessive levels and a refined equilibrium between deficient and toxic concentrations has to be maintained. This is particularly important for very specialized tissues, such as the brain, where metals induce oxidative damage and some of the essential micronutrients, such as Fe, Zn and Cu, have been implicated in etiology and development of different neurological and neurodegenerative diseases (Castellani et al., 2007; Smith et al., 2007). Less obviously, living organisms may find use for non-essential toxic metals in extreme conditions. An elegant example of unexpected biological function of Cd has been recently reported in marine diatoms (Xu et al., 2008).

Despite their potential catalytical activity, toxic metals represent a threat to mammalian cells, principally because their occurrence in air, soil and food has increased significantly as a consequence of recent human industrial activity and specialized cells and tissues have not developed the intrinsic ability to distinguish between physiologically essential and noxious ions. Recent research has gathered an increasing body of evidence for the interaction of toxic metals with intracellular components that lead to cellular injury and defense, but the mechanisms of transport of these metals and metal-containing species across plasma membranes remain to be fully characterized. A special case is transport of these toxicants across tight membrane barriers, such as gastrointestinal and blood-brain barrier, that have developed as protective mechanisms for sensitive organs (Bressler et al., 2007).

Most evidence indicates that toxic metals are capable of replacing or *mimicking* essential metals in the first or early step of transport and metabolism, but then are incapable of mediating subsequent vital functions (Clarkson, 1993; Bridges and Zalups, 2005). The toxic effects are then a consequence of both impairment of transport systems and accumulation into the cell.

Cd and Pb are two highly toxic metals whose harmfulness has been linked to their putative ability to mimic calcium (Ca) and zinc (Zn) at specific binding sites. Due to a close structural similarity to Ca, Cd binds Ca-binding proteins, such as calmodulin (Sutoo, 1994; Ouyang and Vogel, 1998), while Pb binds several intracellular proteins, such as protein kinase C (Markovac and Goldstein, 1988; Long et al., 1994; Tomsig and Suszkiw, 1995), Ca- and Zn- binding (Goldstein, 1993; Kern et al., 2000) and synaptic proteins (Bouton et al., 2001). Both Cd and Pb are potent blockers of voltage-dependent Ca channels (VDCC) by competing with Ca at a cation specific site on the protein; therefore it is not surprising that they could be driven into the cell by depolarization. Voltage-dependent and receptor-operated Ca channels are regarded as the major route of Cd and Pb entry into mammalian cells, although it has become increasingly evident how these channels are frequently not the main pathway of influx of these metals and work in coordination with other transporters.

VDCC mediate Cd influx in excitable cells (Hinkle et al., 1987; Hinkle et al., 1992; Shibuya and Douglas, 1992; Hinkle and Osborne, 1994), including mammalian neurons (Usai et al., 1999) and have been proposed to participate in Cd uptake also in cells from non-excitable tissues (Souza et al., 1997). In previous works of my laboratory, we showed that in certain cells Cd permeation occurs mainly through VDCC of the L-type (Usai et al., 1997; Gavazzo et al., 2005). However, in other studies, the presence of VDCC *per se* did not seem to enhance sensitivity to Cd; for example VDCC expressing (PC12) and non-expressing (PC18) cells showed comparable LD_{50} for Cd (Hinkle and Osborne, 1994) and the role of this pathway in the induction of cell death appears questionable.

The situation is even more complicated for Pb, whose ability to mimic Ca is more debatable. While Clarkson, quoting Simons (Simons, 1985), argues that "the chemical basis for Pb mimicking Ca is not obvious", because the ionic radii of Ca and Pb are "not particularly close" (Clarkson, 1993), Bridges and Zalups (2005) make the opposite assumption stating that "since the ionic radius of Pb is similar to that of Ca, it is possible that Pb mimic Ca at the site of Ca transporter". This contrasting opinion indicates how difficult it is to reach factual conclusions based only on chemical parameters. At odds with observations in other cell types (Simons and Pocock, 1987; Tomsig and Suszkiw, 1991; Legare et al., 1998), in cerebellar granule neurons the greatest part of depolarization-driven Pb influx takes place through DHP-insensitive VDCC, which are characteristic of neuronal tissue, and Pb is also rapidly taken up even in the absence of stimuli that open Ca-permeable channels (Mazzolini et al., 2001). All these observations downsize the importance of mimicking Ca in the transport of lead, although this toxic metal has been shown to interfere heavily with Ca- and Zn-dependent processes.

In this chapter, I will present some general considerations and experimental evidence concerning influx pathways of Pb and Cd across mammalian cell membranes and the relative contribution of VDCC in the uptake of these two toxic metals.

Materials and Methods

Cell Culture

Granule cells were prepared from cerebella of 8-day-old Wistar rats as previously described (Usai et al., 1999), plated on 20mm poly-L-lysine coated glass coverslips and maintained in basal Eagle's culture medium, supplemented with 10% fetal calf serum, 100 μg/ml gentamicin and 25mM KCl, in a humidified 95% air/CO_2 atmosphere at 37°C. Cultures were treated with 10μM cytosine arabinoside from day 1 to minimize proliferation of non-neuronal cells. Experiments were carried out in cultures between 5 and 13 days in vitro.

Wild type Chinese hamster ovary (CHO) cells were obtained from American Type Culture Collection. CHO stably transfected with cDNA encoding for different subunits of voltage-dependent calcium channels (VDCC) were a generous gift of Franz Hofmann and Norbert Klugbauer (Institut für Pharmakologie und Toxikologie, Münich Germany). The cells used in this study are termed CHOCα9β3α2/δ4 (abbreviated CHOCα) to indicate that they express the $\alpha_{1C\text{-}b}$ subunit of VDCC, as well as β3 and α2/δ4 subunits from smooth muscle (Welling et al., 1993).

Both wild type CHO and CHOCα cells were maintained in DMEM medium (Sigma Chemical Co, St. Louis, MO, USA) supplemented with 10% bovine serum, in a 5% CO_2 humidified atmosphere at 37°C. For CHOCα cells, the culture medium was routinely supplemented with antibiotic geneticin G418 20μgr/ml. For the experiments, cells were plated in 12-multiwell trays with or without glass coverslips at a density of $2x10^5$ cells/ml.

Microscopy

Imaging of intracellular Cd and Pb was performed on monolayer cultures plated on glass coverslips. For confocal microscopy, cells were incubated with Oregon Green 488 Bapta-1, a fluorescent indicator, whose fluorescence intensity is increased upon binding Ca, Cd or Pb. Neurons were incubated in 6 μM of the cell-permeant AM ester form of the dye for 45 min at 37°C, and then washed several times with standard saline at room temperature. Other indicators, Fura2 and Indo1, were used as previously described (Usai et al., 1999; Mazzolini et al., 2001; Esposito et al., 2005). Cd and Pb were applied in the absence of Ca and terminated by the addition of the membrane-permeant metal chelator (N,N,N′,N′- tetrakis- 2-pyridyl methyl ethylenediamin, TPEN, 100 μm).

Cells were imaged using a confocal laser scanning microscope Nikon PCM2000 (Nikon Instr., Florence, Italy) with an oil immersion 100×objective (Na=1.3), coupled to a 50 μm confocal pinhole condition (Diaspro et al., 1999).

Cd Treatment

Cells were subjected to 30-60 min treatments with 50 or 100 μM $CdCl_2$ in culture medium without serum. When specified, the medium was supplemented with 30 mM KCl to depolarize the cell membrane and with the dihydropyridines (DHPs) BayK 8644 or nimodipine, which are a L-type VDCC agonist and antagonist, respectively. Stock solutions of nimodipine and BayK were made up at 10 mM in 100% ethanol and diluted in medium to the final concentration. High KCl alone, as well as DHPs (no Cd added), had no effect on cell viability and appearance. After incubation cells were washed three times with PBS containing 1 mM EDTA and fresh medium containing serum was replaced. Cultures were kept in the incubator for another 18-24 hours before testing. Cell viability in each sample was assessed by trypan-blue exclusion assay. All chemicals, culture media and sera were from Sigma-Aldrich Italy.

Capacitance and Current Measurements by Patch-Clamp Technique

Membrane currents were measured from wild type CHO and permanently transfected CHOCα cells in whole-cell clamp configuration by a patch-clamp Axopatch amplifier (Molecular Devices Corporation, Union City, CA). Electrodes were manufactured from borosilicate glass capillaries (Hilgenberg GmbH, Malsfeld, Germany) and had resistance of ≈ 4 MΩ. Voltage stimulation and data acquisition were performed by a PC through a Digidata 1440A interface and Pclamp-10 software (Molecular Devices). Currents were low-pass filtered at 5 kHz and digitized at 10 kHz. Capacitance transients were minimized by analog compensation and the value obtained by this compensation was taken as an estimate of the cell capacitance. All currents traces were further corrected for leak and residual transients by a computer generated P/4 protocol. The holding potential was -80 mV in all the experiments. Current traces were analyzed with Clampfit-10 and Sigma Plot (Jandel Scientific, Erkrath, Germany) software.

Cells were continuously superfused by gravity flow (10 ml/min) with a solution containing (in mM) NaCl 140, KCl 5.4, $CaCl_2$ 1.8, Hepes 5. The pH was 7.4. The internal (pipette) solution contained (in mM): CsCl 20, CsOH 110, Aspartic acid 100, EGTA 5, Hepes 5, with pH adjusted to 7.3 with Trizma base. Calcium currents were recorded in similar external solution containing 130 mM NaCl and 5 mM $CaCl_2$. Wild type CHO cells do not possess prominent voltage-dependent currents, and CHOCα only contained the L-type VDCC under study; therefore there was no need to antagonize or minimize the other conductances to resolve Ca currents. Substitution of internal cesium (Cs) with equimolar potassium (K) had no appreciable effect on the measurements, confirming that CHO cells do not possess any voltage-dependent conductance, including that due to K permeable channels. Application of modifiers, such as agonist and antagonist DHPs, was accomplished by gravity flow; control ion substitution experiments showed that the external bath was completely changed in 10 sec, which was the maximal stimulation rate in these experiments

Apoptosis Detection

The occurrence of apoptosis was evaluated in living cells by Annexin V-FITC apoptosis detection kit (Biovision Research Products, Mountain View ,CA, USA). Monolayer cultures on glass coverslips were incubated with a solution of Annexin V-FITC (0.5 μg/ml), which binds to phosphatidylserine, and propidium iodide (1 μg/ml), which permeates into dead cells, in a buffer containing 10 mM Hepes/NaOH (ph 7.4), 140 mM NaCl and 2.5 mM $CaCl_2$. For evaluation of apoptotic nuclei, cells were incubated with 4,6-diamidino-2-phenylindole (DAPI, Sigma) 1μg/mL in PBS for 5-10 min, as in previous work (Gavazzo et al., 2005). Cells were observed on a Zeiss Axiovert 100 equipped with filters for UV and visible excitation. Nuclei were judged apoptotic when showing segmentation and masses mostly associated with the nuclear envelope or apoptotic bodies. Photographs were collected by a digital camera (Nikon Coolpix 995).

Absorption Spectroscopy Cd Determination

Cells were exposed to Cd in culture medium without added serum. After treatment, cultures were washed three times with PBS containing 1 mM EDTA and harvested immediately. Viable cells were counted in triplicate in each sample. Cells were then washed again three times with the above buffer, re-suspended in distilled water and disrupted by sonication (Sonopuls Ultrasonic Homogenizer, Bandelin) for 2 minutes in an ice bath. The total Cd content of each sample was measured by Flameless Atomic Absorption Spectroscopy (FAAS) using a Perkin Elmer Spectrophotometer (Model 1100 B) equipped with a graphite furnace (Model HGA 700), as in previous work (Gavazzo et al., 2005). The Cd content was normalized to the volume occupied by the cells and molarity was calculated (see results).

Statistical Analysis

Experiments were run in triplicate. Data are presented as mean ± standard error of the mean in at least two experiments. Statistical significance was evaluated by Student-Newman-Keuls multiple comparison test (In Stat, GraphPad Software).

Results

Influx of Toxic Metals into Mammalian Cells

Influx of toxic divalent metal ions into mammalian cells can be monitored in cells pre-loaded with specific fluorescent probes, such as Fura2 (Hinkle et al., 1992; Usai et al., 1997; Usai et al., 1999; Mazzolini et al., 2001), Indo1 (Kerper and Hinkle, 1997a; Esposito et al., 2005), or Oregon green. These fluorescent indicators were developed and are predominantly used for determining intracellular calcium levels, but they have relatively high affinity also for other polyvalent ions. These include Cd and Pb, whose pathways of permeation have been studied by this technique.

Figure 1 shows how exposure to a solution containing a toxicologically significant concentration of Cd or Pb causes a sizeable increase in Oregon Green fluorescence, similar to that seen following Ca influx (Pellistri et al., 2004). This demonstrates feasibility of this probe as toxic metal influx indicator. In this experiment, Cd uptake was triggered by depolarization, indicating the essential contribution of VDCC, while neurons rapidly accumulate Pb even in the absence of a specific stimulus.

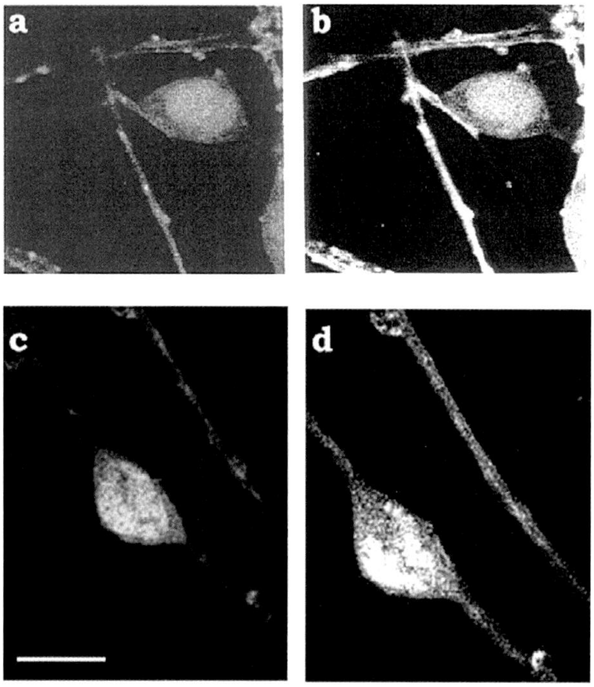

Figure 1. Confocal microscopy images of Cd (a, b) and Pb (c, d) uptake by cerebellar granule cells pre-loaded with the divalent metal-sensitive dye Oregon Green. In (a) and (c), neurons were bathed in a physiological saline. In (b) they have been superfused with a solution containing 0 Ca, 30 mM KCl and 100 μM Cd Cl_2, which caused the dye fluorescence to increase significantly. In (d), external solution contained 0 Ca and Pb 10 μM, which permeates through the neuron membrane even in the absence of a depolarizing stimulus and also caused an increase in the dye fluorescence. Bar = 10 μM

This observation is in agreement with previous findings. In Fura2 loaded neurons (Usai et al., 1999) and insulinoma cells (Usai et al., 1997; Gavazzo et al., 2005), influx of Cd and Pb was evaluated from the fluorescence emission ratio R (E_{380}/E_{340}) and its rate of rise dR/dt. In Ca-free, high-KCl solution, R increased upon application of 100 μM Cd and experiments demonstrated that Cd permeates into these two cell types using Ca-selective pathways, and, in particular, through L-type VDCC. In contrast, in cerebellar neurons, Pb^{2+} (5 - 50μM) determined an increase of R even in the absence of any specific stimulus that open VDCC (Mazzolini et al., 2001; Esposito et al., 2005). In addition, although in Pb dR/dt was increased up to a factor of 5 by depolarizing high-KCl solution, suggesting permeation through VDCC, the effect was neither antagonized by nimodipine, nor enhanced by BayK8644; instead dR/dt decreased in the presence of the Q-type VDCC blocker ω-agatoxin IVA (Mazzolini et al., 2001), suggesting an involvement of non-L type VDCC. Further experiments also proved that glutamate channels of the NMDA type are an additional pathway of both Cd and Pb uptake (Usai et al., 1999; Mazzolini et al., 2001).

Pb caused a rapid saturation of the dye, whose intracellular concentration is ~ 10 μM, indicating that intracellular Pb^{2+} can readily reach concentration in the micromolar range. Parallel experiments showed that non-excitable HeLa cells take up Pb^{2+} much more slowly (Mazzolini et al., 2001), so fast uptake of Pb appears to be a peculiar feature of neurons and may be linked to specific lipidic component of neuronal membranes.

The study of Pb pathways of permeation in rat cerebellar granule neurons was further pursued by taking advantage of a different fluorescent dye, Indo1 (Kerper and Hinkle, 1997a; Esposito et al., 2005). Pb binds Indo1 with high affinity acting as a quencher. Its permeation through the neuronal membrane was indicated by a decrease of the fluorescence emission, which occurred even in resting condition confirming stimulus-independent permeation. Indo1 proved more suitable than Fura2 to study Pb permeation in neurons, because it accumulates into the cell to a concentration estimated 10 times larger than that of Fura2 and, due to a lower affinity for Pb, it is saturated by higher concentration of Pb. In 20 μM Pb, uptake reached a plateau level (≈ 45% of initial fluorescence) in 4 min and was partially antagonized by 25 μM lanthanum (La). Subsequent addition of a membrane permeant ionophore caused a further (> 70%) quenching of the dye, suggesting that previous saturation was due to inactivation of the transport system. Intracellular Pb concentrations were evaluated by comparing the fluorescence intensity of pre-loaded cells with that of cells individually loaded by patch pipette. This estimate indicated that the concentration of free Pb^{2+} sufficient to inactivate the transport system is close to 50 pM (Esposito et al., 2005), a dose of Pb that has been shown to induce a significant effect on the activity of intracellular proteins, such protein kinase C (Markovac and Goldstein, 1988; Long et al., 1994; Tomsig and Suszkiw, 1995), Ca-binding (Goldstein, 1993; Kern et al., 2000) and synaptic proteins(Bouton et al., 2001).

Role of VDCC in Cd-Induced Mortality

Given the prominent role of L-type VDCC in Cd uptake by excitable cells, the involvement of these channels in Cd toxicity was investigated further. Tests of mortality, chromatine condensation and DNA fragmentation showed that a pulsed Cd treatment induces delayed apoptotic cell death in VDCC-containing insulinoma cells and that nimodipine protects against Cd-induced apoptosis and necrosis in these cells, supporting a specific involvement of L-type VDCC (Gavazzo et al., 2005). The same treatments were largely harmless in VDCC-free HeLa cell cultures, in which neither death nor DNA fragmentation was observed. Therefore, excitable cells appear more susceptible than non-excitable epithelial-like cells to Cd accumulation. However, it is not clear whether the mere presence of VDCC can make cells more vulnerable to toxic metal injury. To answer this question we have compared the effects of Cd in two cell types that are virtually identical, except for the expression of VDCC in one of the two. This approach is made possible by the availability of a CHO cell line that has been permanently transfected with VDCC α_1, β and α_2/δ subunits, CHOCα9$\beta_3\alpha_2/\delta_4$ cells (Welling et al., 1993; Cataldi et al., 1999). Prominent voltage-dependent Ca currents were recorded from the permanently transfected cell line, while such currents are completely absent in wild type CHO (figure 2A). The currents are larger with Ba as charge carrier and are sensitive to DHP modulation (figure 2B), as expected for a L-type Ca channel.

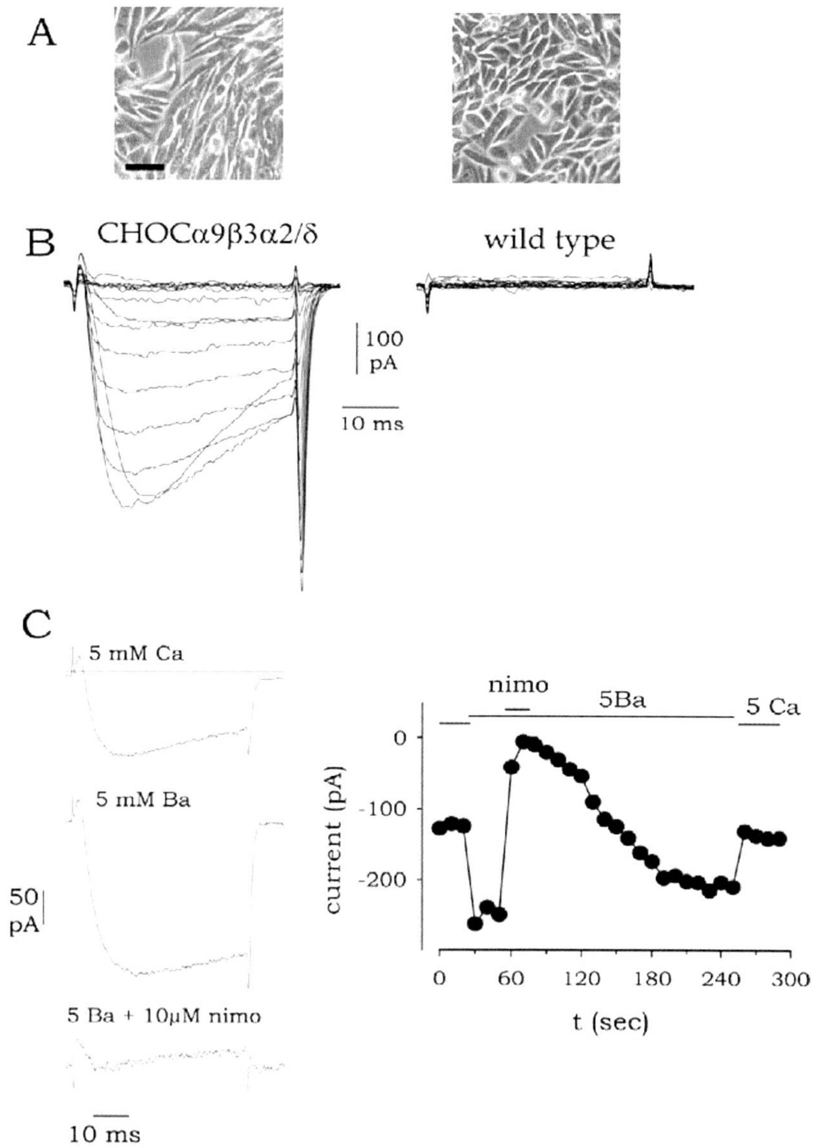

Figure 2. CHOCα cells express L-type VDCC .
(A) Representative microphotographs of monolayer cell cultures of permanently transfected CHOCα cells (left) and wild type CHO cells (right). Bar = 35 μM.
(B) Representative current traces evoked by voltage steps of 40 ms duration from –60 to +60 mV, in 10 mV intervals, from a holding potential of -80 mV in a CHOCα cell (left) and a wild type CHO cell (right). The external solution contained 5 mM $CaCl_2$. This protocol evoked voltage-dependent calcium current in the permanently transfected cell, while no current was present in CHO cell of the wild type.
(C) Characterization of the Ca current in CHOCα cell. The current was increased by more than 50% when the external solution was changed from 5 mM $CaCl_2$ to 5 mM $BaCl_2$ and it was reversibly blocked by 10 μM nimodipine. Current traces evoked by 50 ms depolarizing steps from –80 mV (holding potential) to +10 mV in the three conditions are shown on the left. The graph on the right shows the time course of the experiment.

Both cell types (wild type and CHOCα) were subjected to a 'pulse treatment' with 50 or 100 μM Cd in serum-free culture medium and observed after 18-24 hour after wash. Cell suffering and mortality was negligible immediately after the treatment, but it was evident after at least 16 hours. As previously observed (Galan et al., 2001; Poliandri et al., 2003; Gavazzo et al., 2005), Cd accumulates into the cell during treatment, but the toxic mechanism leading to cell death becomes effective at a later time. Figure 3 shows the appearance of the cell monolayer 24 hours after an incubation of 30 min with 100 μM Cd. Cells exposed to Cd were of smaller size than control cells and had an increased tendency to detach from their neighbors and from the substrate and float. In addition, CHOCα VDCC-expressing cells frequently acquired an irregular shape, with shredded edges (figure 3). The number of viable cells was estimated from the number of adherent, trypan-blue excluding cells and this number was normalized to that of adherent viable cells in control conditions, as in our previous works (Gavazzo et al., 2005). Despite some morphological features, the dose dependence of Cd sensitivity was very similar in the two cell types (figure 3). Therefore, by this approach it is not possible to demonstrate a clear role of VDCC in mediating Cd-induced cell detachment.

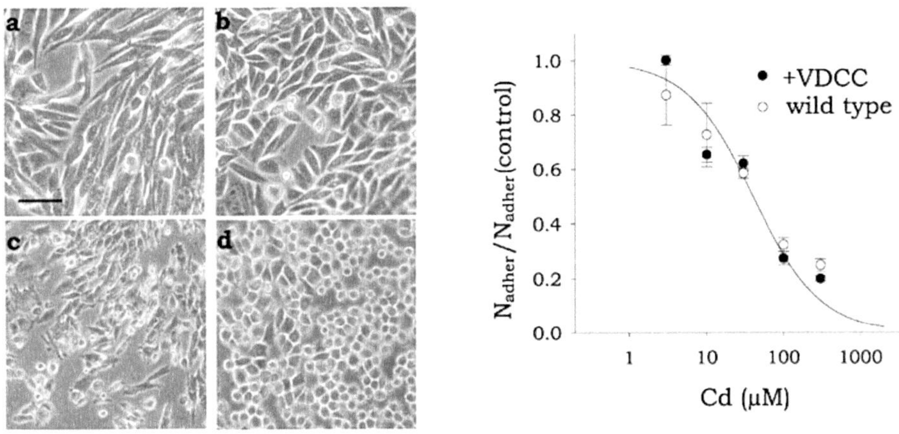

Figure 3. Effect of pulse treatment with Cd in CHOCα VDCC-expressing and wild type CHO cells. Microphotographs of control (a, b) and Cd-treated (c, d) cells: (a, c) VDCC-expressing CHOCα cells and (b, d) wild type CHO cells. Both cell types were incubated in 100 μM Cd for 60 minutes in the presence of 30 mM KCl. Pictures were taken 24 h after wash of the metal. Bar = 35 μM.

The graph shows the effect of a 60 min pulse treatment with Cd on cell adhesion, as a function of concentration. Trypan-blue excluding adherent cells were counted 24 h after wash of the metal. Points are average ± sem of 3 experiments in the same condition in CHOCα (filled circles) and wild type CHO (empty circles) and were best fitted to the function

$$N_{adher}/N_{adher(control)} = 1/(1+([Cd]/ ED_{50}))$$

where $N_{adher}/N_{adher(control)}$ is the number of adherent cells after Cd treatment normalized to the number of adherent cells in control culture; [Cd] is the concentration of Cd and ED_{50} is the concentration of Cd that causes detachment from the substrate of 50% of cells. The best fit yielded ED_{50}=40 μM for CHOCα and 43 μM for wild type CHO. The two curves are overlapped. In contrast with the different appearance, the two cell types were similarly affected by Cd treatment.

In subsequent experiments, we challenged the cells with Cd in the presence of 1 μM DHP modifiers. When VDCC-expressing CHOCα cells were exposed to Cd in the presence of the antagonist (nimodipina) or agonist (BayK 8644), the effect of the metal was clearly modified with respect to cells exposed to Cd alone (figure 4), with a sizeable increase of mortality in the presence of BayK and protection in the presence of nimodipine. DHP modifiers were largely ineffective in wild-type CHO cells exposed to Cd (not shown). Therefore, it seems that the actual contribution of L-type VDCC to Cd uptake is clearly underscored only in the presence of these modifiers.

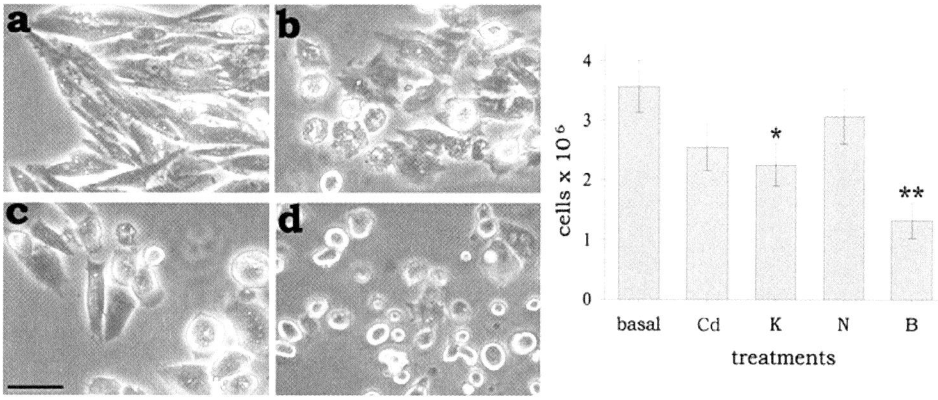

Figure 4. Effect of dihydropyridines on Cd cytotoxicity in VDCC-expressing CHOCα cells. VDCC-expressing CHOCα cells were treated for 30 min in (a) control medium containing 30 mM KCl, (b) 100 μM Cd + 30 mM KCl ; (c) 100 μM Cd + 30 mM KCl + 1μM nimodipine; (d) 100 μM Cd + 30 mM KCl + 1μM BayK8644. Note the partial recovery in shape of the cells treated in the presence of nimodipine and definitive lost of adhesion when the treatment was performed in the presence of BayK. Bar = 25 μM

The graph represents counts of viable adherent cells in the same experiment (average ± sem in 3 samples) in control (basal), following for a 30 min treatment with 100 μM Cd (Cd), Cd+30 mM KCl (K), Cd+30 mM KCl and 1 μM nimodipine (N), Cd+30 mM KCl and 1 μM BayK (B). * indicates significantly different from control with $p < 0.05$, and ** with $p < 0.0001$

In the attempt to reveal any discrepancy in Cd sensitivity between wild type and VDCC-expressing CHO cells, we investigated other features of cellular death caused by transient acute exposure to Cd. Previous work has shown how Cd-induced mortality is basically due to apoptosis in several cell types (el Azzouzi et al., 1994; Hamada et al., 1997; Chrestensen et al., 2000; Lopez et al., 2003) and in CHO cells in particular (Banfalvi et al., 2005). In insulinoma cells, we have shown that apoptosis is the main pathway of cell demise following short (0.5 – 1 hour) treatments with relatively low doses of metal (≤ 100 μM; (Gavazzo et al., 2005). We investigated development of Cd-induced apoptosis by two indicators: exposure of phosphatidylserine, measured by Annexin V binding (Martin et al., 1995) and chromatine condensation (Banfalvi et al., 2005). Although both redistribution of phospholipids and chromatin condensation are regarded as early events in the apoptotic process, we were unable to identify any of these markers before 16 hours after treatment in all the cultures. Therefore initiation of apoptosis did not occur before that time, or before that time it is not possible to foresee the occurrence of apoptosis by straightforward means. In CHO cells that had been treated with 100 μM Cd for one hour and tested for apoptosis occurrence 18-24 hours later, incubation with annexin V revealed green staining in the plasma membrane, indicating translocation of phosphotidylserine to the outer surface of the membrane (figure 5 a-d). Less frequently, green cells also displayed a red staining throughout the nucleus (figure 5 e), indicating loss of membrane integrity, a feature of late apoptosis or necrosis. Figure 6 shows living cells stained with DAPI to visualize nuclear morphology. In cells treated with medium only, nuclei appeared homogeneously bright (figure 6a,b). In cells that had been exposed to Cd, nuclei were smaller, frequently condensed and segmented, as typical of apoptotic cells (figure 6c,d). Again, despite some variation in appearance, counts of stained cells could not reveal any statistically significant difference and reveal a specific sensitivity of CHOCα cells to Cd-induced apoptosis.

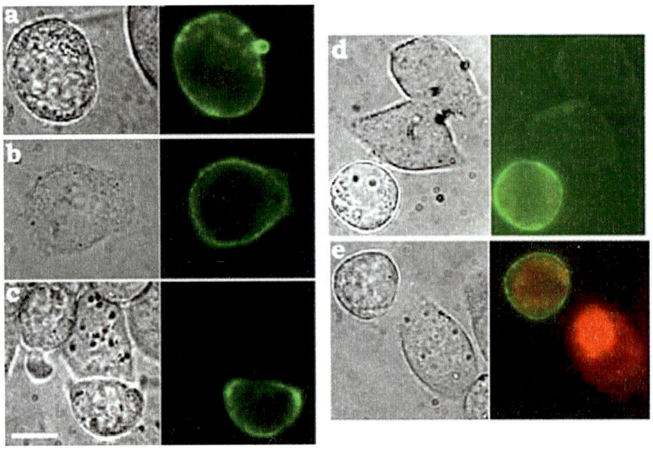

Figure 5. Evidence of Cd-induced apoptosis: membrane staining. Cells were treated with 100 μM Cd for 30 minutes and, 24h after wash of the metal, they were incubated with Annexin V-FITC (0.5 μg/ml), and propidium iodide (1 μg/ml) for 15-45 min. In each panel, the left side is the bright field photography of the cell and the right side is a fluorescent image obtained using a dual filter set for FITC and rhodamine. Panel a through d are from CHOCα cells, while e is from wild type CHO. In a, b and c, exposure of phosphatidylserine on the outer leaflet of the plasma membrane is revealed by the green staining; no red staining was detectable in these cells; in d the green staining is more diffuse into the cells and a pale red staining is also present; in e both cells have lost membrane integrity and taken up the red dye (PI) and are in late apoptotic (top cell) and necrotic state (bottom right cell).

Another indicator of apoptosis is cell shrinkage, which can be quantified by cell capacitance measurements. Both wild type and CHOCα cells showed a decrease in cell capacitance 18-24 hours after Cd treatment. In wild type, cell capacitance was 24 ± 2 pF (n= 15) in control and 15 ± 3 pF (n=6) after incubation with 100 μM Cd, significantly different from control with $p < 0.05$. In CHOCα cells, cell capacitance was 35 ± 3 pF (n= 17) in control and 22 ± 2 pF (n =39) after incubation with 100 μM Cd, significantly different from control with $p < 0.001$. In CHOCα cells, the effect of Cd treatment was partially prevented by nimodipine and the difference between this condition and control condition was not significant (figure 7). This indicates again that the contribution of L-type VDCC in Cd-induced toxicity is evident only in the presence of DHPs.

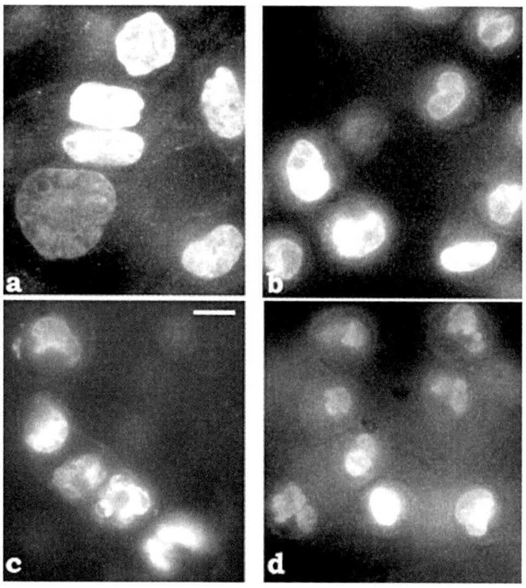

Figure 6. Evidence of Cd-induced apoptosis: nuclear staining. CHOCα (a, c) and wild type CHO (b, d) cells were incubated with medium only (a, b) or 100 μM Cd with 30 mM KCl (c, d) for 30 minutes and stained with DAPI 24h after wash of the metal. Bar = 12 μM. Both cell types were reduced in size by Cd treatment and their nuclei were condensed and segmented.

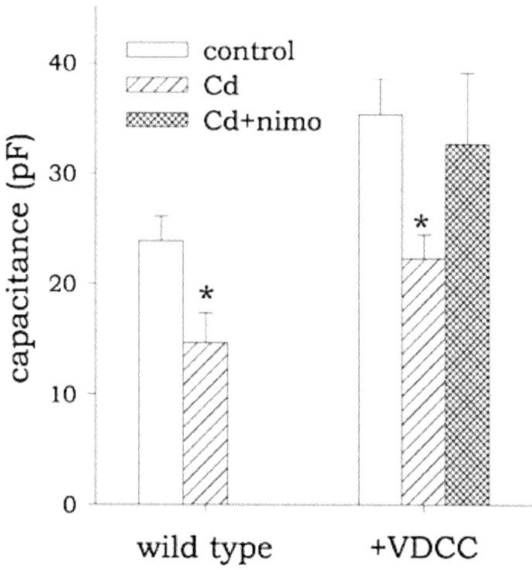

Figure 7. Evidence of Cd-induced apoptosis: change in cell capacitance. Cells were incubated with 100 μM Cd for 60 minutes and electrical measurements were performed 24h after wash. Membrane capacitance was estimated from transient compensation (see methods section). Treatment with Cd caused shrinkage of all cell, as revealed by reduction of the cell capacitance in both wild-type and CHOCα cells. * indicates a significant difference from control with $p < 0.05$. Nimodipine protected CHOCα cells from Cd-induced shrinkage.

Finally, we performed measurements of total Cd accumulation during incubation time by atomic absorption spectroscopy. For these experiments, cells were harvested immediately after the treatment and the total amount of Cd was normalized to the cell volume, estimated from the average single cell capacitance. Assuming that both cell types are approximately spherical and have a specific membrane capacitance of $1\mu F/cm^2$, 10^6 wild type CHO cells comprise a volume of 11μl, while 10^6 CHOCα cells a volume of 19.5 μl. This difference is not negligible and, with this method, results were considerably more reproducible than using a standard normalization to the total protein content. Data shown in figure 8 indicate that the difference in Cd accumulation is barely or not significant when both cell types are incubated in 100 μM Cd with or without elevated KCl. On the other hand, BayK 8644 significantly enhanced metal accumulation in depolarized CHOCα cells treated with 50 or 100 μM Cd, but not in wild type CHO cells, while nimodipine caused a significantly decrease in total metal content during incubation with 50 μM Cd in the permanently transfected cells.

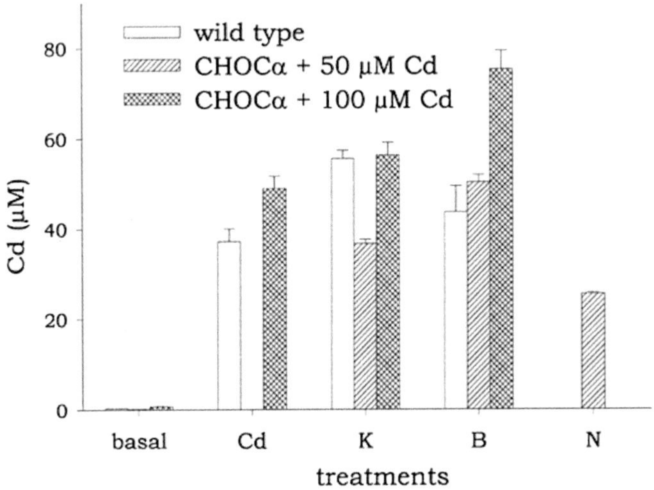

Figure 8.Total Cd accumulation measured by FAAS in wild type CHO and CHOCα cells. Both cell types were incubated for 1 hour in medium containing 0, 50 or 100 μM Cd and other modifiers. Cd determination was normalized to the cell volume (see text). Treatments were as follows: control 0 Cd (basal), 60 min Cd (Cd), Cd+30 mM KCl (K), Cd+30 mM KCl and 1 μM BayK (B), Cd+30 mM KCl and 1 μM nimodipine (N). Depolarization (treatment K) did not increase significantly the intracellular concentration of Cd, with respect to Cd alone ($p > 0.05$). In CHOCα cells, but not in wild type CHO, BayK significantly enhanced and nimodipine significantly reduced Cd accumulation, both with $p < 0.001$, with respect to the same dose of Cd administered alone in depolarization (treatment K).

Discussion

Toxicity of both Cd and Pb is mainly due to their ability to permeate not simply mammalian cell membranes, but specifically tight membrane barriers, such as the epithelial lining in the gastrointestinal tract, luminal membrane of proximal tubules in the kidney and the blood-brain barrier. In early work, the influx of Pb and Cd into mammalian cells was frequently ascribed to permeation through calcium channels, and in particular L-type DHP-sensitive VDCC (Hinkle et al., 1987; Simons and Pocock, 1987; Tomsig and Suszkiw, 1991; Hinkle and Osborne, 1994; Souza et al., 1997). More recently other types of VDCC, different from the classical DHP-sensitive L-type, have been implicated both in Cd (Usai et al., 1999; Leslie et al., 2006) and Pb (Mazzolini et al., 2001) uptake.

In the attempt to identify the clear role of VDCC in toxic metal uptake, it is remarkable that Pb permeation appears to be largely independent of VDCC even in neurons (Mazzolini et al., 2001; Esposito et al., 2005). As far as Cd is concerned, in this chapter I have shown that expression of L-type VDCC is not sufficient to modify Cd uptake to a toxicological significant extent, at least in the case of a brief, intense exposure, as those used in this study. Cd accumulation and sensitivity were comparable in two cell lines whose only difference was the presence of L-type VDCC in one of the two. Indeed, wild-type CHO cells, which do not possess VDCC, are vulnerable to Cd poisoning and take up Cd during relatively brief exposures to an extent sufficient to trigger cell death by apoptosis or necrosis similarly to VDCC expressing CHOCα cells. Although VDCC play an important role in Cd uptake in CHOCα cells, as indicated by the sizeable effect of L-type channel DHP modifiers, CHO cells lacking VDCC must use other transport proteins with similar efficiency and outcome.

It is thus clear that VDCC are not the main responsible of toxic metal permeation through cell membranes and tight barriers. Rapid passive transport of Pb, independent of VDCC, was reported at the brain endothelium (Deane and Bradbury, 1990) and even protein-independent transport of lead has been described (Diaz and Monreal, 1995). Store-depletion activated calcium channels have been suggested to mediate the uptake of Pb in various cells, including a glial cell line (Kerper and Hinkle, 1997a; Legare et al., 1998) and brain capillary endothelial cells (Kerper and Hinkle, 1997b). In a rat astroglial cell line, Cheong et al. (2004) reported kinetics data concerning two putative distinct pH-sensitive transport mechanisms for Pb. VDCC-independent uptake of Pb by cerebellar granule neurons was inhibited by La (Esposito et al., 2005) and this observation can provide some clue as for the mechanism of permeation. The involvement of a La-sensitive store-depletion activated channel, as proposed in other systems (Kerper and Hinkle, 1997a; Evans et al., 2003), is unlikely because Pb influx did not require drainage of the stores. Another possibility is that Pb^{2+} enters the neuron via reverse operation of an exchanger similar (or identical) to the Na-Ca exchanger, which is also very sensitive to La block (Kaczorowski et al., 1989) and whose role in Zn permeation was demonstrated in cortical neurons (Sensi et al., 1997). Also this possibility was discarded because Na-Ca exchanger does not play a significant role when the Na concentrations are close to the physiological value, as in Esposito et al. (2005). It is more probable that Pb is taken up through a La-sensitive ion carrier similar to the Zn transporter found in brain neurons (Colvin, 1998), whose identity and features still wait full definition.

Involvement of other transport proteins has been described. At the epithelial lining of the gastrointestinal barrier, DMT1, the main transporter that absorbs Fe in the brush border membrane of the mammalian intestine has been shown to carry also different toxic metals (Picard et al., 2000; Lecoeur et al., 2002; Bannon et al., 2003; Okubo et al., 2003; Suzuki et al., 2007). The role of this transporter in Cd absorption is underscored by the fact that iron deficiency creates a significant risk for increased cadmium exposure by increasing gastrointestinal absorption from 5% to as much as 20%. In general, a low level of essential metals favor uptake of non-essential (toxic) elements that compete for their site and this stresses the importance of evaluating metal balance, in contrast with metal concentration. In addition, other metal-specific transporters have been identified at the cDNA level (Bressler et al., 2007) and studied in expression systems (Liu et al., 2008).

Research has also driven attention to the importance of active back transport, mediated by active transporter and pumps, which mitigates metal uptake. This role is frequently played by ATP-driven metal pumps (P_{1B}-type ATPases) that actively transport several metals, including toxicant, from a variety of cells (Solioz and Vulpe, 1996; Arguello et al., 2007). Interestingly, a functional link has been postulated between L-type VDCC and the Zn transporter ZnT-1, which would attenuate Cd and Zn influx thus limiting accumulation (Ohana et al., 2006). This is in agreement with other studies that showed how the expression of ZnT-1 inhibits the activity of L-type channels (Beharier et al., 2007) and is a notable example of how different transport systems may act in a concerted manner in the regulation of metal transport.

In the balance between positive and negative factors, the expression of detoxification proteins is also an important component, chiefly in the case of Cd, as revealed by the presence of a strong negative correlation between cellular MT content and the rate of apoptosis induced by Cd (Shimoda et al., 2001).

The present study completes and corrects our previous work (Gavazzo et al., 2005). Data also confirm that short treatments with Cd trigger delayed cell death, chiefly by apoptosis. The latency time is longer than 8 hours, and during this time it was not possible to find any indication of the forthcoming demise. It is probable that the cell, once hit by the toxic insult, tries all available defense strategies to face the insult, until it reaches a point of no return. The time course of this process would in principle be dependent on the noxious stimulus intensity and cell susceptibility. However, this latency could not be shortened even increasing Cd concentration up to 300 μM in any of the model cells investigated (Gavazzo et al., 2005), an observation that requires further study.

Conclusion

Ionic mimicry is a useful framework to study the mechanisms of metal toxicity. Uptake of toxic metals, such as Cd and Pb, by mammalian cells of different tissues occurs through many different pathways, of which VDCC have frequently been regarded as prominent. Because of the abundance of VDCC, excitable cells may be more susceptible to accumulate Cd and Pb than non-excitable cells. In this chapter I have shown that although Cd and Pb can permeate through VDCC, this is not the most relevant mechanism of passive uptake of Pb, while the mere presence of VDCC is not sufficient to make cells significantly more vulnerable to Cd injury. Other factors, including different transport systems and expression of detoxification proteins, may be more relevant in the accumulation process of these toxic metals.

Acknowledgements

The experiments described in this chapter were carried out by the author with the collaboration of several people. Ilaria Zanardi and Irena Baranowska-Bosiacka helped in Cd treatment experiments, while Elisabetta Morelli performed the FAAS measurements. I thank Mauro Robello and Paola Gavazzo for help and suggestions throughout this work. I am also grateful to Franz Hofmann and Norbert Klugbauer (Institut für Pharmakologie und Toxikologie, Münich Germany) for their permission to use the permanently transfected CHO cells and to Maurizio Taglialatela and Mauro Cataldi (Università Federico II, Naples, Italy) for sending me the cells.

REFERENCES

Arguello, J. M., Eren, E. and Gonzalez-Guerrero, M. (2007). The structure and function of heavy metal transport P1B-ATPases. *Biometals*, *20*, 233-48.

Banfalvi, G., Gacsi, M., Nagy, G., Kiss, Z. B. and Basnakian, A. G. (2005). Cadmium induced apoptotic changes in chromatin structure and subphases of nuclear growth during the cell cycle in CHO cells. *Apoptosis*, *10*, 631-42.

Bannon, D. I., Abounader, R., Lees, P. S. and Bressler, J. P. (2003). Effect of DMT1 knockdown on iron, cadmium, and lead uptake in Caco-2 cells. *Am. J. Physiol. Cell Physiol*, *284*, C44-50.

Beharier, O., Etzion, Y., Katz, A., Friedman, H., Tenbosh, N., Zacharish, S., Bereza, S., Goshen, U. and Moran, A. (2007). Crosstalk between L-type calcium channels and ZnT-1, a new player in rate-dependent cardiac electrical remodeling. *Cell Calcium*, *42*, 71-82.

Bouton, C. M., Frelin, L. P., Forde, C. E., Arnold Godwin, H. and Pevsner, J. (2001). Synaptotagmin I is a molecular target for lead. *J. Neurochem*, *76*, 1724-35.

Bressler, J. P., Olivi, L., Cheong, J. H., Kim, Y., Maerten, A. and Bannon, D. (2007). Metal transporters in intestine and brain: their involvement in metal-associated neurotoxicities. *Hum. Exp. Toxicol*, *26*, 221-9.

Bridges, C. C. and Zalups, R. K. (2005). Molecular and ionic mimicry and the transport of toxic metals. *Toxicol Appl. Pharmacol*, *204*, 274-308.

Castellani, R. J., Moreira, P. I., Liu, G., Dobson, J., Perry, G., Smith, M. A. and Zhu, X. (2007). Iron: the Redox-active center of oxidative stress in Alzheimer disease. *Neurochem. Res*, *32*, 1640-5.

Cataldi, M., Secondo, A., D'Alessio, A., Taglialatela, M., Hofmann, F., Klugbauer, N., Di Renzo, G. and Annunziato, L. (1999). Studies on maitotoxin-induced intracellular Ca(2+) elevation in chinese hamster ovary cells stably transfected with cDNAs encoding for L-type Ca(2+) channel subunits. *J. Pharmacol. Exp. Ther.*, *290*, 725-30.

Cheong, J. H., Bannon, D., Olivi, L., Kim, Y. and Bressler, J. (2004). Different mechanisms mediate uptake of lead in a rat astroglial cell line. *Toxicol. Sci.*, *77*, 334-40.

Chrestensen, C. A., Starke, D. W. and Mieyal, J. J. (2000). Acute cadmium exposure inactivates thioltransferase (Glutaredoxin), inhibits intracellular reduction of protein-glutathionyl-mixed disulfides, and initiates apoptosis. *J. Biol. Chem.*, *275*, 26556-65.

Clarkson, T. W. (1993). Molecular and ionic mimicry of toxic metals. *Annu. Rev. Pharmacol. Toxicol.*, *33*, 545-71.

Colvin, R. A. (1998). Characterization of a plasma membrane zinc transporter in rat brain. *Neurosci. Lett*, *247*, 147-50.

Deane, R. and Bradbury, M. W. (1990). Transport of lead-203 at the blood-brain barrier during short cerebrovascular perfusion with saline in the rat. *J. Neurochem.*, *54*, 905-14.

Diaspro, A., Annunziata, S., Raimondo, M., Ramoino, P. and Robello, M. (1999). A single-pinhole confocal laser scanning microscope for 3-D imaging of biostructures. *IEEE Eng. Med. Biol. Mag.*, *18*, 106-10.

Diaz, R. S. and Monreal, J. (1995). Protein-independent lead permeation through myelin lipid liposomes. *Mol. Pharmacol.*, *47*, 766-71.

Duffus, J. H. (2002). "Heavy metals" a meaningless term? *Pure Appl. Chem.*, *74*, 793-807.

el Azzouzi, B., Tsangaris, G. T., Pellegrini, O., Manuel, Y., Benveniste, J. and Thomas, Y. (1994). Cadmium induces apoptosis in a human T cell line. *Toxicology*, *88*, 127-39.

Esposito, A., Robello, M., Pellistri, F. and Marchetti, C. (2005). Two-photon analysis of lead accumulation in rat cerebellar granule neurons. *Neurochem. Res.*, *30*, 949-54.

Evans, T. J., James-Kracke, M. R., Kleiboeker, S. B. and Casteel, S. W. (2003). Lead enters Rcho-1 trophoblastic cells by calcium transport mechanisms and complexes with cytosolic calcium-binding proteins. *Toxicol. Appl. Pharmacol*, *186*, 77-89.

Galan, A., Troyano, A., Vilaboa, N. E., Fernandez, C., de Blas, E. and Aller, P. (2001). Modulation of the stress response during apoptosis and necrosis induction in cadmium-treated U-937 human promonocytic cells. *Biochim. Biophys. Acta*, *1538*, 38-46.

Gavazzo, P., Morelli, E. and Marchetti, C. (2005). Susceptibility of insulinoma cells to cadmium and modulation by L-type calcium channels. *Biometals*, *18*, 131-42.

Goldstein, G. W. (1993). Evidence that lead acts as a calcium substitute in second messenger metabolism. *Neurotoxicology*, *14*, 97-101.

Hamada, T., Tanimoto, A. and Sasaguri, Y. (1997). Apoptosis induced by cadmium. *Apoptosis*, *2*, 359-67.

Hinkle, P. M., Kinsella, P. A. and Osterhoudy, K. C. (1987). Cadmium uptake and toxicity via voltage-sensitive calcium channels. *J.Biol.Chem.*, *262*, 16333-16337.

Hinkle, P. M. and Osborne, M. E. (1994). Cadmium toxicity in rat pheochromocytoma cells: studies on the mechanism of uptake. *Toxicol. Appl. Pharmacol*, *124*, 91-8.

Hinkle, P. M., Shanshala, E. D. and Nelson, E. J. (1992). Measurements of intracellular cadmium with fluorescent dyes - Further evidence for the role of calcium channels in cadmium uptake. *J.Biol.Chem.*, *267*, 25553-25559.

Kaczorowski, G. J., Slaughter, R. S., King, V. F. and Garcia, M. L. (1989). Inhibitors of sodium-calcium exchange: identification and development of probes of transport activity. *Biochim. Biophys. Acta*, *988*, 287-302.

Kern, M., Wisniewski, M., Cabell, L. and Audesirk, G. (2000). Inorganic lead and calcium interact positively in activation of calmodulin. *Neurotoxicology*, *21*, 353-63.

Kerper, L. E. and Hinkle, P. M. (1997a). Cellular uptake of lead is activated by depletion of intracellular calcium stores. *J. Biol. Chem*, *272*, 8346-52.

Kerper, L. E. and Hinkle, P. M. (1997b). Lead uptake in brain capillary endothelial cells: activation by calcium store depletion. *Toxicol. Appl. Pharmacol*, *146*, 127-33.

Lecoeur, S., Huynh-Delerme, C., Blais, A., Duche, A., Tome, D. and Kolf-Clauw, M. (2002). Implication of distinct proteins in cadmium uptake and transport by intestinal cells HT-29. *Cell. Biol. Toxicol*, *18*, 409-23.

Legare, M. E., Barhoumi, R., Hebert, E., Bratton, G. R., Burghardt, R. C. and Tiffany-Castiglioni, E. (1998). Analysis of Pb2+ entry into cultured astroglia. *Toxicol. Sci*, *46*, 90-100.

Leslie, E. M., Liu, J., Klaassen, C. D. and Waalkes, M. P. (2006). Acquired cadmium resistance in metallothionein-I/II(-/-) knockout cells: role of the T-type calcium channel Cacnalpha1G in cadmium uptake. *Mol. Pharmacol*, *69*, 629-39.

Liu, Z., Li, H., Soleimani, M., Girijashanker, K., Reed, J. M., He, L., Dalton, T. P. and Nebert, D. W. (2008). Cd2+ versus Zn2+ uptake by the ZIP8 HCO3--dependent symporter: kinetics, electrogenicity and trafficking. *Biochem. Biophys. Res. Commun*, *365*, 814-20.

Long, G. J., Rosen, J. F. and Schanne, F. A. (1994). Lead activation of protein kinase C from rat brain. Determination of free calcium, lead, and zinc by 19F NMR. *J. Biol. Chem, 269*, 834-7.

Lopez, E., Figueroa, S., Oset-Gasque, M. J. and Gonzalez, M. P. (2003). Apoptosis and necrosis: two distinct events induced by cadmium in cortical neurons in culture. *Br. J. Pharmacol., 138*, 901-11.

Markovac, J. and Goldstein, G. W. (1988). Picomolar concentrations of lead stimulate brain protein kinase C. *Nature, 334*, 71-3.

Martin, S. J., Reutelingsperger, C. P., McGahon, A. J., Rader, J. A., van Schie, R. C., LaFace, D. M. and Green, D. R. (1995). Early redistribution of plasma membrane phosphatidylserine is a general feature of apoptosis regardless of the initiating stimulus: inhibition by overexpression of Bcl-2 and Abl. *J. Exp. Med, 182*, 1545-56.

Mazzolini, M., Traverso, S. and Marchetti, C. (2001). Multiple pathways of Pb(2+) permeation in rat cerebellar granule neurones. *J. Neurochem, 79*, 407-16.

Ohana, E., Sekler, I., Kaisman, T., Kahn, N., Cove, J., Silverman, W. F., Amsterdam, A. and Hershfinkel, M. (2006). Silencing of ZnT-1 expression enhances heavy metal influx and toxicity. *J. Mol. Med, 84*, 753-63.

Okubo, M., Yamada, K., Hosoyamada, M., Shibasaki, T. and Endou, H. (2003). Cadmium transport by human Nramp 2 expressed in Xenopus laevis oocytes. *Toxicol. Appl. Pharmacol., 187*, 162-7.

Ouyang, H. and Vogel, H. J. (1998). Metal ion binding to calmodulin: NMR and fluorescence studies. *Biometals, 11*, 213-22.

Pellistri, F., Cupello, A., Esposito, A., Marchetti, C. and Robello, M. (2004). Two-photon imaging of calcium accumulation in rat cerebellar graule cells. *Neuroreport, 15*, 83-87.

Picard, V., Govoni, G., Jabado, N. and Gros, P. (2000). Nramp 2 (DCT1/DMT1) expressed at the plasma membrane transports iron and other divalent cations into a calcein-accessible cytoplasmic pool. *J. Biol. Chem, 275*, 35738-45.

Poliandri, A. H., Cabilla, J. P., Velardez, M. O., Bodo, C. C. and Duvilanski, B. H. (2003). Cadmium induces apoptosis in anterior pituitary cells that can be reversed by treatment with antioxidants. *Toxicol Appl. Pharmacol, 190*, 17-24.

Sensi, S. L., Canzoniero, L. M., Yu, S. P., Ying, H. S., Koh, J. Y., Kerchner, G. A. and Choi, D. W. (1997). Measurement of intracellular free zinc in living cortical neurons: routes of entry. *J. Neurosci, 17*, 9554-64.

Shibuya, I. and Douglas, W. W. (1992). Calcium channels in rat melanotrophs are permeable to manganese, cobalt, cadmium, and lanthanum, but not to nickel: evidence provided by fluorescence changes in fura-2-loaded cells. *Endocrinology, 131*, 1936-41.

Shimoda, R., Nagamine, T., Takagi, H., Mori, M. and Waalkes, M. P. (2001). Induction of apoptosis in cells by cadmium: quantitative negative correlation between basal or induced metallothionein concentration and apoptotic rate. *Toxicol. Sci, 64*, 208-15.

Simons, T. J. (1985). Influence of lead ions on cation permeability in human red cell ghosts. *J. Membr. Biol., 84*, 61-71.

Simons, T. J. and Pocock, G. (1987). Lead enters bovine adrenal medullary cells through calcium channels. *J. Neurochem., 48*, 383-9.

Smith, D. G., Cappai, R. and Barnham, K. J. (2007). The redox chemistry of the Alzheimer's disease amyloid beta peptide. *Biochim. Biophys. Acta, 1768*, 1976-90.

Solioz, M. and Vulpe, C. (1996). CPx-type ATPases: a class of P-type ATPases that pump heavy metals. *Trends Biochem. Sci*, *21*, 237-41.

Souza, V., Bucio, L. and Gutierrez-Ruiz, M. C. (1997). Cadmium uptake by a human hepatic cell line (WRL-68 cells). *Toxicology*, *120*, 215-20.

Sutoo, D. (1994). Disturbance of brain function by exogenous cadmium. The Vulnerable Brain and Environmental Risks. R. L. Isaacson and K. F. Jensen. New York, Plenum Press. **3:** 281-299.

Suzuki, T., Momoi, K., Hosoyamada, M., Kimura, M. and Shibasaki, T. (2007). Normal cadmium uptake in microcytic anemia mk/mk mice suggests that DMT1 is not the only cadmium transporter in vivo. *Toxicol Appl. Pharmacol*, *17*, 17.

Tomsig, J. L. and Suszkiw, J. B. (1991). Permeation of Pb2+ through calcium channels: fura-2 measurements of voltage- and dihydropyridine-sensitive Pb2+ entry in isolated bovine chromaffin cells. *Biochem.Biophys.Acta*, *1069*, 197-200.

Tomsig, J. L. and Suszkiw, J. B. (1995). Multisite interactions between Pb2+ and protein kinase C and its role in norepinephrine release from bovine adrenal chromaffin cells. *J. Neurochem*, *64*, 2667-73.

Usai, C., Barberis, A., Moccagatta, L. and Marchetti, C. (1997). Pathways of cadmium uptake in excitable mammalian cells: a microspectrofluorimetric study. *Eur. J. Histochem*, *41*, 189-90.

Usai, C., Barberis, A., Moccagatta, L. and Marchetti, C. (1999). Pathways of cadmium influx in mammalian neurons. *J. Neurochem*, *72*, 2154-61.

Welling, A., Bosse, E., Cavalie, A., Bottlender, R., Ludwig, A., Nastainczyk, W., Flockerzi, V. and Hofmann, F. (1993). Stable co-expression of calcium channel alpha 1, beta and alpha 2/delta subunits in a somatic cell line. *J. Physiol*, *471*, 749-65.

Xu, Y., Feng, L., Jeffrey, P. D., Shi, Y. and Morel, F. M. (2008). Structure and metal exchange in the cadmium carbonic anhydrase of marine diatoms. *Nature*, *452*, 56-61.

In: Heavy Metal Pollution
Editors: S. E. Brown and W. C. Welton
ISBN: 978-1-60456-899-8

Chapter 3

HEAVY METAL POLLUTION: TWO STUDYING CASES

N. Calace, N. Cardellicchio, B.M. Petronio[1], M. Pietrantonio and M. Pietroletti

Department of Chemistry, University "La Sapienza", p.le A Moro 5, 00185 Rome, Italy
CNR – Institute for Coastal Marine Environment, Taranto Section, via Roma 3, 74100 Taranto, Italy

Heavy metals are natural constituents of the earth crust; consequently they are present practically in all environments (air, soils, sediments, waters) as a result of erosion and weathering of parent rocks (Sillanpaa, 1972; Bohn et al., 1985). Nevertheless, the metal introduction in ecosystems occurs not only through natural sources, but also through anthropogenic sources, such as smelters, mining, power station industry and the application of metal-containing pesticides, fertilizers and sewage sludge (Sposito et al, 1982; Brimblecomebe, 1994; McGrath et al., 1995; Chuan et al., 1996; Cambier et al., 1997; Akhter and Al-jowder, 1997; Dijkstra, 1998; Ranjbar, 1998; et al., 2000; Pichtel et al., 2000; Odukoya et al., 2000; Robinson et al., 2001; Abdul-Wahab and Yaghi, 2003).

Anthropogenic atmospheric inputs contribute considerably in soil and sediment contamination (Allowey, 1990). Moreover, long-range atmospheric transport is one of the main factors for dispersion and deposition of main pollutants, including Pb, Zn, Cu, Cd and As (Steiness et al., 1997; Loska et al., 2003) and can lead to pollutant deposition even in supposedly pristine areas (De Vries et al., 2002).

At trace levels some heavy metals as Zn, Cu, Ni, and Cr play an important role in organisms and human life (Mokhtar et al., 1991; Juvanovic et al., 1995; Lapitajs et al., 1995) whereas others, such as Cd, Hg and Pb have no known biological and/or physiological function. However, all metals above reported present toxicity at relative low concentrations (Lisk, 1972; Gadd, 1992; Abdul Rida and Bouchè, 1995).

Heavy metals unlike other toxic pollutant are not biodegradable in the environment. As consequence, they are able to accumulate in soils, sediments and organisms and can be considered as a major group of compounds involved in soil and sediment pollution (Yaron et

[1] Corresponding author. Tel. Fax +39-6-49913723
E-mail-address: biancamaria.petronio@uniroma1.it

al., 1996). Heavy metal concentration levels higher than natural values in soil and sediment are sign of a direct potential toxicity to biota and indirect threats to human health from groundwater contamination and the possible entry into the food chain.

In natural waters the introduction of metals can occur in dissolved and particulate form. Depending of physico-chemical conditions, dissolved metal form can precipitate as hydroxides, carbonates and other salts, or form dissolved organic and inorganic complexes. They can also be adsorbed by the iron-manganese oxides and hydroxides or co-precipitate with these (Förstner, 1985; Drever, 1988; Buffle, 1990). In particulate form, associated with suspended matter, metals can be deposited on the bottom. As consequence, sediments act as traps and accumulate the low concentrations of metals in water (Shne et al., 1995). It is know that heavy metals are linked to fine fraction of sediments, especially to clay and organic matter (Rubinstein et al., 1983; Duzzin et al., 1988; Martincic et al, 1990).

Metal content in sediments is generally well above detection limits even when in water it is extremely low, so sediments play an important role in the assessment of metal contamination in natural water (Pardo et al., 1990; Gonçsalves et al., 1994; Lapaquellerie et al., 1995; Borovec, 1996; Wardas et al., 1996). In this way are also eliminate the errors due to the erratic fluctuations often observed in waters (Salomons and Förstner, 1984; Erel et al., 1991).

A large part of metals accumulate in to sediments is not readily available for aquatic organisms, but the variation of some physical and chemical parameters of the overlying water, such as pH, salinity, redox potential, content of organic compounds, etc., may cause the release of metals. Hence under changing environmental conditions sediments can act as a source of metals (Förstner, 1989; Burgess and Scott, 1992; Gibbs, 1993; Izquierdo et al., 1997; Zoumis et al., 2001).

In soils heavy metals coming from anthropogenic sources slowly redistribute among solid-phase components (McGrath and Cegarra, 1992; Han et al., 1995, 2001). The process is characterised by initial fast retention and subsequent slow reactions that redistribute heavy metals from labile forms into more stable forms (McLaren and Ritchie, 1993; Han and Banin, 1999).

The retention or the solubility of metals is affected by various factors, in particular by metal-binding sites such as soil organic matter or clay (Elliot et al., 1986; Kuo, 1986; Ziper et al., 1988) and by soil chemical properties such as pH, cation exchange capacity, redox conditions (Farrah and Pickering, 1977; Yong and Phadungchewit, 1993). Moreover microbial activity is also considered as a factor affecting metal speciation in soils because microorganisms are involved in solubilization, insolubilization and volatilisation of metals (Berthelin and Leyval, 2000).

The mobility, bioavailability and toxicity of the metals in soils are strictly depending from their distribution between solution and solid-phases, and among various solid-phase components (Ham et al., 2001), which can be measured by selective sequential extractions.

Soil heavy metal pollution can have adverse effect on various parameters relating yield, plant growth and plant quality. The possible entry into the food chain with negative effect on human health is documented (Khas et al., 1998). Moreover, pollution of soil by heavy metals can cause change in the size, composition and activity of microbial community (Giller et al., 1998; Huang and Khan, 1998; Kelly, 1999; Shi et al., 2002; Khan and Scullion; 2002, Abaye, 2005), that play an important role in recycling of plant nutrients, maintenance of soil structure, detoxification of noxious chemicals, and the control of plant pest and growth (Robert and Chenu, 1991; Elsgaard et al., 2001; Filip, 2002).

All above reported make evident that informations on total concentrations of metals can evidence the pollution degree of soils and sediments, but alone it is not sufficient to assess they environmental impact because the different chemical forms (easily exchangeable ions, metal carbonates, oxides, sulphides, metal-organic compound complexes, ions in crystal lattices of minerals, etc.) in which heavy metals are present in the solid phases can strongly influence metal mobility, toxicity and bioavailability (Allen et al., 1995; Lopez-Sanchez et al., 1996; Weisz et al., 2000; Yu et al., 2001). So, besides total metal concentration it is also necessary to know their distribution (speciation).

The partitioning of metals by the use of selective extractants is widely used in order to identify the proportion of metals which are present in different chemical forms. Various extraction schemes have been developed (Kersten, 2002).

At a stretch we report two monitoring studies concerning heavy metal presence in two different environmental compartments (soil and sediment).

First case: Marine sediments of the Mar Piccolo in Taranto (Ionian Sea, Southern Italy)

The Mar Piccolo (surface area of 20.72 km^2) is located in the Northern area of the Taranto town (figure 1). It is an inner, semi-enclosed basin with lagoon features connected with Mar Grande through two channels which are very important for water exchange. Tidal range does not exceed 30-40 cm. Mar Piccolo is divided by a rocky promontories into two inlets, called First Inlet and Second Inlet (maximum depth of 13 and 8 m, respectively).

The presence of both 34 submarine freshwater springs (locally called "Citri") and the outfalls of small tributary rivers influence salinity (Caroppo and Cardellicchio, 1995) whereas the low hydrodynamism and the reduced water exchange with the nearby Mar Grande determine, mainly in summer, a high water stratification. The area is characterised by a mean accumulation rate of 0.3 cm/years (Guzzi et al., 2005).

Mar Piccolo basin is subjected to urbanisation, industry, agriculture, aquaculture and commercial fishing. The main source of pollutants are due to 9 pipes discharge sewages from the Northern area of Taranto and nearby towns into the Mar Piccolo, the ship-yard of the Italian Navy with its drydocks (located in the First Inlet), the largest mussel farm (ca. 30,000 tons $y^{-1}$of mussels), the fishing-boat fleet and small rivers and freshwater springs which drain the surrounding agricultural soils in the basin (Lerario et al., 2003; Cardellicchio et al., 2004, 2006). Urban effluents may transport organic matter, nutrients, petroleum hydrocarbons and toxic heavy metals.

The monitoring of heavy metals in sediments of Mar Piccolo is important because this marine ecosystem is devoted to mussel-farming activities. It is know that mollusk are highly tolerant to many pollutants and exhibit high accumulation of them, particularly heavy metals (Beeloy and Eaves, 1983; Gardenfors et al., 1988)

The research was carried out in two times.

In the first time (Calace et al., 2008) surface sediment samples (0-5 cm) were collected in October 2001 with a Van Veen grab (having an opening of 0.045 m^2) in nine stations in Mar Piccolo (five in the first inlet and four in the second inlet) (figure 1). After collection, samples were transferred to a plastic vessel and frozen at -20°C until analysis.

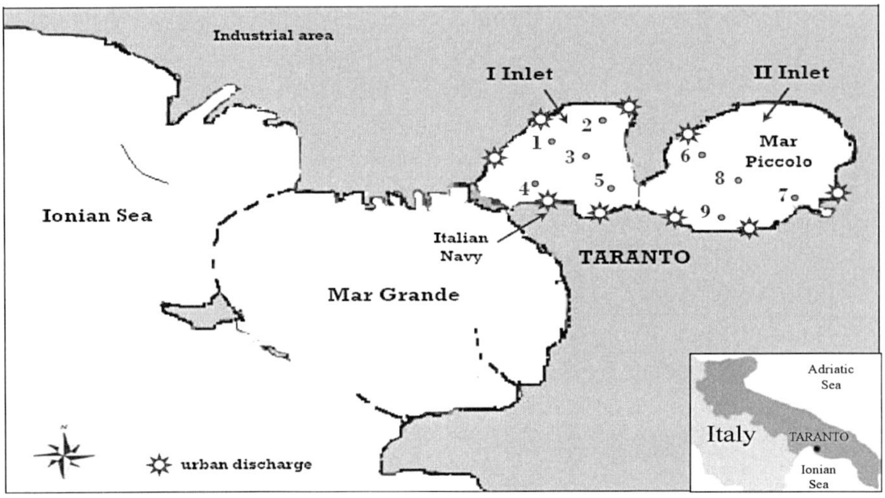

Figure 1. Sampling sites

The primary objectives were to determine:

- the spatial distribution of metals (Cu, Fe, Hg, Mn, Ni, Pb, and Zn) in the study area,
- they distribution in the different chemical forms (speciation) in order to evaluate mobility of metals,
- the content of organic matter (total organic carbon), because organic matter is an important factor influencing metal distribution (Lu et al., 1983). It is known that metals display a different degree of affinity for organic and inorganic compounds.

The results obtained (Calace et al., 2008) have show that organic carbon content, especially in the first inlet, is high respect to data reported by Boetius et al. (1996) for sediments of the Ionic Sea and of the Taranto gulf, probably because of the high trophic status (Jonson and Carman, 1994).

Copper, lead, zinc and mercury concentrations, in stations of first inlet, are higher than those of Taranto Gulf (Buccolieri et al., 2004, 2006; Selli et al., 1977). In particular, in the first inlet the higher copper and lead concentrations are found in stations n.2 and 4 (near an urban discharge and near an other urban discharge and Italian Navy, respectively). Zinc values are high in all stations with the exception of station n.3, located far to anthropogenic sources. Unlike other metals, manganese and nickel display comparable concentration values in both inlets. The relatively low concentrations of manganese could be due to diagenic manganese reduction and subsequent diffusion into overlying water (Lin and Huang, 1993).

Metal speciation, carried out according to the sequence of Campanella et al. (1995) shows different distribution patterns for the six metals, according to different degrees of affinity for organic and inorganic compounds. In all stations, high proportions of copper occur in the humic organic fraction confirming the role of organic matter in enrichment of copper in surface sediments. Manganese-iron oxides as well as organic matter are involved in zinc distribution. Lead is mainly present as exchangeable form and bound to carbonate. The poor amount of lead bound to manganese-iron oxides suggest that Mn oxides play an important role in the enrichment of lead in the surface of sediments. Reduction of manganese oxides, justified by redox potential (Eh) values of sediments (from -180 mV to -230 mV), causes the remobilization and the redistribution in available forms of lead. Also iron and nickel speciation can be related to reducing conditions of sediments.

In conclusion, all the results suggest that reducing conditions in sediments are the major driving forces for manganese, iron, and nickel and lead speciation.

On the basis of results above reported we have identify the first inlet as a contaminated site requiring further examination.

So, a sediment core (21 cm length) was collected in April 2004 in station 2, in which heavy metal concentration in superficial layer is high. After collection, the core was immediately subdivided into twenty-one layers (each layer 1 cm high) in order to obtain a high-depth resolution. Sub-samples were transferred to plastic vessels and frozen at -20°C until analysis.

On the core we have studied the temporal metal distribution, considering both total metal content and metal speciation. Moreover total organic carbon and the hydrophobic fraction of organic carbon was also determined.

EXPERIMENTAL

Total Organic Carbon

Total organic carbon was determined by means of elemental analysis on sediment previously acidified with concentrated HCl to eliminate carbonates. Briefly, 100 mg of sediment were accurately weighed out, acidified with 500 μl of HCl and then dried under an infrared lamp (60°C). The residue was again weighed and analysed. The percentage carbon obtained from elemental analysis was recalculated for the original sediment and expressed as mg/g. The limit of detection (LOD) was 0.03 %.

Organic Carbon Hydrophobic Fraction

To obtain the hydrophobic (Hb) fraction, lyophilised sediment (ca.10 g) was extracted for 24 h at room temperature with a CH_2Cl_2/CH_3OH (2:1 v/v) mixture in a 1:5 (w/v) solid/liquid ratio. Two consecutive extractions of 24 h were carried out in a continuous rotating agitator. After centrifugation for 30 min at 4000 rpm, the supernatants were dried at 60°C in a crucible under vacuum in a desiccator, previously brought to constant weight in a desiccator at 100°C, and then weighed. The organic carbon content of the Hb fraction was determined by elemental analysis.

Total Metal Determination

For total metal determination (Cd, Cu, Fe, Mn, Ni, Pb and Zn), 200 mg of sediment, (vacuum dried, sieved to < 2 mm and homogenised) was digested in a microwave oven (MLS

Mega, 1200 W, 60 bar) with 8 ml of HNO_3 conc., 2 ml of H_2O_2 conc. and 4 ml of HF conc. A four steps digestion program was employed: 250 W for 5 min, 400 W for 5 min, 500 W for 10 min and 0 W for 7 min. After digestion, samples were diluted with ultrapure water and analysed by flame atomic absorption spectrophotometry (F-AAS) using a Varian model Spectr-AA10 spectrophotometer equipped with an air-acetylene flame (acetylene flow 1.5 L min^{-1}). Operating conditions are shown in table 1. Six determinations of each sample were effected.

Table 1. Operational conditions for flame atomic absorption spectrometry

	λ(nm)	Split (nm)	Lamp Current (mA)	Range conc. ($\mu g\ mL^{-1}$)
Cu	324.7	0.5	3.5	2.0-8.0
Fe	372.0/386.0	0.2	5.0	25-100 / 50-200
Mn	279.5/403.1	0.2	5.0	1.0-4.0 /15.0-60.0
Ni	232.0	0.2	3.5	3.0-12.0
Pb	217.0	1.0	5.0	5.0-20.0
Zn	213.9	1.0	5.0	0.5-1.5

Metal Speciation

For metal speciation, the extraction sequence proposed by Tessier et al. (1979) and modified by Campanella et al. (1995) was used. To distinguish metal linked to organic substances from that linked to sulphide, a NaOH 0.5 N extraction solution, after treatment with HCl 0.1 N, was used (step four). A subsequent extraction with HNO_3 8 M for three hours at 80°C was carried out, solubilising metal linked to sulphide. The sequence can be summarised as follow:

-*step one*: $MgCl_2$. Metal present in exchangeable forms (solution A)
-*step two*: NH_4Ac (1M) pH 5. Metals bound to carbonate (solution B)
-*step three*: $NH_2OH.HCl$ (1M) AcOH. Metals bound to manganese-iron oxides (solution C)
-*step four*: HCl (0.1 M). Metals bound to organic matter solubilised in acid medium and acid-base exchangeable forms at pH 1 (solution D)
-*step five:* NaOH (0.5 M). Metals bound to humic and fulvic acids (solution E).
-*step six*: HNO_3 (8 M) 85° C. Metals present as sulphide (solution F).

Three grams of the lyophilised and homogenised sediments were treated and six repetitions of every sample were made. Relative standard deviations were less than 10%.

All reagents were of analytical grade, with very low trace metal contents. Nitric acid, hydrofluoric acid and hydrogen peroxide were Suprapur (Merck, Darmstadt, Germany). Ultrapure water with conductivity < 0.1 μS was obtained from a MILLI-Q® system (Millipore, Bedford, MA, USA). Working standard solutions of metals were prepared daily by serial dilution of stock standard solutions of each element (Merck) containing 1 mg L^{-1} in 0.5 M HNO_3. Normal precautions for trace analysis were always observed.

RESULTS

Organic carbon values and hydrophobic organic carbon content in the different layers of the sediment core are reported in table 2.

The percentage of organic carbon is high in layer 1, than a sharp decrease is observed from layer 1 to layer 7. In the remaining portion of the core the organic carbon content is constant. It is to suppose that the superficial layer is richer of organic matter both to a high nutrient load deriving from anthropogenic sources, and because transformation processes of organic matter are in an early stage. On the contrary, the decrease along the core (layers 2-7) could be ascribed to the increasingly advanced stage of transformation processes of the organic matter associated with loss of CO_2 and/or to an initial lower content of organic matter (poor anthropogenic sources). In the remaining portion of the core (layers 7-21) transformation processes associated with the carbon loss could be considered ended.

Table 2. Total organic carbon and hydrophobic carbon content in sediment layers

Layer	Total organic carbon (mg g^{-1})	Hydrophobic carbon (mg g^{-1})
1	47.13	2.84
2	32.40	3.82
3	34.53	4.99
4	27.50	3.40
5	26.06	2.94
6	22.80	2.51
7	23.00	2.31
8	22.15	2.22
9	22.32	2.35
10	20.87	2.09
11	21.80	2.18
12	20.94	2.11
13	21.15	2.21
14	21.78	2.04
15	20.35	2.12
16	19.04	2.10
17	17.61	1.81
18	19.66	1.88
19	21.30	2.17
20	20.87	2.05
21	22.05	2.09

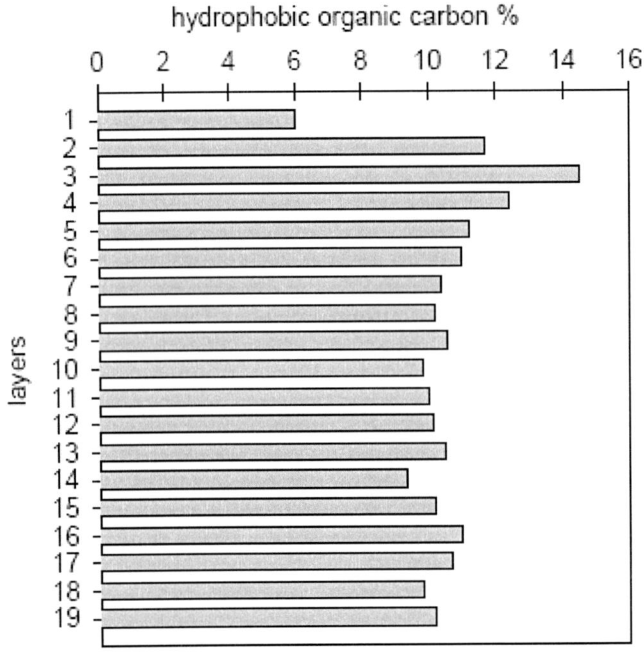

Figure 2. Behaviour of hydrophobic carbon percentage in sediment layers.

On the contrary, the amount of hydrophobic carbon in the upper portion of the sediment core increases until the layer 4. Figure 2 shows the behaviour of hydrophobic carbon percentage in sediment layers.

The high percentage of hydrophobic carbon in layers 2-6 could be related to human pressure. In fact, urban effluents may transport important loads of pollutants, including organic matter, nutrients, petroleum hydrocarbons and toxic heavy metals.

Total metal concentrations are reported in tables 3 and 4. The concentration trend is Fe > Mn > Zn $\geq$ Pb = Cu > Ni

Table 3. Concentration of iron, manganese and nickel in sediment layers

Layer	Fe g kg^{-1}	Mn mg kg^{-1}	Ni mg kg^{-1}
1	25.2 ± 1.3	418.6 ± 27.6	63.5 ± 3.2
2	17.5 ± 1.5	453.7 ± 21.6	65.3 ± 2.5
3	17.0 ± 0.45	340.1 ± 16.8	68.0 ± 1.4
4	17.3 ± 1.1	432.3 ± 27.2	63.5 ± 6.0
5	20.5 ± 1.9	452.0 ± 31.2	61.2 ± 4.3
6	21.8 ± 0.3	464.5 ± 29.7	59.9 ± 5.2
7	27.7± 0.3	360.1 ± 18.5	55.5 ± 1.2
8	33.8 ± 0.5	409.2 ± 23.6	49.6 ± 3.1
9	19.6 ± 0.4	416.3 ± 21.6	53.5 ± 2.7
10	21.2 ± 1.9	397.2 ± 14.7	58.2 ± 3.9
11	22.9 ± 0.4	446.0 ± 19.6	55.5 ± 4.2
12	21.2 ± 2.0	375.9 ± 14.6	56.6 ± 3.5

Layer	Fe g kg-1	Mn mg kg-1	Ni mg kg-1
13	20.5 ± 1.4	476.7 ± 25.7	57.8 ± 4.5
14	22.7 ± 1.8	389.5 ± 18.3	59.2 ± 2.5
15	21.0 ± 1.7	445.6 ± 24.6	58.2 ± 2.9
16	22.7 ± 2.0	401.7 ± 20.4	60.3 ± 3.8
17	19.7 ± 1.8	374.9 ± 18.5	64.6 ± 4.2
18	21.5 ± 1.9	398.3 ± 12.9	61.4 ± 4.8
19	22.3 ± 1.8	406.7 ± 23.9	58.9 ± 5.1
20	22.1 ± 1.9	435,2 ± 23.6	59.3 ± 3.9
21	21.5 ± 1.7	395.7 ± 15.2	60.3 ± 3.7

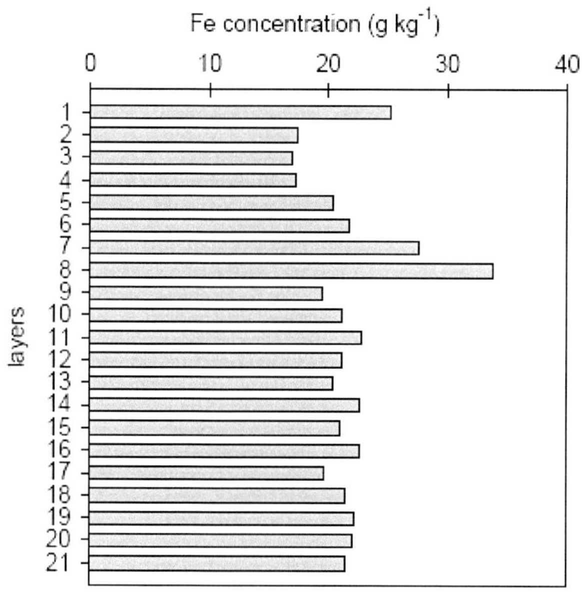

Figure 3. Behaviour of iron along the sediment core.

Iron concentration values (figure 3) along the core, corresponding to natural values in marine sediments, can be divided into two groups: the first including layers from 1 to 8 with a metal concentration variable, the second in which metal concentration is constant.

The differences could be ascribed to a different composition of the sediment. In fact, a good correlation (figure 4) between iron content and grain size percentage between 1 and 2 mm is observed.

Manganese concentration changes along the core: higher values are found in the upper part (table 3). This situation could be related both with a change of sediment composition and with different oxidation states of the system. In anoxic condition, manganese is reduced to a more water soluble forms which can migrate up through the sediment pores and re-precipitate as Mn(II) compounds. The manganese reduction and the subsequent diffusion into the overlying water may have as consequence a not constant metal concentration in sediments.

Ni concentration ranges between natural values and is constant along the core.

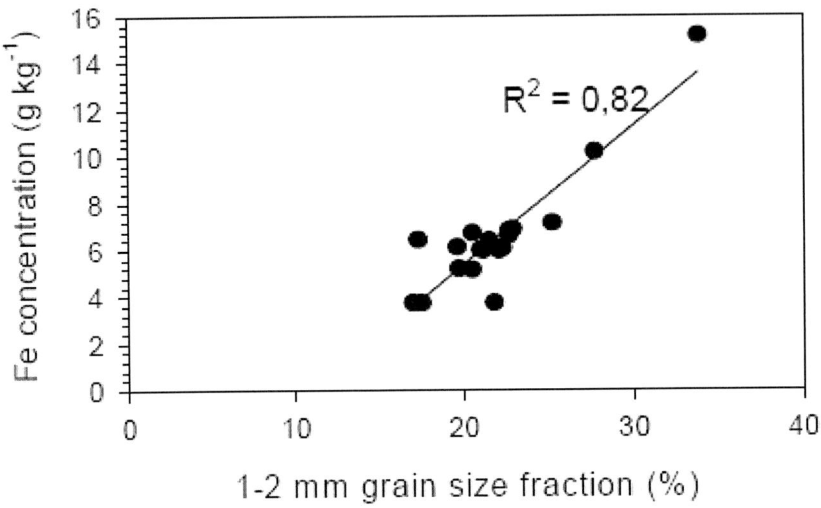

Figure 4. Iron content as function of 1-2 mm grain size percentage.

On the contrary, Zn, Pb, Cu concentrations (table 4) in the upper part of the core are higher and exceed natural values in marine sediments. Moreover, the higher values are in correspondence of layers 7-9. It is hence to conclude that the high Cu, Pb and Zn concentration are due to a change in historical input corresponding to an increase in anthropogenic sources. Moreover, the decrease of metal content in the more superficial layers underlines that the human impact in the last years is reduced. The different behaviour of three metals could indicate different pollution input. Zinc seems to be the metal for which pollution input is older.

Metal distribution underlines that the exchangeable forms (soluble in $MgCl_2$, step 1) are practically absent for all metals.

Table 4. Concentration of lead, zinc and copper in sediment layers

Layer	Pb mg kg^{-1}	Zn mg kg^{-1}	Cu mg kg^{-1}
1	170.9 ± 17.3	300.4 ± 14.4	120.5 ± 4.6
2	165.0 ± 8.5	320.7 ± 10.8	125.3 ± 2.5
3	170.0 ± 6.5	315.0 ± 10.7	152.5 ± 5.4
4	184.8 ± 4.3	340.9 ± 5.2	144.3 ± 10.0
5	162.0 ± 3.6	374.4 ± 15.0	150.8 ± 8.7
6	164.5 ± 3.6	324.5 ± 27.5	177.9 ± 6.2
7	320.5 ± 23.3	355,1 ± 28.4	138.4 ± 10.2
8	366.8 ± 20.5	366.9 ± 32.1	129.6 ± 3.1
9	196.2 ± 15.4	400.4 ± 25.5	120.5 ± 4.2
10	175.4 ±12.5	357.5 ± 25.7	100.2 ± 6.9
11	176.9 ± 3.4	300.9 ± 31.8	79.2 ± 6.8
12	162.4 ± 8.5	253.6 ± 20.5	76.6 ± 3.5
13	140.1 ± 4.2	244,7 ± 11.5	76.8 ± 4.5
14	142.4 ± 8.5	243.2 ± 17.6	80.2 ± 2.5

Layer	Pb mg kg^{-1}	Zn mg kg^{-1}	Cu mg kg^{-1}
15	135..0 ± 7.7	230.6 ± 32.0	84.2 ± 2.9
16	115.7 ± 3.5	235.8 ± 18.8	60.3 ± 3.8
17	98.7 ± 5.1	238.0 ± 23.6	48.6 ± 4.2
18	80.2 ± 5.9	225.9 ± 18.7	51.4 ± 4.8
19	75.3 ± 1.8	230.4 ± 20.5	48.9 ± 5.1
20	66.9 ± 5.9	239.9 ± 17.8	46.3 ± 3.9
21	70.5 ± 1.7	230.8 ± 20.5	50.3 ± 3.7

Fe and Mn speciation shows reducing conditions in the sediments. In fact, more of 70 % of iron is present as sulphide (figure 5), which as example shows iron distribution in layer 3) whereas the prevailing manganese forms are Mn(II) compounds (carbonate and/or hydroxides ~60% and sulphides ~30%) (figure 6, example of metal distribution, layer 6).

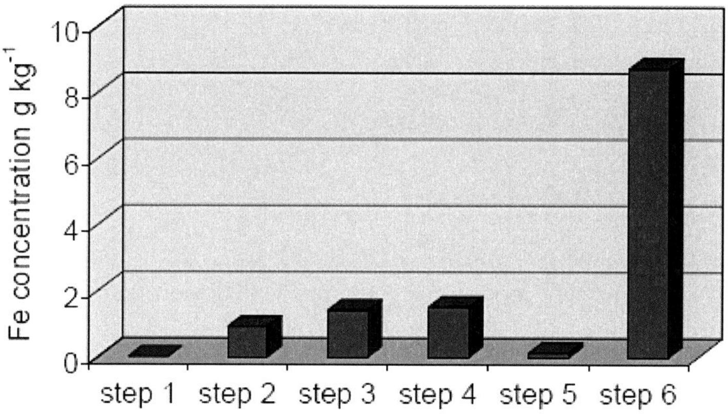

Figure 5. Iron distribution in layer 3.

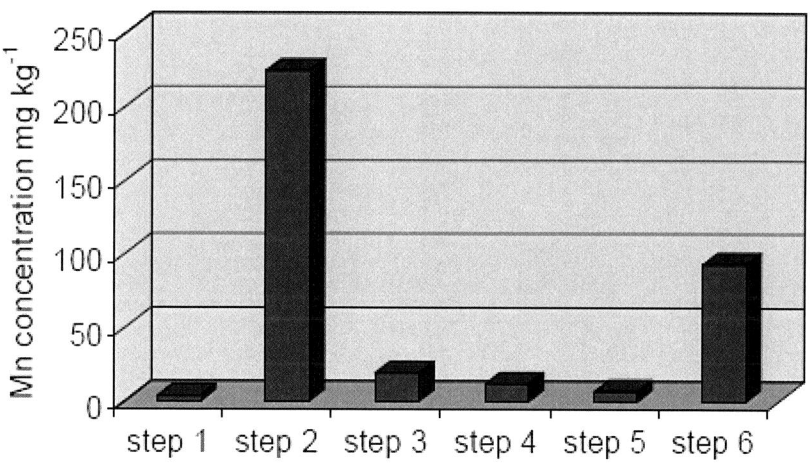

Figure 6. Manganese distribution in layer 6).

A negative correlation (figure 7) is shown between Mn(II) concentrations (step 2 plus step 6) and Mn(IV) (step 3). Ni is in high amount as sulphide.

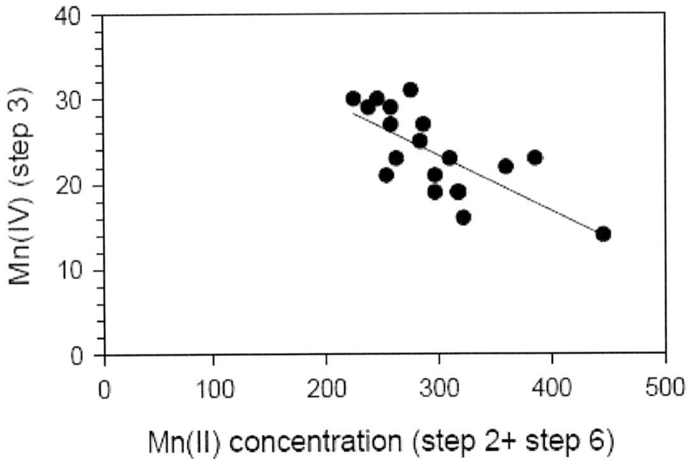

Figure 7. Mn(II) concentrations vs. Mn(IV) concentrations.

Cu, Pb and Zn speciation are among themselves different.

The prevailing forms of copper are mobile forms, copper bound to organic matter and copper as sulphide (figure 8, corresponding to layer 9). In mobile forms (step 2) metal concentration decreases from layer 1 (27 μg g^{-1} corresponding to 21%) to layer 9 (9-10%), than they concentration is constant (figure 9).

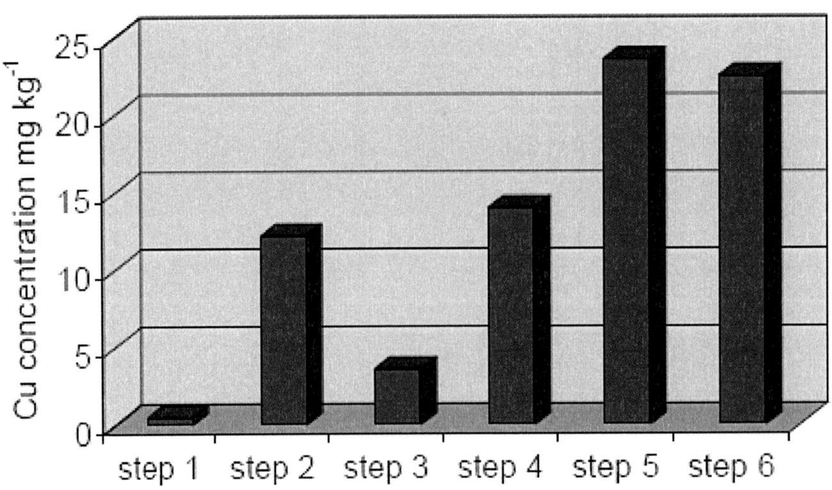

Figure 8. Copper distribution in layer 9.

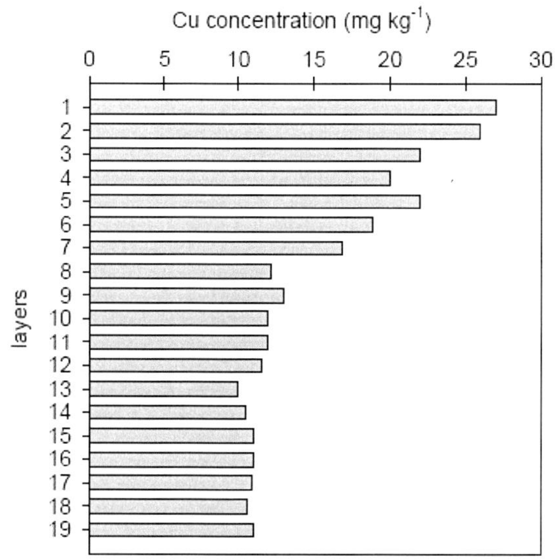

Figure 9. Copper mobile forms along the core.

The behaviour of copper bound to organic matter (figure 10) in the first layers (from 1 to 6) show a behaviour contrary to that of mobile forms (an increase is observed from layer 1 to layer 5), than it is shown a decrease with a subsequent stabilisation indicating in the time the redistribution of the metals from labile forms to more stable forms.

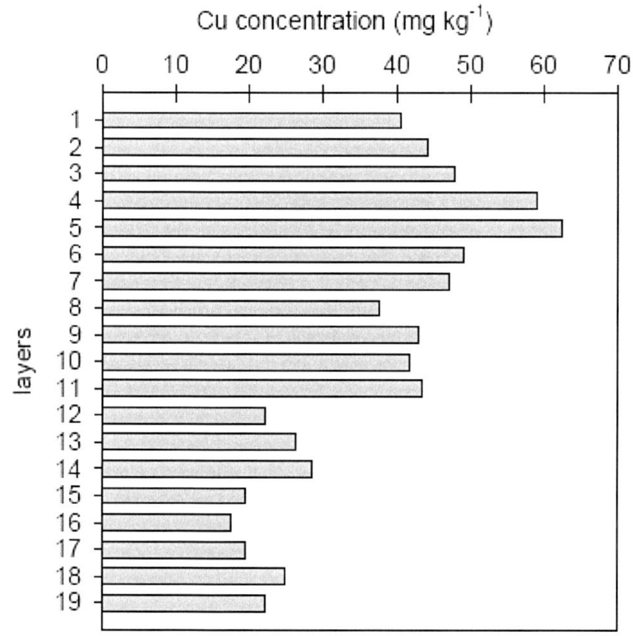

Figure 10. Copper bound to organic matter in the sediment core.

The high affinity of copper for organic matter is known.

Lead (figure 11, corresponding to layer 5) is present in all layers prevalently as mobile forms (~70% in the first layer, 40-45% in others) indicating that equilibrium condition between chemical forms is fast.

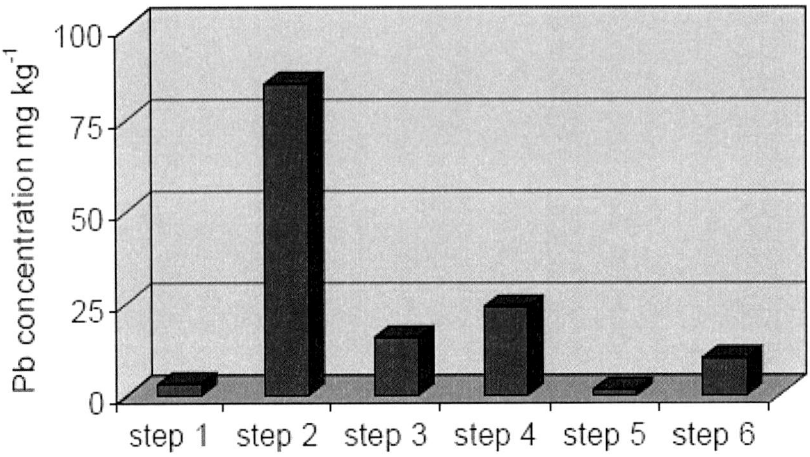

Figure 11. Lead distribution in layer 5.

Also for zinc the mobile forms prevail (figure 12, corresponding to layer 7), but metal as sulphide is also present in not negligible amount. As for lead, zinc bound to humic acids (step 5) is not significant.

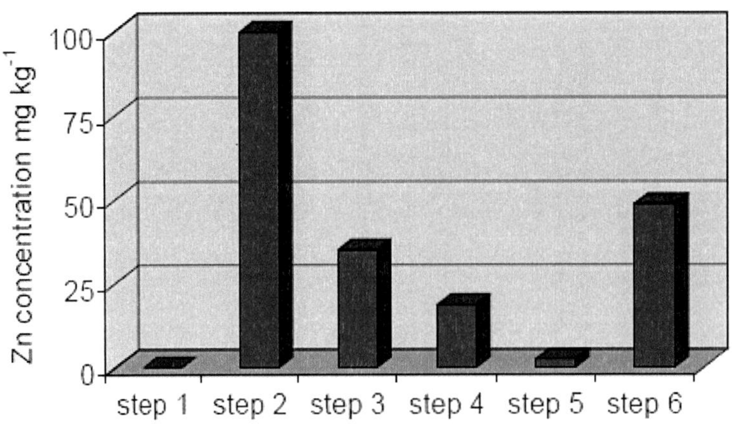

Figure 12. Zinc distribution in layer 7.

In conclusion, results obtained show that in the first inlet of Mar Piccolo sediments contain high amount of contaminants, both organic hydrophobic compounds and heavy metals, in particular copper, lead and zinc. The analysis carried out on the core underline that in the last sixty years the amount of copper and lead is increased. In the latest time a decrease is observed even if metal concentrations are still high. Zinc concentration is high in all layers showing an increase in the latest thirty years. Speciation studies evidence that mobile forms prevail for copper, lead and zinc even if redistribution from mobile forms to more stable forms occurs.

Second case: monitoring of soils of Macchiagrande ecosystem, a WWF oasis near the city of Roma (Italy)

The Macchiagrande oasis (over 280 hectares of land) is in the borough of Fiumicino, on the Roman coastline, a north of Fiumicino Airport (figure 13).

The Oasis is very interesting because it is constituted of a group of various environments including the Mediterranean scrub at the edge of the sand dunes and one of the best-conserved Mediterranean woods on the Roman coastline. The most humid areas of the Oasis are home to an interesting and precious hygrophilous wood, which is one of the largest on the coast.

Also in this case the study can be divided in two parts, the first concerning the monitoring of the total area. Sampling stations (figure 13) were located so that the different environments characterising the oasis are represented. Superficial soil samples were collected with a plastic spoon, than soil parameters (pH, cation exchange capacity, redox conditions, elemental composition, and organic carbon) and heavy metal content (copper, lead, zinc, manganese, nickel and iron) were determined.

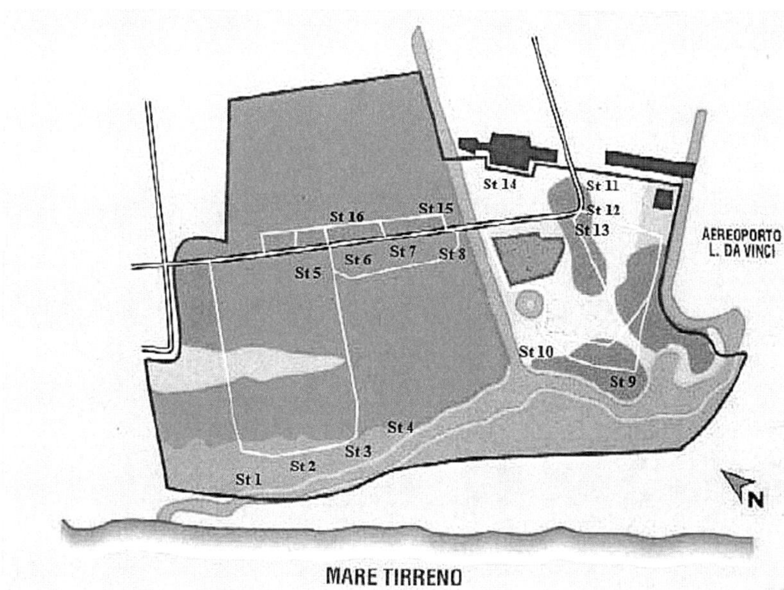

Figure 13. Sampling map of Macchiagrande Oasis.

Experimental

pH values were measured with a Crison model 2022 pH-meter on water surnatant obtained by batching the soil sample with distilled water (1:2.5 ratio w/v) for 15 min.

A Carlo Erba EA11110 model CHNS analyser was employed to measure total carbon and organic carbon, as well as nitrogenous and hydrogen contents, in solid samples. Total organic carbon was determined on samples previously acidified with concentrated HCl to eliminate carbonates. Briefly, 100 mg were accurately weighed out, acidified with 500 μl of HCl and then dried under an infrared lamp. The residue was again weighed and analysed with the CHNS analyser. The percentage carbon obtained from elemental analysis was recalculated for the original sample and expressed as mg/Kg. The limit of detection (LOD) was 0.3 %.

Cation exchange capacity was determined by means of Italian Official Methods for soil (register no. 248 21/10/1999). Briefly, 2 g of soil were placed in a 50 ml centrifuge tube with 25 ml of $BaCl_2$ solution (10% w:v), shaken for 1 min and than centrifuged. Soil was again shaken with a 25 ml of $MgSO_4$ solution (0.1 N) for 5 min and centrifuged too. The magnesium ions remaining in supernatant solution were titrated with ethylenediaminetetraacetic acid disodium salt dihydrate (Na_2EDTA 0.05 N) solution. The difference between the initial concentration of Mg^{2+} ions and the concentration of Mg^{2+} ions remaining in solution gives the cation exchange capacity expressed as meq $100g^{-1}$.

Total metal content and metal speciation were determined as above reported for sediments of Mar Piccolo.

Results

pH values of soil samples are show in figure 14. All samples are characterized by basic values, with differences between samples collected on the sand dunes (more basic) and that corresponding to wooded soils.

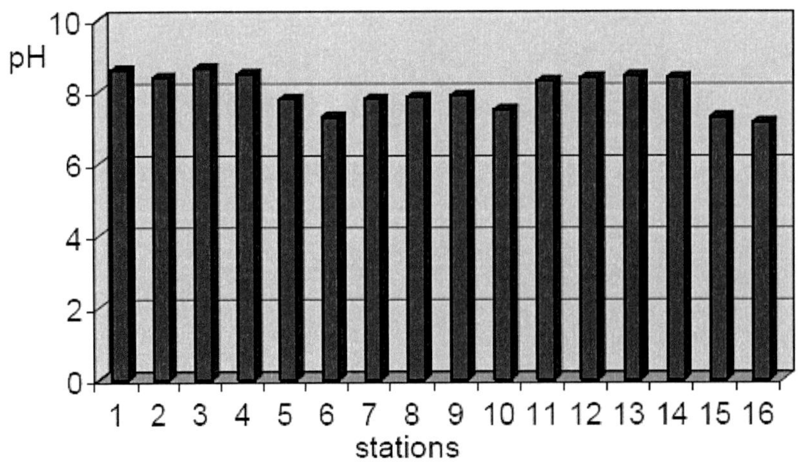

Figure 14. pH values of soil samples.

An opposite behavior is shown from data of cation exchange capacity (figure 15).

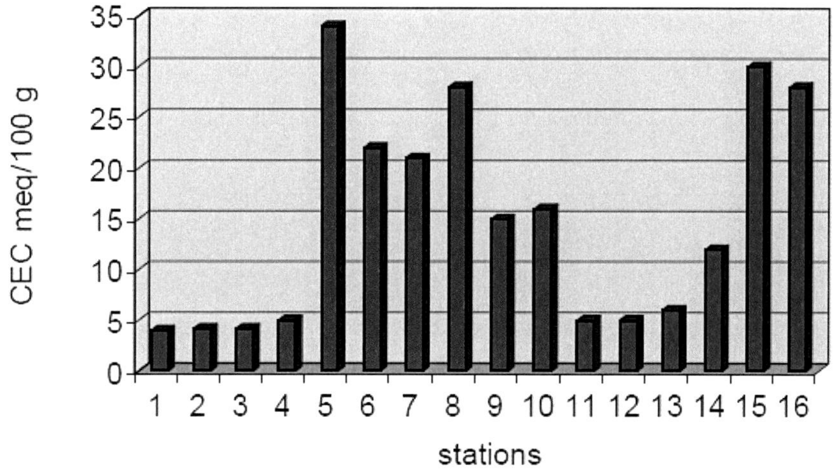

Figure 15. Cation exchange capacity of soil samples.

Redox potential values, ranging from 350 to 400 mV, show that the soil is in oxic conditions. Elemental analysis and organic carbon content (table 5) underline that the soil composition changes with the soil nature: nitrogen, hydrogen and carbon percentages decrease passing from wooded soils to sand dune soils. As regards metals, their total concentrations are shown in table 6.

Table 5. Elemental analysis and organic carbon percentage in soil samples

Stations	N (%)	C (%)	H (%)	C_{org} (%)
1	n.d.	4.33	0.14	0.52
2	n.d.	3.05	0.16	0.37
3	n.d.	3.37	0.13	0.27
4	n.d.	3.00	0.11	0.24
5	0.47	20.35	2.34	18.31
6	0.61	15.47	1.99	8.51
7	0.51	11.55	1.35	8.32
8	0.68	11.91	1.45	7.15
9	0.46	9.37	0.95	7.68
10	0.41	7.81	0.77	3.28
11	0.04	2.46	0.10	1.08
12	0.03	3.06	0.10	0.61
13	0.09	2.90	0.11	0.90
14	0.05	2.30	0.08	0.60
15	0.93	19.04	2.15	11.42
16	0.99	18.00	2.16	14.94

Table 6. Total metal concentration in soil samples

Stations	Ni (mg/Kg)	Cd (mg/Kg)	Cu (mg/Kg)	Pb (mg/Kg)	Zn (mg/Kg)	Mn (mg/Kg)
1	47.9 ± 2.1	4.9 ±0.8	5.1 ± 0.6	76.6 ± 9.1	37.2 ± 3.8	573.7 ± 23.5
2	42.7 ± 2.4	3.9 ± 0.7	6.5 ± 0.2	55.9 ± 6.2	34.6 ± 1.4	608.8 ± 40.9
3	42.9 ±3.5	3.7 ± 0.6	4.7 ± 0.5	54.8 ± 5.5	32.1 ± 2.5	499.1 ±18.8
4	52.5 ± 1.1	3.3 ± 0.4	4.5 ± 0.8	60.7 ± 8.1	40.8 ± 3.9	703.0 ± 30.2
5	31.5 ± 3.4	3.1 ± 0.6	13.7 ± 1.5	71.7 ± 7.3	63.5 ± 3.7	513.6 ± 20.2
6	40.4 ±2.0	4.2 ± 0.4	11.3 ± 0.7	77.2 ± 4.5	54.1 ± 1.6	502.3 ±13.3
7	39.2 ± 2.7	3.6 ± 0.6	10.5 ±1.6	54.5 ±3.0	44.4 ± 2.6	511.2 ±11.3
8	37.5 ± 2.2	3.3 ± 0.3	18.4 ± 1.7	73.1 ± 5.4	46.2 ± 1.6	599.0 ±29.7
9	45.4 ± 1.3	4.8 ± 0.6	20.5 ± 1.4	146.3 ±10.9	52.9 ± 1.9	609.9 ±14.5
10	36.5 ± 1.4	3.1 ± 0.4	8.0 ± 0.9	67.9 ±8.0	41.6 ± 1.7	513.0 ±11.5
11	42.4 ± 3.7	4.0 ± 0.6	20.5 ± 0.9	65.8 ± 7.8	39.6 ±2.3	497.5 ± 20.8
12	38.8 ± 2.1	3.2 ±0.4	20.5 ± 0.7	40.1 ±2.0	37.1 ± 1.3	511.8 ± 20.2
13	50.4 ± 3.3	4.7 ± 0.5	25.8 ± 1.2	66.4 ± 7.6	51.3 ±3.5	545.2 ±15.1
14	52.3 ± 1.9	5.1 ± 0.4	13.9 ± 1.9	63.1 ± 6.9	62.9 ± 5.1	740.4 ±51.4
15	42.7 ± 4.3	3.5 ± 0.5	9.3 ± 1.1	72.3 ± 6.3	53.3 ±3.5	500.7 ±17.6
16	34.8 ± 1.4	2.3 ±0.3	17.5 ± 0.6	91.6 ± 2.7	69.6 ± 7.3	620.6 ±11.6

All values are in agreement with that of not polluted soils, only in station 9 the lead concentration is higher, indicating a possible pollution input. The lead speciation of sample 9 (table 7) shows that 35.7 of the metal are in exchangeable and mobile forms.

Table 7. Lead speciation in station 9

Step	Metal (μg/g)	Metal percentage
1	10.6 ± 0.8	7.4
2	41.7 ± 3.9	28.3
3	12.6 ± 2.4	8.6
4	28.4 ± 1.3	19.4
5	34.1 ±1.8	23.3
6	27.6 ± 0.5	18.9

Consequently the study was gone on doing only the area in which station 9 was located. In this second phase twelve stations (9_1 -9_{12} have been selected (figure 16) in which soil cores were sampled in order to control both the spatial and the temporal metal content.

Values of metal concentration in superficial layers are reported in table 8.

Table 8. Metal concentrations in superficial layer of stations 9_1-9_{12}

Stations	Ni (mg/Kg)	Cu (mg/Kg)	Pb (mg/Kg)	Zn (mg/Kg)	Mn (mg/Kg)
9_1	37.6 ± 1.6	15.4 ± 1.0	122.1 ± 2.4	53.3 ± 2.3	662.4 ± 23.5
9_2	34.3 ± 1.4	5.5 ± 0.3	40.5 ± 1.3	31.7 ± 0.9	472.7 ± 18.3
9_3	24.3 ± 1.1	8.1 ± 0.9	60.7 ± 1.9	38.5 ± 0.8	418.6 ±14.4
9_4	34.8 ± 1.6	9.8 ± 0.6	126.4 ± 15.1	43.6 ± 1.0	487.0 ± 10.2
9_5	24.8 ± 0.7	6.5 ± 0.3	73.7 ± 2.9	35.2 ± 1.0	583.3 ± 27.9
9_6	37.4 ± 2.0	17.8 ± 1.3	572.8 ± 47.1	30.8 ± 1.2	495.1 ± 23.8
9_7	32.2 ± 0.7	13.6 ± 0.4	102.4 ± 7.7	50.3 ± 1.0	425.8 ± 7.3
9_8	25.8 ± 2.1	12.6 ± 0.5	92.4 ± 3.5	46.4 ± 2.2	325.5 ± 9.3
9_9	33.5 ± 0.7	13.6 ± 0.4	164.7 ± 25.9	41.5 ± 0.6	488.6 ± 5.5
9_{10}	31.0 ± 1.4	9.1 ± 0.4	90.3 ± 8.8	35.9 ± 1.2	427.9 ± 15.2
9_{11}	33.9 ± 1.2	9.7 ± 0.6	94.3 ± 7.8	64.8 ± 1.0	450.5 ± 37.8
9_{12}	35.2 ± 1.6	9.1 ± 0.9	80.1 ± 7.7	39.8 ± 1.1	445.2 ± 1.6

The results show that lead is higher than 90 mg/Kg in five stations; therefore metal content was determined also along the cores (layers of 1 cm) to know their temporal distribution (table 9).

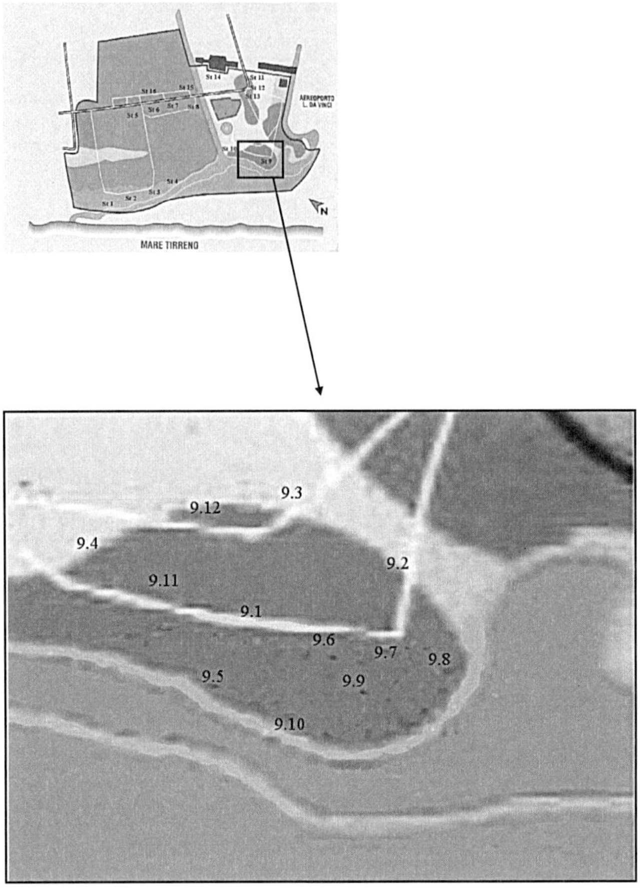

Figure 16. Sampling map of integrated sampling around station 9.

Metal content decreases along the cores. In the station 9_6 Pb concentration is very high in superficial layer. It is to record that the station is located in the central part of the study area where vegetation is quit absent.

Results underline that the pollution is enough recent and that the pollution source is still active.

Finally, the analysis of metal content in leaves sampled of bushes grown on the soil both of the polluted area and of dunes are reported in table 10. In the table are also reported metal values determined in the washing water of the leaves.

The results are in agreement with the high content of lead in the soil of area 9 and show that the metal is present in bioavailable form.

In conclusion the study carried out evidences that the Macchiagrande Oasis is a not metal polluted environment with the exception of the area 9 in which a high content of lead is present in upper layers. The metal contamination is not an old contamination as it is shown from the core analyses. The soil parameters and composition, such as redox potential, pH and organic matter content, make a large part of the metal present in mobile forms, which are partly bioavailable.

Table 9. Metal distribution along soil cores

Core	Layer	Pb (mg/Kg)	Core	Layer	Pb (mg/Kg)
9_1	1	122.1 ± 2.4	9_8	1	92.4 ± 3.5
	2	108.8 ± 8.3		2	135.6 ± 6.5
	3	68.2 ± 8.0		3	145.8 ± 8.1
	4	44.6 ± 3.1		4	119.0 ± 3.1
	5	48.9 ± 2.4		5	70.8 ± 4.3
9_4	1	126.4 ± 15.4	9_9	1	164.7 ± 13.9
	2	135.7 ± 7.5		2	184.0 ± 15.4
	3	97.7 ± 3.3		3	94.5 ± 4.9
	4	72.8 ± 7.6		4	74.8 ± 3.8
	5	60.78 ± 5.7		5	66.7 ± 3.1
9_6	1	572.8 ± 47.1	9_{11}	1	153.4 ± 13.4
	2	375.8 ± 38.3		2	120.3 ± 6.5
	3	91.4 ± 10.6		3	94.3 ± 5.8
	4	66.7 ± 3.1		4	91.1 ± 5.8
	5	64.1 ± 5.7		5	72.1 ± 3.4
9_7	1	102.4 ± 7.7			
	2	150.7 ± 7.5			
	3	139.6 ± 14.3			
	4	92.3 ± 2.3			
	5	79.8 ± 4.4			

Table 10. Metal content in leaves and washing water

Metals	Leaves (μg/g)		Washing water (μg/dm^2)	
	9_6	Dunes	9_6	Dunes
Pb	120.8 ± 10.2	12.6 ± 1.4	1.1 ± 0.2	0.8 ± 0.1
Zn	65.7 ± 5.2	52.4 ± 4.8	2.7 ± 0.3	5.1 ± 0.2
Cu	2.7 ± 0.1	3.9 ± 0.2	1.0 ± 0.1	0.5 ± 0.1
Ni	4.6 ± 0.3	2.9 ± 0.3	0.4 ± 0.03	0.4 ± 0.04
Mn	34.5 ± 2.8	27.7 ± 1.9	2.9 ± 0.1	4.0 ± 0.2

The absence of roads interesting to a high automobile traffic excludes the hypothesis of heavy metal pollution caused by automobiles, both by their emission and by the wear and corrosion of their parts.

It is therefore possible that the source of the lead can be related to the Fiumicino airport. In fact just above the polluted area, that is an open area, the passage of planes is intense.

REFERENCES

Abaye, DA; Lawlor, K.;Hirsch, PR; Brookes, PC. Changes in the microbial community of an arable soil caused by long-term metal contamination. *Eur. J. Soil Sci.,* 2005, 56, 93-102.

Abdul Rida, AMM; Bouché MB. A method to assess chemical biorisks in terrestrial ecosystems. In: Donker, MH., Eijackers, H; Heimbach, F (Eds). Ecotoxicology of Soil Organisms. Lewis, Boca Raton, 1994, 383-394.

Abdul-Wahab, SA; Yaghi, B: Total suspended dust and heavy metal levels emitted from a workplace compared with nearby residential houses. *Atmos. Environ.,* 2003, 38, 745-750.

Akter, MS; Al-Jowder, O. Heavy metal concentrations in sediments from the coast of Bahrain. *Inter. J. Environ. Health Res.*, 1997, 7, 85-93.

Beeby, A; Eaves, SC. Short-term change in Ca, Pb, Zn and Cd concentrations of the garden snail *Helix asperaa* Muller from a central London car park. *Environ. Pollut.*, 1983, A30, 233-244.

Berthelin, J; Leyval, C: Contamination des m ilieux par les elements en trace. Les consequences sur les sols et les eaux superficielles'. *CR Acad. Agric. Fr.*, 2000, 3, 25-37.

Boetius, A; Scheibe, S; Tselepides, A; Thiel, H. Microbial biomass and activities in deep-sea sediments of the Eastern Mediterranean: trenches are benthic hotspots. Deep Sea Res. Part I. *Oceanog. Res. Papers. Part I*, 1996, 43, 1439-1460.

Bohn, HL; McNeal, BL; O'Connor, GA. Soil chemistry. John Wiley, New York, 1985.

Borovec, Z. Evaluation of the concentrations of trace elements in stream sediments by factor and cluster analysis and the sequential extraction procedure. *Sci. Total Environ.*,1996,177, 237-250.

Brimblecomebe, P. Long-term changes in elemental deposition at the earth' surface. *Environ. Pollut.*, 1994, 83, 81-85.

Buffle, J.Complexation Reactions in Aquatic Systems. Prentice Hall, Englewood Cliffs, *N.J.*, 1990

Burgess, ML; Scott, KJ. The significance of in-place contaminated marine sediments on the water column. Processes and effects. In: Burton, GAJr (Ed). Sediment Toxicity Assesment, *Lewis Publ.*, Boca Raton, 1992, 213-240.

Cambier, P. Evaluation of the mobility of toxic elements in contaminated soils. *Anal. Mag.*, 1997, 25, 35-38.

Cardellicchio, N; Annicchiarico, C; Di Leo, A; Giandomenico, S. Mercury in marine sediments and mussel from Taranto Gulf (Ionian Sea, Southern Italy). 7th International Conference on mercury as a Global Pollutant. Lubiana, 2004, June 27-July 2.

Cardellicchio,N; Buccolieri, A; Giandomenico, S; Lerario, VL; Lopez, L; Pizzulli C. Distribution and occurence of polycyclic aromatic hydrocarbons (PAHs) in sediments from the Mar Grande and Gulf of Taranto (Ionian Sea, Southern Italy). *Ann. Chim. (Rome)* 2006, 96, 51-64.

Caroppo, C; Cardellicchio, N. Preliminary study on phytoplankton communities of mar Piccolo in Taranto (Jonian Sea). *Oebalia* 1995, XXI, 61-76.

Chuan, MC; Shu, GY; Liu, JC. Solubility of heavy metals in a contaminated soil: effects of redox potential and pH. *Water Air Soil pollut.*, 1996, 90, 543-556.

De Vriees, W; Vel, E; Reinds, GJ; Deelstra, H; Klap, JM; Leeters, EEJM; Hendriks, CMA; Kerkvoorden, M; Landmann, G; Herkendell, J; Haussmann, T; Erisman, JW. Intensive monitoring of forest ecosystems in Europe 1. Objectives, set-up and evaluation strategy. *Forest Ecol. Manag.*, 2002, 5890, 1-19.

Dijkstra, E. A micromorphological study on the development of humus profiles in heavy metal polluted and non-polluted forest soils under Scots pine. *Geoderma*, 1998, 82, 341-358.

Drever, JI. The Geochemistry of Natural Waters, 2nd edn. Prentice Hall, Englewood Cliffs, N.J., 1988.

Duzzin, B; Pavoni, B; Donnazolo, R. Macroinvertebrate communities and sediments as pollution indicators for heavy metals in the River Adige (Italy). *Water Res*., 1988, 22,1353-1363.

Elsgaard, L; Petersen, SO; Debosz, K. Effects and risk assessment of linear alkylbenzene sulfonates in agricultural soil. 1. Short-term effects on soil microbiology. *Environ. Toxicol. Chem*., 2001, 20, 1656-1663.

Erel, Y; Morgan, JJ; Patterson, CC: Naturals levels of lead and cadmium in a remote mountain stream. *Geochim. Cosmochim. Acta*, 1991, 55, 707-719.

Farrah, H; Pickering, WF. The sorption of lead and cadmium species by clay minerals. *Aust. J. Chem*., 1977, 30, 1417, 1422

Filip, Z. International approach to assessing soil quality by ecologically related biological parameters. *Agric. Ecosyst.Environ*., 2002, 88, 169-174.

Förstner, U. Chemical forms reactivities of metals in sediments. In: Leschber, R; Davis, RD; L'ermite, P (Eds). Chemical Methods for Assessing Bioavailable Metals in Sludges and Soils. Elsevier, London. 1985, 1-30.

Förstner, U. Contaminated sediments. In: Bhattacharji, S; Friedman, GM; Neugebauer, HJ; Seilacher,A (Eds). Lecture notes in earth sciences, Springer-Verlag, Berlin,1989, 21, 157-168.

Gadd, GM. Metals and microorganisms: a problemof definition. *FEMS Microbiol. Lett*., 1992, 100, 197-204.

Gardenfors, U; Westermark, T; Carell, B. Use of land snail shell as environmental archives. *Ambio*, 1988, 17, 347-349.

Gibbs, RJ. Metals in the bottom muds in Townville harbor, Australia. *Environ. Pollut*., 1993, 81, 297-300

Giller, KE;Witter, E; McGrath, SP. Toxicity of heavy metals to microorganisms and microbial processes in agricultural soils: a rewiew. *Soil Biol. Biochem*., 1998, 30, 1389-1414.

Gonçalves, EPR; Soares, HMVM; Boaventura, RAR; Machado, AASC; Silva, JCGE. Seasonal variations of heavy metals in sediments and aquatic mosses from the Cávado river basin(Portugal). *Sci. Total Environ*., 1994, 142, 143-156.

Guzzi L., Martinotti W., Perotti M., Achilli M., Novo A., Cardellicchio N. Ciclo del mercurio: effettuazione di indagini sperimentali (deposizioni, accumulo nelle matrici ambientali). Report SCENARI/EVAMB/WP5/ Milestone 2005; 3, 94-104 (in Italian).

Han, FX; Banin, A. Long-term transformations and redistribution of potentially toxic heavy metals in aride-zone soils. II:under the field capacity regime. *Water Air Soil Pollut*., 1999, 114, 221-250.

Han, FX; Banin, A; Triplett, GB. Redistribution of heavy metals in arid-zone soils under a wetting-drying soil moisture regime. *Soil Sci*., 2001, 166, 18-28.

Han, FX; Hargreaves, J; Kingery, WL; Huggett, DB; Schlenk, DK. Accumulation, distribution and toxicity of copper in soils of catfish pond receiving periodic copper sulfate applications. *J. Environ. Qual*., 2001, 30, 912-919.

Han, FX; Hu, AT; Quin, HY. Transformation and distribution of forms of zinc in acid, neutral and calcareous soils of China. *Geoderma*, 1995, 66, 121-135.

Huang, CY; Khan, KS. Effects of cadmium, lead and their interaction on the size of microbial biomass in a red soil. *Soil Environ.*, 1998, 1, 227-236.

Juvanovic, S; Carrot, F; Deschamps, N; Vukotic, P. A study of the air pollution in the surroundings of an aluminum smelter using Epiphytic and Lithophytic Lichens. *J. Trace Microprobe Tech.*, 1995, 13, 463-471.

Izquierdo, C; Usero, J; Gracia, I. Speciation of heavy metals in sediments from salt marshes on the southern atlantic coast of Spain. *Mar. Poll. Bull.*, 1997, 34, 123-128.

Kelly, JJ; Häggblom, M; Tate, RL. Effects of the land application of sewage sludge on soil heavy metal concentrations and soil microbial communities. *Soil Biol. Biochem.*, 1999, 31, 1467-1470.

Kersten, M. Speciation of trace metals in sediments. In: Ure, AM; Davidson, CM (Eds). Chemical Speciation in the Environment. Blackwell Science, Oxford, 2002, 301-321.

Khan, M; Scullion, J. Effects of metals (Cd, Cu, Ni, Pb or Zn) enrichment of sewage-sludge on soil micro-organisms and their activities. *Appl. Soil Ecol.*, 2002, 20, 145-155.

Khas, KS; Xie, ZM; Huang, CY. Relative toxicity of lead chloride and lead acetate to microbial biomass in red soil. *Chin. J. Appl. Environ. Biol.*, 1998, 4, 179-184.

Lapaquellerie, Y; Jouanneau, JM; Maillet, N; Latouche, C. Cadmium pollution in sediments of the Lot river (France).Estimate of the mass of cadmium. *Environ. Tech.*, 1995, 16, 1145-1154.

Lapitajs, G; Grg, U; Dunemann, L; Begerow, J; Moens, L; Verrept, P. ICP-MS in the determination of trace and ultra trace elements in the human body. *Int. Labor.*, 1995, 5, 21-27.

Lerario V.L., Giandomenico S., Lopez L., Cardellicchio N. Sources and distribution of polyciclic aromatic hydrocarbons (PAHs) in sediments from the Mar Piccolo of Taranto, Ionian Sea, Southern Italy. Ann. Chim. (Rome) 2003; 93, 397-406.

Lisk, D:J: Trace metals in soils, plants and animals. Adv Agron., 1972, 24, 267-325.

Lopez-Sanchez, JF; Rubio,R; Samitier, C; Rauret,G: Trace metal partitioning in marine sediments and sludges deposited of the coast of Barcelona (Spain). *Water Res.*, 1996, 30, 153-159.

Martincic, D; Kwokal, Z; Branica, M. Distribution of zinc, lead, cadmium and copper between different size fractions of sediments. I. The Limski Kanal (North Adriatic Sea). *Sci. Total Environ.,* 1990, 95, 201-215.

McGrath, SP; Chaudri, A M; Giller, KE. Long-terms effects of metals in sewage sludge on soil, microorganisms and plants. *J. Ind. Microbiol.*, 1995, 14, 94-104.

McLaren, RG; Ritchie, GSP. The long-term fate of copper fertilizer applied to a lateritic sandy soil in Western Australia. *Austr. J. Soil Res.*, 1993, 93, 39-50.

Mokhtar, MB; Awaluddin, AB; Fong, CW; Woojdy, M. Lead in blood and air pollution near an operational and a proposed area for copper mining, malaysia. *Bull. Environ. Cont. Toxicol.,* 1991, 52, 149-154.

Odukoya, OO; Arowolo, TA; Bamgbose, O. Pb, Zn and Cu levels in tree barks as indicator of atmospheric pollution. *Environ. Int.*, 2000, 26, 11-16.

Pardo, R; Barrado, E; Pèrez, L; Vega M. Determination and speciation of heavy metals in sediments of the Pisuerga River. *Water Res.*, 1990, 24, 373-379.

Pichtel, J; Kuroiwa, K; Sawyerr, HT. Distribution pf Pb, Cd and Ba in soils and plants of two contaminated sites. *Environ. Poll.*, 2000, 110, 171-178.

Ranjbar, GA. Heavy metal concentration in surficial sediments from Anzali wetland, Iran. *Water Air Soil Poll.,* 1998, 104, 305-312.

Robert, M; Chenu, C. Interactions between soil minerals and microorganisms. In: Stotzky, G; Bollog, IM (Eds.). Soil Biochemistry. Marcel Dekker, New York, 1991, 307-379.

Robinson, B; Russel, C, Hedley, M; Clothier, B. Cadmium adsorption by rhizobacteria: implications for New Zealand pastureland. *Agric. Eco. Environ.*, 2001, 87, 315-321.

Rubinstein, NI; Lores, E; Gregory, NR. Accumulation of PCBs, Hg and Cd by *Nereis virens, Mercenaria mercenaria* and *Palaemonetes pugio* from contaminated harbour sediments. *Aquatic Toxicol.*, 1983, 3, 249-260.

Salomon, W; Förstner, U. Metals in the Hydrocycle. Springer, Berlin, 1984.

Sheppard, DS; Claridge, GGC; Campbell, IB. Metal contamination of soils at Scott base, Antarctica. *Appl. Geochem.*, 2000, 15, 513-530.

Shi, W; Bischoff, M; Turco, R; Konopka, A. Long-term effects of chromium and lead upon the activity of soil microbial communities. *Appl. Soil Ecol.*, 2002, 21, 169-177.

Shne,JP; Ika, RV; Ford, TE. Multivariate statistical examination of spatial and temporal patterns of heavy metal contamination in New Bedform Harbor marine sediments. *Environ. Sci. Technol.*, 1995, 29, 1781-17963.

Sillanpaa, M. Trace elements in soils and agriculture. *FAO Soils Bull.*, 1972, 17, 1-67.

Sposito, G; Lund, LJ; Chang, AC. Trace metal chemistry in arid-zone field soils amended with sewage sludge: I. Fractionation of Ni, Cu, Zn, Cd and Pb in solid phases. *Soil Sci. Soc. Am. J.*, 1982, 46, 260-264.

Wardas, M; Budek, L; Rybicka, EH. Variability of heavy metals content in bottom sediments of the Wilga River, a tributary of the Vistula River (Kraków area, Poland) *Appl. Geochem.*, 1996, 11, 197-202.

Weisz, M;Polyák, K; Hlavay, J. Fractionation of elements in sediment samples collected in rivers and harbors at lake Balaton and its catchment area. *Microchem. J.*, 2000, 67, 207-217.

Yaron, B; Calvet, R; Prost, R. Soil Pollution. Springer-Verlag, Berlin. 1996.

Yong, RN; Phadungchewit, Y. PH influence on selectivity and retention of heavy metals in some clay soils. *Can. Geotech. J.*, 1993, 30, 821-830.

Yu, K-C; Tsai, L-J; Chen, S-H; Ho, S-T. Correlation analyses binding behavior of heavy metals with sediment matrices. *Water Res.*, 2001, 35, 2417-2428.

Zoumis, T; Schmidt, A; Grigorova, L; Calmano, W. Contaminants in sediments: remobilisation and demobilisation. *Sci. Total Environ.*, 2002, 266, 195-202.

In: Heavy Metal Pollution
Editors: S. E. Brown and W. C. Welton
ISBN: 978-1-60456-899-8

Chapter 4

HEAVY METALS POLLUTION ASSOCIATED WITH THERMAL ENERGY PROCESSES

Helena Lopes, Filomena Pinto and Ibrahim Gulyurtlui
INETI, Estrada do Paço do Lumiar, 22, 1649-038 Lisboa, PORTUGAL

ABSTRACT

The presence of heavy metals in fossil fuels or other materials, such as certain types of biofuels and wastes that present enough energy content for its use as fuels, is a problematic issue. There are several national and international programs to control its emissions, namely that of Hg, Cd and Pb, although much more metals are emitted every day from industrial sources. Heavy metals are not destroyed at high temperatures; they suffer complex transformations and different partitioning among the several streams of thermal processes, depending on the operational conditions employed and materials composition. Metals may volatilize at high temperatures and be carried out as vapors or condense as fine particles contributing to atmospheric pollution or may remain attached to the ashes. Atmospheric pollution may be controlled with efficient air pollution control devices or dedicated systems in the case of Hg, but ashes may present high quantities of metals that require appropriate management or reutilization. Management costs depend on their hazardous nature, which are evaluated with appropriate methodologies. Apart from the common use of fossil fuels and Municipal Solid Wastes, two cases are of enormous interest nowadays and will be discussed as case studies: one is the recovery of energy from sewage sludge, either through combustion or gasification, and the other is the use of automobile shredder wastes (ASR) from end life vehicles dismantling. The study of the combustion of ASR in a fluidized bed pilot showed that its energy recover is highly attractive, however, as the residue contains high quantities of heavy metals and chlorine, emissions may be significant, especially that of Hg, thus requiring efficient flue gas treatment. Ash produced retained most of the metals but evaluation of leachability showed that ashes behave as hazardous materials, essentially because of the release of hexavalent Cr, chlorides and sulfates. In the case of sewage sludge, which also contains significant quantities of mineral matter, including mercury, combustion in a fluidized bed pilot led to lower metal emissions because retention in ashes was higher. Evaluation of leachability revealed that metals did not leach, but the release of sulfate and chloride was high in fly ashes. When sewage sludge was gasified to produce syngas, instead of being directly burned, the reducing environment increased the volatilization of some metals, thus the combustible gas produced requires further removal of metals before being used. Metals that were strongly retained in the gasification residues did not leach out and release of sulfates was minimized because gasification process does not favor sulfur oxidation.

Keywords: Heavy Metals, Combustion, Coal, Residues, Emissions, Leachability.

1. Anthropogenic Emissions of Heavy Metals

The term "heavy metals" has a particular significance in environmental sciences. Generally it is applied to high-density metals and its compounds such as: antimony, arsenic, cadmium, chromium, copper, lead, mercury, nickel, selenium, tellurium, thallium, tin, etc. Reinman and Caritat (1998) and Gadd (1992). These high-density metals, with density higher than 4.5 g/cm^3 or 5.0 g/cm^3, are considered as heavy metals and includes a large number of metals (about 65), such as transition metals actinides and lanthanides, with different physicochemical properties.

Heavy metals are natural constituents of the Earth's crust and participate in natural fluxes. Biogeochemical cycles involves emission of heavy metals to the atmosphere through dust, evaporation, volcanic activity, fires, biologic activity, etc. and deposition by the action of natural agents such as win, rain and sedimentation. However, anthropogenic activities, such as mineral extraction, metallurgic industries, road transport and combustion process, with emphasis to incineration, are responsible for the emission of some heavy metals to the atmosphere, both as vapors or as particles of low dimension, with high Enrichments Factors to the earth crust, EF_{crust}, i.e levels much higher than those found in natural materials (Lim, 1979; Swaine, 1998; Clark and Sloss, 1992; Nriagu and Pacyna, 1988). Wadge *et al.* (1986) referred enrichment factors, EF_{crust}, of 132 for Se and 67 for As in particles emitted from coal combustion, but for incinerator particles, the EF_{crust} are much higher, from 1000 to 10000, for As, Cd, Pb and Se.

Previous researchers (Clark and Sloss, 1992 and Nriagu and Pacyna, 1988) revealed that thermal power stations were responsible for the emission of 95% of V, 80% of Ni, 60% of Hg, 65% of Sn and 60% of Se, of the total anthropogenic emissions, being also important the emissions of Cd, Hg, Mn and Sb, from incineration. However, combustion sources and incineration were responsible for the emission of a similar quantity of Hg as the natural sources, being coal combustion responsible by 2/3 of the anthropogenic emissions. In the case of Pb, transports were responsible for 74% of the global emissions due to the use of leaded gasoline in the past.

Comparing anthropogenic emissions with natural sources it can be defined an interference factor, IF (table 1), to express the human contribution, obtained by the ratio between the anthropogenic emissions and natural emissions. This factor evidences the disturbance to the natural equilibrium of the earth ecosystem by human hands, being more pronounced for Cd and for Pb, which shows that these two metals are of most importance. Another metal highly relevant is Hg, as its intensive use and difficulty in controlling its emissions foresees emissions increases of 1.46% to 1.17% by year, respectively, in the north and south hemispheres (Clark and Sloss, 1992).

Table 1. Emissions of some heavy metals to the atmosphere

	Anthropogenic Sources a) $x10^3$ t/year	Natural Sources b) $x10^3$ t/year	IF c) (Interference factor)	Contribution to the total anthropogenic emissions d)	
				Energy (%)	Incineration (%)
As	18.9	12.2	1.6	11.6	1.6
Cd	7.6	1.4	5.8	10.5	10.5
Cr	30.5	43.3	0.71	41.6	2.6
Cu	35.3	28.1	1.25	22.7	4.5
Hg	3.6	2.5	1.4	65.7	34.3
Mn	38.2	316.9	0.09	31.6	21.7
Ni	51.6	29.3	1.9	81.4	0.8
Pb	332.3	12.2	28.0	3.8	0.7
Sb	3.5	2.6	1.5	37.1	20.0
Se	6.3	10.3	0.68	61.9	1.6
Sn d)	5.1	-	-	64.7	15.7
Tl b)	5.1	-	-	21.6	-
V	86.0	27.7	3.1	97.7	1.4
Zn	131.8	44.7	2.9	12.7	4.5

a) Nriagu and Pacyna(1988), b) Nriagu (1989), c) Buat-Ménard(1993), d) Clark and Sloss (1992)

In 1998 a Protocol on Heavy Metals was signed in Aarhus (Denmark), in the framework of the Convention on Long-Range Transboundary Air Pollution (CLRTAP) of UNECE (United Nations Economic Commission for Europe). Lead, cadmium and mercury were considered as priority metals. Later on, in 2001, the United Nations Environment Programme (UNEP) adopted the Programme of International Actions on the restriction of negative impact of mercury on the environment and human health. Since 2005 Hg emissions from coal power stations are already regulated in the USA by the Clean Air Mercury Rule and in European Union, the control of Mercury has already been started (COM(2005) 20 final).

Within the Aarhus Protocol, the European Monitoring and Evaluation of Pollutants (EMEP) is carrying out the assessment of pollution levels of Cd, Pb and Hg in Europe. 64 monitoring stations were installed in Europe for the measurement of Pb and Cd and 19 stations also measure Hg. Local emissions were estimated to vary from 0.1 to more than 5 $kg/km^2/y$ for lead, 5 to 150 $g/km^2/y$ for Cd and 3 to 50 $g/km^2/y$ for mercury. Hg is transported by atmospheric flows all around the globe; hence mercury emission sources located in other continents have a significant impact on pollution in Europe (EMEP Status Report 2/2003).

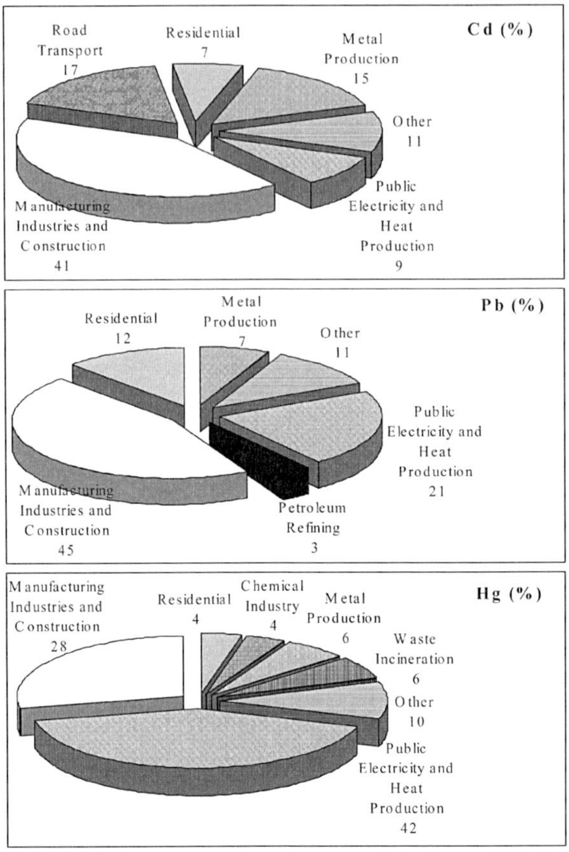

Figure 1. Relative contribution of source categories to total emission of lead, cadmium and mercury from European countries in 2005, (EMEP Status Report 2/2007).

For 2005, data on emissions of lead, cadmium and mercury were actualized. Information was presented by 24 countries (Austria, Belarus, Belgium, Bulgaria, Denmark, Estonia, Finland, France, Germany, Hungary, Ireland, Italy, Latvia, Lithuania, Monaco, the Netherlands, Norway, Poland, the Russian Federation, Slovakia, Spain, Sweden, Switzerland, and United Kingdom). Between 1990 and 2005 total anthropogenic emissions in the EMEP region decreased for all the three metals. Lead decrease was about 87%, cadmium reduction was 50% and for mercury 48% decrease was achieved. Total emissions of lead, cadmium and mercury for the year 2005, were a combination of official data and non-official estimations are were estimated to be around 4700, 245 and 170 t/y, respectively (EMEP Status Report 2/2007).

Analysis of the emissions data by source categories, showed that the public electricity and heating, plus residential sectors (figure 1) contribute to 16% of the total Pb emissions, for Cd, these activities contributed to 34 % and for Hg these two sectors, plus waste incineration, were responsible by 57% of the total emissions reported.

2. Regulations on Heavy Metals Emissions

Most heavy metals are also considered toxic metals. They are assimilated by living organisms, either by inhalation, ingestion or absorption through the skin, and usually they tend to accumulate and affect vital functions (metabolism, reproduction, etc.).

The Sixth Community Environment Action Program, adopted by Decision N° 1600/2002/EC, established the need to reduce pollution to levels which minimize harmful effects on human health, with particular attention to sensitive populations, and the environment as a whole.

Scientific studies revealed that metals like arsenic, cadmium, nickel and mercury are human genotoxic carcinogens and that there is no identifiable threshold below which these substances do not pose a risk to human health. However, through Directive 2004/107/EC, target concentrations in ambient air were set, as shown in table 2.

Table 2. Target values of heavy metals in ambient air (Directive 2004/107/EC)

Pollutant Target value for ambient air (total content in the PM10 fraction averaged over a year.	Ambient air concentration
Arsenic	6 ng/m^3
Cadmium	5 ng/m^3
Nickel	20 ng/m^3

The bioavailability, that is the fraction of a substance that can be absorbed by living organism (Nieboer et al., 1999), depends on its speciation, and in the particular case of solid particles, of its dimension and morphology. The properties of organic fluids are also determinant for the bioavailability extend which as to do with the root of absorption by living organisms. The most fine atmospheric particles (submicron particles) can be breathed by man and reach the lung, where the rate of absorption is higher (50-80%), while the larger particles might be ingested or trapped in the nasal zone and pharynges, being further transported to the stomach, where the absorption efficiency range is between 5% to 15% for most of the metals (Natusch et al., 1973, 1974a, b).

Ontiveros et al. (1989) studied the leachability of fly ashes at pH of 7.4 and of 2, typical for inhalation and ingestion, and found out that more than 90% of Cd was mobilized at both pH values, while 12 % of Pb was mobilized at pH 2, but at pH 7.4 the extraction was lower than 0.5%. Clapp et al. (1988) also verified that for incinerator fly ashes, 15% and 16% of Pb and Cd, respectively, were absorbed by inhalation, while the ingestion represented less than 0.07% for both metals.

Regarding chemical speciation, heavy metals might form a variety of compounds with different oxidation states, depending on the oxy-redox conditions. For example, oxidation states might be; Ag (+1), Sr, Zn, Cd, Be, Ba (+2), Cu (+1 and +2), Ni (+2 and +3), Pb (+2 and +4), Co (+2 and +3), Se (-2,+4 and +6), Cr (+2, +3 and +6), As (-3, +2, +3 and +5), Sb (-3, +3, +4 and +5) and Hg(+1, +2 and +4). Mn exhibits a larger range of oxidation states (+2, +3, +4, +6 and +7).

In the case of Cr, it is recognized that the hexavalent oxidation state (+6) presents higher toxicity. In combustion ashes compounds of trivalent chromium prevail (Clarke and Sloss,

1992), although the typical oxidant conditions and higher temperature may favor the formation of hexavalent chromium compounds. In natural environments at pH values below 6, the organic matter, together with the presence of reducing agents, such as Fe(2+) and sulfides, may promote its reduction. In the case As, the trivalent form, arsenites, is considered more dangerous than the pentavalent form. In combustion ashes As is normally present as arsenate, and during leaching processes the trivalent form tends to oxidize (Clarke and Sloss, 1992). However in natural environments microorganisms may generate methylated forms of As(+3), extremely toxic. Pb is also highly toxic, it can be methylated in natural environments to $(CH_3)_4Pb$, which is a gas at ambient temperature. The use of leaded gasolines in the past, contributed to high environmental contaminations, because of gasoline spills, dispersion of contaminated used oils and air emissions. Both metallic Hg and the major part of Hg compounds are extremely mobile and toxic, due to the high volatility, even at ambient temperature. In combustion systems it may be emitted in the elemental or metallic form or in a oxidized state, in the gaseous form or associated to particles. The toxicity an volatility may increase in environmental systems due to alkylation by microorganisms, (Desauziers et al., 1997), producing extremely dangerous methylated species The living times of Hg in the atmosphere are extremely high, varying from half year to two years, while living times of other heavy metals are smaller (minutes to hours). Hg is found in atmosphere mainly in the elemental form (Clake and Sloss, 1992; Pacyna, 1987). Bergstrom (1986) referred a concentration of Hg of 2 ng/m^3 in the ambient air, mainly in the gaseous metallic form. Mn is an example of indirect toxicity, as with the case of Fe, it catalyzes the formation of H_2SO_4, from SO_2 emissions, leading to the formation of smog.

Table 3. Limit values of heavy metals emissions from waste incineration (Directive 2000/76/EC)

Emission Limits (WID) (mg/m^3)	Co-Combustion Cement kilns 10% O_2	Co-Combustion Other systems 6% O_2	Dedicated Incineration 11% O_2
Hg	0.05	0.05	0.05
Cd+Tl	0.05	0.05	0.05
Sb + As, Pb + Cr + Co + Cu + M n + Ni + V	0.5	0.5	0.5

Traditional incineration of residues was mostly applied to MSW (Municipal Solid Waste) in dedicated incinerators or to hazardous wastes incineration. In 1989, Directive 89/369/EEC and Directive 89/469/EEC started the control of emissions from MSW incinerators, including heavy metals. For hazardous waste incineration and co-incineration, Directive 94/67/CE, set lower limits of pollutants including more 5 heavy metals than those considered for MSW incineration. Nowadays the concept of incineration has changed, and the regulations on emission control became unified (table 3), through Directive 2000/76/EC (WID Directive), which applies both to dedicated incineration and to co-combustion processes, including oxidative processes, gasification, pyrolysis and plasma pyrolysis. WID Directive has also defined limit values for the liquid residues resulting from the cleaning process of waste incineration, such as those obtained with scrubbers (table 4).

Table 4. Limit values for waste waters from treatment of flue gases (Directive 2000/76/EC)

Metals (mg/l)	Limit values for non filtered samples
Hg	0.03
Cd	0.05
TI	0.05
As	0.15
Pb	0.2
Cr	0.5
Cu	0.5
Ni	0.5
Zn	1.5

Ashes produced in thermal processes and incineration are classified in the European Waste Catalogue - EWC (Directive 2000/532/EC), under title 10 and 19. In the EWC Directive wastes marked with an asterisk are considered as hazardous wastes in accordance with Directive 91/689/EEC. One of the criterions for the hazardous nature is the presence of heavy metals above certain limits, whose classification is presented in table 5.

These metals may be associated to health and environmental harmful properties which are classified as: H5 “Harmful”, H6 “Toxic” H7 “Carcinogenic”, H8 “Corrosive”, H9 “Infectious”, H10 “Teratogenic”, H11 “Mutagenic”, H12 the substances that release toxic gases when exposed to ambient conditions; H13 the substances that after disposal generate leachates with any of the metal listed before; H14 “Ecotoxic”, i.e. that present risks to the environment.

Table 5. Classification of hazardous metals

Class	Compounds
C1	beryllium; beryllium compounds
C2	vanadium compounds
C3	chromium (VI) compounds
C4	cobalt compounds
C5	nickel compounds
C6	copper compounds
C7	zinc compounds
C8	arsenic; arsenic compounds
C9	selenium; selenium compounds
C10	silver compounds
C11	cadmium; cadmium compounds
C12	tin compounds
C13	antimony; antimony compounds
C14	tellurium; tellurium compounds
C15	barium compounds; excluding barium sulfate
C16	mercury; mercury compounds
C17	thallium; thallium compounds
C18	lead; lead compounds

In Europe, the IPPC (Integrated Pollution Prevention Control) (Directive 96/61/EC) is one of the most important tools for the control of pollutant emissions from industrial activities, including heavy metals from industries of iron and steel, non ferrous metals, combustion processes, petroleum, cement and glass sectors. Reference Documents on Best Available Techniques (BAT's) were produced for several industrial activities. Of particular interest in this thematic are the BREFs (Reference Document on Best Available Techniques) for the LCP (Large Combustion Plants) and Waste Incineration. BREF for LCP identifies as BATs that contribute to control heavy metal emissions in combustion and co-combustion processes, the flue gas treatment technologies used to control particulate emissions, $PM_{2.5}$ and PM_{10}. These BAT technologies are ESP (Electrostatic Precipitators) and fabric filters. In the case of Hg, it is considered that the use of FGD (flue gas desulfurization) scrubbers used to remove SO_2 emissions may also contribute to reduce Hg emissions. At the same time, it is defined the need to control heavy metals in combustion residues, especially when co-combustion of wastes is performed.

3. Behavior of Heavy Metals in Combustion Systems

Emissions of heavy metals depend on its behavior in the combustion process and also on the type of technology used for flue gas treatments. By behavior is meaning the transformation that metal suffers when the fuel matrix is burned at the high temperatures produced during combustion and also during flue gases cooling. These transformations are influenced by several factors, namely fuel characteristics and type of combustion system (table 6).

Table 6. Factors that influence heavy metals behavior in combustion process

Fuel	Combustion Equipment
Concentration and Affinity of Heavy Metals Levels of S, P, Cl and other halogens Nature of organic fraction Level of ashes Size of Particles Moisture Blending of Fuels	Technology for ash removal Fuel feeding and charge regime Use of adsorbents Cooling surfaces Excess air Temperature Pressure

3.1. Heavy Metals Concentrations in Coals

The elements present in coals may be classified according to their concentration; as "major elements" (above 1%), "minor elements" (from 1% to 0.1%) and vestigial (below 0.1%). Heavy metals are usually present in vestigial concentrations, which vary depending on the origin of coal (Raask, 1985). Table 7 presents typical concentrations of vestigial elements normally found in coals.

Table 7. Vestigial elements in coals (Raask, 1985)

Above 50 mg/kg	Ba, B, F, Mn, P, Sr, Ti, Zn
10 a 50 mg/kg	As, Ce, Br, Cr, Cu, Pb, Li, Ni, Rb, V, Zr
1 a 10 mg/kg	Sb, Be, Cd, Ce, Co, Ge, I, La, Mo, Nb, Sc, Se, Tl, Sb, U
below 1 mg/kg	Hg, Ag, Ta

Table 8. Impacts of heavy metals usually present in coals

Most concerning elements	As, B, Cd, Hg, Mo, Pb, Se
Moderately concerning elements	Cr, Cu, Ni, V, Zn
Less concerning elements	Ba, Co, Ge, Li, Mn, Sb, Sr
Radioactive elements	Rn, Th, U
Concerning elements, present in very low concentrations	Be, Sn, Te, Tl

As heavy metals are usually associated to toxic effects, they have been also classified according to their environmental impact (Vuthaluru et al., 1994), as presented in table 8.

Smith (1980) referred that in combustion processes, even equipped with electrostatic precipitators, metals such as As, Sb, Pb, Cd, V, Mo, Zn, Ga, Cr and Cu are volatilized and further condensed in submicron particles enriched with the volatilized elements up to 10 times related to other non volatilized metals, which increases the probability of lung deposition in 10 to 1000 times. For species that do not condense at the combustion and exhaust conditions such has Cl, Se, B and Sb, lung deposition may reach 1000 to 10000 times higher that the non volatilized species. Based on coal consumption in the USA and considering 98% retention of metals in ashes captured in the system, volatile metals even present in such low concentrations as 1 mg/kg in the fuels would be emitted at a rate of 20 t/year. For Hg, considering a lower retention of 95%, amounts emitted could reach 950 t/year. Happily, most coals exhibit Hg concentrations below 1 mg/kg.

3.2. Ash Formation

The behavior of heavy metals is intimately connected with the production and partitioning of ashes. Ashes are mainly composed of Si, Al, Ca, Si, Fe, K, Na, Mg, Ti, S, P and Cl. These elements may be present in several forms in coals (Vassilev and Vassileva, 1997). The most relevant minerals are silica, quartz, pyrites, silicates, clays, carbonates, oxides, oxyhydroxides, sulfates, phosphates and chlorides. The thermal behavior of these minerals depends on the temperature and atmosphere in the combustion process. Sulfides and pyrites are oxidized forming sulfates. Carbonates suffer calcination and are transformed into oxides, releasing CO_2. Clays are dehydrated and deshydroxilized, and may suffer phase changes. The oxides, sulfates and phosphates may undergo physicochemical transformations and interact.

Ashes differ substantially in amounts and quality for different thermal power system. Usually in combustion systems two main ash streams may be distinguished, bottom ash and fly ashes. In the case of rotary cement kilns or other kilns that produce ceramic materials, most constituents that usually form bottom ash are incorporated in the clinker produced.

Bottom ash, also called slag, usually remain in the bottom of boilers and are continuously collected in quenching systems. Fly ashes correspond to particles below 1000 μm that are transported with the flue gases and may be divided in several ash streams depending on the point of collection, e.g. boiler, cyclones, electrostatic precipitators and bag filters or may be emitted. MSW combustion generates large quantities of slags (bottom ash and grate siftings produced in grate systems) and fly ashes (boiler ash and electrostatic precipitators or bag house filters ashes).

Fly ashes may be distinguished as a function of their size; the gross fraction or super micron, present particulate diameters between 1 μm and 1000 μm, and the fine fraction or submicron (aerosols) have sizes below 1 μm. Nowadays, alternative classifications that are connected with impacts on human beings are PM_{10} (particles with sizes below 10 μm), $PM_{2.5}$ (particles with sizes below 2.5 μm) and PM_1 (particles with sizes below 1 μm).

The development of new techniques for the measurement of extremely low particulate sizes turned possible the study of particles arising from combustion between 0.04 μm and 0.15 μm (Kauppinen and Pakkanem, 1990a, b; Markowski and Filby, 1985). The works of Kauppinen and Pakkanem (1990b) revealed that the fraction of submicron particles emitted was much more important in waste incinerator processes than in pulverized coal combustion processes, representing about 30% of the total emitted particles. Lind *et al.* (1994) verified that the fine particulates emitted from a fluidized bed system presented higher sizes than those of pulverized combustion systems.

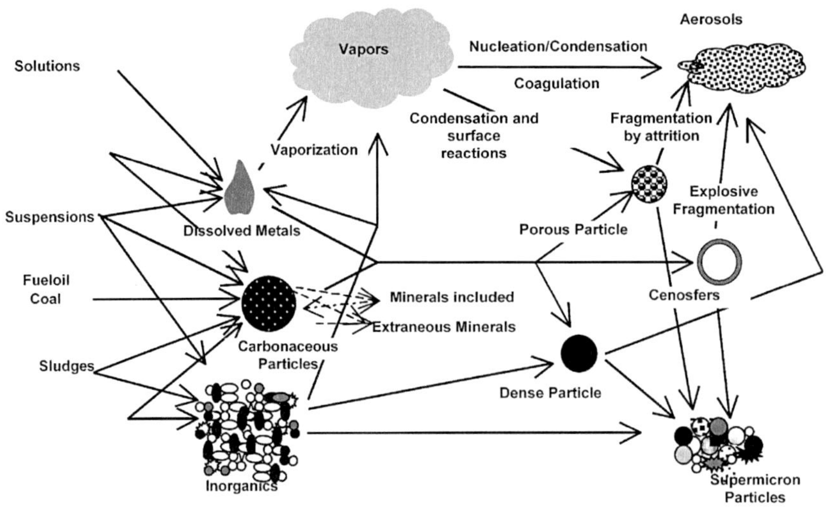

Figure 2. Formation of particles (adapted from Linak and Wendt, 1993)

The types of ashes that are formed in combustion processes depend on the nature of minerals present in the fuels, particle size of fuels, and design of the system. Refractory minerals such as silicates, aluminum silicates and iron tend to remain in bottom ashes, especially if they are present as extraneous matter in the fuels. They may be also carried out as fly ashes with the combustion gases, if they are present in fuels as dispersed and finely divided matter. Other minerals may be preferentially associated with the fly ashes if they are finely dispersed in the fuel matrix or associated to the organic compounds. Figure 2 presents a scheme of the mechanisms involved in the formation of ashes which is similar for the behavior of metals, except that the most volatile metals may also be emitted in the gaseous form as is the case of Hg.

3.3. Transformations of Heavy Metals During Combustion

Several mechanisms may be involved in the transformations of heavy metals during combustion, namely:

- Retention in bottom ashes
- Transport with the gross fly ashes in the combustion gases
- Volatilization without condensation
- Volatilization with condensation
- Explosion of cenosfers
- Chemical reactions with the gaseous species
- Physical or chemical adsorption in the fly ashes
- Encapsulation by other minerals
- Vitrification

The retention of heavy metals in bottom ashes depends on the partition of ashes, which is influenced by the system characteristics and fuel properties. The transport of heavy metals depends on the quantity and nature of mineral matter present in the fuels, as well as on the fluid dynamics of the thermal equipment. In pulverized coal systems the particles transported present normally diameters between 1 μm e 100 μm. The quantity of transported particles may be influence by the charge of the installation (Klika et al., 2001) and this influence the concentration and partition of particles.

There are a large number of studies about the fate of heavy metals in combustion systems for coal: Natusch et al., 1973, 1974a, b; Davison et al., 1974; Lee et al., 1975; Klein et al., 1975a, b; Gladney et al., 1976; Campbell et al., 1978; Coles et al., 1979; Smith et al., 1979; Smith, 1980; Hayes et al., 1982; Hansen and Fisher, 1980; Hansen et al., 1984; Markowski and Filby, 1985; Kauppinen and Pakkanen, 1990a; Clarke, 1993; Linak and Wendt, 1994; Querol et al., 1995 and Meij, 1993. For waste incineration there are also works such as those of Clapp et al., 1988; Pluss and Ferrell, 1991; Norton et al., 1998, 1986; Reimann, 1989; Kauppinen and Pakkanen, 1990b and Linak and Wendt, 1993. All of them agree that the enrichment of submicron particles or emission as vapors is produced because of the volatility of metals.

The volatility of metals may be defined as the temperatures to which corresponds a vapor pressure, P_v, of 1.10^{-6} atm (Barton et al., 1990). These temperatures differ for each metal and even for different specific chemical species.

The degree of volatilization of heavy metals increases with the temperature. Variation of P_v with the temperature may be expressed with a parametric equation of the type:

$$log \quad P_v = A/T + B \log T + C \tag{1}$$

in which A, B e C, are different constants for each chemical species and thermodynamic state, and T is the temperature in K (Knacke et al., 1991).

The intensity of emissions and speciation of metals depend on the stability of their vapor phase, when the gas cools down till the exit of the stack. The metals with higher tendency for emission in the vapor phase are Hg, Se and Tl, and light elements such as the halogens. However, interaction between the several species present in the gas might lead to the formation of new species of different volatility. The vapor phases may also be removed from the flue gas using adsorbing techniques and reactions with other compounds present in the particles.

Natusch *et al.* (1973, 1974a, b) proposed the so called volatilization/condensation mechanism as being the responsible for the higher enrichments verified for Pb, Tl, Sb, Cd, Se, As, Ni, Cr, Zn, Be and Mn in fly ashes, as the size of particles diminish. These authors claim that volatilization of metals, such as Cr and Ni, considered less volatile, might be due to their presence as sulfides and formation of carbonyls in reducing zones of the furnace. Carbonyls are extremely volatile, being in gas phase even at 50°C (Habashi, 1986). Several compounds may be formed, such as $Ni(CO)_4$, $Fe(CO)_5$, $Cr(CO)_6$, $Mo(CO)_6$, $W(CO)_6$ and $Mn_2(CO)_{10}$. This phenomenon may generate pure metallic compounds when carbonyls are further decomposed at high temperatures.

The model of Volatilization/Condensation comprises several phenomena:

- Volatilization that depends on the original speciation of the metals, temperature and internal atmosphere.
- Homogeneous condensation, which involves formation of condensation nucleus of extremely low dimensions (0.01 μm) that may undergo coalescence and grow. This occurs when P_v exceeds the saturation pressure vapor of the metal, either because of increase of vapor phase concentration of the metal during cooling, or as a consequence of formation of new compounds of lower vapor pressure. This phenomenon occurs preferentially in high temperature systems with low concentration of particles in the gaseous phase.
- Heterogeneous condensation is the condensation of metals vapors at the surface of other particles carried out with the gas phase. It originates non homogeneous particles in which the concentration varies with the diameter, $C_X = f(1/D)$ (Davison *et al.* 1974).

Smith et al. (1979) found out that the variation of concentration with the diameter could be even higher ($1/D^2$). Because of this phenomenon, usually particles of lower diameters present higher concentrations of heavy metals and they are more available. However, studies of particles below 0.15 μm (Smith et al., 1979) and 0.037 μm (Markowsky and Filby, 1985) showed bimodal concentration profiles with lower enrichments in the submicron range than what could be expected. These authors also verified the presence of lithophilic species such as Si, Al, Mg and Ca in the submicrometric range. Afterward, it was proposed another mechanism for submicrometer particles, involving explosion of cenosferes that were formed in the early phases of fuel desvolatilization at high temperature. The fragments, typically in the range of 0.01 μm, might coalesce, growing to sizes between 0.1 μm and 1 μm. In agreement with this theory the different heavy metals enrichments found for diverse installations might be attributed to condensation phenomenon occurring in different stages, before or after the explosion of particles.

In addition to simple condensation, the preferential deposition at the particles surface could also be due to absorption and possible reaction with the matrix of particles, as for example in the case of calcium arsenates (Haynes et al., 1982).

Another important phenomenon consists in the migration of metals to the core of fused silica aluminates particles, producing surfaces enriched in light metals that after cooling encapsulates the metals Rink *et al.* (1995) and Kozinsky *et al.* (1995).

3.4. Volatility and Affinity

Several authors classified elements according to their volatility in coal combustion system, such as Natusch *et al.* (1973), Klein *et al.* (1975a), Campbell *et al.* (1978), Coles *et al.* (1979), Smith (1980), Clarke (1993), Pires *et al.* (1997) and Aunela-Tapola *et al.* (1998). Meij (1993) proposed a classification based on the relative enrichments, RE.

The enrichment of a certain element in the ashes may be defined as "Relative Enrichment", RE, i.e. the ratio between the concentration of the element (C_x), present in the ash (i), to concentration of the same element in the fuel (C_{xFuel}), considering the ash base.

$$RE = \left[\frac{C_{x\,i}}{C_{x\,Fuel}} \right] \times \frac{Ash\ in\ the\ Fuel}{100} \qquad (2)$$

In table 9 is presented the classification of elements according to volatility. Aunela-Tapola *et al.* (1998) included Mn in class I and referred that Hg and Se also appear highly enriched in both fly ashes and bottom ashes, especially for coals with high concentrations of Ca. They also referred that Cr and Ni might be classified in class I, as in many cases their presence could be associated with contaminations from the combustion system.

Table 9. Classification of elements according to volatility

Class I	Elements not volatilized, (ER≈1): Al, Ca, Fe, K, Na, Mg, Ti, Ce, Cs, Eu, Hf, La, Sc, Sm, Si, Sr and Th
Class II	Elements volatilized but condensed in the smallest particles.
IIa	ER > 4: As, Cd, Pb, Zn, Ge, Mo, Sb and Tl
IIb	2 < ER < 4: Co, Cu, Ni, V, P, Be, U and W
IIc	1.3 < ER < 2: Cr, Mn, Na, Ba, Rb and Sr
Class III	High volatile elements preferentially emitted as vapors: RE<<1: C, Cl, N, S, B, Br, F, I, Se and Hg

These authors stated that the behavior of elements was influenced by the nature of coals and association tendency of metals. The litophilic elements, with higher tendency to associate with silica, could be included in class I. The calcophilic, i.e with tendency to associate with Ca, could be classified in class II. Besides this, the chemical speciation of metals in coals also determines their behavior.

Mineral matter present in coals and other combustible materials, may be classified as inherent, when it is linked to the organic matrix, fine particles spread in the organic matter, as clay particles (Smith, 1980), or be present as discreet minerals or extraneous minerals, as is the case of natural mineral materials, such as sand, pyrites, sulfides, chlorides, limestone, dolomite, clay and sulfates. They may also be present as liquid solutions (Vassilev and Vassileva, 1997; Querol et al., 1995).

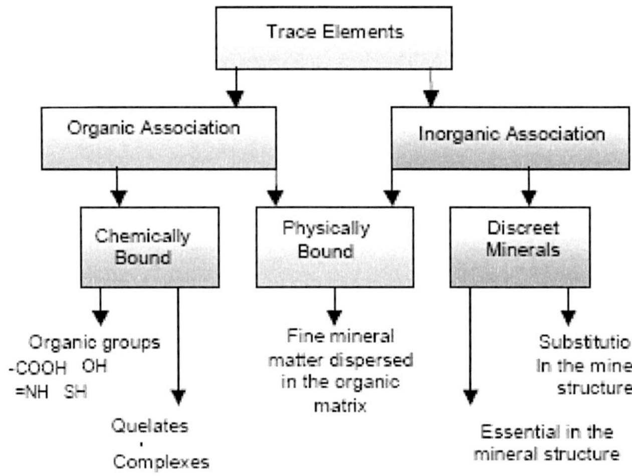

Figure 3. Association of trace elements in coals.

The association of metals to the organic phase increases its volatility as they are released to the vapor phase when the organic matter is consumed. Smith (1980) verified that in coal combustion systems, the enrichment of fly ashes was variable, e.g. 7 to 15 for Pb, from 2.5 to 7 for Cu, 2.5 to 5 for Ni, 7 to 21 for Zn, and from 4 to 6 for Cr and Mn, although the boiling point of most of these pure metals is superior to 1800°C, except in the case of Pb. This may be justified by the association of theses metals to the organic forms.

Association of metals has been extensively studied, as it is important for the control of emissions and also pre-treatment of coals. Clarke and Sloss (1992) defined the associations of trace elements in coals, as presented in figure 3.

Heavy metals may also appear associated to clays, in consequence of its retention in interstitial water retained in the early stages of coal formation. Other elements, like F may be incorporated in inclusions such as apatite (*monazites* and *xenotime*). Mercury may appear as sulfide or organically bound. There are many possibilities of combinations. In table 10 several examples are presented.

Table 10. Associations of several elements in coals

	(Raask, 1985)	(Querol et al., 1995)	(Pires et al., 1997)
Organic Matter	Sb, Ga, P, Ti, V	N, S, Be, B, Ge, V, W, Zr	B, Be, Br, Ge, V, Co
Sulfides	Fe, As, Cd, Co, Cu, Cr, Mn, Hg, Mo, Ni, Se, Zn, Pb	Fe, Ni, Co, Cu, Zn, As, Se, Mo, Cd, Sb, Hg, W, Pb, Tl	As, Cd, Co, Cu, Dy, Hg, Lu Mo, Ni, Pb Se,W, Zn,
Sulfates and phosphates	Ca, F, rare elements	Ca, Fe, Ba	
Clays	Na, K, Mg, Li, Rb, Cs	Na, K, Mg, P, Ti, Li, Cr, Ni, Cu, Ga, Rb, V, Sr, Y, Sn, Cs, Ba, Ta, Pb, Bi, Th, U	Ba, Cr, Cs, Ga, Mn, Rb, Sb, Sn, Sr, Ta, U, Zr, Ni, Cu
Carbonates	Ca, Fe, K, Na, Mg	Ca, Mn, Co	
Interstitial Water	Na, Ca, K, Ba, Sr, Ce, Br, I		

The partition between organic phase and sulfides, besides depending on the type of metals, is also variable from coal to coal (Raask, 1985). The process responsible for the inclusion of heavy metals and enrichment relatively to other common geologic materials (Vassilev, 1994; Vassilev and Vassileva, 1997) is linked to the origin of coals, i.e. the original humic material that exhibits high affinities for metals. During the carbonization process the reducing atmospheres favor the formation of sulfides, either through mineralization of the organic sulfur and associated metals, or through reduction of the metallic sulfates. Raask (1985) correlated the degree of metal bondage to pyritic sulfur and the presence of pyritic itself with the increasing quality of coals.

The works of Klika *et al.* (2001) about partitioning of heavy metals in a circulating fluidized bed system using coal and lignite and those of Miller *et al.* (1999), that studied coal and biomass in a laboratorial fluidized bed system, showed that volatility of metals is not the same for different fuels. However, they also found out that the behavior of metals is also determined by the quantity of ashes formed in the case of biomass and coals.

3.5. Thermodynamic Predictions

Thermodynamic calculations have been applied to understand the behavior of heavy metals in combustion system. Calculations are useful as they allow predicting the formation of chemical species, however, they may deviate from reality, as the behavior is affected by several phenomena that are not taken into account in mathematical models. Mathematical models assume thermodynamic equilibriums that might not be attained in combustion systems and there may be also kinetic limitations.

Among other limitations of mathematical models, the original chemical and physical speciation of metals in fuels, reactions between the several species and the adsorption and chemical reactions with the solid particles are not usually taken into account.

Another gap is the difficulty in reproducing the non isothermal conditions and varying atmospheres to which the metals are subjected during combustion processes, as they may change from reducing to oxidant conditions in fractions of seconds.

Yan *et al.* (2001) made the thermodynamic calculations of some heavy metals for coal combustion, concluding that for high contents of mineral mater the reducing conditions may be extremely important. Helble *et al.* (1996) modeled a gasifying system and predicted an intense volatilization of As, Se, Sb, Pb e Hg, and to a less extend of Cs, Cr, Co, Mn, Ni, U, e Th. Reed *et al.* (1997) also predicted gasification of sewage sludge, finding high volatilities of Hg, Se, As and Cd, that may lead to its emission in the gaseous phase at temperatures below 400°C. Sn, Pb and Mn, might be retained in the solid phase at 400°C and Zn, Cu and Mo at 600°C. Co, Ni, Cr and V, should be kept in the solid phase till 1000°C.

In table 11 is shown the volatilization and boiling temperatures of some heavy metals species, as stated by Clark and Sloss, 1992.

When chlorine is present in the fuels the formation of chlorides is favored, which increase metals volatility. Wu and Biswas (1993) verified that metals exhibit different affinities to Cl, following the order: Pb>Cr>Hg>As>Sn>Cd. When the levels of Cl are low, as in most coals, the competition for Cl is important.

Table 11. Volatilization and boiling temperatures of some heavy metals species (Clark and Sloss, 1992)

	Species	Volatilization Temperature (°C)	Boiling Temperature (°C)
As	As	-	613 (sublimation)
	As_2O_3	77	465
Cd	Cd	157	-
	CdO	604	-
Co	Co	-	2870
	CoO	1551*	1800 (decomposition)
Cr	Cr	-	2670
	Cr_2O_3	-	3000-4000
Cu	Cu	-	2570
	CuO	975*	-
Hg	Hg	47*	357
Mn	Mn	-	1960
	MnO	1628*	-
Ni	Ni	-	2730
	NiO	1485*	-
Pb	PbO_2	299	-
	Pb	488	-
	PbO	943**	-
	Species	Volatilization Temperature (°C)	Boiling Temperature (°C)
V	V	1482	-

	VO	1010	-
Zn	Zn	-	907
	ZnO	1057*	1725 (sublimation)

*para Pp $\approx 10^{-5}$ atm and **para Pp$\approx 10^{-3}$ atm

Durlak *et al.* (1997) showed that the formation of chlorides during MSW incineration is influenced by moisture and Na in the system, because of the high affinity of H and Na to Cl, according to the reactional equilibriums (3) to (5).

$$M + Cl \leftrightarrow MCl \quad (3)$$

$$MCl + Cl \leftrightarrow MCl_2 \quad (4)$$

$$MCl + OH \leftrightarrow MO + HCl \quad (5)$$

Hydroxides may be also formed in the presence of water vapor (Buekens et al., 1995), according to reaction (6).

$$MCl_x + xH_2O \leftrightarrow M(OH)_x + xHCl \quad (6)$$

The presence of S simultaneously with Cl might be important, interfering with the formation of chlorides. This effect is more important for metals such as Cd, Zn and Pb. In table 12 is presented the volatilization temperatures of heavy metals in a waste with 3% of Cl, as reported by Verhulst et al., 1996.

The proportion between Cl and S may influence the volatilization of metals, either through chemical equilibriums between them and the metals or reactions between S and Cl.

Table 12. Volatilization temperatures of heavy metals in a waste with 3% of Cl (Verhulst et al., 1996)

	Volatilization Temperature (°C) Without S		Volatilization Temperature (°C) With S	
	initial	final	initial	final
As	500	550	500	550
Cd	300	400	500	600
Cu	600	800	600	800
Hg	-	<100	-	<100
Mn	>1100	-	>1100	-
Ni	1000	>1000	1000	>1000
Pb	350	500	700	800
Zn	300	>1000	600	>1000
Fe	>1100	-	>1100	-
Al	>1100	-	>1100	-

HCl forms Cl_2 through the Deacon reaction, (7), under excess air. The reaction might be catalyzed by CuCl and $CuCl_2$, as well as by ashes (Liu et al., 2000). Then, Cl_2 converts SO_2

into SO_3, favoring the formation of HCl, through reaction (8). SO_3 favors the formation of metal sulfates:

$$4HCl + O_2 \leftrightarrow 2Cl_2 + 2H_2O \quad (7)$$

$$Cl_2 + H_2O + SO_2 \leftrightarrow SO_3 + 2HCl \quad (8)$$

Frandsen *et al.* (1994) and Yan *et al.* (2001) classified the metals according to the most probable reactions in the combustion processes. Considering a generic metal M, and assuming that sulfates are the most stable species at low temperatures, the most probable reactions are of type A, B, C, D, E and F, as shown in chemical reactions (9) to (14).

In oxidative thermal processes, metals tend to lower their oxidation state with the increase of temperature; hence at the highest temperatures used, the most probable forms should contain the metals in the lowest oxidant state.

Hansen *et al.* (1984) assumed that metals in silicate matrix should present the lowest oxidative states, namely Cu(+1), Cr(+3), Co(+2), Mn(+2) e Mn(+3), because they are trapped at high temperatures, while the metals deposited at the particulate surfaces may present higher oxidant states, as Cu(+2), because they condense at relatively low temperatures.

Decomposition of Sulfates into Oxides

Equilibrium A) $$MSO_{4(s)} \leftrightarrow MO_{(s)} + SO_{2(g)} + 1/2\, O_{2(g)} \quad (9)$$

Conversion of Oxides and Chlorides, in the presence of Cl (F, Br, I):

Equilibrium B) $$MO_{(s)} + 2HCl_{(g)} \leftrightarrow MCl_{2(g)} + H_2O_{(g)} \quad (10)$$

Volatilization of Oxides:

Equilibrium C) $$MO_{(cr)} \leftrightarrow MO_{(g)} \quad (11)$$

Interaction of vapor phase with the matrix compounds (ex. Al_2O_3, SiO_4, Fe_2O_3,, etc.) or other metals, forming refractory compounds:

Equilibrium D) $$MO_{(g)} + RO_{y(s)} \leftrightarrow MRO_{y+1(s)} \quad (12)$$

MRO_{y+1} being: silicates or aluminates:
$aMO.bSiO_2$; $aMO.bAl_2O_3$; $aMO.bSiO_2.cAl_2O_3$ or spinels, e.g.:
$ZnO.Fe_2O_3$, $CuO.Fe_2O_3$, $NiO.Fe_2O_3$, $CdO.F_2O_3$, $FeO.Cr_2O_3$

Conversion of Oxides in the vapor phase to the atomic gas phase:

Equilibrium E) $$MO_{(g)} \leftrightarrow M_{(g)} + 1/2\, O_{2(g)} \quad (13)$$

Conversion of Sulfates to Chlorides
Equilibrium F)

$$MSO_{4(s)} + Cl_{2(g)}(2HCl \ ou \ 2Cl) \leftrightarrow MCl_{2(s,g)} + SO_{2(g)} + O_{2(g)} \tag{14}$$

According to Frandsen *et al.* (1994), metals may be divided in two groups. The fist one, metals that form sulfates at low temperature, include Be, Cd, Co, Cr, Hg, Ni, Pb, Sb, Sn, V and Zn, some forming hydrated sulfates. The second group, those forming oxyhydroxides, include As, B, Ga, Ge, P, Se and Ti. For the group that form sulfates, the reaction sequence that leads to volatilization includes reactions A, C and E.

The major part of sulfates is instable in combustion systems as can be seen in table 13. In the presence of Cl they may form chlorides, by mechanism F, increasing the volatility of the metals.

Table 13. Decomposition temperatures of metallic sulfates (Habashi, 1986)

Sulfates	Decomposition Temperatures (°C)	Products
$CdSO_4$	825-850	$2CdO.CdSO_4$
	890	CdO
$CoSO_4$	720-770	Co_3O_4
$CuSO_4$	650-670	$CuO.CuSO_4$
	710	CuO
$HgSO_4$	450-500	$HgO.HgSO_4$
	>500	$2HgO.HgSO_4$
$MnSO_4$	700-800	Mn_3SO_4
$NiSO_4$	700-765	NiO
$PbSO_4$	640-700	$PbO.PbSO_4$ / $2PbO.PbSO_4$ / $4PbO.PbSO_4$
	960	PbO
$VOSO_4$	400	V_2O_5
$ZnSO_4$	700-720	$2ZnO.ZnSO_4$
	780	ZnO
$Al_2(SO_4)_3$	540-640	Al_2O_3
$FeSO_4$	500-535	$Fe_2O(SO_4)_2$
	650	Fe_2O_3

Yan *et al.* (2001) modified the classification of Frandsen, taking into account the reactions of metals between each other and with other species.

Table 14. Classification of reactivity of heavy metals in oxidant conditions

Group I	Elements that form sulfates, unstable above 700°C Reactions of type A; B; C; E and F.
Ia	Elements that do not react with other elements. Cd, Pb, Se and Te
Ib	Elements that react with other metals, after decomposition. Reactions type D. Co, Cr, Cu, Mn, Ni, Tl and Zn
Groupo II	Elements that form preferentially oxides, halides e arsenates. As, Hg, Sb, Sn and V

3.6. Retention of Heavy Metals in Ashes

Retention of heavy metals in residual solids produced in thermal system is one of the most important processes to prevent heavy metals from being emitted. These phenomena may occur if the fuels containing species capable of adsorbing or reacting with metals, or may be promoted by the use of adsorbents.

In low temperature systems, adsorbents such as calcium oxides or activated carbon may be used. In fluidized bed systems it is common to use limestone or dolomites to control sulfur oxides. These sorbents may contribute to retain metals. Also the high quantities of particles inside the fluidized bed reactors may also contribute to retain the heavy metals.

The fixation of heavy metals in solids may occur through several mechanisms.

- Condensation,
- Physical adsorption at the surface or inside porous particles,
- Chemical adsorption with reaction,
- Scavenging in the vitrified structure,
- Isomorphic substitution of crystalline structure.

Immobilization in vitrified ashes may occur only in systems of very high temperatures. In fluidized bed systems this type of phenomena is not probable due to the moderate working temperatures that are not enough to fuse silica, alumina, silicoaluminates and iron oxide. However, it may occur some melting at particles surface or sintering which may contribute to trap metals.

In a study of combustion of wastes in a fluidized bed, at 700°C, Litt and Tewksbury (1984) optimized the retention of metals in the silica bed material, using small amounts of Na_2CO_3 that produced controlled sinterization.

Uberoi and Shadman (1990, 1991) investigated the retention of Pb and Cd chlorides with alumina, silica, kaolin, bauxite, emathlite and limestone between 700°C and 800°C, in a fluidized bed reactor. They verified that limestone was not efficient in metals retention. The retention by silica and alumina was found to be considerable and kaolin and bauxite were the most efficient, being kaolin more efficient with Pb, due to the formation of $PbO.Al2O_3.2SiO_2$. In the case of bauxite, $PbO.Al_2O_3$, was the product formed. Cd was strongly retained in alumina and bauxite, forming $CdO.Al_2O_3$, being kaolin more efficient due to the blocking of porous with $CdO.Al_2O_3.2SiO_2$.

Scotto *et al.* (1992, 1994) investigated retention of Pb in kaolin, at 1320°C, verifying that retention diminished with the increase of Cl. The material obtained appeared to be fused, although its fusion temperature is higher (2083°C) than the operational conditions.

Linak *et al.* (1995) have made the same type of studies, using kaolin and calcium hydroxide to retain Pb, Cd e Ni. They verified that Pb and Cd were efficiently captured by kaolin, involving fusion, while bauxite captured Cd without evidences of melting. Calcium oxide was able to capture Cd, forming eutectic species with low melting temperature, which was helped by the presence of Cl. Ni was also adsorbed in presence of Cl.

Eddings *et al.* (1994) used clays to retain metals in a rotary kiln, and verified that Cd, Pb and Cr presented different behavior with the temperature increase. At temperatures below 800°C, retention of metals that were volatilized was due to physical adsorption in the clays. Above that temperature, volatilization decreased and it was found that Pb and Cd were trapped in the silicates. In the case of Cr, the mechanism of retention was different; as it appeared that Cr was incorporated in the structure of clays.

Ho *et al.* (1992, 1993) verified that in fluidized bed systems, sand and limestone worked as adsorbents for Cd and Pb, producing low solubility products. Limestone showed to be more efficient than sand, being higher the retention at 750°C. However, at higher temperatures, the high volatilization inhibited the scavenging of metals. Pb retention by limestone reached 95%, while in presence of sand it was only 67%. Limestone Cd retentions reached 90%, but were lower than 30% in presence of sand. It was suggested that the mechanisms of retention of heavy metals should include partial fusion and coagulation of aerosols with the adsorbent particles.

Chen *et al.* (1998) investigated retention of Cr, Cu, Pb and Cd, with silica and limestone in a fluidized bed and found out that the optimal temperature for Cd and Pb retention was 600°C. For Cr, maximum retention occurred at 700°C and for Cu at 800°C. Cr appeared to be retained by chemical and physical adsorption, while for Cu chemical adsorption was more important. For Pb and Cu physical adsorption was more important.

Mojtahedi (1989) found out that limestone used in a fluidized bed retained As, Cd, Pb and Se. However, Hughes and Littlejohn (1987) observed higher emissions of Cd, Mn e Pb, when limestone was used, relating this with the higher emissions of limestone.

Ramesh and Kozinsky (2001) investigated the capture of Cr, Pb and Cd with aluminosilicates, *mulite* and *corundum*, at 1000°C and 1500°C, verifying that different mechanisms were responsible for the retentions of those metals. Cr^{+3} with an ionic radius and charge similar to Al^{+3}, could migrate through the matrix of *corundum* and *mulite*, making isomorphic substitutions in the aluminates. Pb^{2+} and Cd^{2+}, with higher ionic radius were captured by a different mechanism, occupying the empty spaces left by oxygen in *mulite* or being incorporated in the vitreous matrix.

There are other possibilities of isomorphic substitutions of heavy metals. The phenomenon appears to be dictated by the similarity of ionic radius with the same oxidation number, table 15, as referred by Vehlow (1993).

Biswas e Wu (1998) referred that mineral matter exhibits a high tendency for oxides of heavy metals, hence addition of sorbents must be done in an early phase to the formation of metals aerosols, or before the occurrence of sulfating and chlorination reactions.

Table 15. Identification of some cations for isomorphic substitutions

Na^+	K^+	Ca^{2+}	Fe^{2+}	Fe^{3+}
Cu^+	In^+ Tl^+	Ti^{2+} Cd^{2+}	Mg^{2+} Co^{2+} Ni^{2+} Cu^{2+} Zn^{2+}	Cr^{3+} Mn^{3+} Co^{3+} Ni^{3+}

Addition of sorbents to capture heavy metals, besides reducing the quantity of metals volatilized, may suppress the nucleation process, decreasing the formation of submicron particles enriched in heavy metals. This contributes to improve the dedusting process of flue gases, as most technologies are less efficient in removing submicron particles from the flue gases.

3.7. Behavior of Heavy Metals in Fluidized Bed Systems

In fluidized bed systems the tendency for heavy metals to be emitted is generally lower than in pulverized coal combustion (Tillman, 1984; Lim, 1979), as exemplified in table 16.

Table 16. Emission rates of heavy metals in coal fluidized bed system and pulverized combustion (Lim, 1979)

% of metal emitted	Pulverized combustion	Fluidized bed combustion
Hg	90	80
As	50-60	15
Pb	0-60	0-20
Cr	0	25
Co	10-20	0-20

This tendency is related to the specific features of fluidized bed systems, such as the low temperatures, low partial pressures in pressurized systems and the use of fuels with higher granulometry. Another effect is the possibility of metals adsorption in particles, because in fluidized systems there are great quantities of porous particles in the gas phase (Mojthahedi, 1989). The use of cyclones to remove particles also favors the capture of metals, because they increase the residence time between the gas phase and the solid particles, at moderate temperatures.

The use of limestone and other sorbents in fluidized bed systems may also contribute to diminish emissions, although they may also introduce metals in the system (Lind et al., 1994). However, the transport of fine particles of sorbents containing heavy metals and the increase of bed depth due to the use of sorbents may increase metals emissions. The effect of using sorbents is complex, because in addition to the possibility of metals adsorption, they also modified the atmosphere, increasing CO_2 and absorbing SO_2 and HCl.

The concentrations of heavy metals in the several ash streams of coal fluidized bed systems is highly variable as can be seen in table 17. The RE values are much lower than in pulverized systems, as shown in table 18.

Table 17. Concentrations and relative enrichments (RE) of heavy metals in ashes from coal fluidized bed combustion (Mojtahedi et al., 1990)

mg/kg	Bed ashes	Cyclone ashes	Filter ashes	RE of filter ashes
As	137-139	237-257	356-378	2.1
Cd	6.7-7.4	11-13	14-18	1.3
Co	18-20	33-36	58-97	1.8
Cu	438-696	618-777	724-802	0.9
Hg	6.6-8.4	8-9.4	10-12	1.0
Ni	51-53	75-86	147-241	2.3
Pb	2.8-4.6	33-41	72-75	0.8
V	66-84	122-135	210-302	1.7

Table 18. Concentrations and relative enrichments (RE) of heavy metals in ashes of pulverized coal combustion (adapted from Meij, 1993)

mg/kg	Bottom ash	ESP ash	Flyash	RE of flyash
As	2.6	34	221	6.2
Cd	0.4	1.0	8.0	7.6
Co	37	55	134	2.4
Cr	132	132	212	1.6
Cu	68	142	335	2.4
Ge	9	19	144	7.8
Hg	0.09	0.3	3.0	2.4
Mn	444	444	741	1.7
Mo	14	30	131	4.3
Ni	76	103	338	3.3
Pb	45	82	394	4.8
Sb	2	8	46	5.8
Se	<0.2	12	276	13.5
V	196	281	845	3.0
Zn	117	234	1307	5.6

Mojtahedi (1989) and Mojtahedi *et al.* (1990) research works compared the behavior of metals in atmospheric and pressurized fluidized bed systems. It was verified that emissions of Cd, Pb, Hg and Zn were higher in the pressurized systems, Pb emissions increased from 1% to 3%, Hg ones from 7.5% to 16%, Cu emissions rose from 0.5% to 1.2% and Cd and As ones changed from 4% to 5%, respectively. These authors also verified that Cr, Cu, Ni e Zn concentrations also increased in the fine ashes, corresponding to 11% of the inlet metals, while there were ashes retentions of 25% and 64% in the two first cyclones. They reported emissions of Cd between 0.2 $\mu g/m^3$ and 0.9 $\mu g/m^3$, Hg from 0.4 $\mu g/m^3$ to 6 $\mu g/m^3$ and those of Pb from 2 $\mu g/m^3$ to 3 $\mu g/m^3$ for a coal burning circulating fluidized bed system of 14 MWe.

Lind *et al.* (1994) studied fly ashes emissions in fluidizing system of 80 MW and found values in the range of 0.01 μm to 70 μm, verifying that the submicron range represented less than 1% and that none of the trace elements presented preferential concentrations in this range, showing that the mechanism of volatization/condensation was not relevant in fluidized bed combustion systems.

4. Heavy Metals Leachability in Ashes

Leachability may be defined as the mobilization, extraction and washout of soluble compounds contained in a solid phase, when subjected to contact with an extracting fluid. The process occurs naturally in the environment, when residues enter in contact with the natural fluids such as rain and hydrologic flows.

The evaluation of leachability of heavy metals is one of the most important parameters for the assessment of environmental risks associated to waste management. (IAWG, 1997). Directive 2000/76/EC, for example, obliges to the characterization of leachability of ashes produced in waste incineration to allow decisions about their elimination.

Landfilling operations (Directive 1999/31/CE) requires evaluation of leachability of wastes. European Regulation (Decision 2003/33/CE) classifies the wastes to be landfilled, based on some characteristics and also on the leachability behavior. This Decision defines the leachability standards that shall be used, so that values obtained allow classification in different hazardous classes: inert, non-hazardous and hazardous residues. One of the recommended standards which are most commonly applied to the study of ashes is the method EN 12457. It is a batch leaching test, in which waste samples e are put in contact with deionised water, with liquid to solid ratios of 10L per kg, at room temperature, in continuous agitation during 24h. The leachates are filtered and analyzed for pH, conductivity and several contaminants determinations, as presented in table 19.

Leachability of ashes presents some typical characteristics which are related to their composition. Ashes contain high quantities of Ca, Mg, Na and K oxides, which suffer quickly hydrolysis when in contact with water. This justifies that ashes behave usually as alkaline materials during leaching process, generating leachates with high pH values. For ashes obtained from fluidized bed systems, in which limestone is used to control SO_2, alkalinity may attain extremely high pH values, above 12. Fly ashes may present lower pH values, because of the possibility of acid adsorption or condensation of H_2SO_4 and HCl during flue gas cooling in the exhaust system.

Table 19. Residues classification for landfilling (Decision 2003/33/CE)

Leachability mg/kg	Inert Wastes	Non hazardous wastes	Hazardous Wastes
As	0.5	2	25
Ba	20	100	300
Cd	0.04	1	5
Cr	0.5	10	70
Cu	2	50	100
Hg	0.01	0.2	2
Mo	0.5	10	30
Ni	0.4	10	40
Pb	0.5	10	50
Sb	0.06	0.7	5
Se	0.1	0.5	7
Zn	4	50	200
Cl^-	800	15000	25000
F^-	10	150	500
SO_4^{2-}	1000	20000	50000
DOC	500	800	1000
TDS	4000	60000	100000

The leachability of heavy metals present in ashes is generally very low or insignificant at the normal pH values of leachates generated, probably because metals that may dissolve from the ash matrix, forming cations, tend to combine as hydroxides or carbonates that precipitate in the pH range between 8 and 12. However, some heavy metals may form oxyanions, such as chromates, vanadates, arsenates and molibdates, which are soluble under alkaline pH, hence leachability of these metals may be significant in the typical pH range of ashes. In the case of amphoteric metals, such as Al, Pb and Zn, excessive alkalinity may also favor solubilization.

Fly ashes present normally higher leachability than bottom ashes, which can be explained by the formation process of fly ashes, as they include condensed compounds that are more soluble and pH values are usually acid. Total dissolved solids in bottom ashes from incinerators are usually lower than 10% (Hartman, 1991), while in fly ashes leachability of solids may reach 30%. This is essentially due to the dissolution of calcium compounds, sulfates and chlorides. Sawell et al. (1988) verified that leachability of bottom ashes from waste incineration was lower than 0.5% of total ash mass. However, for fly ashes leachability attained 100% for Cd, 63% for Zn and 36% for Cu.

Buchholtz and Landsberger (1993), applied the US leachability test, TCLP, which is performed with acetic acid solutions, to fly ashes from incineration of MSW and verified that the levels of leachability depend on the diameter of particles (between 20 μm and 250 μm), relating this to different pH values. Leachability of Cd, Cu, Mn, Ni and Zn increased for ashes with higher sizes that presented lower pH values. In the case of Hg and Pb, leachability was higher for the lower particles that were more alkaline.

For coal ashes from fluidized bed systems, the species that leach more are SO_4, Ca, K, Na, and Cl. The leachability of most heavy metals that form cations is generally very low, typically below 0.1 mg/l or even 0.001mg/l, depending on the type of metal or kind of coal, due to alkaline pH of leachates. The metals that form oxianions, such as: Mo, V, Cr, As, Ti, B and Se, may leach in alkaline conditions, however, generally in levels below 1 mg/l. For more alkaline pH values, it may be found a higher leachability of Pb, Ba, Cs and Sr (Morgan II et al., 1997; Lecuyer et al., 1996; Vassilev e Vassileva,1997; Lopes et al., 1999).

Fly ashes from pulverized coal combustion, present generally similar behaviors to those of MSW for most metals, although in lower levels (Van der Sloot, 1997). In the case of Cu, minimums of leachability occur in the range of pH 5 to 12, but lower levels, less than 10 mg/kg, are leached than in the case of ashes from MSW incineration. Cd leachability in coal fly ash, at pH below 9, also increase relatively to the corresponding bottom ashes, being, however, lower (0.001 mg/kg) than in MSW fly ashes (0.01 mg/kg). For Pb and Zn, there is also a lower mobility in coal ashes, as they do not leach above 0.01 mg/kg, for pH values above 7, while in MSW, the minimum leachability of 0.1 mg/kg for Pb, or slightly higher for Zn, are located between pH 10 e 12.

Lundberg et al. (1997) studied leachability of ashes from combustion of MSW in a fluidized bed and verified that bed ashes and ashes captured in cyclones presented lower leachability than ashes obtained in grate systems. However, the ashes captured in the flue gas treatment filter of the fluidized bed system were highly leachable, hence their landfiling had to be performed in special cells of sanitary landfills, or required inertization before being landfilled.

Differences between several ashes may be related, not only due to different pH values and amounts of heavy metals, but also due to the presence of chlorides or other anionic species and residual organic matter, usually present in waste ashes, as they turn metallic species more labile and act as complexing agents in the liquid media produced during leaching conditions (Van der Sloot, 1997).

5. Heavy Metals in Wastes to Be Used as Fuels

Many biofuels and wastes are gaining importance to produce energy as an alternative to fossil fuels utilization; however, wastes may present higher degrees of heavy metals contamination than that found during coal utilization. Besides, the greenhouse gases pressure issue, the restriction on the landfilling of residues containing organic materials works as a driving force to the utilization of wastes for energy production (Directive 1999/31/EC). However, this may be a problematic issue, as heavy metals behavior is still a current environmental problem. Stringent control has to be implement in order to guarantee that regulated emissions are not exceed, however the issue of ash management will continue to be a challenge.

Wastes that are traditionally utilized for energy production are for example tyres (in cement kilns), MSW and sewage sludge. The politic on sustainable use of resources and recycling of wastes has brought other waste materials to the energy valorization subject, for example, ASR, i.e part of the autoshredder materials from end-of-life vehicles dismantling, residues from electric and electronic industry, and other specific residues.

The need to recycle in a more sustainable manner, and also the necessity to control fuel properties, both to assure good technical conditions and to control emissions, led to the SRF concept. SRF (Solid Recovered Fuels) are produced from MSW, biomass and non-hazardous residues. Its production requires separation of biodegradable fractions, milling, drying and compactation in pellets, resulting in a more durable and homogeneous material which is important for storage and combustion stability.

SRF may be burned in different combustion systems for which emission factors differ, depending on the technology and operational parameters used, as e.g. temperature, and type of air pollution control device. Hence, a wide range of properties may be accepted, which led to the definition of SRF classes (CEN/TS 15359). The classification of SRF in classes was initially proposed to cover 7 key properties (Net Caloric Value, moisture and ash contents, Cl, Hg, Cd+Tl concentrations and sum of heavy metals), but was restricted to three properties NCV, Cl and Hg+Cd content. However, the other heavy metals are important and may be requested as indicative values, or established by the final industry that intends to consume SRF.

Definition of acceptable heavy metals concentration in SRF, which may be tolerable in certain industrial processes, is based on the behavior of metals in combustion systems. This concept was defined as "transfer factor" (ERFO), i.e. "the proportion of the input of a material in a process that is emitted to air" (after the gas being cleaned). It was established based on experimental results collected in several German and Dutch industries.

In table 20 are presented the "transfer ratios", TF, of metals for different thermal processes in which SRF might be used. Mercury is the metal, which presents the higher transference to the gaseous emissions, because of its high volatility. Cd (Tl usually is present in insignificant concentrations) has a transfer ratio between 0.5 and 10% and the rest of the metals, namely Sb, As, Pb, Cr, Co, Cu, Mn, Ni, Sn and V present low transfer ratio, below 1%.

In cement kilns, most of the metals are incorporated in the clinker and even Cd and Pb presents lower emission ratios than in pulverized coal systems. Hg, and also Tl vaporize in the kiln and may start condensing at 450-550°C over fine particles that will be collected by electrostatic precipitators. Hg removal from emissions might not be an efficient process, as this metal may appear associated to particles or in the form of metallic vapor, which is not trapped in the dedusting systems. In pulverized coal combustion systems, although temperatures may be lower than in cement kilns, higher transfer ratios are verified for all metals, even though it may vary with the type of coal used and furnace design.

Emission control is usually achieved with electrostatic precipitators, and because of that Hg presents high TF. Regarding to fluidized bed systems, significant TF appear for all the metals, although the lower working temperatures. Metals transferences to the gaseous emissions are essentially due to the mass transfer associated with the high particles load inside the furnace system and to the use of electrostatic precipitators, however treatment of flue gases with activated carbon decrease significantly TF.

Table 20. Transfer ratios of heavy metals present in SRF to emissions (ERFO)

TF (%)	Cement kilns (wet process)	Cement kilns (dry process)	Coal fired power plants	Fluidized bed systems	Fluidized bed systems with activated carbon
Hg	20	10-40	20-50	60	6.5
Cd+Tl	0.48	0.48	1.4-6.7	2.7	0.02
Sum HM	0.02	0.02	0.1-0.5	1.0	0.2

Acceptable heavy metals concentration (in mg/MJ) was hence defined, as presented in table 21, so that emissions limit values defined in the WID (Waste Incineration Directive) are obeyed, table 21.

Table 21. Heavy metals concentrations in SRF to obey emissions limit values of WID

Maximum possible concentrations of HM in SRF (ERFO)	Cement kilns (wet process)	Cement kilns (dry process)	Coal fired power plants	Fluidized bed systems	Fluidized bed systems with activated carbon
Hg mg/MJ	0.17	0.082-0.33	0.034-0.085	0.028	0.26
Cd+Tl mg/MJ	6.90	6.9	0.25-1.21	0.063	85
Sum HM mg/MJ	1650	1650	34-170	17	850
Cl (% wt)	0.5-1.0	3.0	0.4/2.0	0.4/1.4	0.4/1.4

Besides, the definition of the range of acceptable heavy metals, chlorine acceptable contents was also included. The inclusion of Cl is mostly linked to the control of high temperature corrosion and fouling, on the other hand, Cl is extremely important in determining the behavior of heavy metals, because of the possibility of forming metal chlorides with low volatilization temperatures.

6. Case Studies

6.1. Combustion of Automotive Shredder Residues

European Directive 2000/53/EC obliges to the recycling of end-of-life vehicles (ELVs). It is predictable that the number of ELVs will increase from 8 million in 1996 to 11 million in 2015 in Europe. About 75% of ELVs are subjected to dismantling, which generates about 2 to 3 million tones of Auto Shredder Residues, ASR, per year, in Europe (Nourridine, 2007). Its content in organic matter, like plastics, rubber and fibrous materials, may reach 40 to 65% of its weight (Mirable et al, 2002). Although this residue is highly heterogeneous, its heating value could range from 17 to 30 MJ/kg (Nourridine, 2007), thus it is highly attractive for energy recover, avoiding the necessity of its landfilling.

However, ASR usually presents significant levels of heavy metals, as shown in table 22, which bring environmental concerns to the combustion process, due to the possibility of metal emissions and generation of contaminated ashes that have to be eliminated without endanger the environment.

In order to evaluate problems associated with the energetic valorization of ASR, a study was performed in an atmospheric fluidized bed pilot installation of 0.3 MW. The ASR was supplied by a Portuguese company that recycles end-of-life vehicles. The original ASR had to be sorted out in order to remove gross metallic material, like metallic pieces, then was sieved to remove fine inorganic materials below 0.5 mm and prepared as pellets to be fed into the combustor. Composition of the ASR burned is presented in table 23.

Table 22. Heavy metals in ASR

mg/kg	Roy et al, 2001	Mirable et al, 2002	Abelha et al, 2005
Cu	6200-39000	1200	10000
Cr	230-660	800	200
Cd	50-70	-	50
Ni	180-930	700	300
Pb	224-6450	200	2300
Zn	7200-9400	19000	4900
Mn	-	-	2200
Hg	-	-	0.90

Table 23. Characterization of the combustible ASR tested

a.r – as received	Combustible ASR
Low Heating Value *(MJ/kg a.r.)*	12.3
Moisture (% wt a.r.)	4.2
Ash (% wt a.r.)	48.0
Volatile Matter *(% wt a.r.)*	47.8
C (% wt a.r.)	31.1
H (% wt a.r.)	4.1
N (% wt a.r.)	1.0
S (% wt a.r.)	0.43
Cl (% wt a.r.)	0.73
F (% wt a.r.)	<0.02

The system, (Abelha et al., 2005), works at about 850°C (bed temperature) and has two cyclones, working between 650 and 400°C, to collect particles. Fly ashes leaving the system, through the stack were sampled isokinetically and analyzed for metal emissions.

Three types of ashes were collected in the system; bed ashes collected with sand from the bottom of the fluidized bed system and two streams of fly ashes, collected in the first and in the second cyclones. Levels of heavy metals in ASR were significant, especially those of Cu, Pb, Zn and Hg; hence ashes produced presented high concentrations of metals, as it is presented in figure 4. Bed ashes had lower concentrations, due to the incorporation of sand from the fluidized bed system. About 90% of the metals present in ASR were retained in ashes collected in the system, especially in fly ashes collected in the two cyclones. The highest metals contents were collected in the finer particles of the second cyclone. Relative enrichments in the finer ashes varied from 1.1 to 4.8, although higher values were obtained in the case of Ni, which might be due to contamination from the refractory steel of the fluidized bed system.

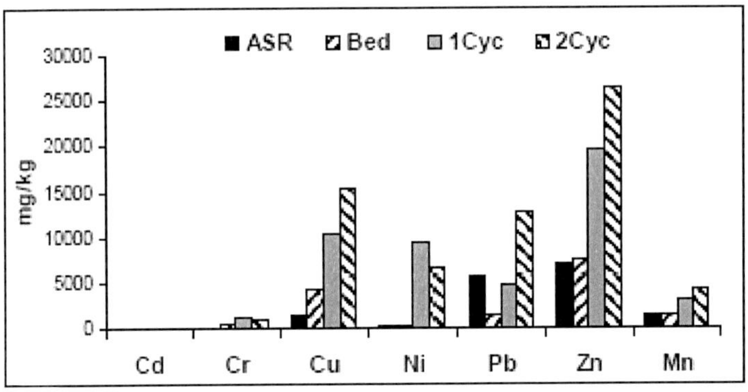

Figure 4. Heavy Metals in ASR and ashes (1Cyc - first cyclone, 2Cyc - second cyclone).

As it may be observed in figure 5, Hg was relatively high in the ASR (1.4 mg/kg). Hg concentration in bed ashes was insignificant and in the cyclone ashes was found to be lower than in the ASR, being higher in the second cyclone than in the first one. Global retention in ashes was about 32% of the inlet amounts. Most of the Hg was released with the flue gas, exceeding allowable limits of waste incineration, as can be seen in table 24.

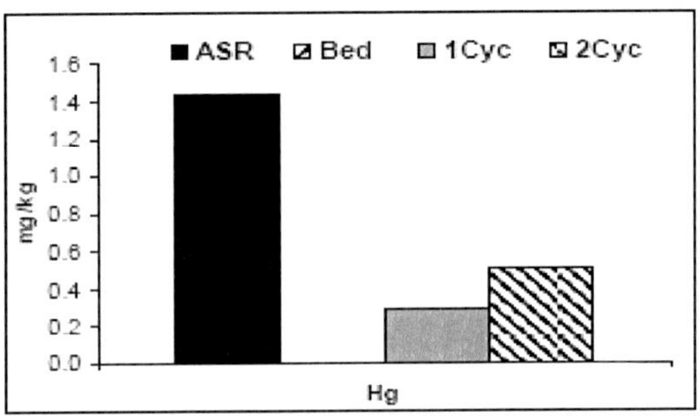

Figure 5. Mercury in ASR and Ashes.

Other metals emissions of were also relatively high, which may be related with the presence of Cl in the ASR, as it promotes the formation of metallic chlorides that present high volatilities. The used pilot-scale combustion installation does not have other flue gas treatment than particles retention by cyclones, but full scale combustion systems will have to be equipped with efficient flue gas treatments equipments in order to remove mercury and particles enriched with heavy metals from the flue gas stream.

Table 24. Heavy metals emission from ASR combustion

Emissions	mg/Nm^3 at 11% O_2
Particles	534
Cd	1.2
Cr	0.47
Cu	87
Ni	0.12
Pb	26
Mn	0.16
Hg	0.17

Besides the need to control emissions from energetic valorization of ASR, ash leachability has to be evaluated in order to allow their correct management. Table 25 shows the results obtained for leachability performed in accordance with EN 12457. The ashes presented alkaline nature, producing high pH leachates. Although ashes contained high concentrations of heavy metals, leachability of metals was insignificant, except in the case of chromium, which may be due to the formation of hexavalent chromium under the oxidant combustion conditions and presence of Cl that may favor formation of $CrCl_6$ and $CrOCl_4$ (Linak and Wendt, 1994). In addition, ashes also leached significant levels of chlorides and sulfates. According to the landfilling classification (Decision 2003/33/CE), the three ash streams produced behaved as highly hazardous materials, hence requiring pretreatment before disposal.

Table 25. Leachability of ashes from ASR combustion

	Bed ash	1° Cyclone ash	2° Cyclone ash
pH	11.0	10.4	10.4
Conductivity (mS/cm)	2.2	13.8	21.8
Cl^- mg/kg	1030	44990	75530
SO_4^{2-} mg/kg	11090	16200	16900
Pb mg/kg	<0.3	<0.3	3.4
Cd mg/kg	<0.2	<0.2	<0.2
Cr mg/kg	54.0	183.0	22.4
Cu mg/kg	<0.03	<0.03	0.09
Ni mg/kg	<0.3	<0.3	<0.3
Mn mg/kg	<0.3	<0.3	<0.3
Zn mg/kg	<0.2	<0.2	<0.2
Hg mg/kg	<0.03	<0.03	<0.03

6.2. Combustion and Gasification of Sewage Sludges

Production of sewage sludge from waste water treatments facilities was estimated to be 6.5 millions tones (d.m.) in 1990. Due to more efficient waste water treatment requirements (Directive/271/EEC) it was estimated that the amounts of sewage sludge produced could grow to 10 millions tones in 2005 Davis (1995). As sewage sludge contains significant quantities of organic matter, calorific value of dry sewage sludge may vary between 11.5 and 17.3 MJ/kg (Reimann, 1990), which makes sewage sludge attractive for energetic valorization. About 11% of sewage sludge amounts produced in Europe in 1994 was burned, but it is estimated that combustion will grow to 40% (Davis, 1995; ISWA, 1995), because of restriction on landfilling of organic wastes. In the USA, production of sewage sludge was estimated to be 7 millions tons in 1990 (Werther and Ogada, 1999). In Japan 7.5 millions tons were produced in 1991, growing to 70% till 2000.

One of the problems associated with the combustion of sewage sludge is the presence of heavy metals; especially Hg. Table 26 presents the contents of the main heavy metals found in sewage sludge. Other metals may be also present, such as Ag (nd-960 mg/kg), Ba (0-3000 mg/kg), V (0-2100 mg/kg), Sr (0-2230 mg/kg) and Se (10-180 mg/kg) (Dewling, 1980). Table 27 presents the characterization of the sludge.

Table 26. Heavy metals in several sewage sludges

mg/kg (m.s.)	Portugal a)	Denmark b)	Belgian c)	UK d)	Germany e)	USA g)
Cd	2.1-17	5.0-23	0.37	3.3	3.2-7.8	nd-1100
Cu	50-1860	179-1120	278	568	280-960	45-16030
Pb	46-460	226-1350	165	221	242-720	80-26000
Ni	15-182	31-265	31	57	40-106	nd-2800
Zn	325-8150	1033-2282	1067	792	875-1400	nd-28360
Cr	21-291	111-831	52.5	157	24-91	22-30000
Hg	2.3-7.4	-	1.0	2.4	2.7-22.1	0.1-89
As	-	-	-	-	11-15 [f)]	10-50
Co	2.0-54	-	-	-	-	nd-800
Mn	68-1225	-	-	-	-	100-8800
B	5.5-52	-	-	-	-	200-1430

a) (Domingues, 1999) b) (Peterson, 2001) c) (Huybrechts, 2001) d) (Smith, 2001) e) (Reimann, 1990) f) (Albrescht and Schelhaas, 1997) g) (Dewling, 1980)

Table 27. Characterization of the combustible sewage sludge

a.r. – as received	Combustible Sewage Sludge
Low Heating Value *(MJ/kg a.r.)*	13.1
Moisture (% wt a.r.)	6.6
Ash (% wt a.r.)	42.8
Volatile Matter *(% wt a.r.)*	49.8
C (% wt a.r.)	30.9
H (% wt a.r.)	3.8
N (% wt a.r.)	3.7
S (% wt a.r.)	0.96
Cl (% wt a.r.)	0.07
F (% wt a.r.)	<0.02

Combustion studies of dry sewage sludge were performed in the same installation used for ASR. Bed temperatures were slightly lower; around 750°C in the bed zone and cyclones were operated below 400 °C, due to the lower calorific value of the sludge that was burned without auxiliary fuel.

Figure 6 compares the levels of heavy metals present in sewage sludge to those obtained in ashes. More than 98% of inlet heavy metals amounts were retained in the bed ashes and in the ashes collected in the two cyclones. Bed ashes were separated from the bed sand, because they maintained their granular form and are easily segregated from the sand. This allowed verifying that ashes retained in the bed present similar concentration of metals as the fly ashes collected in cyclones. Relative enrichments of ashes were usually lower than 1.1, except in the case of Cd that reached 1.8, and in the case of Cr and Ni, which may be attributed to contamination with elements present in the refractory steel of the combustion system.

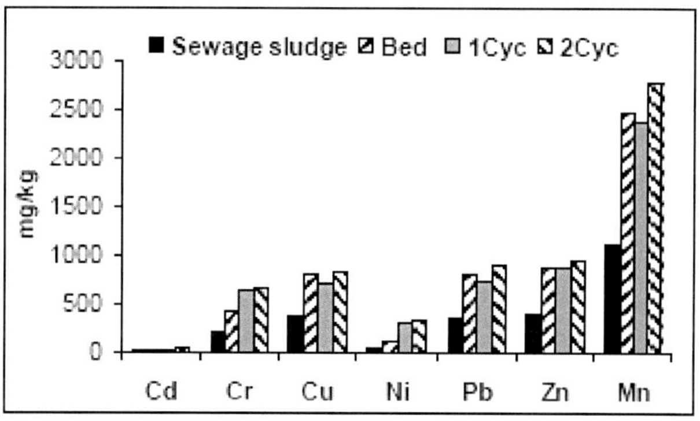

Figure 6. Heavy Metals in dry sewage sludge and ashes.

Hg content found in the sewage sludge was significant, 2.8 mg/kg. In this case, although Hg has been completely volatilized (bed ashes had no Hg), Hg concentrations in fly ashes collected in cyclones were very high due to retention of mercury. Enrichments found in cyclone ashes reached 4.2 related to the inlet concentration for second cyclone ashes. Higher retention of Hg in the solid phase for sewage sludge than those obtained for ASR, was not expected, as Cl level in sewage sludge was lower than in the ASR. It is recognized that the presence of Cl may favor the formation of $HgCl_2$ (Frandsen *et al.*, 1994), which is stable below 430°C, favoring its capture in the solid phase. When Cl is low in fuels, vapors of metallic Hg are preferentially formed and are extremely difficult to capture requiring special techniques. The higher retention in the solid phase might be explained by the lower operational temperatures of the cyclones during sewage sludge combustion that may have favored condensation of oxidized Hg. Nevertheless, retention in the ashes accounted for only about 60% of the inlet Hg, being emitted the remaining.

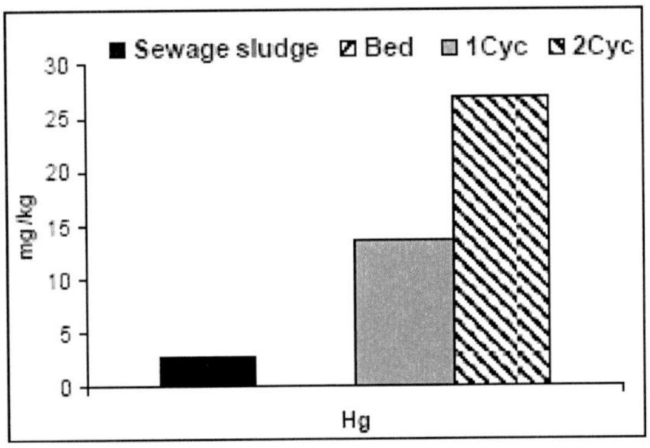

Figure 7. Mercury in dry sewage sludge and ashes.

As a consequence of the high retention of most heavy metals, emissions were very low, as can be seen in table 28, except in the case of Hg, that presented higher emissions than the limit values for waste incineration, as it happen for ASR. The low levels of metal emissions may be also explained by the low emissions of particles, which is a particular feature of sewage sludge mineral mater that tend to remain mostly as bed ashes, retaining heavy metals.

Leachability of ashes was evaluated in accordance with EN 12457 to classify ashes for landfilling. All the ashes showed alkaline nature, producing leachates with high pH values. Moderate values of conductivity, as in the case of ASR ashes, were probably due to solubilization of sulfates and chlorides of alkaline metals. Leachability of heavy metals was insignificant for most of the metals and detected in very low concentrations in the case of Cu and Cr. Comparing with the limit values for landfilling, cyclone ashes present hazardous nature, mainly due to the release of chlorine, as although sulfate release is high, levels remained in the non-hazardous levels.

Table 28. Heavy metal emissions from sewage sludge combustion

mg/Nm^3 at 11% O_2	Sewage Sludge
Particles	49
Cd	<0.001
Cr	<0.001
Cu	0.021

Ni	<0.002
mg/Nm^3 at 11% O_2	Sewage Sludge
Pb	0.009
Mn	0.012
Hg	0.19

Table 29. Leachability of ashes from sewage sludge combustion

	Bed ash	1° Cyclone ash	2° Cyclone ash
pH	10.1	9.3	9.4
Conductivity (mS/cm)	23.4	33.7	35.0
Cl^- mg/kg	1700	19700	27000
SO_4^{2-} mg/kg	15100	18900	19000
Pb mg/kg	<0.3	<0.3	<0.3
Cd mg/kg	<0.2	<0.2	<0.2
Cr mg/kg	<0.3	0.4	0.4
Cu mg/kg	<0.3	0.3	0.4
Ni mg/kg	<0.3	<0.3	<0.3
Mn mg/kg	<0.3	<0.3	0.3
Zn mg/kg	<0.2	<0.2	<0.2
Hg mg/kg	<0.03	<0.03	<0.03

Sewage sludge was also subject to gasification in a pilot fluidized bed gasification installation, working at 850°C, in presence of steam and air. Solid residues were collected from the bed zone and from the cyclone installed to remove particulates from the syngas produced. Figure 8 compares the concentration of heavy metals in the sludge and in the solids collected in the bed and in the cyclone. All the heavy metals were enriched in the residues compared with the parent sludge, although residues still contained about 8% of unburned carbon. Relative enrichment varied from 2.2 to 6.1, the highest values were obtained for Pb, being higher than in the case of sewage sludge combustion. This reveals that volatilization was more pronounced in the case of gasification.

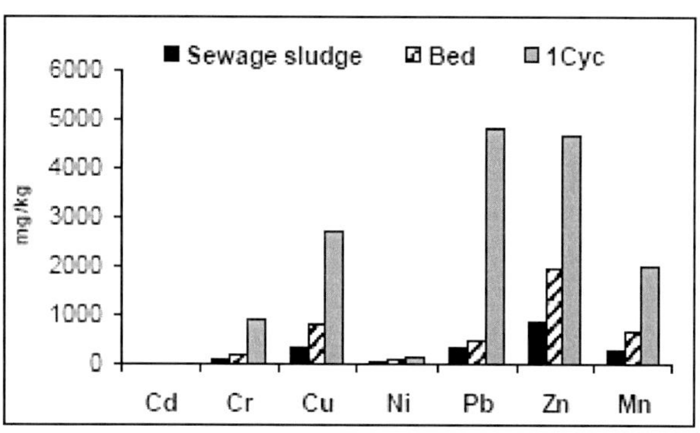

Figure 8. Heavy metals in dry sewage sludge and gasification residues.

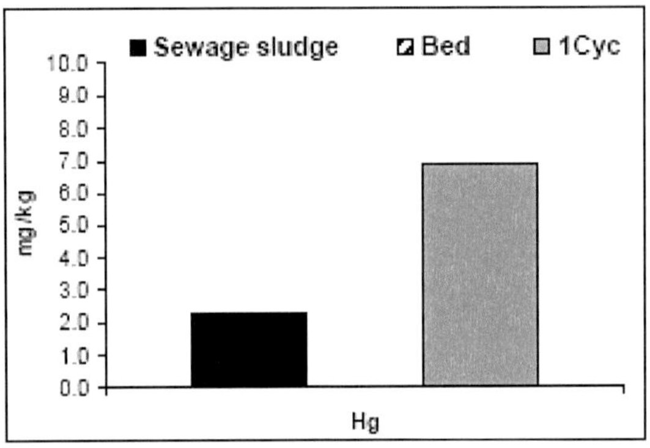

Figure 9. Mercury in dry sewage sludge and gasification residues.

In the case of Cd and Hg (figure 9), enrichments were lower in residues collected in the cyclone, 0.4 and 1.2 respectively, which indicates that these metals were transferred to the gas phase and thus the syngas produced must be further cleaned to be used as a fuel.

Leachability of residues from gasification was evaluated through EN 12457. The residues produced alkaline leachates as in the case of ashes obtained in combustion, as can be seen in table 30. Conductivity was lower than in the case of ashes, which mean that these residues are less soluble. Levels of Cl released by both residues were lower than those presented by ashes formed in combustion. The same applies to sulfates as their levels are much lower in residues from gasification, because conversion of sulfides to sulfates is not favored during gasification. Leachability of heavy metals was not detected. Reducing conditions of gasification also prevents oxidation of Cr, thus no hexavalent species are released under the alkaline pH.

Table 30. Leachability of residues from sewage sludge gasification

	Bed residue	1° Cyclone residue
pH	10.1	10.8
Conductivity (mS/cm)	1.5	3.2
Cl^- mg/kg	55	530
SO_4^{2-} mg/kg	556	1500
Pb mg/kg	<0.3	<0.3
Cd mg/kg	<0.2	<0.2
Cr mg/kg	<0.3	<0.3
Cu mg/kg	<0.3	<0.3
Ni mg/kg	<0.3	<0.3
Mn mg/kg	<0.3	<0.3
Zn mg/kg	<0.2	<0.2
Hg mg/kg	<0.03	<0.03

7. CONCLUSIONS

The problems of heavy metals associated to thermal energy process, including e.g. combustion and gasification, are very complex. Since the sixties it is know that anthropogenic activities contribute to increase heavy metals in the atmosphere and since then intensive investigation has been carried out. It was found that thermal processes involved in energy production from coal are one of the major sources of heavy metals, because of the high temperatures involved. Metals are emitted as very small particles or aerosols that are extremely toxic for human beings and can be spread into the atmosphere crossing continents and prevailing for long periods of time. For that reason, international programs and regulations have been launched to measure and control heavy metal pollution. In the new era; the problem of greenhouse gases and climate change, the scarcity of raw materials and the protection of the environment by avoiding landfilling of organic wastes, reinforced the valorization of wastes as a source of energy. However, one of the major drawbacks is the presence of heavy metals in wastes that present new challenges in the area of control of heavy metals emissions, especially of the most volatile metals, such as mercury. Control of emissions results in the production of liquid or solid effluents highly contaminated and of high quantities of ashes enriched in heavy metals requiring appropriate management. Hence, utilization of wastes for power production requires a detailed control of both gaseous emissions of heavy metals and ashes produced. Among the several thermal processes, those that work at lower temperatures diminish the emission of heavy metals. Studies of thermal valorization of wastes containing heavy metals whose landfilling has to be avoided, such as sewage sludge and auto shredder residues were carried out in a fluidized bed combustion system to investigate the fate of heavy metals The results showed that ashes produced in these types of systems contribute to control metal emissions at moderate temperatures, but still flue gas treatment has to be performed. Ashes have to be dealt with care to prevent environmental pollution, because although most metals are immobilized in alkaline ashes, oxidant environments tend to oxidize metals that may form oxyanions, such as heaxavalent Cr, especially in the presence of chlorine. Oxidant combustion of wastes also generates ash that may leach high quantities of sulfates and chlorides that turn them into hazardous wastes. The reducing environments used in gasification processes may decrease this problem, but increase volatilization of some metals, hence requiring gas treatment.

8. REFERENCES

Abelha, P.; Gulyurtlu, I.; Lopes, H.; Cabrita, I. (2005). The burning of Automotive Shredder Residue using Fluidized Bed Technology. *Proceedings of the 18th International Conference on Fluidized Bed Combustion*, Ed.Preto, ASME, CD-ROM edition (FBC 2005-78023)

Albrecht, J.; Schelhaas, K.P. (1997). Investigation in NOx-Formation at Combustion of Sewage Sludge in a Commercial Scale Bubbling Fluidized Bed. *Proceedings of the 14th International Conference on Fluidized Bed Combustion*, Ed.Preto, F.D.S.-ASME, 997-1005.

Aunela-Tapola, L.; Hatanpaa, E.; Hofftren, H.; Laitinem, T.; Larjava, K.; Rasila, P. and Tolvanen, M. (1998). A Study of Trace Element Behaviour in Two Modern Coal-Fired Power Plants II – Trace Elements Balances in Two Plants Equipped with Semi-Dry Flue Gas Desulphurisation Facilities. *Fuel Processing Technology, 55*, 13-34.

Barton, R.G.; Clark, W.D.; and Seeker, W.R. (1990). Fate of Metals in Waste Combustion Systems. *Combustion Science. and Technology, 74*, 327-342.

Bergstrom, J.G.T. (1986). Mercury Behaviour in Flue Gases. *Waste Management and Research, 4(1)*, 57-64.

Biswas, P.; Wu, C.Y. (1998). Control of Toxic Metal Emissions from Combustors Using Sorbents: A Review. *Journal of Air and Waste Management Association*, 48, 113-127.

BREF LCP – Reference Documents on Best Available Techniques for Large Combustion Plants, July 2006, European Commission

BREF WI - Reference Documents on Best Available Techniques for Waste Incineration, August 2006, European Commission.

Buat-Ménard, P. (1993). Global Change in Atmospheric Metal Cycles. In: C.N. Hewitt and W.T.Sturges (Eds.), *Global Atmospheric Chemical Change* (pp 470). London, Elsevier Applied Sciences

Buchholz, B.A.; Landsberger, S. (1993).Trace Metal Analysis of Size-Fractionated Municipal Solid Waste Incinerators Fly Ash and its Leachates. *Journal of Environmental Science Health, A28(2),* 423-441.

Buekens, A.G.; Verhulst and Dimova, C. (1995). Heavy Metals in MSW Incineration Thermodynamic and Experimental Study. *Proceedings of the International Conference on Solid Waste Management Thermal Treatment and Waste-to-Energy Technologies (VIP-53) 1995, Air and Waste Management Association – EPA*, 461-473.

Campbell, J.A. ; Laul, J.C. ; Nielson, K.K. ; Smith, R.D. (1978). Separation and Chemical Characterization of Finely-Sized Fly-Ash Particles. *Analytical Chemistry 50(6)*, 1032-1040.

CEN/TS 15359:2006, Solid recovered fuels - Specifications and classes, CEN, Management Centre: rue de Stassart, 36 B-1050 Brussels

Chen, J.; Wey, M., Liu; Y; Chiang, B. (1998). Dynamic Adsorption of Heavy Metals Under Various Incineration Temperatures. *Journal of Environmental Engineering*, *August 1998*, 776-779.

Clapp, T.L.; Magee II, J.F.; Ahlert, R.C.; Kosson, D.S. (1988). Municipal Solid Waste Composition and the Behavior of Metals in Incinerator Ash. *Environmental Progress, 7(1),* 22-30.

Clarke, L.B.; Sloss, L.L. (1992). Trace Elements Emissions from Coal Combustion and Gasification, In: IEA Coal Research (pp 111), Gemini House, London.

Clarke, L.B. (1993). The Fate of Trace Elements during Coal Combustion and Gasification: an overview. *Fuel, 72*(6), 731-736.

Clean air act mercury rule. 2007. 07.13. Available from: http//www.epa.gov/air/mercuryrule/rule.htm.

Coles, D.G.; Raaini, R.C.; Ondov, J.M.; Fisher G.L.; Silberman, D.; Prentice, B.A.; (1979). Chemical Studies of Stack Fly ash from a Coal-Fired Power Plant. *Environmental Science and Technology, 13(4)*, 455-459.

{SEC(2005) 101}, Communication from the Commission to the Council and the European Parliament, Community Strategy Concerning Mercury. Brussels, 28.01.2005. available at: http://eur-lex.europa.eu/LexUriServ/site/en/com/2005/com2005_0020en01.pdf

Davis, R.D. (1995). Legislative Developments and Sewage Sludge Disposal Strategy in Europe. In: Rhenipal and Dirk European Holdings Limited (Eds.), *Symposium on Sewage Sludge Disposal* (21 pp). Prague, England.

Davison, R.L.; Natusch, D.F.S.; Wallace, J.R. (1974). Trace Elements in Fly Ash: Dependence of Concentration on Particle Size. *Environmental Science and Technology, 8(13),* 1107-1113.

Decision N° 1600/2002/EC of the European Parliament and of the Council of 22 July 2002, OJEC, N°L 242, 10.9.2002

Decision 2003/33/CE of the Council of 19 Dec. 2002, OJEC, N°L 11, 16.01.2003.

Desauziers, V.; Castre, N.; LeCloirec, P. (1997). Sorption of Methylmercury by Clays and Mineral Oxides. *Environmental Technolology, 18,* 1009-1018.

Dewling, R.T. (1980). Sewage-Sludge Incineration Raises Air Pollution Concerns. *Water and Sewage Works, Oct. 1980*, 28-43.

Directive 2004/107/EC of the European Parliament and of the Council of 15 December 2004, OJEC N° L 23, 26.11.2005

Directive 89/369/EC of the European Parliament and of the Council of 8 June 1989, OJEC N.° L 163,14.06.1989.

Directive 89/429/EC of the European Parliament and of the Council of 21 de June 1989, OJEC N.° L 203, 15.07.1989.

Directive 94/67/EC of the European Parliament and of the Council of 16 December 1994, OJEC N.° L 365, 31.12.1994.

Directive 2000/76/EC of the European Parliament and of the Council of 4 December 2000, OJEC N.° L 332/91, 28.12.2000.

Directive 2000/532/EC of the European Parliament and of the Council of 3 May 2000, OJEC N.° L 226, 06.09.2000.

Directive 91/689/EEC of the European Parliament and of the Council of 12 December of 1991, OJEC N.° L 337/20, 31.12.91.

Directive 1999/31/EC of the Council of 26 April 1999, OJEC, N ° L 182, 16.07.1999 .

Directive 91/271/EEC of the Council of 21 May 1991, OJEC n.° L 135 de 30.05.1991.

Directive 96/61/EC of the Council of 24 September 1996, OJEC, N° L 257, 10.10.1996.

Directive 2000/53/EC of the European Parliament and of the Council of 18 September 2000, OJEC, N°L 269, 21/10/2000.

Chen, J.; Wey, M., Liu, Y; Chiang, B. (1998). Dynamic Adsorption of Heavy Metals Under Various Incineration Temperatures. *Journal of Environmental Engineering, August 1998*, 776-779.

Domingues, H.L.B.R.C. (1999). Comportamento de Metais Pesados (Cd, Cr, Cu, Ni, Pb e Zn) em solos Tratados com Lamas Residuais Urbanas. *PhD Dissertation in Hydrosphere Ecology* FCT-UNL, Portugal

Durlak, S.K.; Biswas, P.; and Shi, J. (1997). Equilibrium Analysis of the Effect of Temperature, Moisture and Sodium Content on Heavy Metal Emissions from Municipal Solid Waste Incinerators. *Journal of Hazardous Materials, 56,* 1-20.

Eddings, E.G. ; Lighty, J.S. ; Kozonsky, J.A. (1994). Determination of Metal Behavior During the Incineration of a Contaminated Montmorillonite Clay. *Environmental Science and Technology 28*, 1791-1800.

EMEP Status Report 2/2003, Heavy metals: Transboundary Pollution of the Environment, I. Ilyin, O. Travnikov, W. Aas, H.Th. Uggerud, UN-ECE's Convention on Long-Range Transboundary Air Pollution. 2005.06.03. Available at: www.emep.int; www.msceast.org/countries.

EMEP Status Report 2/2007, Heavy metals: Transboundary Pollution of the Environment, 2008.03.03. Available at: http://www.nilu.no/projects/ccc/reports/emep4-2007.pdf

ERFO - European Recovered Fuel Organization, Classification of Solid Recovered Fuels. 2008.03.03. Available at: http://www.erfo.info/fileadmin/user_upload/erfo/documents/ standardisation/ Classification_report.270205.pdf,2005

Frandsen, F.; Dam-Johansen, K.; and Rasmussen, P. (1994). Trace Elements from Combustion and Gasification of Coal-an Equilibrium Approach. *Progress on Energy Combustion Science, 20*, 115-138.

Gadd, G.M. (1992). Heavy Metal Pollutants: Environmental and Biotechnological Aspects, *Encyclopedia of Microbiology* (Vol.2, pp 35), Academic Press, Inc.

Gladney, E.S.; Small, J.A.; Gordon, G.E.; Zoller, W.H. (1976). Composition and Size Distribution of In-Stack Particulate Material at Coal-Fired Power Plant. *Atmospheric Environment, 10*, 1071-1077.

Habashi, F. (1993). Métallurgie Extractive. *A Textbook of Hydrometallurgy* (689 pp), Québec Enr.

Hansen, L.D.; Fisher, G.L. (1980). Elemental Distribution in Coal Fly Ash Particles. *Environmental Science and Technology 14(9)*, 1111-1117.

Hansen, L.D.; Sillberman, D., Fisher; G.L.; Eatough, D.J. (1984). Chemical Speciation of Elements in Stack-Collected, Respirable-Size, Coal Fly ash. *Environmental Science and Technology 18(3)*, 181-186.

Hartman, R.M. (1991). An Overview of the fate of Metals during Combustion at the Mid-Connecticut Resource Recovery Facility. *Proceedings of the Second Annual International Conference Municipal Waste Combustion - Air and Waste Management Association*, 347-362.

Haynes, B.S.; Neville, M.; Quann, R.J.; Sarofim, A.F. (1982). Factors Governing the Surface Enrichment of Fly ash in Volatile Trace Species. *Journal of Colloid and Interface Science, 87(1)*, 266-278.

Helble, J.J.; Mojtahedi, W.; Lyranem, J.; Jokiemi, J.; Kauppinen, E.; (1996). Trace Element Partitioning During Coal Gasification, *Fuel, 75(8)*, 931-939.

Ho, T.C.; Lin, C.K.; Lee, H.T.; Hopper, J.R.; (1993). Characterization of Lead and Cadmium Capture by Sorbent during Fluidized Bed Combustion. *Proceedings of the 12th International Conference on Fluidized Bed Combustion. Ed.Rubow, L.N.-ASME*, 31-40.

Ho, T.C.; Chen, C.; Hoper, J.R.; Oberacker, D.A. (1992). Metal Capture During Fluidized Bed Incineration of Wastes Contaminated with Lead Chloride. *Combustion Science and Technology 85*, 101-116.

Hughes, I.S.C.; Littlejohn, R.F. (1987). Trace Element Emissions from AFBC. *Proceedings of the 9th Int. Symposium on Fluidized Bed Combustion, Ed. Mustonen, J.P.*, 906-915.

Huybrechts, D. (2001). Best Available Techniques for the Treatment and Disposal of Sewage Sludge in Flandres. *Proceedings of the International Symposium on the Recycling and Reuse of Sewage Sludge, Ed. Dhir, R.K., Limbachya, M.C. and McCarthy, M.J.Thomas Telford,* 58-68.

IAWG, (1997). Municipal Solid Waste Incinerator Residues. In: Chandler, A.J., Eighmy, T.T.Hertlén, J., Hjelmar, O., Kosson, D.S., Sawell, S.E., Van der Sloot, H.A., Vehlow (Eds.) *Studies in Environmental Science* 67 (pp.974), London, Elsevier Applied Sciences

Kauppinem, E.I.; Pakkanem, T.A. (1990b). Mass and Trace Element Size Distribution of Aerosols Emitted by a Hospital Refuse Incinerator. *Atmospheric Environment, 24A(2),* 423-429.

Kauppinen, E.I.,; Pakkanen, T.A. (1990a). Coal Combustion Aerosols: A Field Study. *Environmental Science and Technolology, 24,* 1811-1818.

Klein, D.H.; Andren, A.D.; Bolton, N.E.; (1975b). Trace Element Discharges from Coal Combustion for Power Production. *Water Air and Soil Pollution, 5*, 71-77.

Klein, D.H.; Andren, A.D.; Carter, J.A.; Emeny, J.F.; Feldman, C.; Fulkerson, W.; Lyon, W.S.; Ogle, J.C.; Talmi, Y.; Hook van R.; Bolton, N.E.; (1975a). Pathways of Thirty-Seven Trace Elements through Coal-Fired Power Plant. *Environmental Science and Technology, 9(10),* 973-979.

Klika, Z.; Bartonova, L.; Spears, D.A. (2001). Effect of Boiler Output on Trace Element Partitioning During Coal Combustion in Two Fluidized-Bed Power Stations. *Fuel, 80,* 907-917.

Knacke, O.; Kubaschewski, O.; Hesselmann, K. (1991). Thermo-Chemical Properties of Inorganic Substances, Vol.I e II, (pp.87), Springer-Verlag.

Kozinski, J.A.; Rink, K.K.; Lighty, J.S. (1995). Combustion of Sludge Waste in FBC. Distribution of Metals and Particle Sizes. *Proceedings of the 13th International Conference on Fluidized Bed Combustion. Ed.Heinschel,* K.J.-ASME, 139-147.

Lecuyer, I.; Bicocchi, S.; Ausset, P.; Lefevre, R. (1996). Physico-Chemical Characterization and Leaching of Desulphurization Coal Fly Ash. *Waste Management and Research ,14,* 15-28.

Lee, R.E.; Crist, H.L.; Riley, A.E.; Macleod, K.E., (1975). Concentration and Size od Trace Metals Emissions from a Power Plant, a Stell Plant, and a Cotton Gin. *Environmental Science and Technology, 9(7)*, 643-647.

Lim, M.Y. (1979). Trace Elements from Coal Combustion - Atmospheric Emissions", Report Number ICTIS/TR05, IEA Coal Research, London.

Lind, T.; Kaupinen, E.I.; Jokiniemi, J.K. (1994). A Field Study on the Trace Metal Behaviour in Atmospheric Circulating Fluidized-Bed Coal Combustion. *Proceedings of the Twenty-Fifth International Symposium on Combustion.* The Combustion Institute, 201-209

Lind, T.; Kauppinen, E.I.; Nilsson, K.; Sfiris, G.; Maenaut, W.; Huggins, F.E. (1999). Heavy Metal Bahviour During Circulating Fluidized Bed Combustion of Willow (Salix). *Proceedings of the 15th International Conference on Fluidized Bed Combustion,* Ed.Reuther, R.B. 201-209.

Linak, W.P.; Wendt J.O.L. (1993). Toxic Metal Emissions from Incineration: Mechanisms and Control. *Progress in Energy Combustion Science, 19*, 145-185.

Linak, W.P.; Wendt, J.O.L. (1994). Trace Metal Transformations Mechanisms during Coal Combustion. *Fuel Processing Technology, 39,* 173-198.

Linak, W.W.; Srivastava, R.K.; Wendt, J.O.L. (1995). Sorbent Capture of Nickel Lead, and Cadmium in a Laboratory Swirl Flame Incinerator. *Combustion and Flame, 100*, 241-250.

Litt, R.D.; Tewksbury, T.L. (1984). Trace Metal retention When Firing Hazardous Waste in a Fluidized Bed Incinerator. In: EPA Office of Research and Development. *Technical Report Data. EPA-600/2-84-198*, Research Triangle Park, NC 27711.35-37.

Liu, K.; Pan, W.-P.; Riley, J.T. (2000). A Study of Chlorine Behaviour in a Simulated Fluidized Bed Combustion System. *Fuel, 79,* 1115-1124.

Lopes, H.; Trindade, T.; Gulyurtlu, I.; Cabrita, I. (2001). Chracterization of FBC Ashes from Co-Combustion of Coal with Oily Residues. *Fuel, 80,* 785-793.

Lopes, H.; Trindade, T.; Gulyurtlu, I.; Cabrita, I. (1999). Characterization of Heavy Metals in Ashes from a FBC Burning Coal and Industrial Residues. *Proceedings of the 1999 Int. Ash Utilization Symposium. Un. Kentucky-CAER*, 176-187.

Lundberg, M.; Hagman, U.; Anderson, B-A; Olofsson, J. (1997). Environmental Performance of the Kvaerner BFB Boilers for MSW Combustion-Analysis of Gaseous Emissions and Solid Residues. *Proceedings of the 14th International Conference on Fluidized Bed Combustion, Ed.Preto, F.D.S.-ASME*, 7-13.

Markowski, G.R.; Filby, R. (1985). Trace Element Concentration as a Function of Particle Size in Fly Ash from a Pulverized Coal Utility Boiler. *Environmental Science and Technology 19(9),* 796-804.

Meij, R. (1993). The Fate of Trace Elements at Coal-Fired Power Plants. Proceedings of the Conference on Thermal Power Generation and the Environment, IEA, Hamburg. 1-4.

Miller, B.B.; Richaud, R.; Dugwell, D.R.; Kandiyoti, R. (1999). Partitioning of Trace Elements During Co-combustion of Coal and Biomass in a Suspension-firing Reactor: Preliminary Findings. *Proceedings of the International Workshop on Fluidized Bed Processing of Unconventional Fuels, Ed. Ciemat, Cabanilhas, A.C., Micio, M., Madrid,* 45-65.

Mirabile, D.; Pistelli, M.I.; Marchesini, M.; Falciani, R.; Chiappelli, L.(2002). Thermal valorisation of automobile shredder residue: injection in blast furnace. *Waste Management, 22*, 841-851.

Mojtahedi, W. (1989).Trace Metals Volatilization in Fluidized-Bed Combustion and Gasification of Coal. *Combustion Science and Technology, 63,* 209-227.

Morgan II, R.P.; Holtsmaster, W.; Meagher, K.; Pegg, M.L.; O'Connor, J. (1997). Chemical Assessment of Fluidized Bed Ash. *Fuel, 76(8),* 717-725.

Natusch D.F.S.; Davison, R.L.; Lamb, R.E.; Wallace, J.R.; Evans, C.A. (1973). Trace Metals in Airborne Particles. *Proceedings of the Third International Clean Air Congress, Dusseldorf, FRG. VDI. Verlag Gmbh*, C12-C14.

Natusch, D.F.S.; Wallace, J.R.; Evans, C.A. (1974a). Toxic Trace Elements: Preferential Concentration in Respirable Particles. *Science; 183*, 202-204.

Natusch, D.F.S.; and Wallace, J.R.; (1974b). Urban Aerosol Toxicity: The Influence of Particle Size. *Science, 186,* 695-699.

Nieboer, E.; Fletcher, G.G.; Thomassen, Y. (1999). Relevance of Reactivity Determinants to Exposure Assessment and Biological Monitoring of the Elements. *Journal of Environmental Monitoring, 1*, 1-14.

Norton, G.A.; Dekalb, E.L.; Malaby, K.L. (1986). Elemental Composition of Suspended Particulate Matter from the Combustion of Coal and Coal/Refuse Mixtures. *Environmental Science and Technology, 20(6)*, 604-609.

Norton, G.A.; Malaby, K.L.; Dekalb, E.L. (1988). Chemical Characterization of Ash Produced during Combustion of Refuse-Derived Fuel with Coal. *Environmental Science and Technology, 22(11)*, 1279-1283.

Nourreddine.M.(2007). Recycling of auto shredder residue. *Journal of Hazardous Materials, A139*, 481-490

Nriagu, J.O.; Pacyna, J.M. (1988). Quantitative Assessment of Worldwide Contamination of Air, Water and Soils by Rrace Metals. *Nature, 333*, 134-139.

Nriagu, J.O. (1989). Natural Sources of Trace Metals in the Atmosphere" *Nature, 338*, 47-49.

Ontiveros, J.L.; Clapp, T.L.; Kosson, D.S. (1989). Physical Properties and Chemical Species, Distribution within Municipal Waste Combuster Ashes *Environmental Progress, 8(3)*, 200-206.

Pacyna, J.M. (1987). Atmospheric Emissions of Arsenic Cadmium, Lead and Mercury from High Temperature Process in Power Plant Generation and Industry. In: T.C.Hutchinson and K.M.Meema (Eds.), *Lead, Mercury. Cadmium and Arsenic in the Environment* (pp. 69-87). John Wiley and Sons.

Petersen, J., (2001). Review of Danish Field Experiments with Sewage Sludge and Composted Household Refuse as a Fertilizer Source. *Proceedings of the International Symposium on the Recycling and Reuse of Sewage Sludge. Ed.Dhir, R.K., Limbachya, M.C. and McCarthy, M.J.Thomas Telford,* 175-189.

Pires, M.; Fiedler, H.; Teixeira, E.C. (1979). Geochemical Distribution of Trace Elements in Coal: Modelling and Environmental Aspects. *Fuel, 76(14/15)*, 1425-1437.

Pluss, A.; Ferrell, R.E. Jr. (1991). Characterization of Lead and other Heavy Metals in Fly Ash from Municipal Waste Incinerators. *Hazardous Waste and Hazardous Materials, 8(4)*, 275-292.

Protocol on Heavy Metals, UNECE, 2008.03.12. Available at: http://www.unece.org/env/lrtap.

Querol, X.; Fernandez-Turiel, J.L.; Lopez-Soler, A. (1995). Trace Elements in Coal and their Behaviour During Combustion in a large Power Station. *Fuel, 74(3)*, 331-343.

Raask, E., (1985). The Mode of Occurrence and Concentration of Trace Elements in Coal. *Progress in Energy and Combustion Science, 11*, 97-118.

Ramesh, A.; Kozinsky, J.A. (2001). Rearrangements in Metals Environment of Inorganic Particles Combustion and Solidification, *Combustion and Flame, 125*, 920-930.

Reed, G.P.; Brain, S.; Cahill, P.; Fanton, I. (1997). Control of Trace Elements in Gasification: A Measurement Methodology to Validate Thermodynamic Predictions. *Proceedings of the 14th International Conference on Fluidized Bed Combustion; Ed. Preto, F.D.S.-ASME*, 1285-1293.

Reimann, D.O. (1989). Heavy Metals in Domestic Refuse and Their Distribution in Incinerator Residues. *Waste Management and Research, 7*, 57-62.

Reimann, D.O. (1990). Heavy Metals and Heat Value in Sewage Sludge. *Water Science Technology, 22(12)*, 85-92.

Reinman and Carita, (1998). *Chemical Elements in the Environment*, (pp.35-36), London, Springer.

Rink, K.K.; Kozinsky, J.A.; Lighty, J.S. (1995). Biosludge Incineration in FBCs: Behavior of Ash Particles. *Combustion and Flame, 100*, 121-130.

Roy, C; Chaala, A. (2001). Vacuum Pyrolisis of Automobile Shredder Residues, *Resources Conservation and Recycling, 32*, 1-27

Sawell, S.E.; Bridle, T.R.; Constable, T.W. (1988). Heavy Metal Leachability from Solid Waste Incineration Ashes. *Waste Management and Research, 6*, 227-238.

Scotto, M.K.; Peterson, T.W.; Wendt, J.O.L. (1992). Hazardous Waste Incineration: *The In-Situ Capture of Lead by Sorbents in a Laboratory Down-Flow Combustor. Proceedings of the Twenty-Fourth Symposium (Int.) on Combustion, The Combustion Institute*, 119-1117.

Scotto, M.K.; Uberoi, M.; Peterson, T.W.; Shadman, F.; Wendt, J.O.L. (1994). Metal Capture by Sorbents in Combustion Process. *Fuel Processing Technology, 39*, 357-372.

Smith, S.R. (2001). The Scientific Basis to Recycling Sewage Sludge in Agriculture. *Proceedings of the International Symposium on the Recycling and Reuse of Sewage Sludge, Ed.Dhir, R.K., Limbachya, M.C., and McCarthy, M.J., Thomas Telford*, 141-157.

Smith, R.D. (1980). The Trace Element Chemistry of Coal during Combustion and the Emissions from Coal-Fired Plants. *Progress in Energy Combustion Science, 6*, 53-119.

Smith, R.D.; Campbell, J.A.; Nielson, K.K. (1979). Concentration Dependence upon Particle Size of Volatilized Elements in Fly Ash, *Environmental Science and Technology, 13(5)*, 553-558.

Swaine, D.J. (1998). Why Trace Elements are Still Important. *Proceedings of the 1998 Conference on Air Quality – Mercury, Trace Elements, and Particulate Matter*, Virginia. 1-3.

Tillman, D.A. (1994). *Trace Metals in Combustion Systems* (pp. 273), Academic Press.

TCLP - Method 1311, SW-846 - Test Methods for Evaluating Solid Waste, Physical/Chemical Methods, U.S. Environmental Protection Agency, Office of Solid Waste, Washington, D.C. 1996.

Uberoi, M.; Shadman, F. (1991). High-Temperature Removal of Cadmium Using Solid Sorbents. *Environmental Science and Technology, 25(7)*, 1285-1289.

Uberoi, M.; Shadman, F. (1990). Sorbents for Removal of Lead Compounds from Hot Flue Gases. *AIChE Journal. 36(2)*, 307-309.

Van der Sloot, H.A.; Heasman, L.; Quevauviller, Ph. (1997). Harmonization of Leaching Extraction Tests.(pp.281), *Studies in Environmental Science 70*, Elsevier.

Vassilev, S.V. (1994). Trace Elements in Solid Waste Products from Coal Burning at some Bulgarian Thermoelectric Power Stations. *Fuel, 73(3)*, 367-374.

Vassilev, S.V.; Vassileva, C.G.; (1997). Geochemistry of Coals, Coal Ashes and Combustion Wastes from Coal-Fired Power Stations, *Fuel Processing Technology, 51*, 19-45.

Vehlow, J. (1993). Heavy Metals in Waste Incineration. *DAKOFA Conference: AffaldsforbrAEnding Under Nye Betingelser*, 1993. 1-9.

Verhulst, D.; Buekens, A.; Spencer, P.J.; Eriksson, G. (1996). Thermodynamic Behaviour of Chlorides and Sulfates under the Conditions of Incineration Furnaces. *Environmental Science and Technolology, 30*,.50-56.

Vuthaluru, M.B.; Enkhoorn, S.; Kiel, J.H.A; Veringa, H.J. (1994). Trace Element Emissions – Literature Review, *ECN-C-94-096*, ECN, 43 pp.

Wadge, A., Hutton, M.; Peterson, P.J. (1986). The Concentration and Particle Size Relationships of Selected Trace Elements in Fly Ashes from U.K. Coal-Fired Power Plants and Refuse Incinerator. *The Science of the Total Environment, 54*, 13-27.

Werther, J.; Ogada, T. (1999). Sewage Sludge Combustion", *Progress in Energy and Combustion Science, 25,* 55-116.

Wu, C.Y.; Biswas, P. (1993). An Equilibrium Analysis to Determine the Speciation of Metals in a Incinerator. *Combustion and Flame, 93*, 31-40.

Yan, R.; Gauthier, D.; Flamant, G. (2001). Partitioning of Trace Elements in the Flue Gas from Coal Combustion. *Combustion and Flame, 125*. 942-954.

In: Heavy Metal Pollution
Editors: S. E. Brown and W. C. Welton
ISBN: 978-1-60456-899-8

Chapter 5

FROM FIELD STUDIES TO LABORATORY EXPERIMENTS FOR ASSESSING THE INFLUENCE OF METAL CONTAMINATION ON RELATIVE SPECIFIC GROWTH RATES OF PERIPHYTIC DIATOMS

Soizic Morin**, *Michel Coste and François Delmas
Cemagref, UR Réseaux, Epuration, Qualité des Eaux, 50 avenue de Verdun, F33600 Cestas cedex, France

ABSTRACT

This study reports field and laboratory experiments in which the impact of cadmium on the growth characteristics of several diatom taxa from multispecific natural biofilms was assessed under non-limiting nutrient availability. The diatom community structures were determined for each time and each treatment, and coinertia analysis with the physicochemical characteristics of the water revealed the strong influence of many "uncontrolled" factors (differentiating field assemblages and microcosms), but also of metal contamination and duration of exposure. Under non-contaminated conditions, the inverse relationship between specific growth rates (μmax) and cell sizes was in accordance with the allometry studies reported by several authors. Metal contamination influenced the growth responses of the diatoms but, under these toxic conditions, the growth rates were not necessarily correlated with biovolumes.

Microcosm and *in situ* results showed marked differences; examination of the standardized variations [ratio of (treated μmax) - (control μmax) over control μmax] revealed that most of the species considered were saprophilous and metal-tolerant, and the various taxa had slightly increased growth rates under cadmium contamination. These trends, similar in both laboratory and *in situ* experiments, underline the fact that microcosm experiments may be valuable alternatives to field studies, when considered with caution.

* soizic.morin@bordeaux.cemagref.fr

Introduction

Aquatic environments are often exposed to various pollutants like pesticides, oil hydrocarbons, heavy metals or radioactive pollution, released from industry, agriculture or households. Emissions of heavy metals in the ecosystem may induce severe problems in the food web – from plants and animals to man. There are a number of standardized tools for assessing aquatic ecosystem health from physical and chemical analyses to biomonitoring, *i.e.* the use of living organisms that are likely to reveal the environmental conditions of their habitat. Among the major sources of primary productivity in freshwater systems, diatom communities are composed of a large number of species with various ecological preferences [1-3], which confer to these organisms a high relevance among biological indicators.

An attempt to determine growth responses of freshwater diatoms from the Riou-Mort watershed [4] has been followed on here by a study of the influence of cadmium contamination on their growth characteristics. Species-specific studies of growth responses have already been performed on monocultures of diatoms exposed to metal contamination [5-8]; however particular growth patterns in natural multispecific populations may be influenced by enriching elements [9] or reduced by competitive exclusion between organisms [10-12] or contaminations with toxicants [6, 13], and may also vary among species depending on individuals' sizes. Indeed, it has been demonstrated under unpolluted conditions that large diatom species from natural biofilms have longer cell cycles than smaller ones [4, 14, 15], but little can be said about the influence of metal contamination on their growth characteristics.

This study attempts to elucidate the effects of cadmium on an integrated metabolic parameter, *i.e.* growth rate, in many diatoms grown in multispecific cultures, with regards to their biovolumes. The shifts in benthic diatom species composition are also monitored, to determine the sensitivity or tolerance to metal for several species. The contribution of both laboratory and *in situ* relative specific growth rates were investigated to provide valuable information on the potential value of using microcosms assays for river biomonitoring.

Materials and Methods

Data for this study were collected from 2 sites in the Riou-Mort stream, South West France (44°N/2°E). The stream receives metal discharge from a small tributary, the Riou-Viou stream, that carries seepage from a former zinc factory. One site was placed upstream of this discharge and another downstream (figure 1). The growth rate experiments using samplers made of artificial substrate were carried out in March 2005 over a 20-day period. This survey was complemented with a laboratory assay conducted in March 2006, the experimental systems being inoculated with natural biofilm sampled from the upstream site. Periphytic communities in both studies were composed of thin biofilms dominated by diatoms.

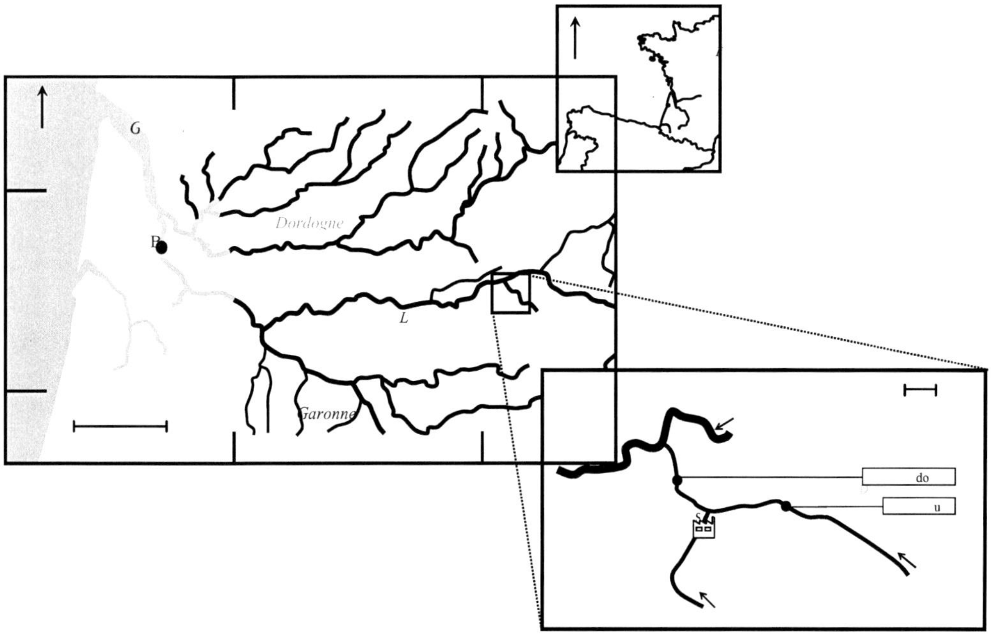

Figure 1. Location of the study sites.

Field Sampling

Two caged, artificial samplers loaded with six vertically aligned glass slides [see 16] were installed at each site for algal collection. Depth, velocity and irradiance were standardized as closely as possible because of their potential influence on growth rates.

The samples (three replicate slides per sampling date) were removed after 4, 7, 14 and 20 days for processing. On each sampling occasion, temperature, pH, conductivity and dissolved oxygen were measured in the stream (WTW, Weilheim, Germany) and stream water samples were simultaneously collected and brought back to the laboratory for nutrient measurements according to French and international standards (NF T90-023 and NF EN ISO 13395). Metal (cadmium, zinc) concentrations were analyzed by the GEMA team, University Bordeaux 1.

Laboratory Assays

Diatoms sampled from the Riou-Mort river (March 2006) were inoculated into freshwater microcosms [17, 18]. In these experimental systems, the periphytic algae were allowed to grow on glass slides similar to those used in the field to maintain their benthic growth features and collected after 1, 2, 4 and 6 weeks colonization. The culture medium used for the experiment was 4-fold diluted Woods Hole medium, without EDTA and supplemented with silica [17, 18]. For all experiments, the units were placed in an air-conditioned room where temperatures were kept constant. A constant light intensity of approximately 70 $\mu mol \cdot s^{-1} \cdot m^{-2}$ illuminated the cultures with a 10:14 hour light-dark cycle, and the systems were under continuous water movement (approximate speed: 0.4 $cm \cdot s^{-1}$).

Cultures were grown in triplicate and allowed to settle on the slides and acclimate to their treatments (control; treated: 10 μgCd·L^{-1}) for one week. The concentration for the treated experimental unit was chosen because of its ecological relevance in the watershed where the diatoms were collected: it corresponds to average concentrations recorded downstream the Riou-Mort river in Winter [19]. In each experimental unit and at each sampling date, physical and chemical parameters were determined according to the procedures used for field assays. Depending on the results of the analyses, culture medium was added as required to compensate for the decrease of nutrient concentrations due to algal uptake.

Diatom Counts and Identification

Periphyton films were removed from the artificial substrates by scraping with a cutter blade and diluted to a standard volume of 100 mL to form three replicate periphyton samples per site and per sampling date. Suspensions of the material were subsampled for various analyses. Diatom cell density was estimated for each replicate sample by enumeration of the total number of cells exhibiting chloroplasts in 10 fields (1.25 μL each, 0.5 mm depth) of a Nageotte counting chamber, using light microscopy at 400× magnification (photomicroscope Leica DMRB, Wetzlar, Germany). Data were recorded as cells per unit area of sampled substrate (number of cells·cm^{-2}).

Aliquots assigned to taxonomic analysis of diatom assemblages were prepared according to the procedure described in ANSP protocols [20]. After digestion of the formalin-fixed samples in boiling hydrogen peroxide (30% H_2O_2) and hydrochloric acid (35%), permanent slides were prepared by mounting the cleaned diatom frustules on a glass microscope slide in Naphrax© (Northern Biological Supplies Ltd, UK), a high refractive index (1.74) medium. At least 400 frustules were counted per replicate slide at 1,000× magnification and identified following the Süβwasserflora [21] classification and recent nomenclatural updates. Cell volumes were calculated for each species using a set of geometric shapes and mathematical equations for biovolume calculations [22] from literature linear dimensions.

Absolute abundances of each species, in individuals·cm^{-2}, were calculated after total diatom densities and relative abundances of each species identified during taxonomic determinations.

Calculation of Specific Growth Rates

For each experimental condition, growth rates (μmax, expressed as divisions·day^{-1}) and generation - doubling - times (G) of the most abundant diatoms from the three replicates were calculated according to Guillard [23]. Differences in growth rates between the diatoms collected over the two experiments were investigated among the species found in all the samples and representing at least 2% relative abundance.

The number of divisions in a day was calculated from the slope of the linear portion of a curve showing log cell number as a function of time: $\mu max = \dfrac{\log(x2/x1)}{d2 - d1}$, where x1: cell number at day d1 (cell·cm^{-2}) and x2: cell number at day d2 (cell·cm^{-2}). In this study, instead of the needed two countings, three or four dates (depending on the results obtained for μmax) were sampled during the exponential growth phase and regression coefficients were calculated by regression analysis of the logarithm of the cell counts against the time in days, in order to increase the precision of the results. From μmax, generation times (in hours) were computed by: $G = \frac{24}{\mu max}$.

Data Analyses

Correlations between growth rate and cell sizes were analyzed using linear regressions (MS-Excel) performed on log10 transformed biovolume data. *t*-statistics were used to test the regression coefficient (R^2) for significance, at the 0.01 α-level.

A coinertia analysis [24] was used to study the relationships between individual species and water physicochemical parameters. The coupling of both data matrices produced coinertia axes by projecting in a new factorial map the 10 environmental variables and the 20 species representing more than 1% relative abundance. Coinertia analysis and plots were performed using ADE-4 [25] package for the R statistical environment [26].

Results

Laboratory and Field Colonization Conditions

The physical and chemical parameters measured in the sampled waters for both experiments are shown in table 1.

pH (7.7 ± 0.1) and oxygen saturation (> 75%) were very comparable between treatments. The temperature imposed in the microcosms were quite stable (17.9 ± 0.8 °C) but fluctuated in the field from 2.7 to 13.9 °C during the experimental period. Of the four treatments, nutrient concentrations were quite similar but nitrate concentrations were markedly weaker at the upstream site. The conductivities were very different between experiments: in laboratory microcosms, values were around 200 $\mu S \cdot cm^{-1}$, whereas mean values exceeding 1500 $\mu S \cdot cm^{-1}$ were recorded in the field study.

Although biofilm settlement in the field experiment took place under higher cadmium water concentrations (two-fold), the contamination imposed in the treated experimental unit exceeded thresholds commonly used for diatom toxicity assays [5].

Growth Experiments in Field Conditions

146 species distributed over 45 genera were recorded in March 2005; however only 15 taxa were found in cumulative abundances reaching more than 1% of the community and allowed growth rate calculations. Taxonomic composition and relative abundances of diatom communities differed between sites: upstream assemblages were dominated by *Ulnaria ulna* (Nitzsch) Lange-Bertalot (23%), *Nitzschia palea* (21%), *Navicula gregaria* (11%) and

Gomphonema parvulum (7.5%), and by *Ulnaria ulna* (24%), *Surirella angusta* (12.5%) and *Nitzschia palea* (8.5%) downstream. The species *Eolimna minima* and *Encyonema minutum* (Hilse ex Rabhenhorst) D.G. Mann were also well represented downstream.

Table 1. Summary of water physical and chemical data at the 2 sites up- and downstream of the metal discharge and for the 2 experimental units (SE: standard errors, dl: detection limit: 1 μgCd·L^{-1} and 10 μgZn·L^{-1})

		Field study		Laboratory experiment	
		upstream	downstream	control	treated
pH	mean	7.7	7.7	7.6	7.7
	± SE	0.1	0.2	0.0	0.1
temperature (°C)	mean	7.7	6.5	18.8	17.2
	± SE	2.2	2.2	0.0	0.1
oxygen saturation (%)	mean	76.5	79.5	94.5	89.7
	± SE	4.9	4.2	0.5	0.5
conductivity (μS·cm^{-1})	mean	2546.0	1451.6	173.2	233.6
	± SE	533.1	210.3	2.8	8.0
dissolved cadmium (μg·L^{-1})	mean	< dl	18.2	< dl	9.1
	± SE		0.3		0.3
dissolved zinc (mg·L^{-1})	mean	< dl	1.6	-	-
	± SE		0.2		
nitrate (mg·L^{-1})	mean	4.8	27.0	25.0	26.3
	± SE	0.3	5.0	1.9	2.2
dissolved inorganic nitrogen	mean	2.1	8.5	5.9	6.3
(mg·L^{-1})	± SE	0.3	1.4	0.4	0.3
orthophosphate (mg·L^{-1})	mean	0.3	1.0	1.9	2.1
	± SE	0.0	0.3	0.3	0.2

Comparing the growth of the species from both sites showed a large spectrum of μmax values varying from 0.04 to 0.18 div·day^{-1} upstream, and from 0.13 to 0.31 div·day^{-1} downstream of the metal inputs (table 2). With the exception of three taxa – *Navicula veneta*, *Nitzschia acicularis* (Kützing) W. M. Smith and *Nitzschia linearis* (Agardh) W. M. Smith var. *tenuis* (W. Smith) Grunow – that did not grow downstream, all the major species remained dominant and presented faster growth rates under contaminated conditions. The species *Eolimna minima, Nitzschia dissipata* and *Surirella angusta,* that were almost absent upstream, were found in sufficient amounts to allow the calculation of growth rates in the samples collected downstream, values that were included in the range of growth constants found for the other species.

Allometry analyses revealed that growth rates were not always correlated with cell volume (figure 2). In upstream samples, growth rates correlated significantly with the log of cell size (μmax = 1.07VOL$^{-0.33}$; $R^2 = 0.61$; $p < 0.01$) but, in the metal-contaminated site, no significant correlation was observed ($R^2 = 0.039$; $p = 0.24$).

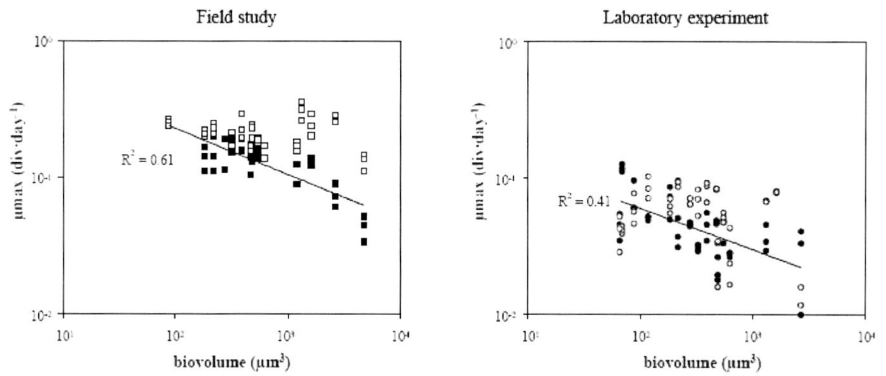

Figure 2. Relationships between growth rate and cell biovolume (logarithmic sizes) of the diatoms inventoried in the samples collected from the field (■: upstream, □: downstream) and in laboratory experiments (●: control, ○: treated).

Growth Experiments in Indoor Microcosms

Jaccard similarity index values (*ca.* 0.95) expressed very similar taxonomic composition between the diatoms inoculated in the microcosms and those sampled at week 1 in both experimental units. Richness in the systems was of 68 species representing 26 genera, but growth rates were assessed for 18 of the 20 main species (*i.e.* the ones reaching a relative abundance > 1% of the total raw species data) that were recorded almost at each date. Communities from both experimental units were characterized by an association of *Navicula gregaria* Donkin (mean relative abundances: 19%), *N. lanceolata* (Agardh) Ehrenberg,(18%), *Surirella brebissonii* Krammer and Lange-Bertalot (15%) *Nitzschia palea* (Kützing) W. Smith (11%) and *Nitzschia dissipata* (Kützing) Grunow (7%) at week 1, and then suffered a significant decrease ($p < 0.05$) in relative abundances. At the end of the experiment, control communities were dominated by *Navicula gregaria* (21%), *N. lanceolata* (15%) and *Nitzschia palea* (13%), whereas in the treated units they were represented by higher amounts of *Nitzschia palea* (20%), *Nitzschia acidoclinata* Lange-Bertalot (17%), *Eolimna minima* (Grunow) Lange-Bertalot (16%) and *Planothidium frequentissimum* (Lange-Bertalot) Lange-Bertalot (9%).

Most of the diatoms reached their exponential phase of growth in the first three weeks of experimentation. The specific growth rates (table 2) ranged from 0.02 to 0.12 div·day^{-1} in the control unit, and from 0.01 to 0.09 div·day^{-1} in the treated one. Growth rates were species-specifically influenced by cadmium contaminations. For some of the taxa studied (*Craticula molestiformis* (Hustedt) Lange-Bertalot, *Gomphonema parvulum* (Kützing) Kützing, *N. palea*, *Planothidium frequentissimum*, *P. lanceolatum* (Brebisson ex Kützing) Lange-Bertalot, *Surirella angusta* Kützing), μmax values increased significantly in response to metal exposure. Only *Navicula(dicta) seminulum* (Grunow) Lange Bertalot and, to a lesser extent, *Eolimna minima* and *Surirella brebissonii* showed a significant decrease in specific growth rates. During the course of the experiment, the species *Navicula gregaria*, *N. lanceolata* and *Surirella brebissonii* almost disappeared in the treated units. Generation times were below 9 days for the fastest growing species (*Navicula(dicta) seminulum* in the control unit).

Table 2. Individual biovolumes (in μm^3) and growth rate characteristics (μmax in div·day^{-1}, G in h) for species common to both field and laboratory assays. Calculations were based upon: [1] the first three dates, [2] the four dates, [3] the last three dates.

		Field study		Laboratory experiment	
		upstream	downstream	control	treated
Mayamaea atomus	biovolume	66			
var. permitis	mean μmax ± SE			0.04 ± 0.01[1]	0.04 ± 0.01[2]
	mean G ± SE	-	-	556.1 ± 88.7	624.6 ± 137.4
	R^2			0.61	0.62
Navicula(dicta)	biovolume	69			
seminulum	mean μmax ± SE			0.12 ± 0.00[1]	0.04 ± 0.00[2]
	mean G ± SE	-	-	204.7 ± 8.3	582.5 ± 33.8
	R^2			0.93	0.80
Eolimna minima	biovolume	88			
	mean μmax ± SE		0.25 ± 0.02[1]	0.07 ± 0.01[1]	0.06 ± 0.01[3]
	mean G ± SE	-	95.6 ± 5.8	349.3 ± 58.8	422.4 ± 75.9
	R^2		0.68	0.90	0.99
Craticula	biovolume	119			
molestiformis	mean μmax ± SE			0.05 ± 0.00[1]	0.09 ± 0.01[1]
	mean G ± SE	-	-	476.1 ± 11.1	288.5 ± 36.9
	R^2			0.86	0.99
Nitzschia pusilla	biovolume	185			
	mean μmax ± SE	0.14 ± 0.02[3]	0.21 ± 0.01[3]	0.07 ± 0.01[1]	0.06 ± 0.01[1]
	mean G ± SE	176.3 ± 24.9	115.8 ± 6.2	365.6 ± 76.3	389.5 ± 33.6
	R^2	0.86	0.90	0.77	0.80
Planothidium	biovolume	219			
frequentissimum	mean μmax ± SE	0.14 ± 0.01[2]	0.23 ± 0.02[1]	0.04 ± 0.01[2]	0.09 ± 0.00[1]
	mean G ± SE	178.9 ± 17.4	105.8 ± 8.5	637.5 ± 108.9	263.3 ± 10.8
	R^2	0.98	0.80	0.75	0.98
Navicula veneta	biovolume	279			
	mean μmax ± SE	0.15 ± 0.04[3]		0.05 ± 0.00[1]	0.07 ± 0.00[2]
	mean G ± SE	169.0 ± 43.4	-	525.3 ± 13.2	363.6 ± 15.3
	R^2	0.94		0.71	0.96
Gomphonema	biovolume	326			
parvulum	mean μmax ± SE	0.18± 0.02[3]	0.20 ± 0.02[1]	0.03 ± 0.00[2]	0.07 ± 0.01[1]
	mean G ± SE	136.0 ± 14.0	123.1 ± 10.5	791.3 ± 29.5	363.0 ± 59.7
	R^2	0.94	0.87	0.73	0.83
Nitzschia palea	biovolume	391			
	mean μmax ± SE	0.17 ± 0.02[1]	0.24 ± 0.03[1]	0.04 ± 0.01[2]	0.08 ± 0.01[1]
	mean G ± SE	144.2 ± 12.1	104.6 ± 14.3	555.1 ± 95.3	299.7 ± 39.1
	R^2	0.97	0.89	0.81	0.94
Planothidium	biovolume	475			
lanceolatum	mean μmax ± SE	0.14 ± 0.04[3]	0.17 ± 0.01[3]	0.05 ± 0.00[1]	0.08 ± 0.00[1]
	mean G ± SE	183.5 ± 47.0	144.4 ± 6.4	509.8 ± 9.7	292.3 ± 1.5
	R^2	0.88	0.92	0.96	0.99
Navicula gregaria	biovolume	485			
	mean μmax ± SE	0.14 ± 0.01[3]	0.22 ± 0.02[1]		
	mean G ± SE	170.3 ± 11.5	109.3 ± 9.5	-	-
	R^2	0.94	0.84		

Gomphonema micropus	biovolume	488			
	mean μmax ± SE mean G ± SE R²	-	-	0.02 ± 0.00[2] 1161.6 ± 156.4 0.85	0.03 ± 0.01[3] 983.6 ± 326.0 0.91
Nitzschia capitellata	biovolume	549			
	mean μmax ± SE mean G ± SE R²	0.15 ± 0.01[1] 211.3 ± 39.4 0.97	0.19 ± 0.01[2] 128.9 ± 3.9 0.83	0.04 ± 0.01[1] 577.6 ± 93.4 0.85	0.05 ± 0.00[1] 472.0 ± 36.1 0.99
Nitzschia dissipata	biovolume	625			
	mean μmax ± SE mean G ± SE R²	-	0.15 ± 0.01[3] 163.2 ± 14.0 0.68	0.03 ± 0.00[1] 877.2 ± 26.4 0.85	0.03 ± 0.01[1] 1009.6 ± 313.1 0.75
Navicula lanceolata	biovolume	1227			
	mean μmax ± SE mean G ± SE R²	0.11 ± 0.01[2] 219.1 ± 32.2 0.84	0.17 ± 0.01[3] 142.2 ± 8.1 0.80	-	-
Surirella angusta	biovolume	1315			
	mean μmax ± SE mean G ± SE R²	-	0.31 ± 0.03[1] 78.9 ± 8.7 0.89	0.04 ± 0.01[2] 685.0 ± 104.5 0.81	0.07 ± 0.00[1] 354.9 ± 1.6 0.93
Nitzschia linearis	biovolume	1624			
var. linearis	mean μmax ± SE mean G ± SE R²	0.13 ± 0.01[2] 184.0 ± 9.1 0.77	0.24 ± 0.03[1] 100.8 ± 12.7 0.87	-	0.08 ± 0.00[1] 303.7 ± 3.7 0.96
Surirella brebissonii	biovolume	2654			
	mean μmax ± SE mean G ± SE R²	0.07 ± 0.01[2] 331.4 ± 13.3 0.66	0.27 ± 0.01[1] 89.0 ± 3.9 0.80	0.03 ± 0.01[1] 1236.2 ± 708.8 0.72	0.01 ± 0.00[3] 4012.2 ± 2744.8 0.63
Ulnaria ulna	biovolume	4724			
	mean μmax ± SE mean G ± SE R²	0.04 ± 0.01[2] 574.1 ± 88.6 0.85	0.13 ± 0.01[1] 187.1 ± 18.4 0.80	-	-

Specific growth rates were also strongly correlated with cell biovolume (figure 2) under control conditions ($\mu max = 0.24 VOL^{-0.30}$; $R^2 = 0.41$, $p < 0.01$); the allometric relationship was not so meaningful but did express size dependence of the growth rates under metal contamination ($\mu max = 0.16 VOL^{-0.20}$; $R^2 = 0.10$; $p = 0.018$).

Comparison of Field with Laboratory Data

Both experiments (field and laboratory) revealed comparable diatom community species composition, with Jaccard similarity indices (determined on a dataset using the main species) between all the samples reaching at least 75%. The factorial map deduced from the coinertia analysis of the co-structure between species data and water physicochemical characteristics (figure 3a) revealed 4 separate groups. The diatom assemblage structure of the laboratory experiment was discriminated from field samples on the first coinertia axis, which explained 20.4% of the variability. The second axis, accounting for 5.0% of the variability, separated samples from Cd-polluted and non-polluted conditions, the samples from the Cd-treated units at the first two dates being associated with those of the control units. The superposition of the corresponding correlation circle of water physicochemical variables (figure 3b) onto the species factorial map (figure 3c) revealed that the discrimination of the laboratory samples was mainly explained by higher temperatures and orthophosphate concentrations, correlated with higher relative abundances of the species *Mayamaea atomus* var. *permitis* (Hustedt) Lange-Bertalot, *Navicula lanceolata*, *Navicula(dicta) seminulum*, *Nitzschia dissipata* and *N. pusilla*; higher nitrate concentrations in the Cd-treated unit were related with the presence of *Surirella brebissonii*, *Nitzschia acidoclinata* and *Eolimna minima* in higher amounts at weeks 4 and 6. On the factorial maps for species and environmental data, field samples were strongly correlated to higher conductivities upstream, inducing a shift of diatom assemblages towards elevated relative abundances of the species *Ulnaria ulna*, *Nitzschia linearis* and *Gomphonema parvulum*. Higher ammonia, dissolved oxygen and cadmium concentrations separated downstream samples, associated with *Surirella angusta*, *Encyonema minutum* and *Planothidium lanceolatum*, from upstream communities with larger proportions of *Nitzschia acicularis*.

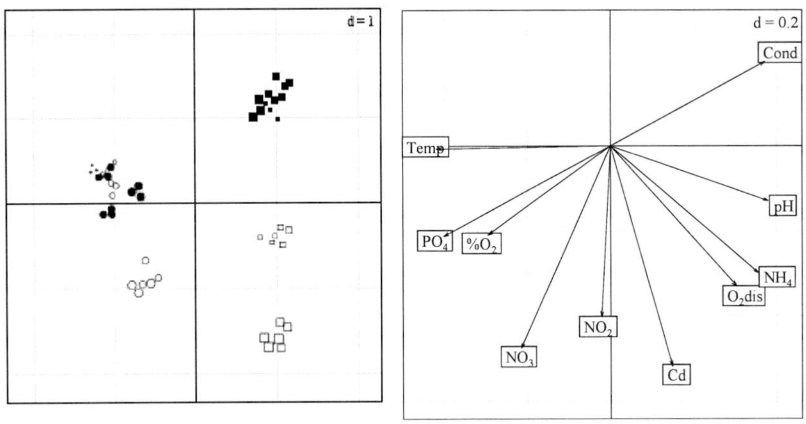

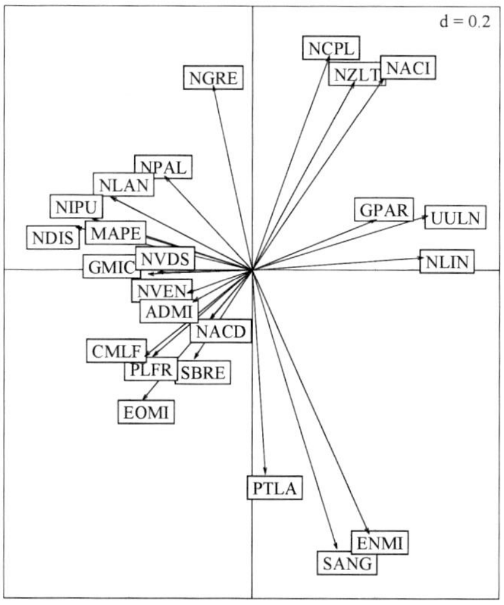

Figure 3. Coinertia factorial map of the samples (a), and first factorial plots of the physicochemical variables (b) and the species relative abundances (c). Labels corresponding to field study (■: upstream, □: downstream) and laboratory experiment (●: control, ○: treated) increase in size with duration of colonization. ADMI: *Achnanthidium minutissimum*, CMLF: *Craticula molestiformis*, ENMI: *Encyonema minutum*, EOMI: *Eolimna minima*, GMIC: *Gomphonema micropus*, GPAR: G. *parvulum*, MAPE: *Mayamaea atomus* var. *permitis*, NGRE: *Navicula gregaria*, NLAN: *N. lanceolata*, NVEN: *N. veneta*, NVDS: *Navicula(dicta) seminulum*, NACI: *Nitzschia acicularis*, NACD: *N. acidoclinata*, NCPL: *N. capitellata*, NDIS: *N. dissipata*, NLIN: *N. linearis*, NPAL: *N. palea*, NIPU: *N. pusilla*, PLFR: *Planothidium frequentissimum*, PTLA: *Planothidium lanceolatum*, SANG: *Surirella angusta*, SBRE: *S. brebissonii*, UULN: *Ulnaria ulna*.

Table 2 illustrates the variability in growth rates for experiments carried out in the field and in the laboratory: in the upstream site, field growth rates were 2- to 5-fold those found in laboratory conditions, and variations were higher when exposed to cadmium. However, variations in specific growth rates between treated (respectively downstream) and control (resp. upstream) conditions were examined through a plot of standardized variations calculated as follows: $\Delta\mu = \dfrac{\mu(\text{treated}) - \mu(\text{control})}{\mu(\text{control})}$ (figure 4). For five of the seven diatom species used, there was an increase in growth rates between metal-free and contaminated conditions in both experiments, and variation of growth characteristics and breadth of ranges were comparable between field and laboratory studies; nevertheless qualitative results for *Surirella brebissonii* and, with a less marked difference, *Nitzschia pusilla* were totally opposed: specific growth rates were found to be reduced by metal contamination in microcosm experiments and enhanced in field samples.

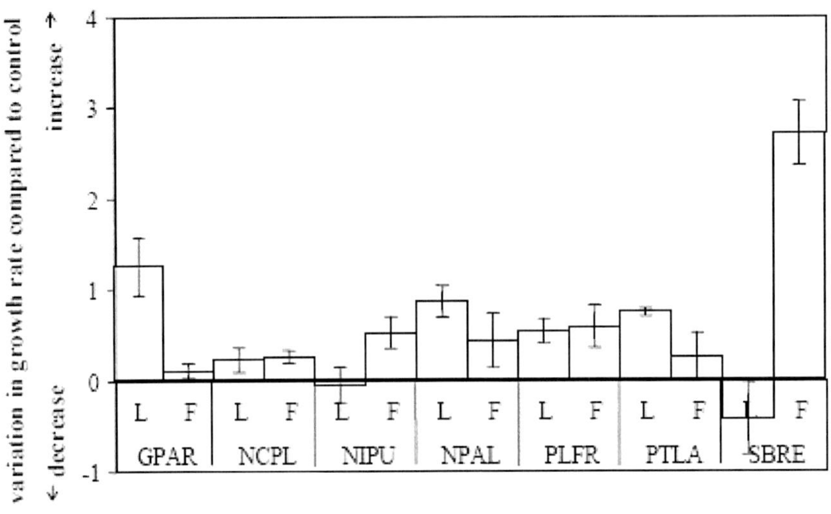

Figure 4. Growth rate variations between contaminated and control samples with their error bars for the taxa inventoried both in laboratory (L) and field (F) experiments. Codes: see species correspondence in Figure 3.

DISCUSSION

Community Responses to Metal Contamination

We determined the growth responses of 19 diatom species using natural community experiments in the field and the laboratory, and compared the kinetic data of the taxa common to both experiments. The findings are presented in table 2. As neither field nor laboratory conditions were nutrient-limiting, the main species found are αmeso- to polysaprobous species [3] and are fairly common in disturbed assemblages in this region [27, 28]. The patterns noted in the growth responses and the succession of diatom species under natural and artificial conditions without metal pollution were studied in a previous work [4]. They allowed us to propose certain ecological strategies, in which small-sized opportunistic species reproduce rapidly (e.g. *Craticula molestiformis, Eolimna minima, Mayamaea atomus* var. *permitis*) while slow-growing larger species settle on pioneer diatoms mats and utilize resources more efficiently (e.g. *Surirella angusta, S. brebissonii*).

The similar growth ability of many of the diatoms studied under cadmium contamination is consistent with the observations of Payne and Price [29] that growth rates of some diatom species maintain maximum values at relatively high cadmium concentrations. Many reports [30, 31] suggest antagonistic interactions of nutrients and metals; in the present study there appeared to be trade-offs in relative resource utilization among species exposed to cadmium, some of the competitive taxa (e.g. *Navicula gregaria, Nitzschia palea*) found in much (two-fold) lower relative abundances under metal contamination than others that were consequently probably less submitted to competitive exclusion than in non-contaminated treatments. However, due to species-specific differences, *in situ* and laboratory growth rates were not exclusively determined by cell size. Thus, the allometric relationships between growth constants and cell biovolumes (figure 2) were quite clearly observable in the treated experimental unit but this did not hold for the diatoms collected in the field where a substantial number of species did not respond with a growth-size relationship. However, for both field and laboratory experiments, we found evidence that metal contamination alters the quality of the allometric relationship.

Dominant species composition and responses led us to hypothesize that the species with good nutrient uptake abilities were also in some way stress-tolerant and, for this reason, the exposure to the metal did not cause a significant replacement of the sensitive algal species by more tolerant ones, as described by Blanck et al. [32]. The predominance of strong competitors for nutrients was sometimes observed in both metal-polluted and unpolluted sites [33], complicating data interpretation; this resilience was mostly explained by organic inputs, stimulating growth kinetics of some of the benthic organisms involved.

Responses of diatom communities to toxicants can be measured through different parameters, among which growth characteristics are not necessarily the best indicators. Species-specific responses of algal photosynthetic [34] or metabolic [35] activity have also been demonstrated, and can be valuably used, particularly through PICT [32]. However, even these methods are likely to be biased in such cases where saprophilous species are also found to be tolerant to toxic substances [36]. Whatever the method selected, under elevated resource availability, the toxic effects of metals is probably more weakly observable due to nutrient-stimulated growth. For this reason, it would be interesting to compare, in a complementary study, the responses of diatoms exposed to metals under oligotrophic conditions.

Sensitivity / Tolerance of Diatom Species

Most of the dominant species identified in this study have already been described as metal-tolerant. *Eolimna minima* was found to be very resistant to heavy metals in several reports in the literature [27, 37-39], as well as *Gomphonema parvulum* [40-42] and *Nitzschia palea* [41, 43, 44]. Szabó et al. [39] also reported the species *Navicula lanceolata*, *Nitzschia dissipata* and *Planothidium lanceolatum* in heavy metal polluted sites in Central Europe. The coinertia analysis (figure 3) however expressed differences in the specific structure of the communities from metal-polluted sites, with regards to duration of exposure, compared to non-polluted ones. Using cadmium concentrations and species relative abundances at the last two dates of both experiments (*i.e.* for mature communities), we demonstrated a positive correlation between dissolved cadmium concentrations and abundances of *Surirella angusta* ($R^2 = 0.54$, $p < 0.01$), *S. brebissonii* ($R^2 = 0.49$, $p < 0.01$), *Encyonema minutum* ($R^2 = 0.33$, $p < 0.01$) and *Planothidium lanceolatum* ($R^2 = 0.17$, $p = 0.04$), suggesting that these species are metal-tolerant. On the other hand, species like *Nitzschia acicularis* ($R^2 = 0.16$, $p = 0.05$) or *N. palea* ($R^2 = 0.20$, $p = 0.03$) were negatively correlated to cadmium concentrations and can be regarded as more sensitive to metal contamination although they were still present in the samples (but in lower proportions). The most sensitive species were those that grew in unpolluted conditions but almost totally disappeared in the contaminated samples, like *Navicula(dicta) seminulum* in the laboratory experiment (32% in the control unit vs. 1% in the treated one, which contrasts with other studies [33, 38, 45]) or *Nitzschia acicularis* in the field study (7% upstream vs. 0.5% downstream). The extinction of sensitive taxa therefore constitute a strong signal of metal contamination that could be used for biomonitoring purposes. There is thus a need for further studies in order to identify a number of highly sensitive species which could then be used as valid indicators for the assessment of metal pollution.

Comments on the Use of Microcosms and Artificial Substrates

A number of *in situ* [16, 33, 46, 47] and experimental [17, 18, 41, 48, 49] studies have underlined the impact of combined nutrients and metals on diatom community structures. In this study their effects on specific growth characteristics have also been demonstrated. The *in vitro* experiment was set up to describe biofilm development under constant nutrient load and variable metal concentrations using natural diatom assemblages and was intended to mimic the field conditions observed in March 2005. However, our results indicate that many "uncontrolled" factors still interact and affect µmax and G of the 19 diatoms studied. The discordance between field and laboratory results (table 2) suggest that the growth rates of algal species are not exclusively determined by cell size, and not related in the same way *in situ* as in laboratory cultures. The influence of various determinants, alone or in interaction, on algal growth rates in "control" conditions has been discussed elsewhere [4]. Variations in resource supply (continuous vs. sporadic inputs of N, P, Si) may modulate competitive availability [9, 12, 50, 51], and differences in many physical parameters (daily and seasonal fluctuations vs. stable conditions) like temperature [52], light [53], flow rate [54], etc. probably influenced diatom settlement and species growth responses. Whereas the diatom succession (variations in species composition and specific growth characteristics) may differ under field and laboratory conditions, it is likely that the species-composition trajectory was above all determined by immigration of diatom cells ([55] and figure 3a herein). The natural biofilm inoculated in the microcosms provided the only source (in terms of species likely to develop), and species like *Surirella angusta* or *Encyonema minutum*, which were both found to represent at least 0.5% of the relative abundances in the upstream site and in the inoculum, represented up to 21% and 4.5% downstream, but did not grow in the cadmium-treated microcosms. It can be hypothesized that this site may have benefited from continuous (although weak) inputs from upstream and from the Riou-Viou tributary (accounting for 3.5% and 5.4% of the relative abundances for the two above species in March 2005, unpublished data), allowing a marked development of these species downstream, whereas they were not available in sufficient amounts in the inoculum to settle in the treated experimental units. Another point is that species that may be quite sensitive to metals and even "excess" tolerant species transported by the flow may have colonized the substrates downstream, their settlement being favored by the mats that had already developed and possibly by the protective organic layer, whereas microcosms were only fed by the initial pool of species inoculated.

Our approach was to compare growth responses of diatom species under cadmium contamination, in natural biofilms grown *in situ* and *in vitro*. Laboratory-based experiments may be useful to focus on the particular impact of toxicants, since biofilm communities in nature are often difficult to investigate [56]. The present experiment displayed low variability between replicates in both field and laboratory studies; however the conditions within the experimental units were remote from environmental conditions, and lead to differences in diatom community structure and in individuals' growth kinetics. Nevertheless, the standardized variations in growth rates between untreated and treated conditions showed high profile similarity for most of the species. It can be argued that the unclearly identified differences between field and laboratory conditions that induced the divergences in growth rate did not modify some unidirectional processes of species growth after metal contamination, leading to a comparable final stable community. Microcosm studies may thus be useful to monitor the effects of disturbances on growth kinetics but require particular caution for data interpretation.

CONCLUSION

Diatom growth responses were found to be strongly size-dependant, but also influenced by metal contamination. Allometric relationships were observed under conditions without metal pollution, and an impairment of the relationships occurred for diatom communities exposed to Cd. The assemblages were dominated by saprophilous species, owing to the high nutrient availability in both studies. Moreover, these species were suspected to be tolerant to metals, efficiently using the resources and being more competitive in impacted environments; the changes in the allometry curves however underline the effects of Cd on diatom growth responses. Further investigations under nutrient-limited or low-resource conditions would be necessary to confirm these results and / or to determine the responses of oligosaprobous communities and individuals to metal contamination.

ACKNOWLEDGEMENTS

The authors would like to thank M. Bonnet, M. Boudigues, H. Bouillard, V. Duflo and T.T. Duong for their precious help during both experiments. They are grateful to the GEMA team, University Bordeaux 1, who analyzed cadmium concentrations in field water samples.

REFERENCES

[1] Lange-Bertalot H. Pollution tolerance of diatoms as a criterion for water quality estimation. *Nova Hedw.* 1979;64:285-304.

[2] Steinberg C, Schiefele S. Biological indication of trophy and pollution of running waters. Z Wasser Abwasser Forsch. 1988;21:227-34.

[3] van Dam H, Mertens A, Sinkeldam J. A coded checklist and ecological indicator values of freshwater diatoms from the Netherlands. *Neth. J. Aquat. Ecol.* 1994;28(1):117-33.

[4] Morin S, Coste M, Delmas F. A comparative study between laboratory and field assays using relative specific growth responses of periphytic diatoms. (submitted to Hydrobiologia)

[5] Conway HL, Williams SC. Sorption of cadmium and its effects on growth and the utilization of inorganic carbon and phosphorus of two freshwater diatoms. *Journal of Fisheries Research Board of Canada.* 1979;36(5):579-86.

[6] Fisher NS, Jones GJ, Nelson DM. Effects of copper and zinc on growth, morphology, and metabolism of *Asterionella japonica* (Cleve). *J. Exp. Mar. Biol. Ecol.* 1981;51:37-56.

[7] Gensemer RW. Role of aluminium and growth rate on changes in cell size and silica content of silica-limited populations of *Asterionella ralfsii* var. *americana* (Bacillariophyceae). *J. Phycol.* 1990;26(2):250-8.

[8] Ivorra N, Barranguet C, Jonker M, Kraak MHS, Admiraal W. Metal-induced tolerance in the freshwater microbenthic diatom *Gomphonema parvulum*. Environ Pollut 2002;116(1):147-57.

[9] Tilman D, Mattson M, Langer S. Competition and nutrient kinetics along a temperature gradient: An experimental test of a mechanistic approach to niche theory. Limnol Oceanogr. 1981;26(6):1020-33.

[10] Tilman D. Resource competition between planktonic algae: An experimental and theoretical approach. *Ecology.* 1977;58(2):338-48.

[11] Tilman D, Kiesling R, Sterner R, Kilham SS, Johnson FA. Green, bluegreen and diatom algae: taxonomic differences in competitive ability for phosphorus, silicon and nitrogen. *Arch. Hydrobiol.* 1986;106(4):473-85.

[12] van Donk E, Kilham SS. Temperature effects on silicon- and phosphorous-limited growth and competitive interactions among three diatoms. *J. Phycol.* 1990;26:40-50.

[13] Visser F, Gerringa LJA, van der Gaast SJ, de Baar HJW, Timmermans KR. The role of the reactivity and content of iron of aerosol dust on growth rates of two Antarctic diatom species. *J. Phycol.* 2003;39(6):1085-94.

[14] Tang EPY. The allometry of algal growth rates. *J. Plankton. Res.* 1995;17(6):1325-35.

[15] Kagami M, Urabe J. Phytoplankton growth rate as a function of cell size: an experimental test in Lake Biwa. *Limnology*. 2001;2(2):111-7.

[16] Morin S, Vivas-Nogues M, Duong TT, Boudou A, Coste M, Delmas F. Dynamics of benthic diatom colonization in a cadmium/zinc-polluted river (Riou-Mort, France). Fundam Appl Limnol. 2007;168(2):179-87.

[17] Gold C, Feurtet-Mazel A, Coste M, Boudou A. Effects of cadmium stress on periphytic diatom communities in indoor artificial streams. Freshwat Biol. 2003;48:316-28.

[18] Morin S, Duong TT, Herlory O, Feurtet-Mazel A, Coste M. Cadmium toxicity and bioaccumulation in freshwater biofilms. Arch Environ Contam Toxicol. 2008;54(2):173-86.

[19] Morin S, Duong TT, Dabrin A, Coynel A, Herlory O, Baudrimont M, Delmas F, Durrieu G, Schäfer J, Winterton P, Blanc G, Boudou A, Coste M. Long term survey of heavy metal pollution, biofilm contamination and diatom community structure in the Riou-Mort watershed, South West France. Environ Pollut. 151(3);532-42.

[20] Charles DF, Knowles C, Davies RS. Protocols for the analysis of algal samples collected as part of the U.S. Geological Survey National Water-Quality Assessment Program.: Patrick Center for Environmental Research, The Academy of Natural Sciences, Philadelphia, PA; 2002.

[21] Krammer K, Lange-Bertalot H. Bacillariophyceae 1. Teil: Naviculaceae. 876 p.; 2. Teil: Bacillariaceae, Epithemiaceae, Surirellaceae, 596 p.; 3. Teil: Centrales, Fragilariaceae, Eunotiaceae, 576 p.; 4. Teil: Achnanthaceae. Kritische Ergänzungen

zu *Navicula* (Lineolatae) und *Gomphonema*. 437 p. Stuttgart.: G. Fischer Verlag.; 1986 - 1991.

[22] Hillebrand H, Dürselen CD, Kirschtel D, Pollingher U, Zohary T. Biovolume calculation for pelagic and benthic microalgae. J Phycol. 1999;35(2):403-24.

[23] Guillard RRL. Division rates. In: Stein JR, editor. Culture methods and growth measurements Handbook of physiological methods: University Press, Cambridge; 1973. p. 290-311.

[24] Dolédec S, Chessel D. Co-inertia analysis: an alternative method for studying species-environment relationships. Freshwat Biol. 1994;31(3):277-94.

[25] Thioulouse J, Chessel D, Dolédec S, Olivier JM. ADE-4: A multivariate analysis and graphical display software. Stat Comput. 1997;7(1):75-83.

[26] Ihaka R, Gentleman R. R: A language for data analysis and graphics. *J. Comput. Graph. Statist.* 1996;5:299-314.

[27] Gold C, Feurtet-Mazel A, Coste M, Boudou A. Field transfer of periphytic diatom communities to assess short-term structural effects of metals (Cd,Zn) in rivers. Water Res. 2002;36(14):3654-64.

[28] Tison J, Park YS, Coste M, Wasson JG, Ector L, Rimet F, Delmas F. Typology of diatom communities and the influence of hydro-ecoregions: A study on the French hydrosystem scale. Water Res. 2005;39(14):3177-88.

[29] Payne CD, Price NM. Effects of cadmium toxicity on growth and elemental composition of marine phytoplankton. J. Phycol. 1999;35(2):293-302.

[30] Barranguet C, Plans M, van der Grinten E, Sinke JJ, Admiraal W. Development of photosynthetic biofilms affected by dissolved and sorbed copper in a eutrophic river. Environ. Toxicol. Chem. 2002;21(9):1955-65.

[31] Lozano RB, Pratt JR. Interaction of toxicants and communities - the role of nutrients. Environ Toxicol Chem. 1994;13(3):361-8.

[32] Blanck H, Wängberg SA, Molander S. Pollution-Induced Community Tolerance - A new ecotoxicological tool. Functional Testing of Aquatic Biota for estimating Hazards of Chemicals. Philadelphia: Cairns J. Jr, Pratt J.R. (Eds); 1988. p. 219-30.

[33] Ivorra N, Hettelaar J, Tubbing GMJ, Kraak MHS, Sabater S, Admiraal W. Translocation of microbenthic algal assemblages used for *in situ* analysis of metal pollution in rivers. Arch Environ Contam Toxicol. 1999;37(1):19-28.

[34] Radchenko IG, Il'yash LV. Growth and photosynthetic activity of diatom *Thalassiosira weissflogii* at decreasing salinity. *Biol. Bull.* 2006;33(3):242-7.

[35] Admiraal W, Barranguet C, van Beusekom SAM, Bleeker EAJ, van den Ende FP, van der Geest HG, Groenendijk D, Ivorra N, Kraak MHS, Stuijfzand SC. Linking ecological and ecotoxicological techniques to support river rehabilitation. *Chemosphere*. 2000;41(1-2):289-95.

[36] Guasch H, Admiraal W, Blanck H, Ivorra N, Lehmann V, Paulsson M, Real M, Sabater S. Use of lotic periphyton communities as indicators of sensitivity to certain toxicants. In: Prygiel J, Whitton BA, Bukowska J, editors. Use of algae for monitoring rivers III; 1999. p. 245-52.

[37] Feurtet-Mazel A, Gold C, Coste M, Boudou A. Study of periphytic diatom communities exposed to metallic contamination through complementary field and laboratory experiments. *J. Phys. IV*. 2003;107:467-70.

[38] Pérès F, Coste M, Ribeyre F, Ricard M, Boudou A. Effects of methylmercury and inorganic mercury on periphytic diatom communities in freshwater indoor microcosms. *J. Appl. Phycol.* 1997;9(3):215-27.

[39] Szabó K, Kiss KT, Taba G, Ács É. Epiphytic diatoms of the Tisza River, Kisköre Reservoir and some oxbows of the Tisza River after the cyanide and heavy metal pollution in 2000. *Acta Bot.Croat.* 2005;64(1):1-46.

[40] Admiraal W, Ivorra N, Jonker M, Bremer S, Barranguet C, Guasch H. Distribution of diatom species in a metal polluted Belgian-Dutch river : an experimental analysis. In: Prygiel J, Whitton BA, Bukowska J, editors. Use of algae for monitoring rivers III. Douai; 1999. p. 240-4.

[41] Gold C, Feurtet-Mazel A, Coste M, Boudou A. Impacts of Cd and Zn on the development of periphytic diatom communities in artificial streams located along a river pollution gradient. *Arch Environ. Contam. Toxicol.* 2003;44:189-97.

[42] Shehata SA, Lasheen MR, Kobbia IA, Ali GH. Toxic effect of certain metals mixture on some physiological and morphological characteristics of freshwater algae. Water Air Soil Pollut. 1999;110(1-2):119-35.

[43] De Filippis LF, Pallaghy CK. Heavy metals : Sources and Biological effects. In: Rai LC, Gaur JP, Soeder CJ, editors. Algae and water pollution. Stuttgart, Germany: E. Schweizerbart'sche Verlagsbuchhandlung; 1994. p. 31-77.

[44] Medley CN, Clements WH. Responses of diatom communities to heavy metals in streams: The influence of longitudinal variation. *Ecol. Appl.* 1998;8(3):631-44.

[45] Hirst H, Jüttner I, Ormerod SJ. Comparing the responses of diatoms and macroinvertebrates to metals in upland streams of Wales and Cornwall. Freshwat Biol. 2002;47(9):1752-65.

[46] Gomez N, Licursi M. Abnormal forms in *Pinnularia gibba* (Bacillariophyceae) in a polluted lowland stream from Argentina. *Nova Hedw.* 2003;77(3-4):389-98.

[47] Ivorra N, Bremer S, Guasch H, Kraak MHS, Admiraal W. Differences in the sensitivity of benthic microalgae to Zn and Cd regarding biofilm development and exposure history. *Environ. Toxicol. Chem.* 2000;19(5):1332-9.

[48] Interlandi SJ. Nutrient-toxicant interactions in natural and constructed phytoplankton communities: results of experiments in semi-continuous and batch culture. Aquat Toxicol. 2002;61(1-2):35-51.

[49] Ivorra N, Hettelaar J, Kraak MHS, Sabater S, Admiraal W. Responses of biofilms to combined nutrient and metal exposure. *Environ. Toxicol. Chem.* 2002;21(3):626-32.

[50] Kilham SS. Kinetics of silicon-limited growth in the freshwater diatom *Asterionella formosa*. *J. Phycol.* 1975;11(4):396-9.

[51] Tilman D. Mechanisms of plant competition for nutrients: the elements of a predictive theory of competition. In: Grace J, Tilman D, editors. Perspectives on Plant Competition. New York: Academic Press; 1990. p. 117-41.

[52] Montagnes DJS, Franklin DJ. Effect of temperature on diatom volume, growth rate, and carbon and nitrogen content: Reconsidering some paradigms. Limnol Oceanogr. 2001;46(8):2008-18.

[53] Nicklisch A. Growth and light absorption of some planktonic cyanobacteria, diatoms and Chlorophyceae under simulated natural light fluctuations. *J. Plankton. Res.* 1998;20(1):105-19.

[54] Ghosh M, Gaur JP. Current velocity and the establishment of stream algal periphyton communities. *Aquat. Bot.* 1998;60(1):1-10.

[55] Stevenson RJ, Peterson CG. Emigration and immigration can be important determinants of benthic diatom assemblages in streams. *Freshwat. Biol.* 1991;26(2):279-94.

[56] McCormick PV, Cairns J. Algae as indicators of environmental change. *J. Appl. Phycol.* 1994;6(5):509-26.

In: Heavy Metal Pollution
Editors: S. E. Brown and W. C. Welton
ISBN: 978-1-60456-899-8

Chapter 6

EFFECTS OF HEAVY METALS ON LABORATORY REARED *TIGRIOPUS FULVUS* FISCHER (COPEPODA: HARPACTICOIDA)

Luigi Pane[1], Gian Luigi Mariottini[1], Alessandra Lodi[2] and Elisabetta Giacco[1]

[1]Dipartimento di Biologia, Università di Genova, Viale Benedetto XV 5, I-16132, Genova, Italy
[2]Dipartimento di Ingegneria Chimica e di Processo "G.B.Bonino", Università di Genova Via all'Opera Pia 15, I-16145, Genova, Italy

ABSTRACT

Heavy metal pollution is a growing environmental problem in marine coastal waters; for this reason the monitoring of water pollution needs the availability of test-species able to furnish reliable and cheap results. From this point of view, copepods are important test organisms owing to their wide distribution and for their key-position within the food web. As *Tigriopus fulvus* seems to be a promising target-species in ecotoxicology, in this paper the toxicity of different concentrations of four heavy metals (chrome, zinc, cadmium, mercury) was evaluated. Treatments were performed after 24, 48 and 72 hours on nauplii and after 24, 48, 72 and 96 hours on adult *Tigriopus fulvus*; the percent survival and LC_{50} were evaluated and a method to obtain same-aged nauplii from detached egg sacs from ovigerous females is described. The results showed that mercury is the most dangerous metal for *T. fulvus* nauplii (LC_{50} 72hrs=0.39±0.02) and adults (LC_{50} 96hrs=0.34±0.13). Zinc and cadmium exhibited similar toxicity and chrome was the least toxic metal. All metals affected seriously egg sac and nauplii production of females with a 30% reduction of nauplii produced per egg sac; cadmium and mercury highly affected survival, reproduction and hatching of nauplii. On the whole, the data were repeatable and *Tigriopus fulvus* showed a satisfactory sensitivity to tested metals. Therefore, for its sensitivity and its biological and ecological characteristics, as well as for its adaptability to laboratory conditions and for its easy and cheap maintaining *Tigriopus fulvus* can be eligible as a test-species highly suitable in aquatic ecotoxicology.

INTRODUCTION

Trace metal pollution is a growing environmental problem in marine coastal waters, because heavy metals persist in the environment and can be toxic for living beings (Amiard *et al.*, 1987; Pinto *et al.*, 2003). Low concentrations of some metals are essential to organisms being enzyme components, but at high concentrations they can induce serious toxic phenomena; furthermore, for some of them, such as mercury and lead, a specific biological function is not known, thus they are toxic also at low concentrations. For these reasons the monitoring of water pollution needs the availability of test-species, able to furnish reliable and cheap results, to evaluate the action of pollutants at the individual and ecosystem level. From this point of view, harpacticoid copepods can be useful test-organisms owing to their wide distribution and key-position within the food web, and because they are easier to rear than pelagic copepods; thus they have been indicated to be suitable for toxicity trials (Hutchinson *et al.*, 1994; Miliou *et al.*, 2000). In particular, the Mediterranean harpacticoid *Tigriopus fulvus* seem to be a promising target-species in ecotoxicology (Todaro *et al.*, 2001; Faraponova *et al.*, 2003; Pane *et al.*, 2005). This species has been extensively studied from the biological (Carli and Fiori, 1977; Carli *et al.*, 1989a), biochemical (Carli *et al.*, 1989b; Pane *et al*, 2003) and ecological point of view (Carli *et al.*, 1993; Pane *et al.*, 1996, 2000). On this basis, aim of this study was to evaluate the influence of four metals (chromium, zinc, cadmium, mercury) on nauplii and adults of *T. fulvus* after short-term (24, 48, 72, 96hrs) and long-term (30 days) bioassays to assess their toxicity and to establish the LC_{50} values. Furthermore, a simple technique useful to obtain same-aged nauplii for ecotoxicity tests by detachment of egg sacs is here described.

MATERIALS AND METHODS

Adult *T. fulvus* were collected in the rockpools of Genova-Nervi and acclimatized in the laboratory for several generations. Specimens were maintained in plastic tanks containing 10L of 0.22 μm filtered Instant Ocean® (IO) artificial seawater at 18 PSU salinity and 18.0±0.5°C temperature, with light-dark cycle 12-12 hrs. Copepods were fed once a week by a suspension of cultured *Chlorella minutissima* and *Saccharomyces cerevisiae* cells. A sample of ovigerous females was isolated to obtain a suitable amount of same aged nauplii for toxicity tests. Hatching was stimulated by detaching egg sacs using fine needles after immobilization of females by soft filtration on GF/F Whatman filters (0.47 μm) leaving a thin water film to avoid any damage (Pane *et al.*, 2005). After detachment, egg sacs were transferred into 12-well plates in 4 ml seawater (18 PSU) and maintained at 18±0.5°C for 24 hours. Newborn nauplii (I-II stage) hatched by detached eggs and adult females were utilized to evaluate the acute toxicity of four heavy metal salts (chromium, zinc, cadmium and mercury). Nauplii were exposed to the following metal concentrations, chosen after preliminary trials: 10.0, 20.0, 30.0, 40.0 mg/L $CrCl_3$ (Sigma), 1.0, 2.0, 3.0 mg/L $ZnCl_2$ (Merck), 1.0, 2.0, 3.0, 4.0, 5.0 mg/L $CdCl_2$ (Sigma) and 0.3, 0.4, 0.5, 0.6, 0.7 mg/L $HgCl_2$ (Merck) diluted in IO. Eight replicates for each concentration and a control test were prepared, exposing 1819 nauplii in multiwell plates (Costar Corning) provided with polystyrene inserts (Millicell® PCF, polycarbonate membrane pore size 12μm). The organisms were not fed during the experiment. The survival was evaluated after 24, 48 and 72 hours exposition for each concentration.

Adult females were exposed to the following metal concentrations, chosen after preliminary trials: 75.0, 100.0, 125.0, 150.0 mg/L $CrCl_3$ (Sigma), 2.5, 5.0, 10.0, 20.0 mg/L $ZnCl_2$ (Merck), 2.5, 5.0, 10.0, 20.0 mg/L $CdCl_2$ (Sigma) and 0.19, 0.37, 0.75, 1.50 mg/L $HgCl_2$ (Merck) diluted in IO. Three replicates and a control test were prepared for each concentration, exposing 315 females in multiwell plates (Costar Corning) provided with polystyrene inserts (Netwell^{TM}, membrane mesh size 74 μm). The survival was evaluated after 24, 48, 72 and 96 hours exposition.

In order to evaluate the possible metal precipitation at the adopted conditions and consequently to know the effective metal concentration to which the organisms were exposed, blank tests without organisms were carried out at the same metal concentrations utilized in the experiments. Metal content in solution was determined by spectrophotometer (Varian mod. 240 FS) equipped with the specific lamp for each element.

LC_{50} values and the 95% lower and upper limits were calculated by the Trimmed Spearman Karber method (Hamilton *et al.*, 1977).

The chronic toxicity (survival and influence of metals on brood production) was evaluated exposing ovigerous females to sub-lethal concentrations (lower than the NOECs after 96 hours) calculated after the acute test: 5.0 mg/L $CrCl_3$, 0.4 mg/L $ZnCl_2$, 0.3 mg/L $CdCl_2$ and 0.2 mg/L $HgCl_2$. Twelve ovigerous females per each tested metal, against control test, were exposed for 30 days in multiwell plates (Costar Corning) provided with polystyrene inserts (Netwell^{TM}, membrane mesh size 74 μm). The specimens were fed with *Chlorella minutissima* and *Saccharomyces cerevisiae*; the solutions were changed every other day. The number of egg sacs and nauplii produced per female, the number of hatched nauplii per egg sac and female survival were evaluated. Results have been validated with a non-parametric test (r_s Spearman).

RESULTS

The blank tests (results not shown) evidenced that, at the experimental adopted conditions, only zinc and cadmium do not precipitate, whilst chromium and mercury concentrations decreased about 50% within the first 24 hours, reaching 96-98% of precipitated metals at the end of each test. Thus the effective metal concentrations at which organisms were exposed during the acute test were the following: 1.95, 3.90, 5.85, 7.80 mg Cr^{3+}/L; 0.47, 0.95, 1.41 mg Zn^{2+}/L; 0.61, 1.22, 1.83, 2.44, 3.05 mg Cd^{2+}/L and 0.22, 0.29, 0.37, 0.44, 0.52 mg Hg^{2+}/L for nauplii and 14.60, 19.50, 24.30, 29.25 mg Cr^{3+}/L; 1.17, 2.35, 4.70, 9.40 mg Zn^{2+}/L; 1.52, 3.05, 6.10, 11.20 mg Cd^{2+}/L and 0.14, 0.27, 0.55, 1.11 mg Hg^{2+}/L for adult females. For the chronic test the effective metal concentrations were 0.97 mg Cr^{3+}/L; 0.19 mg Zn^{2+}/L; 0.18 mg Cd^{2+}/L and 0.15 mg Hg^{2+}/L. Then, the results discussed afterwards have to be referred to the actual concentration for zinc and cadmium and to the nominal concentration for chromium and mercury.

Detached clear-orange egg sacs with evident naupliar eye of embryos yield viable *T. fulvus* nauplii; on the contrary, green-coloured immature egg sacs showed no or very low production. With this care, almost all detached egg sacs produced viable larvae; the average production was 13.61±3.51 viable nauplii/egg sac. Detaching of nearly mature egg sacs hastens the hatching and yields nauplii of the same age utilizable for experiments. Most of specimens born from detached egg sacs hatched rapidly (30-40 minutes) and some of them developed precociously. Figure 1 (a-d) shows the survival of nauplii as a function of metal concentration after the acute test. Mercury was the most toxic metal and no viable specimen was found after 48 hours when treated with 0.70 mg/L $HgCl_2$ and after 72 hours when treated with 0.60 mg/L $HgCl_2$; 3 mg/L $CdCl_2$ and 2 mg/L $ZnCl_2$ caused 100% mortality after 72 hours. A similar result was obtained with 4 mg/L $CdCl_2$ after 48 hours. Chromium induced minor effects and 100% mortality was observed at high doses (40 mg/L $CrCl_3$ after 24 hours; 30 mg/L after 72 hours). Nauplii were high sensitive to $HgCl_2$ (Table 1). The LC_{50} for $ZnCl_2$ resulted lower than that for $CdCl_2$ after 24 and 48 hours, but after 72 hours their values were similar. LC_{50} for $CrCl_3$ was always quite high.

Figure 1 (e-h) shows the survival of adult females as a function of metal concentration after the acute test. Mercury caused dramatic decrease of adult female survival and LC_{50} values for zinc and cadmium showed close results (table 2). Chromium was the least toxic metal and 100% mortality was never observed.

Table 1. LC_{50} (mg/L), lower (L.C.) and upper (U.C.) confidence limits after 24, 48 and 72 hours treatment of *T. fulvus* nauplii with given concentrations of $CrCl_3$, $ZnCl_2$, $CdCl_2$ and $HgCl_2$

Hours	Parameter	$CrCl_3$	$ZnCl_2$	$CdCl_2$	$HgCl_2$
24	LC_{50}	29.83 ± 1.28	2.66 ± 0.21	4.39 ± 0.30	0.57 ± 0.02
	95% L.C.	26.37 ± 1.69	2.14 ± 0.31	3.56 ± 0.52	0.53 ± 0.02
	95% U.C.	33.77 ± 1.28	3.10 ± 0.36	5.31 ± 0.88	0.61 ± 0.02
48	LC_{50}	24.12 ± 0.97	1.66 ± 0.17	2.24 ± 0.32	0.45 ± 0.02
	95% L.C.	22.43 ± 0.83	1.28 ± 0.14	1.49 ± 0.41	0.42 ± 0.02
	95% U.C.	26.68 ± 0.84	2.16 ± 0.27	3.43 ± 0.49	0.47 ± 0.02
72	LC_{50}	16.02 ± 2.57	0.95 ± 0.05	0.96 ± 0.05	0.39 ± 0.02
	95% L.C.	10.44 ± 2.07	0.87 ± 0.01	0.78 ± 0.10	0.36 ± 0.01
	95% U.C.	24.84 ± 4.92	1.08 ± 0.15	1.22 ± 0.31	0.43 ± 0.02

All considered parameters were affected by chronic exposition of *T. fulvus* females to sublethal concentrations of metals (Table 3): in particular, the number of produced nauplii and the number of nauplii per egg sac, as well as female survival, decreased remarkably after $HgCl_2$ and $CdCl_2$ treatment.

A noticeable decrease of brood production by females treated with $HgCl_2$ has been emphasized; a 7-fold decrease of produced nauplii, a 3-fold decrease of nauplii/egg-sac and a remarkable reduction of produced egg-sacs was observed. The results were statistically significant in comparison to the control, with $r_s = 0.253$ and $P = 0.417$ for egg sac production, and $r_s = -0.087$ and $P = 0.783$ for the production of nauplii. Almost a halved egg sac production in comparison to the control ($r_s = 0.339$; $P = 0.266$) and more than 4-fold lower number of born nauplii ($r_s = 0.306$; $P = 0.319$) was caused by $CdCl_2$. Furthermore, 16% mortality between the 7th and the 16th day treatment was observed in Hg- and Cd-treated females. The results for $ZnCl_2$ and $CrCl_3$ showed egg sac production was similar to controls

(r_s = 0.287, P = 0.354 for $ZnCl_2$; r_s = -0.012, P = 0.716 for $CrCl_3$), but the number of born nauplii was lower (r_s = 0.129, P = 0.667 for $ZnCl_2$; r_s = -0.477, P = 0.115 for $CrCl_3$).

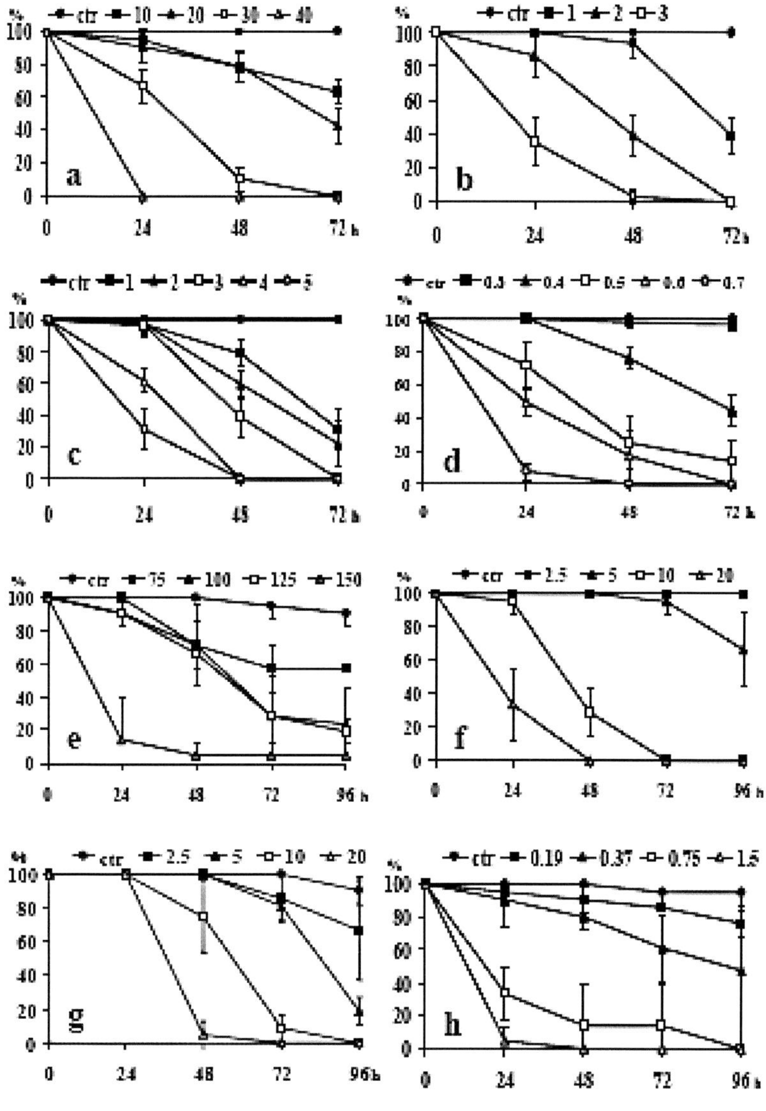

Figure 1. Percent survival (mean ± SD) of *Tigriopus fulvus* nauplii treated (mg/L) with $CrCl_3$ (a), $ZnCl_2$ (b), $CdCl_2$ (c), $HgCl_2$ (d), and adult females treated with $CrCl_3$ (e), $ZnCl_2$ (f), $CdCl_2$ (g), $HgCl_2$ (h).

Table 2. LC_{50} (mg/L), lower (L.C.) and upper (U.C.) confidence limits after 24, 48, 72 and 96 hours treatment of *T.fulvus* females with given concentrations of $CrCl_3$, $ZnCl_2$, $CdCl_2$ and $HgCl_2$. n.c.: not calculable

Hours	Parameter	$CrCl_3$	$ZnCl_2$	$CdCl_2$	$HgCl_2$
24	LC_{50}	137.02 ± 7.36	16.80 ± 2.78	n.c.	0.63 ± 0.13
	95% L.C.	125.91 ± 1.46	n.c.	n.c.	0.46 ± 0.12
	95% U.C.	149.31 ± 14.78	n.c.	n.c.	0.86 ± 0.13

Table 2. Continued

Hours	Parameter	CrCl3	ZnCl2	CdCl2	HgCl2
48	LC50	128.16 ± 2.06	8.65 ± 0.86	12.36 ± 1.82	0.52 ± 0.11
	95% L.C.	113.23 ± 1.56	6.88 ± 0.43	8.34 ± 0.37	0.37 ± 0.06
	95% U.C.	145.40 ± 4.53	10.88 ± 1.48	15.52 ± 2.39	0.75 ± 0.26
72	LC50	86.99 ± 12.19	6.85 ± 0.39	6.54 ± 0.66	0.47 ± 0.17
	95% L.C.	62.37 ± 16.31	n.c.	5.31 ± 0.67	0.32 ± 0.09
	95% U.C.	140.58 ± 9.84	n.c.	8.06 ± 0.61	0.69 ± 0.31
96	LC50	85.25 ±2.34	5.65 ± 0.83	3.32 ± 0.92	0.34 ± 0.13
	95% L.C.	63.93 ± 11.36	4.53 ± 0.83	2.81 ± 0.84	0.25 ± 0.09
	95% U.C.	116.93 ±27.17	7.07 ± 0.80	4.98 ± 0.69	0.46 ± 0.18

Table 3. Chronic exposition of *Tigriopus fulvus* females to 5.0 mg/L $CrCl_3$, 0.4 mg/L $ZnCl_2$, 0.3 mg/L $CdCl_2$ and 0.2 mg/L $HgCl_2$

	control	$CrCl_3$	$ZnCl_2$	$CdCl_2$	$HgCl_2$
exposed organisms	12	12	12	12	12
survival %	100	100	91.67	83.33	83.33
produced egg sacs (n)	7.0±0.7	5.7±0.6	6.4±1.4	4.2±1.1	2.8±0.7
produced nauplii (n)	96.4±14.0	44.4±11.7	41.8±14.5	22.5±7.8	13.0±2.9
nauplii/egg sac (n)	13.7±1.5	7.8±1.8	6.4±1.8	5.3±1.2	4.8±1.5
nauplii/female/day	3.2±0.5	1.5±0.4	1.4±0.5	0.7±0.2	0.4±0.1

On the whole, among non-essential metals $HgCl_2$ was more toxic than $CdCl_2$ and, among essential metals, $ZnCl_2$ was more toxic than $CrCl_3$. *T. fulvus* showed a good resistance to $CrCl_3$, resulting in a LC_{50} 30-fold greater than that for $HgCl_2$ after a 3-hour treatment. Zinc and cadmium showed similar results after 72 hours, while the former was more toxic after 24 and 48 hours for both larvae and adults; this is not easily explainable, because a biological function of cadmium is not known whereas zinc being an essential metal is involved in several metabolic processes. Anyhow, higher toxicity of essential metals in comparison to non-essential ones was reported (Barka *et al.*, 2001).

Barka et al. (2001) reported a lower resistance of adult *Tigriopus brevicornis* to $HgCl_2$ (0.053 mg/L LC_{50}) after 96 h treatment; as concerns cadmium, Todaro *et al.* (2001) reported the LC_{50} 48h was 4 mg/L for *T.fulvus*, while Faraponova *et al.* (2003) reported higher sensitivity after 24 and 48h treatment, but values similar to ours after 72 and 96h treatment. Our results are not in accordance with LC_{50} 96h from Forget *et al.* (1998) and Barka *et al.* (2001), who reported lower resistance of *T. brevicornis*.

CONCLUSION

Results here reported show that mercury and cadmium highly affect all considered reproductive parameters as well as survival. Zinc- and chromium-treated females showed no effect on egg sac production, but the number of born nauplii was highly reduced; this could demonstrate that even though these metals do not affect the reproduction rate, they highly influence egg development. It should be noted that obtained LC_{50} values were higher than the mean concentrations recorded in the natural living environment of *T. fulvus* (Pane *et al.*, 2001) and in surrounding coastal zones (Carli *et al.*, 1991).

To date growing metal amounts rise from anthropogenic sources; thus, this pollution seems to play an increasing environmental role (Pinto *et al.*, 2003). In addition, also metals occurring in air, soils and freshwaters can be carried to seawaters and drained into coastal zones where they persist and exert toxicity on coastal organisms; for this reason, considering also the bioaccumulation processes, the monitoring of such effects by toxicity tests is a useful tool to determine the effects of pollutants on the whole coastal ecosystem. Therefore, the availability of sensitive test-species is required to define pollutant tolerable levels. In this connection copepods, for their wide distribution and position in the food web, can be suitable test species. Available data indicate that copepods are more sensitive or as sensitive as other crustaceans to pollutants (Forget *et al.*, 1998); in particular, harpacticoids are easier to rear than pelagic calanoids such as *Acartia tonsa*, which is a well-known test-organism (ISO, 1999); owing to their sensitivity to pollutants they have been indicated to be useful bioindicators for the monitoring of pollutant effects (Bengtsson, 1978; Forget and Bocquené, 1999). Therefore, for its sensitivity and euryaline characteristics, small dimensions, relative easy finding during the whole year along several Mediterranean coasts, adaptability to laboratory conditions, reduced costs of maintaining and execution of tests, as well as for the simple obtaining of same-aged nauplii using the egg-sac detachment technique here described (Pane *et al.*, 2005), as already observed for *Tigriopus japonicus* (Kahan *et al.*, 1988), *Tigriopus fulvus* can be an eligible test-organism for toxicity trials in aquatic ecotoxicology.

REFERENCES

Amiard, J.C., Amiard-Triquet, C., Berthet, B. and Metayer, C. (1987). Comparative study of the patterns of bioaccumulation of essential (Cu, Zn) and non-essential (Cd, Pb) trace metals in various estuarine and coastal organisms. *Journal of Experimental Marine Biology and Ecology*, *106*, 73-89.

Barka, S., Pavillon, J.F. and Amiard, J.C. (2001). Influence of different essential and non-essential metals on MTLP levels in the copepod *Tigriopus brevicornis*. *Comparative Biochemistry and Physiology C*, *128*, 479-493.

Bengtsson, B.-E. (1978). Use of a harpacticoid copepod in toxicity tests. *Marine Pollution Bulletin*, *9*, 238-241.

Carli, A. and Fiori, M.A. (1977). Sviluppo larvale del *Tigriopus fulvus* Fischer. *Atti IX Congresso S.I.B.M.*, Lacco Ameno d'Ischia, 19-22/5/1977, 181-190.

Carli, A., Colacello, F. and Valente, T. (1983). Variazioni stagionali degli acidi grassi in popolamenti di *Tigriopus fulvus* delle pozze di scogliera della costa ligure (Copepoda Harpacticoida). *Atti della Società Italiana di Scienze Naturali e Museo Civico di Storia Naturale Milano*, *124*, 11-20.

Carli, A., Mariottini, G.L. and Pane, L. (1989a). Reproduction of the rockpools harpacticoid copepod *Tigriopus fulvus* (Fischer 1860), suitable for aquaculture. Deuxième Congrès International d'Aquariologie (1988), Monaco, *Bulletin de l'Institut Océanographique de Monaco, spécial 5*, 295-300.

Carli, A., Balestra, V., Pane, L. and Valente, T. (1989b). Rapporto di composizione percentuale degli acidi grassi nel *Tigriopus fulvus* delle pozze di scogliera della costa ligure (Copepoda Harpacticoida). *Bollettino della Società Italiana di Biologia Sperimentale*, *55*, 421-427.

Carli, A., Di Cintio, R. and Pane, L. (1991). Il plancton della baia di Riva Trigoso (Mar Ligure): presenza di alcuni elementi in traccia. *Bollettino della Società Italiana di Biologia Sperimentale. 67*, 423-430.

Carli, A., Pane, L., Casareto, L., Bertone, S. and Pruzzo, C. (1993). Occurrence of *Vibrio alginolyticus* in Ligurian coast rock pools (Tyrrhenian Sea, Italy) and its association with the copepod *Tigriopus fulvus* (Fischer 1860). *Applied Environmental Microbiology*, *59*, 1960-1962.

Faraponova, O., Todaro, M.A., Onorati, F. and Finoia, M.G. (2003). Sensibilità sesso ed età specifica di *Tigriopus fulvus* (Copepoda, Harpacticoida) nei confronti di due metalli pesanti (Cadmio e Rame). *Biologia Marina Mediterranea*, *10*, 679-681.

Forget, J. and Bocquené, G. (1999). Partial purification and enzymatic characterization of acetylcholinesterase from the intertidal marine copepod *Tigriopus brevicornis*. *Comparative Biochemistry and Physiology B*, *123*, 345-350.

Forget, J., Pavillon, J.F., Menasria, M.R. and Bocquené, G. (1998). Mortality and LC_{50} values for several stages of the marine copepod *Tigriopus brevicornis* (Muller) exposed to the metals arsenic and cadmium and the pesticides Atrazine, Carbofuran, Dichlorvos, and Malathion. *Ecotoxicology and Environmental Safety*, *40*, 239–244.

Hamilton, M.A., Russo, R.C. and Thurston, R.V. (1977). Trimmed Spearman-Karber method for estimating median lethal concentrations in toxicity bioassays. *Environmental Science and Technology*, *11*, 714-719; Correction 12, 417 (1978).

Hutchinson, T.H., Williams, T.D. and Eales, G.J. (1994). Toxicity of cadmium, hexavalent chromium and copper to marine fish larvae (*Cyprinodon variegatus*) and copepods (*Tisbe battagliai*). *Marine Environmental Research*, *38*, 275-290.

ISO 14669 (1999). Water quality. Determination of acute lethal toxicity to marine copepods (Copepoda, Crustacea).

Kahan, D., Berman, Y. and Bar-El, T. (1988). Maternal inhibition of hatching at high population densities in *Tigriopus japonicus* (Copepoda, Crustacea). *Biological Bulletin*, *174*, 139-144.

Miliou, H., Verriopoulos, G., Maroulis, D., Bouloukos, D. and Moraitou-Apostolopoulou, M. (2000). Influence of life history adaptations on the fidelity of laboratory bioassays for the impact of heavy metals (Co^{2+} and Cr^{6+}) on the tolerance and population dynamics of *Tisbe holothuriae*. *Marine Pollution Bulletin*, *40*, 352-359.

Pane, L., Feletti, M. and Carli, A.M. (1996). Fattori ambientali e fluttuazioni della popolazione del copepode *Tigriopus fulvus* delle pozze di scogliera di Genova-Nervi (Mar Ligure). *S.It.E. Atti*, *17*, 317-320.

Pane, L., De Nuccio, L., Pruzzo, C. and Carli, A. (2000). Adhesion of bacteria and diatoms to the exoskeleton of the harpacticoid copepod *Tigriopus fulvus* in culture: electron and epifluorescent microscope study. *Journal of Biological Research*, *76*, 37-43.

Pane, L., Capelli, R., Paganelli, L. and Carli, A.M. (2001). Inorganic mercury in the rock pool environment (Ligurian Sea). *Journal of Biological Research*, *77*, 7-12.

Pane, L., De Nuccio, L. and Franceschi, E. (2003). Contenuto energetico del copepode *Tigriopus fulvus* (Fischer 1860): applicazione della DSC (Differential Scanning Calorimetry). *Biologia Marina Mediterranea, 10*, 593-596.

Pane, L., Giacco, E. and Mariottini, G.L. (2005). Acute and chronic heavy metal bioassay on *Tigriopus fulvus* Fischer (Copepoda: Harpacticoida). *13th Symposium PRIMO 13. "Pollutant responses in marine organisms"*. Alessandria – Italy. June, 19-22 2005, 113.

Pinto, E., Sigaud-Kutner, T.C.S., Leitao, M.A.S., Okamoto, O.K., Morse, D. and Colepicolo, P. (2003). Heavy metal-induced oxidative stress in algae. *Journal of Phycology, 39*, 1008-1018.

Todaro, M.A., Faraponova, O., Onorati, F., Pellegrini, D. and Tongiorgi, P. (2001). *Tigriopus fulvus* (Copepoda, Harpacticoida) una possibile specie-target nella valutazione della tossicità dei fanghi portuali: ciclo vitale e prove tossicologiche preliminari. *Biologia Marina Mediterranea, 8*, 869-872.

In: Heavy Metal Pollution
Editors: S. E. Brown and W. C. Welton
ISBN: 978-1-60456-899-8

Chapter 7

SYNTHESIS OF ALKYL (LAURYL) POLY-GLUCOSIDE AND EFFECT OF TEMPERATURE, CATALYST AND TIME ON REACTION PRODUCT

Mu. Naushad* and Pallavi Jain
Department of Chemistry, Inst. of Sci. and Tech., SRM University, Modinagar, India

ABSTRACT

The goal of this study was to synthesize a new set of Lauryl poly-glucoside as an economic surfactant, which is produced exclusively with renewable sources having good dermatological properties and absence of eco-toxic properties. We have synthesized alkyl glucoside surfactant by direct glucosidation reaction by reacting glucose and n-dodecanol. The suitable conditions for the reaction were obtained, i.e., reaction temperature (120°C), mole ratio of catalyst (glucose to n-dodecanol 0.05:1:5) and reaction time (5 h). The yield of the product was found 54%.

Keywords: alkyl polyglucoside, carbohydrate, direct synthesis, oligomers

INTRODUCTION

Alkyl glucosides are the highly effective nonionic surface active agents, which are produced by the reaction of glucose with fatty alcohols. These surfactants are good emulsifiers, especially for polar molecules such as natural oils and fats and also gives good wetting and foam profile.

The alkyl glucosides have much number of applications in the cosmetic product sector, as auxiliaries in crop protection formulations and as surfactants in industrial cleansing agents. In combination with other surfactants, alkyl polyglucosides shows unique properties. For example, using alkyl polyglucosides in light duty detergent or shampoo formulation reduces

* palli24@yahoo.com, shad_chem@rediffmail.com

the total amount of surfactant without sacrificing any performance. During the washing process, alkyl polyglucoside in other combination, produce stable and fine foam, which protects sensitive textiles. Investigations regarding toxicity of alkyl glucosides in different toxicological and ecological laboratory produced favorable results. Alkyl glucosides are also completely biodegradable, both aerobically and anaerobically [1].

1.1. Alkyl Glucosides

Alkyl glucosides are the compounds derived by the replacement of the hydrogen atom of the cyclic hemiacetal by an organic radical [2].

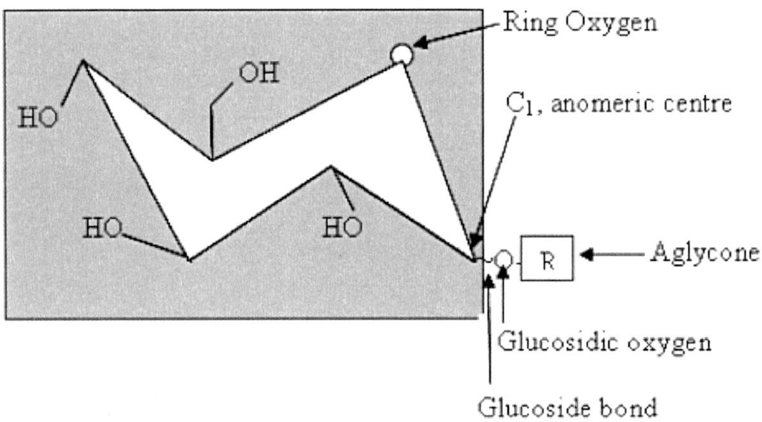

Glucose reacts intramolecularly, where in the carbonyl group combines with an alcohol group of the same glucose to form a hemiacetal. This yields a ring structure and a new chiral center at the original location of the carbonyl group. Glucosides are easy to form with acid catalyst but the reaction is reversible and for a good yield of glucoside, water must be removed as it is produced [3].

1.2. History of Alkyl Glucoside

Emil Fisher in 1893 synthesized and identified the first alkyl glucoside [4], by an acid catalyzed reaction of glucose with alcohols, which is known as Fisher glucosidation.

The products of Fisher glucosidation process are complex, mostly equilibrium mixtures of α/β anomers and pyranoside/furanoside isomers which also comprise randomly linked glucoside oligomers [5].

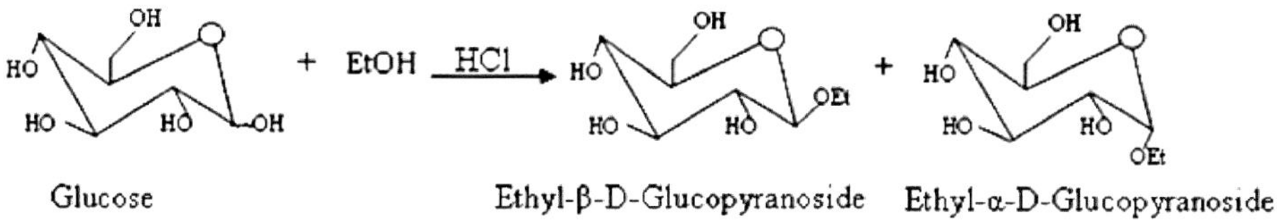

As a result individual molecular species are difficult to isolate from fisher reaction mixtures, so KOEINGS –KNORR SYNTHESIS adopted by Fisher [6]. In koeings –knorr reaction, reacting glycosyl halides with alcohols in the presence of silver oxide or silver carbonate, which is a stereo selective reaction, produces alkyl glucosides.

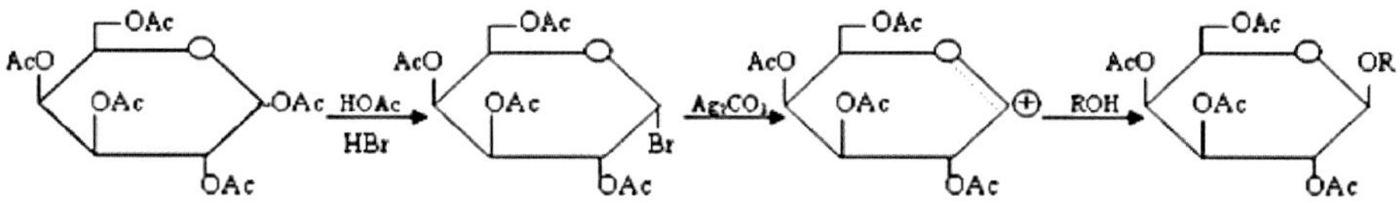

E. Fisher and B. Helferich in 1911 was the first to report the synthesis of long chain alkyl glucoside having surfactant properties [7] by using this stereo selective glucosidation process introduced by W.Koennigs and E.Knorr in 1901.

1.3. Chemistry of Glucosidation

- Protonation of the anomeric hydroxyl group.
- Expulsion of water, with aid from the ring oxygen atom
- Formation of an oxonium ion or carbocation.

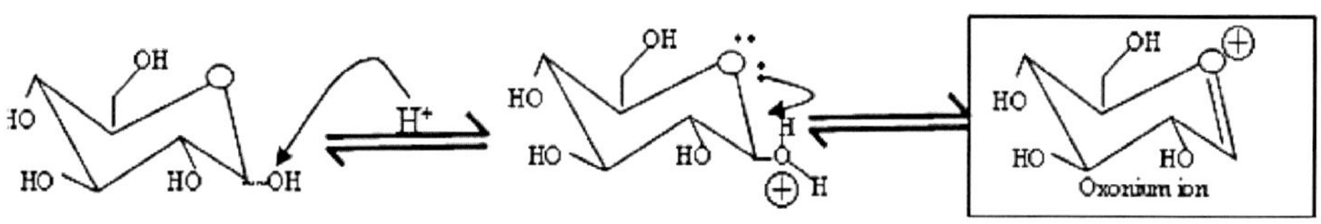

- Attack by the solvent on oxonium ion, say MeOH, from either the top β-face or the lower α-face.

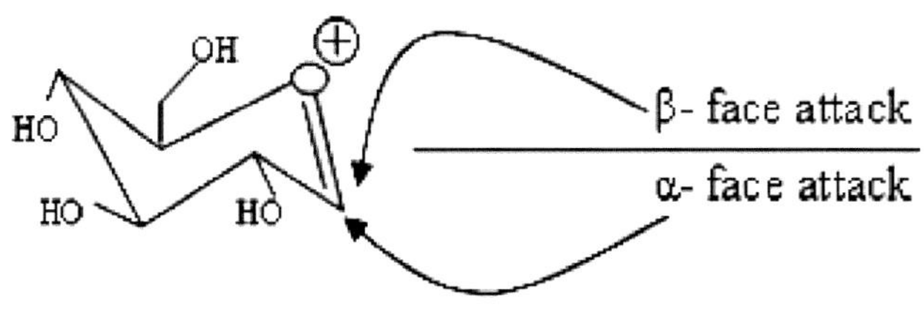

- Production of the protonated methyl glucoside
- Loss of hydronium ion to give the product(s).

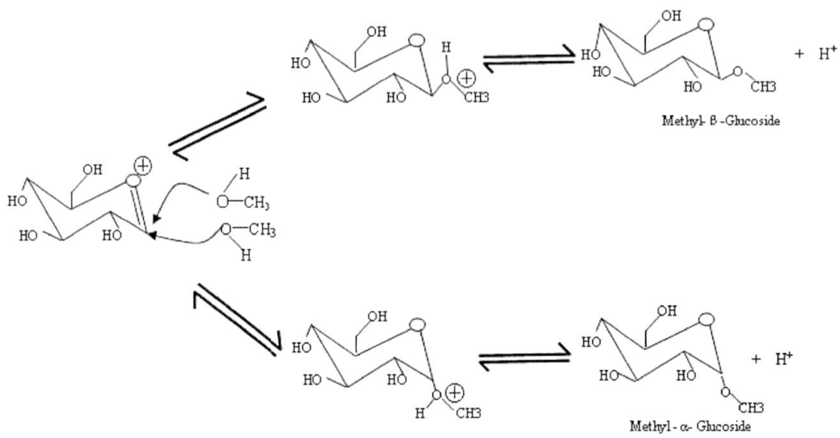

1.4. Raw Materials

In the synthesis of alkyl glucoside by glucosidation process, fatty alcohol and carbohydrate source are required in addition with the catalyst.

1.4.1. Fatty Alcohols

In alkyl glucoside synthesis, fatty alcohols are used to build the hydrophobic part of the molecule. Synthetic fatty alcohols from petrochemical sources or natural fatty alcohols from natural, renewable sources, such as fats and oils can be used in the process.

The natural fatty alcohols are obtained after transesterification and fractionation of fats and oils (triglycerides), which results in the formation of corresponding fatty acid methyl esters and subsequently hydrogenation.

1.4.2. Carbohydrate Source

In alkyl glucoside synthesis, carbodrates are used to build the hydrophilic part of the molecule. Both polymeric and monomeric carbohydrates are used as raw material.

Polymeric → starch or glucose syrups with low degradation level

Monomeric → water free glucose, glucose monohydrate (dextrose) or highly degraded glucose syrup.

1.5. Degree of Polymerisation

Degree of polymerization (DP) is reffered as the average number of glucose units linked to an alcohol group.

The acid-catalyzed fisher reaction results an oligomer mixture in which on average more than one glucose unit is attached to an alcohol molecule, because of the polyfunctionality of the carbohydrate partner.

Any change in the ratio of glucose to alcohol in the reaction mixture automatically changes the concentration of individual oligomers (mono-, di-, tri-glucoside). The average DP is an important characteristic with regard to the physical chemistry and application of alkyl glucosides. In an equilibrium distribution, the DP for a given alkyl chain length correlates well with basic product properties such as polarity, solubility, etc. The content of individual species in the oligomer mixture decreases with increasing degree of polymerization [8].

Thus a wide range of HLB can be obtained giving a wide range of properties. If the polar character predominates due to big hydrophilic head and the middle alkyl chain (8-14 atoms), glucosides are water-soluble and if lipophilic character predominates due to long alkyl group, emulsifiers are obtained.

1.6. Production of Alkyl Glucosides

Any two processes can produce alkyl glucosides by the Fisher synthesis by reaction of carbohydrate and alcohol, namely:

- Direct synthesis
- The transacetalization process

1.6.1. Direct Synthesis

- Involves simple plant equipment [9-11].
- One-stage process.
- Direct rection of carbohydrate with fatty alcohol to form long-chain alkyl glucoside/alkyl polyglucoside.

To minimize side reactions, the carbohydrate is often dried before the actual reaction, which takes place in the presence of water.

1.6.2. Transacetalization Process

- Involves more equipment than the direct synthesis.
- Two-stage process

Butanolysis is the first stage in which the carbohydrate reacts with a short chain alcohol (e.g., n-butanol or propylene glycol) and optionally depolymerizes.

Transacetalization is the second stage in which the short chain alkyl glucoside is transacetalized with a relatively long chain alcohol ($C_{12/14}$ OH) to form required alkyl glucoside. When oligo- and polyglucoses (e.g., starch) are used then transacetalization process is applied [12].

Secondary products such as polydextrose [14], ethers and colored impurities are formed in the acid catalyzed synthesis of alkyl glucosides. Polydextrose are formed in the course of the synthesis through the polymerization of glucoses and have undefined structure. The process parameters such as temperature, pressure, reaction time, catalyst etc. decides the type and concentration of the substance formed.

The loss of alkyl glucoside active substance in the reaction mixture correlates well with the formation of polydextrose, the reaction mixture in the case of overreacted systems gradually becoming heterogeneous again through precipitating polydextrose. Accordingly, product quality and product yield are critically influenced by the time at which the reaction is terminated. Starting with solid dextrose, alkyl glucosides low in secondary products are obtained, providing other polar constituents (polydextrose) are filtered off together with the remaining carbohydrate from reaction mixture that has not fully reacted [15,16].

1.6.4. Temperature

The reaction parameter temperature and pressure are closely related in the Fisher glucosidation.

- Low reaction temperatures (<100°C) → alkyl glucoside low in secondary products.
- High reaction temperatures (>100°C, typically 110-120°C) → changes in color of the carbohydrates.

The acetalization equilibrium is shifted to the product side by removing lower boiling reaction products (water in direct synthesis, short chain alcohols in the transacetalization process) from the reaction mixture. Removal of water from the reaction mixture minimizes secondary reaction (particularly the formation of polydextrose), which takes place in the presence of water.

1.6.5. Pressure

Typical reaction pressure in the transacetalization and direct synthesis variants are between 20-100 mbar.

1.6.6. Catalyst

Sulfuric acid, paratoluene- and alkyl benzene sulfonic acid and sulfosuccinic acid are the acids, which can be used in glucosidation process. The acidity and the concentration of the acid decide the reaction rate.

After the reaction, a suitable base is used to neutralize the acidic catalyst for e.g., NaOH or MgO_2 .The neutralized reaction mixture contains 50-80% fatty alcohol and is yellow in color. Greater molar ratio of carbohydrate to fatty alcohol leads to high fatty alcohol content. To obtain a specific DP for the technical alkyl glucosides, the ratio is adjusted. The excess fatty alcohol is removed by vacuum distillation. The end product after distillation is an almost pure alkyl glucoside.

The so formed alkyl glucoside, after removal of fatty alcohol is directly dissolved in water to obtain a highly viscous 50-70% alkyl glucoside paste.

In successive refining steps, this paste is worked up into an acceptable quality in accordance with performance related requirements. These refining steps may include bleaching of the product, the adjustment of product characteristic, such as pH value, active substance content and microbial stabilization.[17]

1.7. Analytical Chacterization of Alkyl Glucosides

H.Waldhoff, J.Scherler and M.Schmitt [18] were published analytical methods of alkyl polyglucosides and by other groups in 1995 and 1996 [19-22]. Commercial alkyl polyglucosides are complex mixtures of species, which differ mainly in the degree of polymerization and in the length of the alkyl chains.

Alkyl monoglucosides are the main group of components with a content of >50%, followed by the diglucosides and higher oligomers up to heptaglucosides. More highly glucosidated species are also present in small amounts. Species with a degree of glucosidation >5 are not normally determined in routine analysis because the amounts involved are too small. The most important analytical techniques routinely used for the characterization of main and trace components in commercial alkyl polyglucosides are high-performance liquid chromatography (HPLC), thin layer chromatography (TLC) and gas chromatography (GC).

1.7.1. Alkyl Polyglucoside Determined by GC

The high temoperature gas chromatography technique (HTGC) has proven to be suitable for the analysis of alkyl mono- and oligoglucosides. The high temoperature gas chromatography technique uses temperatures of up to 400°C, enables oligomeric alkyl polyglucosides up to the very high boiling heptaglucosides to be analyzed. In this technique the hydroxyl groups in alkyl polyglucosides have to be converted into silyl ethers before analysis to prevent sample decomposition.

1.7.2. Alkyl Polyglucoside Characterization by HPLC

Alkyl polyglucosides characterization by HPLC is regularly performed using an isocratic reversed -phase system. In most cases, no particular sample preparation is necessary; after dissolution in the eluent, the sample solution is filtered and directly injected into the system.

The retention time corresponds to the lipophilicity of the substances separated. The individual species are identified and quantified by the external standard method using commercially available alkyl glucosides for calibration. The alkyl monoglucosides are separated cleanly enough to allow sufficiently accurate quantification. Detailed determination of individual oligoglucosides and separation into α and β anomers, pyranosides and furanosides are only possible with a more polar mobile phase that requires tediously long reaction time. The analysis of alkyl glucoside in commercial alkyl polyglucoside products by HPLC provides good results for analytical tasks, which do not require high resolution of a broad spectrum of components. Typical application includes raw material identification, comparative alkyl polyglucoside analysis, quantifications and calculations solely on the basis of the alkyl monoglucoside contents.

1.7.3. Alkyl Polyglucosides Determined by TLC

TLC can separate the industrial alkyl polyglucosides mixture according to the alkyl chain length on RP_{18} reversed phase plate.

The substabces were made visible as black spots on TLC plates by spraying with dilute H_2SO_4 and heated subsequently. The separated substances identified by online coupling of TLC with secondry-ion mass spectrometry (SIMS). The solvent used is $CHCl_3$ / CH_3OH in a ratio of 9:1.

1.8. Physicochemical Properties of Alkyl Polyglucosides

1.8.1. Surface Tension

The surface tension of alkyl (poly) glucoside was investigated as a function of the alkyl chain and the degree of polymerization (DP) using samples of different in composition.

Shinoda et al.[23] and Boeker et al.[24] determined the CMC and the surface tension values at the CMC from surface tension/concentration curves. A few selected values are set out in the table. [25].

Table. CMC values from surface tension measurements

Substance	β-Database-C_8G_1	β-C_8G_1	β-Database-$C_{16}G_1$	β-Database-$C_{12}G_1$
Temp. (^{0}C)	25	25	25	25
CMC (mol/L)	$2.5*10^{-2}$	$1.8*10^{-2}$	$2.2*10^{-3}$	$1.9*10^{-4}$
References	12	13	12	12

The CMC values of pure alkyl polyglucosides and the technical alkyl polyglucosides are comparable with those of typical nonionics surfactants and decrease distinctly with increasing alkyl chain length. As the figures set out in the table show, the alkyl chain length has a far stronger influence on the CMC by comparison with the number of glucoside groups of the alkyl polyglucosides.

1.8.2. Phase Behavior

The nonionic class of alkyl polyglucosides differs from fatty alcohol ethoxylates by its characteristic structure, which considerably effects the association of molecules in solution, the phase behavior and interfacial activity. The hydrophilic head of the glucose ring, associated with the surrounding water molecules by hydrogen bonds, is rather more as compared to the alkyl chain. However, the hydration is low as compared to fatty alcohol ethoxylates. Therefore, basic phenomena like cloud point, thermal phase inversion, or gel formation in medium concentration of pure solutions cannot be observed in case of alkyl polyglucosides.

The phase behavior of alkyl polyglucosides in aqueous solution gives important hints for the handling and formulation of a product in regard to viscosity, flow behavior and phase stability.

1.8.3. Microemulsion Phases

The solubilization of comparably small amounts of oil components in rinse and shampoo formulations demonstrates the basic emulsion properties which alkyl polyglucosides should be expected to show as nonionic surfactants [25].

In general, the interfacial activity of alkyl polyglucosides is determined by the carbon chain length and, to a lesser extent, by the degree of polymerization (DP). Interfacial activity increases with the alkyl chain length and is at its highest near or above the CMC with a value below 1mN/m. At the water/mineral oil interface, $C_{12/14}$-APG shows lower surface tension than $C_{12/14}$ alkyl sulfate [26]. Medium-chain alkyl polyglucosides may be used as emulsifiers for o/w emulsions in combination with hydrophobic coemulsifiers [27].

Alkyl polyglucosides differ from ethoxylated nonionic surfactants in that they do not undergo temperature-induced phase inversion from oil-in-water (o/w) to water-in-oil (w/o) emulsions. Instead, their hydrophilic/lipophilic properties can be balanced by mixing with a emulsifier, such as glycerol monooleate (GMO) or sorbitan monolaurate (SML).

1.9. Applications of Alkyl Polyglucoside

- $C_{12/14}$ alkyl polyglucosides and DP 1.4 are particularly useful in manual dishwashing detergents (MDDs).[28]
- $C_{8/10}$ APG and DP of about 1.5 are particularly useful in the formation of all-purpose and specialty cleaners. Today, alkyl polyglucoside containing products are found both in all-purpose cleaners and in special cleaners, such as bathroom cleaners, toilet cleaners, window cleaners, kitchen cleaners and floor care products [29]. Alkyl polyglucoside themselves have an excellent cleaning performance which can be determined, for e.g., in accordance with the IPP quality standard [30].
- $C_{12/14}$ alkyl polyglucosides and DP 1.4 are used in laundry detergent formulation. Laundry detergents are the products which is used daily to bring soiled clothing back into the state fit for use. These are marketed in various forms for example powder, paste or liquid detergents whose particular formulation determined by the type of soil, by consumer requirements and by the textile and its washing instruction [31].
- In the cosmetic field alkyl polyglucosides represent a new class of surfactants, which combine properties of conventional nonionics and anionics. By far the largest proportion of commercial products is represented by C_{8-14} alkyl polyglucosides for cosmetic cleansing formulations [32].
- APGs also used in soaps, shampoos, shower bath, body bath and syndets.
- Because of their excellent skin compatibility, they are attractive alternative to conventional syndet formulations based on alkyl sulfates.
- Alkyl polyglucoside/fatty alcohol sulfate combinations show improved mildness toward oral mucous membrane and produce rich foam.
- $C_{12/14}$ –APG was found to be an effective booster for special antibacterial agents, such as chlorohexidine.
- In the presence of APG, the quantity of bacterial agent can be reduced to 1/4 without loosing their bacterial activity.
- In permanent-wave formulations, the use of alkyl polyglucosides provides for an improved permanent-wave effect and at the same time reduces hair damage.

1.10. Toxicology of Alkyl Polyglucosides

On the basis of the information available, alkyl polyglucosides are not considered as toxic or harmful in acute toxic tests, but in high concentrations have to be classified as irritating to the skin and eyes. In addition, sensitizing effects are unlikely to occur.

1.11. Ecological Properties of Alkyl Polyglucosides

Alkyl polyglucosides have been extensively investigated with regard to their fate and effect in the environment [33, 34] and the conclusion is that even under unfavorable conditions, no environmental risk is involved in the use of alkyl polyglucosides.

2. Review of Literature

A lot of work has been done to synthesize and characterize the alkyl glucoside/ polyglucoside. A brief review of the work done is given below:

2.1. Preparation

Various technologists have worked on preparing the different alkyl glucoside/polyglucosides or other detergent compositions containing alkyl polyglucosides by studying the various aspects of the detergent components. Direct synthesis and transacetalization process are the most satisfactory method for the preparation of alkyl glucoside/alkyl polyglucoside, and much of the technical literature deals with innovations with and without solvents and with reaction conditions designed for efficient conversions. A few preparations are given here:

2.1.1. Chemical Synthesis

Lueders et al. [35] prepared alkyl oligoglucoside surfactants with bright color from saccharide and alcohols by glucosidation or transglucosidation in the presence of inorganic acid (e.g., $Na_5P_3O_{10}$ or NaF) or a polycarboxylic acid(e.g., malonicacid or EDTA) or its salt or esters, optionally treating an alcohol solution of the product with an adsorbent such as active carbon, and after distilling the alcohol; bleaching the product with a peroxide,esp.H_2O_2.

M.Biermann et al. [36] prepared surface-active glucosides that have good color stability in alkaline media, have a high content of alkyl mono glucosides and preferably contain polysaccharide or polysaccharide and butyl glucoside. The preparation involves the use of 4-10 mol BuOH/mol saccharide (esp. glucose). A reactor is charged with ~50% of the BuOH and an acid catalyst (e.g., p-toluene sulfonic acid), the mixture is heated to reflux temperature and a suspension of glucose in the remainder of the BuOH is added slowly to maintain a clear reaction mixture while a BuOH-water mixture is distilled off. The preheated fatty alcohol (2-20 mol/mol glucose) is added slowly with vacuum distillation of BuOH to give a reaction mixture containing ≤30-mol% residual butyl glucoside. The mixture is cooled to <95°C, neutralized to pH ≥ 8, filtered and distilled to remove excess fatty alcohol giving a surfactant containing <5% fatty alcohol.

Lueders et al. [37] prepared alkyl glucoside surfactants with bright color by treating the starting lower alkyl glucoside in alcohol solution with activated carbon, distillating the solvent from the product, and bleached the distilled residue or its aqueous solution with peroxide. A mixture of butyl oligosaccharide in BuOH was prepared, mixed with Alfol 1012 and a BuOH-H_2SO_4 mixture; and heated at 110°C in vacuum with distillation of BuOH to give a solution of C_{10-14} alkyl oligoglucosides, which were neutralized with NaOH at 80°C, stirred 1 hr at 100°C with activated carbon, filtered, heated at 150°C/1-2 mbar with removal of alcohols; mixed with 50% water, and heated (100g) with 6.7 mL 30% H_2O_2 solution at 80°C for 1 hr.

Lueders et al. [38] prepared alkyl oligoglucoside surfactants with bright color from saccharide and alcohols by glucosidation or transglucosidation in the presence of a polymeric complexing agent having HLV value 10-20, such as polyethylene glycol, a cyclic ether, or poly(vinyl pyrrolidinione),treatment of an alcohol solution of the product with an adsorbent such as active carbon, and, after distilling the alcohol; bleaching the product with a peroxide, esp. H_2O_2.

K.Hill et al. [39] prepared C_{12-18} alkyl glucoside with high space time yield and high content of C_{12-18} alkyl monoglucoside, low content of butyl glucoside, and, good color stability in alkaline medium by adding an acidic solution of butyl glucoside in butyl alcohol to a preheated (100-125°C) mixtures of C_{12-18} primary alcohol containing an acid catalyst with distillation of butyl alcohol.

W.Wuest et al. [40] prepared surface active glucosides from starch or starch degradation products such as glucose syrup by reaction with a lower alcohol, esp. butyl alcohol, and transacetalization with C_{8-22} alcohols (esp. natural) in surfactant second step. The glucosides have good brightness, color stability, and alkali resistance.

Stefen et al. [41] introduced alcohol process for the work-up of fatty alcoholic solutions of alkyl polyglucosides. The title process involved the adjusting of C_{8-24} alcohol solutions of the C_{8-24} alkyl polyglucosides to pH 10-14,maintained the solutions at 60-120°C for atleast 2hr,and vacuum distilled the alcohols at 7-9. E.g., 147.4kg fatty alcohols (containing n-dodecanol 67%, n-tetradecanol 27%, and n-hexadecanol 6%), 18kg anhydrous glucose and 1.29kg p-$MeC_6H_4SO_3H$. H2O were reacted at199° for 6hr, the reaction mixture was cooled to 70°,and adjusted to pH 11 with 25% aqueous NaOH. The mixture was maintained at 70° for 24hr, allowed to settle for 12hr (pH remaining at9), and then filtered. The fatty alcohols were distilled; he residual alkyl polyglucoside was dissolved in water and treated with H_2O_2 at 80° and pH 7; the solution was stirred at 80° for 6hr;the water quantity was so chosen that the for-run of the bleaching process would result in alcohol 50% aqueous solution; the iodine color index of this solution was 7. The surfactants thus obtained result better color and color stability.

H. Rossmaier et al. [42] introduced a method for the preparation of surface-active glucosides. In this process starch or partially degraded starch was added to a preheated (100-300°C) reaction medium containing propylene glycol and acidic catalyst and optionally C_{8-18} alkanols for transacetalization of fatty alcs. The fatty alc./ propylene glycol molar ratio is 1.The process gives fatty alkyl glucosides having a high content of C_{8-18} alkyl monoglucoside and showing good color stability in alkaline media propylene glycol and fatty alc. were recovered by distillation for recycle.

H. Sawada et al. [43] manufactured alkyl glucosides with good color and fragrance for nonirritating surfactants by treating sugars with higher alcohols or with lower alcohols and then with higher alcohols, bleaching the resulting glucosides with H_2O_2, and contacting with ≥1 compound selected from MnO_2, Pt-group metals, peroxidase, and ascorbic acid or its salts. Thus stirring decyl alcohol 72.0, glucose 18.0, and p-$MeC_6H_4SO_3H.H_2O$ 0.5mol at 95°C and 40mm Hg and neutralization with NaOH gave 4270g glucoside, a 50% reddish black aqueous solution of which was adjusted to pH 9, stirred with aqueous H_2O_2 at 45°C for 30min.,mixed with MnO_2 at room temperature for 3hr and the solution was filtered to give aqueous glucoside with residual H_2O_2 0%.Gardner color no. 2 and no bad odor for ≥120hr, vs. 0.14, 4, and aldehyde odor, resp., without MnO_2 treatment.

C.Pickens et al. [44] prepared C_{12-14} alkyl glucoside from 1 part of anhydrous dextrose and 6 part C_{12-14} alcohols in the presence of p-$MeC_6H_4SO_3H$,was neutralized. In the manufacture of glucoside surfactant, undesired hydrophilic materials such as polysaccharide byproducts are removed from the crude C_{12-14} alkyl glucoside containing unreacted hydrophobic alcohols by mixing with 2% water and heating 30 min. at 105°C and the resulting aqueous phase containing hydrophilic materials was separated. This process reduced the concentration of dextrose and polydextrose in the product from 9.76% to 3.60%.

H.Kamiya et al. [45] prepared glucoside $Y(K)_n$ (Y=residue of fructofurranose, fructopyranose or glucopyranose; K=OCH2CH(OH)CH2OR, OCH(CH2OH)CH2OR, OCH2CHROHOCHRCH2OH; R=C6-24 alkyl or alkenyl; and foam stability, useful as detergents, emulsifiers, dispersants, wetting agents, solubilizers etc by hydrolysis of ZK_m (Z=sucrose residue;m=1-8) in the presence of acid catalyst. Sucrose was treated with glycidyl decyl ether in DMSO in the presence of NaOH to give sucrose decyloxyglyceryl ether, which was hydrolyzed by aqueous HCl in isopropyl alcohol to give glucosides.

G.Borsotti et al. [46] used the quinoline salt of 2, 4, 6 trimethyl benzene sulfonic acid, glucosidation catalyst comprising a strong organic acid and weak organic base having ka between 10^{-8} and 10^{-1} for the preparation of APGs having a high concentration of alkyl monoglucoside and a light color, by reacting 1-dodecanol with glucose, giving a product containing 75% dodecyl monoglucoside and 14.8%dodecyl glucoside.

G Wolf et al. [47] prepared alkyl glucoside for use as surfactant, esp. in detergent compositions by reacting reducing saccharides such as glucose and galactose with a dialkyl sulfate such as didecyl sulfate.

M.Weuthen et al. [48] prepared surface active glucosides by acetalization of saccharides such as glucose with fatty alcohols and used sulfo-carboxy compounds containing 2-8 C (e.g., sulfoacetic acid, sulfopropionic acid, or sulfobenzoic acid) as catalyst.

M.Gibson et al. [49] introduced a process for the preparation of alkyl glucoside surfactants. In the title process, a slurry of a hydrous saccharide source (e.g., dextrose monohydrate) in a first portion of an aliphatic alcohol (esp. C_{7-22} alcohol) is added to a second portion of he alcohol maintained at an elevated temperature and an acid catalyst is added and the aliphatic alcohol is reacted with the saccharide source to form the glucoside.

P.Schulz et al. [50] prepared lower alkyl glucoside having a bright color and a low content of polyglucose and other byproducts by using an in-line mixer for the addition of an aqueous glucose syrup to a hot mixture of a lower alcohol (esp. BuOH) and an acid catalyst (e.g., p-toulene sulfonic acid) while removing water by azeotropic distillation and continuing the reaction after addition of the syrup was complete. The syrup was dispersed uniformly in the alcohol and does not adhere to reactor surfaces.

G.Wolf et al. [51] prepared good color and low content of alkyl oligo-and polyglucosides from a 10-80% aq. glucose solution and C_{8-30} primary aliphatic alcohols by using 2-10mol alcohol/mol glucose, using an acidic emulsifier (e.g., dodecylbenzene sulfonic acid) as the catalyst, using an alkyl glucoside as an additional emulsifier, heating the aq. glucose solution to 50-90°C before emulsification with the alcohol-catalyst-emulsifier mixture, carrying out the glucosidation at 100-150°C/10-100mbar with distillation of H_2O,adding base to give pH 8-10, distilling excess alcohol at 0.01-10mbar, converting the product to an aq. paste containing 30-70% alkyl glucoside, and bleaching with a peroxygen bleach.

P.Schulz et al. [52] prepared alkyl glucoside, showing good detergency and useful as surfactants in laundry detergents etc by reacting 1mol sugar having 5-6 carbon (e.g., glucose) with 6-20mol C_{4-11} alkanol and/or alkenol (e.g., C_{8-10} alkanol) in the presence of acid catalyst (e.g., p-$MeC_6H_4SO_3H$).

H.Wulff et al. [53] prepared improved APG surfactant mixtures by mixing ≥2 binary surface active APG (e.g., an APG having C_8 and C_{10} alkyl groups and an APG having C_9, C_{10} and C_{11} alkyl groups) to give a predetermined, selected average alkyl chain length gives surfactant compositions for use un hard surface cleaners, laundry detergents, shampoos, facial cleaners, toothpaste, mouth washes, antiperspirants foam baths etc.

Ingo Wegener et al. [54] added carbonate esters $R^1O(C_xH_{2x}O)_nCO(OC_yH_{2y})_pOR^2$ ($R^{1\text{-}2}$=$C_{4\text{-}22}$ alkyl;x,y=2-4;n, polyglucoside=2-20) as defoamers to alkyl glucoside surfactant solutions(e.g., textile treatment baths) to suppress foaming.

Patrick M. McCurry et al. [55] prepared C_{8-20} alkyl polyglucoside by reacting a saccharide and an alcohol in the presence of an acid catalyst and added an oxide such as MgO to the reaction product and removed the unreacted alcohol in a thin film evaporator at ≤240°C in vacuo, optionally with sparging with an inert gas. The presence of the oxide improves the color of the product.

M.Weuthen et al. [56] introduced a modified two-step manufacturing process for long chain alkyl polyglucosides C_{16-18} (APG-1618) on the basis of the direct synthesis, consists of the glucosidation of fatty alcohols with glucose (mole ratio of 2.8) at 105°C upto a conversion degree of 75%.This reaction was stopped by neutralization of the catalyst. The residual and polyglucosides were filtered. The white, colorless solid product consists of approximately 70 wt% fatty alcohols and 30wt% glucosides, which were efficient nonionic emulsifiers. The compound APG-1618 is particularly appropriate to cosmetic lotions and creams because it is readily biodegradable and shows good skin compatibility in dermatological investigation according to the international std.test (OECD).

Bergfeld et al. [57] introduced a continuous manufacturing process of alkyl glucosides. In this process, a monosaccharide (e.g., glucose), a fatty alcohol (e.g.,Lorol S), and an acid catalyst (e.g., dilauryl 4-sulfopthalate) were added continuously to a stirred reactor and reacted under stationary conditions to give 95-100% conversion of monosaccharide while water was removed and the alkyl glucoside product was withdrawn.

M.Weuthen et al. [58] prepared surface-active oligoglucosides with bright color by acetalization of saccharides such as glucose with fatty alcohols (e.g., C_{12-14} alcohols), and neutralized the product in the presence of a reducing agent (e.g., NaH_2PO_2) and bleached. The method gives glucosides with good color and fatty alcohols recovered by distillation and had low carbonyl number.

Tietze et al. [59] prepared surface-active oligoglucosides by reacting fatty alcohols with glucose having particle size1-30μm. This method gave high reaction rate and a high monoglucoside content in the product.

M.Weuthen et al. [60] prepared surface-active alkyl glucosides with bright color, a high content of oligo glucoside by acetalization of 1mol glucose with 2-8mol of primary alcohols (e.g., C_{16-18} fatty alcohols) to give 60-98% conversion and neutralized the product and repeated the filtration until the filtrate was clear.

Yongqiang et al. [61] prepared dodecylglucoside with dodecanol and glucose as raw material at 2.76-10.7kPa and 80-160°C in the presence of an acid catalyst. Properties of dodecyl glucoside was determined and compared with other industrial surfactants. Dodecyl glucoside showed good surface activity, dispersibility, emulsification ability and foaming property.

Rainen et al. [62] manufactured alkyl oligoglucoside by direct acid acetalization of glucose with direct acid acetalization of glucose with excess C_{6-22} fatty alcohols containing 0-3 double bonds. Suspension of aqueous glucose syrup in fatty alcohol was used, for e.g., heating of 75°/35mbar a suspension obtained by dispersing 114g of 70% aqueous glucose syrup in 340g molten (40°C) coco fatty alcohol and containing 1.1g dodecylbenzene sulfonic acid catalyst gave glucose acetals containing 3.6% polymeric products and 0.75% residual glucose.

Woo et al. [63] made pale colored and transparent alkyl glucosides by reacting glucose with fatty alcohol in he presence of an acid catalyst and then neutralized, distilled, diluted and bleached. The reaction step was controlled by alcohol product color and an amount of unreacted glucose <3%, an alcohol/glucose mol ratio 3-5, acid content0.002-0.007 mol/mol glucose, reaction temperature 110-125°C, the neutralizing step was controlled by alkali metal oxide powders having alcohol specific surface area >30m^2/glucoside and the product color; distilling step was controlled by unreacted residual fatty alcohol; and the diluting step was controlled by refined ion water and sealed vessel without O. Further, the process provides alcohol method of recycling collected fatty alcohol without being refined.

Xupeng et al. [64] synthesized dodecyl glucoside from glucose and lauryl alc. in the presence of self-made composite catalyst (no data) by direct glucosidation. The effects of catalyst, temp., the ratio of glucose to lauryl alc., and premixing of raw materials on the reaction were studied. The optimum reaction parameters were: pressure 4.05kPa, temp 120□C, mol ratio of lauryl alc. to glucose 5:1, and wt. ratio of catalyst to glucose 0.008:1.

Zhang et al. [65] synthesized n-dodecyl glucoside from glucose and n-dodecanol by direct method, and influencing factors of the synthetic process, such as reaction temp and time, mole ratio of catalyst to glucose and n-dodecanol, etc. were also studied. Finally, the suitable conditions for the reaction were obtained, i.e., reaction temp. 110□C, mole ratio of catalyst to glucose to n-dodecanol 0.015:1:4, reaction pressure 4kPa and reaction time 4h. Yield of the product (n = 1.44) was about 84%.

Yuqin et al. [66] synthesized alkyl glucosides by reaction of glucose with alc. (at a ratio of 1-10) in the presence of catalyst at 80-150□C. The catalyst was the mixt. of C1-20 aliph. carboxylic acid and sulfonic acid or inorganic acid (at a molar ratio of 1:1-20).

2.1.2. Enzymatic Synthesis:

David et al. [67] prepared surface active-6 (C8-22 acyl)-1-(C3-12 alkyl) glucosides, esp. useful as emulsifiers in cosmetics, pharmaceuticals, foods etc., by the esterification of an alkyl glucoside with alcohol fatty acid in the presence of a lipolytic enzyme in a medium containing 30% water. Stirring 7g 1-butyl glucoside (I; 72% solids) with 11.9g oleic acid and 0.33g immobilized lipolytic enzyme of Candida cylindraceae for 96hr gave 5.7g 6-oleic acid ester of I.

Chaiya Panntrarux et al. [68] introduced β-glucosidase-catalyzed synthesis of n-alkyl-β-glucosides through the condensation of glucose and n-alcohols with alkyl chain length of 6-12 was carried out in a biphase system, where the organic phase consisted only of an alcohol substrate. All the alcohols tested were condensed with glucose to give the corresponding n-alkyl-β-glucosides. An equation was formulated to predict the equilibrium yields of n-alkyl-β-glucosides under various conditions, and its validity was experimentally verified. The equilibrium constant for the formation of n-alkyl-β-glucosides was independent of the alkyl chain length and was 1.90 at 60^0C. The volume ratio of the organic to the aqueous phase was a factor significantly affecting the equilibrium yield. It was also shown that the synthesis at lower initial glucose concentrations effectively yielded alkyl glucosides with less β-glucobiase formation.

Gabin Vic et al. [69] used almond-β glucosidase to catalyze alkyl-β-glucoside by reacting glucose and the alcohol in organic media and studied the influence of five different solvents and the thermodynamic water activity on the reaction. The best yields were obtained in 0 or 90% (v/v) t-butanol, acetone or acetonitrile where the enzyme was very stable. In this enzymatic synthesis under thermodynamic control, the yield increases, as the water activity of the reaction media decreases. Enzymatic preparative-scale synthesis was performing in a t-butanol-water mixture, which was found to be the most appropriate medium. 2-hydroxybenzyl-β-glucopyranoside was obtained 17% yield using a 0:10(v/v) t-butanol-water mixture.

Bertho et al. [70] introduced the preparation of glucoside surfactants from waste materials. In this process, plant materials (e.g., bagasse, corn, and sorghum by products) were treated with aqueous acids at 20-150°C or with hydrolytic enzymes at 20-90°C to give sugar syrups, which were separated from solids and treated with alcohols to give solution of alkyl glucoside surfactants. 231g crude barley bran (5% moisture) heated with 770g 3% H_2SO_4 at 130°C for 30minutes, 19g $CaCO_3$ flour was added at 60°C, filtered and then the juice was separated (17% based on solids), treating the juice with strong cationic and weak anionic ion exchangers, and concentrated to 71% solids, gave 130g syrup containing 29.8% pentoses based on solids. 125g BuOH containing 2.8g H_2SO_4 was added to this syrup over 90minutes at 102°C with azeotropic distillation of water, 328g $C_{12\text{-}14}$ fatty alcohols were added over 2h at 90°C, and the excess as were removed by thin film distillation which gave 147g residue containing 3% fatty alcohols which was decolorized as a 50% aqueous solution with H_2O_2.

M.Markropoulon et al. [71] used fusarium oxysporum-β-glucosidase catalyze the production of alkyl-β- glucosides from various disaccharides, based on the transglucosidation reaction, in the presence of primary, secondary and tertiary alcohols of glucosyl acceptors. Primary alcohols were found to be the best acceptors. The influence of the glucosyl donor concentration, as well as the enzyme specificity towards the cleaved glucosidic bond and the aglucone part of the donor, had also been investigated. The enzyme did not exhibit regiospecificity and seems to be unspecific towards the aglucone part. The specificity of the β-linkage had been confirmed by proton nuclear magnetic resonance (^{1}H NMR) analysis.

Yan T-R et al. [72] catalyzed the synthesis of methyl-β-glucoside and ethyl-β-glucoside by an extracellular-β-glucosidase II of Aspergillus niger with 5% (v/v) cellobiose as glucosyl donor in a biphasic media containing 20% (v/v) methanol and 30% (v/v) ethanol, respectively. The maximum yield of methyl-β-glucoside and ethyl-β-glucoside was 83%(mol/mol; 12mg/ml) and 53 % (mol/mol; 5.5mg/ml), based on cellobiosed consumed.

Takashi Kobayashi et al. [73] Synthesized hexyl, heptyl and octyl glucosides, galactosides and fucosides separately through the β-glucosidase-catalyzed condensation of glucose, galactose and fucose with the corresponding alcohols in an aqueous-organic biphasic system, where one of the substrates, alcohol, was the organic phase, at various vol. Ratios. An equation was proposed for predicting the equilibrium yield of an alkyl glucoside and experimentally determined the parameters involved in the equation. The equilibrium constant k for the alkyl glucosides formulation was then evaluated. Although the k-value slightly depends on both the saccharide moiety and the alkyl chain length, it was approx. 2.

Takashi Kobayashi et al. [74] Synthesized hexyl-β-glucoside at 1mM using immobilized β-glucosidase from 2.0M glucose and 1-hexanol. The continuous production of hexyl glucoside was achieved by its extractive recovery with 1-hexanol, and 4.9g of product was obtained during a week operation using a 26.5ml reactor.

2.1.3. Chemical Composition Containing APG

Ando et al. [75] introduced alkylpolyglucoside degreasing detergent compositions with low foaming properties. Title compositions comprise alkyl polyglucosides ROG_n (I; R=$C_{6\text{-}10}$ alkyl; G=$C_{5\text{-}6}$ glucose residue; n=1-3) containing ≥2 polyglucosides selected from (A) I(R=C10 alkyl), (B) I(R=C8 alkyl), and (C)I (R=C6 alkyl) at wt. ratio A/B 1/1000-50/50, A/C 1/1000-50/50, B/C 1/99-99/1, and A/(B+C) 1/1000-50/50.Thus,a biodegradable compositions containing 90% octyl polyglucoside and 10% decyl polyglucoside showed good degreasing property for hydrocarbon oils and polybutene.

Hill et al. [76] used 1 part of dodecyl glucoside (av.DP~1.4) with one part Plurafac RA40 (ethoxylated propoxylated fatty alcohol) in water, which gave a low-foaming compound having good wetting properties. These surfactant compositions are especially useful for automatic dishwashing, metal degreasing and cleaning of hard surfaces.

Baur et al. [77] prepared the glucosides $(G_mR^1)R^2_n$ (where G= glucose unit; R^1=$C_{8\text{-}18}$ alkyl acetal; R^2=$C_{1\text{-}4}$ alkyl or aryl methyl in ether linkage; av. m=1-10; av. n=0.1m-2m),from the glucosides G_mR^1(I) and the compounds R^2X (X=nucleophilic group).A solution of 700g I(av. m=2.6-2.8; R^1=$C_{10\text{-}12}$ alkyl) in 300g and 100g 50% NaOH were mixed at 44° with 0.80 mol. $PhCH_2Cl$ over 1h and stirred 6h to give a solution of substituted glucoside (degree of benzylation 0.9) with greater surface activity than the starting glucoside.

Giesen et al. [78] prepared detergent paste containing 13.5g coco alkyl glucoside and 4.5g α–sulfo $C_{16\text{-}18}$ tallow fatty acid di-sodium salt as well as 1.0% CM-cellulose (thickener) and citric acid (to give pH 7.2).This paste is useful for laundry and for washing hard surfaces, e.g., dishes.

Alfred Oftring et al. [79] prepared alkyl glucoside aminocarboxylates as complexing agents for heavy metals and alkaline earth metals and degreasing agents. These glucoside derivatives I (≥2 R^1=$COCH_2$:$CHLCO_2X$ and the other R=H; L=$N(CH_2CO_2X)2$, $NHCH(CO_2X)CH_2CO_2X$ etc; X=H, alkali metal, etc; R^2=H, Me) 0.5mol 2-hydroxyethyl glucoside and 3g NaOAc at 120°C, stirred for1h at 120°C and154g water was added at 100°C which gave an aqueous solution of pentamaleate with saponification no. 779 and acid no.464. the solution (132g) was added during 45 min. to a mixture of iminodiacetic acid 66.5%, 40% NaOH 82, Na_2CO_3 26.5 and water 70g at 10-15°C with addition of NaOH to maintain pH 10, stirred for 2.5hr at 25°C, acidified at pH 7, and spray dried to give a product which had Ca^{+2} sequestering capacity 175mg $CaCO_3$/g at 20°C and pH 11 and good dispersing power for kaolin in water.

Katsuhiko Deguchi et al. [80] introduced neutral liquid detergent compositions containing alkyl glucosides 3-60, C_{18-14} fatty alcohol 0.001-0.5 and water soluble organic and for inorganic salt(e.g., Na or K sulfate, chloride, phosphate, malate, lactate, p-toulene sulfonate, edetate, borate, citrate or benzoate) 0.001-8%. The compounds had good foaming properties, mildness to skin and hair and rinseability, impart a good feel to hand after use and caused rapid draining of water from tableware etc. The composition contained C_{9-11} alkyl glucoside 23, octyl alcohol 0.1,Na_2SO_4 1.0 and water 75.9%.

Balzer et al. [81] prepared low foaming detergent for machine laundering, which comprised liquid or powder compounds containing alkyl polyglucosides 5-30, and other surfactants 0-3%. A liquid composition contained carboxymethyl ether of ethoxylated (4 mol) C_{12-14} alcohols 15, C_{12-14} alkyl polyglucoside 15, coco fatty acid triethanol amine salt 10,triethanol amine 6, ethyl alcohol 12; 1,2-propanediol 6 and water 36%.

Alfred Oftring et al. [82] prepared alkyl mono - and polyglucoside ether carboxylates RO $(CHR^1{:}CH_2O)_y\ [A(OR^2)_z]_x$(R=alkyl, alkylphenyl,hydroxy alkyl, (hydroxy=alkyl) phenyl, $CH(CO_2X)\ CH_2CO_2X$, $CH_2CH(CO_2X)CH_2CO_2X$,H; R^1=H,Me; $R^2=CH(CO_2X)CH_2CO_2X$, $CH_2CH(CO_2X)CH_2CO_2X$; X=H, alkali metal, NH_4 etc; A=mono- or poly glucose residue, x=1-10, y=0-10, z=1-4) by etherification of $R^3C(CHR^1{:}CH_2O)_y\ [A(OH)_z]_x$ [R3=alkyl, alkylphenyl, hydroxyalkyl(hydroxy=alkyl)phenyl, H] with maleic and/or itaconic acid in the presence of alkaline earth metal ions at pH 9-13 and 50-150°C. The compounds are used in phosphate free or phosphate poor laundry detergents as clay soil dispersants and in crustation inhibitors.

Hans Andree et al. [83] discussed the preparation, properties and the uses of mono-C_{12-14}_alkyl polyglucosides (av.dp 1.4). The synergistic effects of these nonionics surfactants with other surfactants, their favorable ecological behavior were noted. The main fields of applications are in manual dishwashing detergents and in liquid detergents.

Georg Meine et al. [84] gave an aqueous liquid detergent composition, which was the mixture of anionic and nonionic surfactants and a thickener comprised of C_{8-12} alkyl polyglucoside having dp 1.35-2.50.

Manfred Weuthen et al. [85] prepared storage-stable aqueous compositions containing alkyl and/or alkenyl oligoglucoside surfactants. A crystallization inhibitor was selected from alcohols polyethylene glycol, glucose Fe^{+3} ions and alkyl oligoglucosides based on short chain primary alcohols; Guerbet alcohols, or polyols were added to an aqueous compound containing a glucoside surfactant $R^1O(G)_p$ ($R^1=C_{6-22}$ alkyl or alkenyl; $G=C_{5-6}$ glucose residue; p=1-10) to give a liquid compositions showing good pumpability even after prolonged storage at ~30°.

Kiewert et al. [86] prepared low foaming cleaner which containing water, a C_{8-10} alkyl poly glucosides, a C_{6-16} fatty alcohol alkoxylate with HLB>10, a C_{8-22} fatty acid and additive forms little or no foam and shows good detergency.

Hermann Hensen et al. [87] introduced a mild surfactant mixture which was mild to skin and especially useful in MDDs and cosmetics, containing 10-40% C_{12-22} alkyl and/or akenyl oligoglucoside and 20-80% $R(OCH_2CH_2)_nOSO_3M$ ($R=C_{6-22}$ alkyl and/or alkenyl; M=alkali metal, NH^4 etc; n=1-20).

Wolfgang Poly et al. [88] prepared low foaming aqueous detergent mixture which was useful in cleaning compositions, hair and skin conditioners etc, containing glucoside derivatives(R^1OG_p)(R^2)$_n$ (R^1=aliphatic C_{6-22} group containing 0-3 double bond; R^2=C_{1-8} alkyl or benzyl; G=C_{5-6} sugar residue; p=1-10; n=0.1-2p) as well as anionic, nonionic and amphoteric or zwitterionic surfactants.

Frank Wangemann et al. [89] prepared sulfated polyhydroxy compounds from epoxy alkane glucoside reaction products by reacting an alkyl glucoside with a long chain epoxide (e.g., 1,2-epoxytetradecane) were sulfated and neutralized to prepare surfactants.

Lothar Pioch et al. [90] prepared powdered and flowable pumpable surfactant compositions containing alkyl glucoside, an alkyl sulfate surfactant, an alkali silicate having SiO_2/alkali oxide molar ratio ≥ 2, and water was spray-dried or drained in a fluidizing apparatus with a gas stream heated at <250°C which gave a powdered surfactant compound which is resistant to caking useful in detergent formulation containing no alkylbenzene sulfonate surfactant.

Guo et al. [91] prepared by the acid catalyzed reaction of Cl-containing alcohols with carbohydrates to form chloroalcohol glucosides and treating the glucosides with higher t-amines. Thus, a surfactant was prepared from a glycerin-epichlorohydrin reaction product, glucose, and dimethyldecylamine.

Boiteux et al. [92] prepared hemisulfosuccinate of alkyl polyglucoside $RO(S)_xCOCHR^1CHR^2CO_2M$ (R=C_{6-32} alkyl or alkenyl; S=sugar residue; M=alkali metal; 1 of $R^1$and R^2=SO_3M and the other=H; x=1-10) by esterification of polyglucoside $RO(S)_xH$ with maleic anhydride sulfating the maleate double bond. The compounds are useful as surfactant in detergents, cosmetics etc.

Weuthen et al. [93] prepared alkyl and/or alkenyl oligoglucoside sulfo-succinate surfactants by the rapid and complete transesterification of a mono- or di- alkyl sulfosuccinate with an alkyl and/or alkenyl oligoglucoside without partial hydrolysis of the glucoside.

Weuthen et al. [94] used an alkaline compound such as Na_2CO_3 as a catalyst for the etherification of a glucoside such as dodecyl glucoside with an isethionate salt (e.g., Na or ammonium) at 100-200°, which gave products useful as surfactants.

2.2. Purification of Alkyl Polyglucosides

Wulff et al. [95] separated the mono glucosides from alkyl poly glucosides which wereaccomplished by molecular distillation; giving a purified alkyl monoglucoside product and an alkyl poly glucoside product showing better surfactant properties than the starting mixture.

McCurry et al. [96] continuously bleached an alkyl poly glucoside which was substantially free of alcohol and had a dark color with peroxide (e.g., H_2O_2) at controlled pH and temperature, esp. in the presence of MgO, to give a product having a light color.

Paul Schulz et al. [97] claimed a procedure for the bleaching of surface-active alkyl glucosides with peroxide. Stabilizers such as Alcohol sulfate, Na silicate and potassium gluconate were used with H_2O_2 in the bleaching of the glucosides at high pH(e.g., 9-10), giving transparent products showing good resistance to loss of transparency during storage.

Anja Kruse et al. [98] separated alkyl poly glucoside by reversed phase HPLC and two detdction methods were tested, evaporative light scattering detection and electro chemical detection at a good electrode. The results of both were discussed.

Juergen Gruetzke et al. [99] separated fatty alcohols, which remain unreacted in the manufacturing of surface-active alkyl glucosides, which had good color, by using ≥1 falling film evaporator and a final molecular distillation step.

Gerda Grund et al. [100] improved the color of alkyl poly glucoside surfactant solutions by bleaching at an alkaline pH with H_2O_2 in the presence of a decomposition inhibitor (e.g., inorganic Mg compound) for H_2O_2, followed by decomposition of unreacted H_2O_2 in the presence of a inhibitor and a transition metal or transition metal compound.

Karl Schmid et al. [101] improved the color of alkyl oligoglucoside by treating the solutions (40-70% solid) of C_{4-22} alkyl(oligo)glucoside with bleaching compositions containing an active O-compound and an alkali earth compound at pH 9-11.

2.3. Analysis of Alkyl Polyglucosides

Matsumura et al. [102] synthesized and evaluated surface active properties of n-alkyl α- andβ-glucopyranosides, with alkyl chain having 8-12 carbon atoms, such as static surface tension, CMC, occupation area of mol,, dynamic surface tension, foaming properties, biodegradability and antimicrobial activities. Alkyl glucosides containing C_{8-12} C-chains showed surface activities and CMC. Among the alkyl glucosides, α-anomers were less hydrophilic than β–anomers. Alkyl glucosides were readily biodegraded by activated sludge of a municipal sewage plant. N-alkyl glucosides containing C_{8-12} alkyl chains showed a broad spectrum of increasing antimicrobial activity.

Balzer et al. [103] discussed the interfacial activity and miscelle structure, phase diagrams, foaming and wetting power and interfacial tension and illustrated manual dishwashing compositions; liquid fabrics washing compositions; and personnel care formulations.

Nickel et al. [104] characterized alkyl monoglucosides and alkyl polyglucosides in terms of their interfacial properties which exhibit a dependence on alkyl chain length and d.p.and drawn a comparison to other nonionic and anionic surfactants in hard water, clouding phenomena were obtained for $C_{8/10}$ and $C_{12/14}$ alkyl polyglucosides. The CMC of the alkyl glucosides were similar to those of other nonionic surfactants and showed a decrease with increasing alkyl chain length. The d.p. has only a minor influence on the CMC. Above the CMC, alkylglucosides exhibit very low interfacial tension against mineral oil compare to $C_{12/14}$ fatty alcohol sulfate (FAS). In contrast to $C_{12/14}$ FAS and $C_{12/14}$ fatty alcohol hexaglycol ether, resp; $C_{12/14}$ APG shows remarkable high dynamic surface tensions even at a concentration far above the CMC. For mixtures of $C_{12/14}$ APG and $C_{12/14}$ FAS the static surface tensions and the CMC lie close to those of $C_{12/14}$ APG, whereas the dynamic surface tensions behavior is demonstrated by the anionic surfactants.

Waltermo et al. [105] studied the interactions at the air / liquid and the liquid/solid interface for octyl β- glucoside (I). With the thin film balanced the thickness of the equilibrium liquid films was determined as a function of surfactants concentration, inert salt concentration, and pH. At high surfactant concentration and /or at low pH black films ~ 4.6 nm thick were formed. The stabilizing forces in black films were related to the interactions with in and between the layers. The forces acting between I surfactant layers adsorbed on hydrophobized mica were investigated with an interferometric type surface force apparatus. The forces measured between the sugar head groups like dimethylamine oxide and monoglycerides. Considerably more long range repulsion was observed between surfactants with oligo ethylene oxide head groups.

Wodarczak et al. [106] separated alkyl polyglucosides with different alkyl chain lengths by TLC on RP_{18} with $MeOH/H_2O$ 9:1(vol) and visualized by spraying with diluted H_2SO_4 and heated. The detection limit was 600ng. for dodecyl monoglucoside. Other than ethoxylates, APG gave no coloration with Dragendorff reagent. Quantitative determination of APG was performed with anthrone and photometry at 620nm. The detection limit was ~ 10g. The determination was interfered by CM- cellulose, which can be removed by column chromatography.

Buschmann et al. [107] proposed a method for the colorimetric determination of APGs surfactants based on reaction with anthrone. The calibration of showed a linear relationship between concentration and absorbance from 30μg-150μg APG/mL sample. The relative deviation was 1.1(48.5μg APG/mL, n=6). The calibration curve was stable and used for determinations for several months. Cellulose derivatives interface with the determinations substances, which were usually present in formulations do not interfere. The APG concentration was determined for some com.-available APG – containing detergent formulations.

Yoon et al. [108] studied interfacial properties of APG, surface and interfacial tension, CMC, foaming and effectiveness according to degree of polymerization of glucose unit (d.p. = 1.2~ 1.8) and alkyl chain length, the former affected hardly but the later affected greatly on them. The samples with lowest interfacial tensions and maximum adsorption head a great detergency.

Bushmann et al. [109] proposed three analytical methods for the separation of alkyl polyglucosides according to the alkyl chain length by reversed phase TLC, a phometeric determination of the total concentration with the anthrone reagent and separated by HPLC, using an evaporative light scattering detector. The first two methods allows a very fast and easy analysis of APG with minimum instrumentations and well suited for a control of incoming bulk chems. The HPLC methods yielded quantitative determination of the alkyl chain length distribution. The methods were characterized by reproducibility, detection limit and interferences.

2.4. Applications

Johnson et al. [110] claimed ionic derivatives of alkyl mono- and poly- glucosides containing anionic groups (e.g., carboxylates, sulphates) and are useful as surfactants, which do not require a hydrotope to keep them in aqueous suspension. Dodecyl glucoside (d.p 1.3) in DMF was sulphated with SO_3Me_3N to prepare a surfactant having ~ 1 SO_4Na groups / glucosyl group.

Fabry et al. [111] introduced surfactant mixture containing alkyl glucosides and hydroxyl sulfonates, having good biodegradability and skin compatibility and useful in non- and low – phosphate detergents containing alkyl glucosides $RO(G)_x$[R=C≥8 aliphatic groups; Gl=residue of $C_{5\text{-}6}$ reducing saccharide; x=1-10] and hydroxyl sulfonates $Me(CH_2)_yCHR^1(CH_2)_2CH_2O(C_nH_{2n}O)_mH$; (y,z =0-18; p=0.2; y+z+p=4-18; maximum=0-30;

n=2-4; one of R1 and R2 is OH and the other is SO_3M; Maximum=H, alkali metal etc). A detergent contained C_{12-14} alkyl glucoside (average degree of oligomerization 1.3) 7.5, an adduct of 5 mol. ethylene oxide and 1 mol C_{18} diol sulfonate (prepared by sulfonating ocenol 90/95 comprising mainly oleyl alcohol) 7.5, Na_2SO_4 41.0, Soda 7.0, sokalan CP5 3-5, zeolite NaO 20.0, coco fatty acid sodium salt 1.5, water glass 3.0%.

Hill et al. [112] prepared a mixture of 1-20 parts alkyl glucosides ROG_n and 1-9 parts alcohol ROH (R=linear or branched C_{16-28} alkyl; G=Glucose residue; n=1-10) is useful as an antifoaming agent, especially in the food industry. A mixture of 18% behenyl glucoside and 82% behenyl alcohol was mixed (10 parts) with 90 parts paraffin oil and used as an antifoaming agent for a sugar solution.

Balzer et al. [113] used alkyl polyglucosides in low foaming liquid or powder detergents which show good detergency, ROG_n and $R^1OG^1_m$ (R=C_{8-11} alkyl; R^1=C_{12-18} alkyl;G, G^1= residue of hexose and/or pentose; n,m=1,3) 3-30,ethoxylated fatty alcohols 3-30,salt 5-30 and other surfactants 0-5%.

Tagata et al. [114] had made liquid detergent compositions which is useful for wool fabrics, contain $R^1(OR^2)_xG_w$ [I;R^1=linear or branched C_{8-18} alkyl, alkenyl, alkylphenyl; R^2=C_{2-4} alkylene; G=C_{5-6} reducing sugar residue; x=0-5; w=1-10);R^3R^4,$R^5R^6N^+X^{-[}$II; R^{3-6}=C_{14-20} alkyl, C_{1-5} alkyl, hydroxy alkyl, $(CH_2CH_2O)_mH$; average mol.no. of added ethylene oxide 2-30 per mol.;$\geq$ 1of R^3-R^6 are C_{14-20} alkyl; X=halogen, $MeSO_4$, $EtSO_4$) 0.5-10, amino modified silicones (derivatives) 0.05-5, and $\geq$1 anionic surfactants selected from alkyl benzene sulfonates, alkyl ether sulfates and polyoxy ethylene ether sulfates 0.5-10%. Thus, 20% I (R^1=C_{9-11} alkyl; n=0; w=1.35; G=Glucose residue) was mixed with II (R^3=C_{16-18} beef tallow alkyl; R^4=R^5=R^6=Me; x=Cl)1, $Me_3SiO(SiMe_2O)_m$ $[SiMe(CH_2)_3NH(CH_2)_2NHCO(CH_2)_3(C_2H_4O)_4$-$C_8H_{17}O)_nSiMe_3$ 1, C_{10-13} alkyl benzene sulfonic acid sodium salt 1.5, ethyl alcohol 5%, and balanced water to give a detergent, which when used to washing a wool fabric showed amount of fallen cuticle 27mg/100 g. wool.

Hees et al. [115] used nonionic surfactants comprising glucoside ROZ_x (R=C_{8-18} alkyl; Z=glucose residue; x=1-2) with anionic surfactant (not petrochemical derived), builders, water-solution solvents or solubilizers and additives in liquid compounds, which show good detergency in the cleaning of hard surfaces.

Lomax et al. [116] used alkyl polyglucosides for personnel care, laundry and liquid detergent applications and introduced oligomeric (d.p $\leq$3) glucoside alkyl ethers and there nonionic surfactant properties. There benefits are that they contain ether (not ester) linkage, and therefore are resistant to alkyl hydrolysis.

Urfer et al. [117] used alkyl glucosides in aqueous disinfecting cleaners containing quaternary ammonium compounds and reported that the germicidal efficiency of an aqueous disinfecting cleaner containing quaternary ammonium compounds (e.g., Barquat 425 OZ) is increased by adding an alkyl glucoside (e.g., Glucopon 425).

Knaut et al. [118] claimed alkyl polyglucosides as a new category of mild, highly biodegradable surfactants. Due to their performance spectrum and special properties, they are suitable for use in a variety of applications and different industrial branches. Manufactured from carbohydrates and fatty alcohols; alkyl polygluocosides offer advantages to formulators and users, particularly from the economical, ecological and dermatological points of view. Special motion can be made their synergy with standard surfactants, ecological properties and extraordinarily good skin compatibility.

Giesen et al. [119] introduced mixtures of 5-12% alkyl glucosides $R^1O(Z)_p(R^1=C_{6-12}$ alkyl; Z=glucose residue;p=1.3-1.8) and 88-95% alkyl glucosides $R^2O(Z)_p(R^2=C_{10-18}$ alkyl; Z=glucose residue;p=1.3-1.8) show good mildness to skin and are useful in personal cleansing compositions.

Johnsson et al. [120] used branched alkyl glucosides for use in cleaning composition. Alkyl glucosides $RCH_2O(G)_xH$(R=C8-12 alkyl containing 2-4 groups CHMe in the chain; G=glucose residue; x=1-4) are useful in compositions which give good cleaning of hard surfaces and have acceptable foaming properties.

Giesen et al. [121] introduced mixtures of 48-52% alkyl glucosides $R^1O(Z)_p(R^1=C_{6-12}$ alkyl; Z=glucose residue;p=1.3-1.8) and 48-52% alkyl glucosides $R^2O(Z)_p(R^2=C_{6-12}$ alkyl; Z=glucose residue;p=1.3-1.8) show good mildness to skin and are useful in personal cleansing compositions.

Fallon et al. [122] used alkyl glucoside in personnel care products and claimed that alkyl polyglucosides are mild, biodegradable, and offer excellent value and performance when compared to conventional surfactants. Mildness place a vital role in products formulated for bodily use. Testing of demonstrate this characteristic, is therefore increasingly important. Other important factor in cleaning and washing products include foam, foam characteristics and viscosity building in the final formulated product.

Milius et al. [123] used decylpolyglucosides in dishwashing liquid formulation for higher concentration, increased mildness, and greater compatibility with the environment and product clarity. Decylpolyglucosides give either 40% active hand dishwashing liquids, which perform better than 40% active com. brands, or 30% active dishwashing liquids comparable in performance to 40% active com. brands. Therefore decylpolyglucosides permit to use less active ingredients in formulations thus reducing environmental impact. Decylpolyglucosides are mild and decrease skin irritation of anionic surfactants. Decylpolyglucosides are clear raw materials, which permit to prepare clear formulations.

REFERENCES

[1] Karlheinz H., Fats and Oils As Raw Materials, Cogins DurtSclland GmbH, Dusseldorf, Germany, *Pure Appl. Chem.*, Volume 12(7), *2000,* 1255.

[2] Agarwal O.P., Chemistry of Organic Natural Products, Goel Publishing house, Meerut, Volume 1, 73.

[3] Fennema O.R., Food Chemistry, Dept of food science, University of Wisconsin Madison, Madison Wisconsin, IInd edition.

[4] Fisher E., Ber. Dtsh. Chem. Ges, 26, 1893, 2400.

[5] Capon B., *Chem Rev,* 69, *1969*, 389 and Ferrier RJ., *Fortschr Chem. Forsch,* 14, *1970,* 389.

[6] Koenigs W. and Knorr E., *Ber. Dtsh. Chem. Ges,* 34, *1901,* 959.

[7] Fisher E. and Helferich B., *Justus Liebigs Ann. Chem,* 386, *1911,* 68.

[8] Flory PJ., *J Am. Oil Chem. Soc*, 74, 1952, 2718.

[9] EP 0437461 B1, *Henkel, 1988.*

[10] EP 049174, *Huls, 1991.*

[11] EP 0617045 A2, *Akzo, 1994.*

[12] EP 0357969 B1, *Henkel, 1988.*
[13] Lai YZ. and Shafizadeh F., *Carbohydr Res*, 38, *1970,*177.
[14] Ponder GR. and Richards GN., *Cardohydr Res,* 208, *1974,* 93.
[15] WO 90/06933, *Henkel, 1990.*
[16] EP 0492397, *Kao, 1990.*
[17] Friedli F. E., Detergency of Specialty Surfactants, *Surfactant science series,* 98, *2001,* 10.
[18] Waldhoff H., Scherler J. and Schmitt M., Proceedings 4th World Surfactant Congress, Barcelona,1, 1996, 507.
[19] Spilker R., Menzebach B., Schneider U. and Venn I., *Tenside Surf Det,* 33, *1996,* 21.
[20] Buschmann N., Kuse A. and Wodarczak S., *Agro Food Industry, Hi-Tech, 1996,* 6.
[21] Buschmann N. and Wodarczak S., *Tenside Surf. Det* 32,*1995,* 336.
[22] Bruns A., Waldhoff H. and Winkle W., *Chromatographia,* 27, *1989*, 340.
[23] Shinoda K., Yamaguchi T. and Hori R., *Bull. Chem. Soc. Jpn,* 34, *1961*, 347.
[24] Boecker T. and Thiem J., *Tenside Surf. Det,* 26,*1989*, 318.
[25] [Ref.] F.E. Friedli, Detergency of speciality surfactants, 95, 2001, Marcel Dekker, New York.
[26] Forster T., Hensen H., Hofmann R. and Salka B., *Cosmetic toiletries* 110, *1995*, 223; *Parfumerie Kosmetik* 76, *1995*, 763.
[27] Nickel D., Nitsch C., Kurzendorfer CP. and Rybinski W., *Prog. Colloid. Polym Sci* 89, *1992,* 249.
[28] Fukuda K., Sodermann O., Lindman B. and Shinoda K., *Langmuir*, 9, *1993*, 2921.
[29] Putnik CF. and Borys NF., *Soap Cosmet Chem. Spec,* 86, *1986*, 34 and Hughes FA. and Lew BL., *J. Am. Oil Chem. Soc,* ,47, *1970*, 162.
[30] Holdt B. D., Jeschke P., Menke R. and Soldanski JD., *Seifen Ole Fette Wachse J,* 120, *1994*, 42.
[31] IPP Quality Standard, *Seifen Ole Fette Wachse J,* 112, *1986*, 371.
[32] Upadek H. and Krings P., *Seifen Ole Fette Wachse J,* 117, *1991*, 55.
[33] Matthies W., Krachter H-U., Steiling W. and Weuthen M., 18th IFSCC, Venice, Poster Vol 4, *1994,* 317.
[34] Steber J., Guhl W., Sterlter N. and Schroder FR,. *Tenside Surf Det*, 32, *1995*, 515.
[35] Gerike P., Holtmann W. and Jasiak W., *Chemoshere* 13, *1984*, 121.
[36] Harald L. and Peter H., (Huels A.-G.) Eur. Pat. Appl. EP 306,651 (Cl. C07H15/04), *1989*, DE Appl. 3,729,843, *1987,* 7pp, *Chem. Abstr,* 111(16), *1989*, 136480z.
[37] Manfred B., Karlheinz H., Willi W., Rainer E., Josef W., Andreas B., Guenter H., Karl-Heinz O., Walter W. and Klaus W., (Henkel kGaA) Ger. Offen. DE 3,723,826 (Cl. C07H15/04), *1989*, Appl. *1987*; 9pp, *Chem. Abstr,* 110(22), *1989,* 195187m.
[38] Harald L., (Huels A.-G.) Eur. Pat. Appl. EP 306,650 (Cl. C07H15/04), *1989*, DE Appl. 3,729,842, *1987*, 4pp, *Chem. Abstr,* 111(8), *1989*, 60020b.
[39] Harald L. and Peter H., (Huels A.-G.) Eur. Pat. Appl. EP 306,652 (Cl. C07H15/04), *1989*, DE Appl. 3,729,844, *1987*; 7pp, *Chem. Abstr,* 111(16), *1989*, 136479z.
[40] Karlheinz H. and Manfred B., (Henkel kGaA) Ger. Offen. DE 3,842,541 (Cl. C07H15/04), *1990*, Appl. *1988,* 4pp, *Chem. Abstr,* 113(24), *1990*, 214295a.
[41] Willi W., Rainer E., Josef W., Karlheinz H., Manfred B, (Henkel kGaA) Ger. Offen. DE 3,842,534 (Cl. C07H3/02), *1990,* Appl. *1988,* 8pp, *Chem. Abstr,* 113(4), *1990,* 25983n.

[42] Stefan S., (Huels A. -G.) Ger. Offen. DE 4,234,241 (Cl. C07H15/04), *1994*, Appl. *1992,* 3pp, *Chem. Abstr,* 120(24), *1990*, 301654e.

[43] Henry R. and Manfred B, (Henkel kGaA) Ger. Offen. DE 3,900,590 (Cl. C07H15/04), *1990*, Appl. *1989,* 6pp, *Chem. Abstr,* 113(24), *1990*, 214298d.

[44] Hiroki S., Hiroshi N., Makoto A. and Akira Y., (Kao corp.) Jpn. Kokai Tokkyo Koho JP 02,264,789[90,264,789] (Cl. C07H15/04), *1990*, Appl. 89/86,072, *1989*, 5pp, *Chem. Abstr,* 114(14), *1991*, 124950b.

[45] Carl P. and Patrick M., (Henkel kGaA) PCT Int. Appl. WO 9007,515 (Cl. C07H15/04), *1990*, VS Appl. 290,535, *1988,* 30pp, *Chem. Abstr,* 114(6), *1991*, 45526q.

[46] Hiroshi K., Katsumi K. and Yoshiaki F., (Kao corp.) Jpn. Kokai Tokkyo Koho JP 0454,192[9254,192] (Cl. C07H15/04), *1992*, Appl. 90/163,241, *1990*, 10pp, *Chem. Abstr*, 117(4), *1992*, 29239e.

[47] Giampietro B., Claudio S., Luigi N. and Tullio P., (Enichem S.P.A;Eniricerche S.P.A) Eur.Pat. Appl. EP 570,056(Cl.C07H15/04), *1993*, IT Appl. 92/mic 1157, *1992*, 9pp, *Chem. Abstr,* 120(24), *1994,* 301649g.

[48] Gerhard W., Alfred O., Guenter O., Richard S. and Wolfgang K., (BASF A.-G.) PCT Int. Appl. WO 9321,197 (Cl.C07H15/04) *1993*, DE appl. 4,213,016, *1992,* 16pp, *Chem. Abstr,* 121(4), *1994*, 38083j.

[49] Manfred W., Karlheinz H. and Paul S., (Henkel kGaA) Ger. Offen. DE 4,137,636 (Cl. C07H15/04), *1993*, Appl. *1991*, 4pp, *Chem. Abstr*, 120(2), *1994*, 10765p.

[50] Michael G. W., Patrick M. M. and carl P.E., (Henkel kGaA) PCT Int. Appl. WO 9324,504 (Cl.C07H15/00) *1993*, US Appl. 891,305, *1992*, 28pp, *Chem. Abstr,* 121(8), *1994*, 86282v.

[51] Paul S., Herbert E. and Rainer E., (Henkel kGaA) Ger. Offen. DE 4,140,334 (Cl. C07H15/04), *1993*, Appl. *1991,* 5pp, *Chem. Abstr,* 120(2), *1994*, 10767r.

[52] Gerhard W., Alfred O., Georg S., Helmut W., Rudolf A., Heinrich B. H., Dieter H., (BASF A.-G.) Ger. Offen. DE 4,212,080 (Cl.C07H15/04) *1993*, 16pp, *Chem. Abstr,* 120(10), *1994*, 110006k..

[53] Paul S., Rainer E, (Henkel kGaA) Ger. Offen. DE 4,235,647 (Cl. C07H15/04), *1994*, Appl. *1992,* 6pp, *Chem. Abstr*, 121(10), *1994,* 112000y.

[54] Wulff, Harald W.P., Paul S. A., Patricia B. E., Barry S. A., Michael C. W., Robert A. A, (Henkel Corp.) PCT Int. Appl. WO 9307,249 (Cl.C11D1/66) *1993*, US Appl. 774,430, *1991,* 61pp, *Chem. Abstr,* 120(2), *1994,* 10774r.

[55] Ingo W., Brigitte G, (Henkel kGaA) Ger. Offen. DE 4,209,38 (Cl. B01D19/04), *1993*, Appl. *1992*; 7pp, *Chem. Abstr,* 120:194535t, No.16, *1994*.

[56] Patrick M. M., Jr., Brian A., Carl P. E. and Janet V. R., (Henkel kGaA) PCT Int. Appl. WO 9424,139 (Cl.C07H1/06) 27 Oct *1994*, US Appl. 47,217, 14 Apr *1993*; 20pp, *Chem. Abstr,* 123:59614q, No.6, *1995*.

[57] Weuthen M., kawa, R., Hilland K. and Ansmann A., (Henkel kGaA, D-40191 Duesseldorf, Germany), Fett Wiss Technol. *1995*, 7(6), 209-11, (Eng), *Chem. Abstr,* 123(14), *1995,* 173555w.

[58] Manfred B. J. and Juergen S., (Akzo Nobel N.V.) Eur. Pat. Appl. EP 617,045 (Cl.C07h15/04), *1994*, EP Appl. 93/104,599, *1993*, 8pp, *Chem. Abstr,* 122(12), *1995,* 217198t.

[59] Manfred W., (Henkel kGaA) Ger. Offen. DE 4,316,601 (Cl.C07H15/04), *1994*, Appl. *1993*, 5pp, *Chem. Abstr,* 122(22), *1995,* 268685g.

[60] Friedjan L., Kai B., Manfred W., Karlheinz H. and Paul S., (Henkel kGaA) Ger. Offen. DE 4,321,838 (Cl. C07H15/04), *1995*, Appl. *1993*; 4pp, *Chem. Abstr,* 122(26), *1995,* 317436n.

[61] Manfred W., Detra R. and Irmgard H., (Henkel kGaA) Ger. Offen. DE 4,137,089 (Cl.B01F17/56), *1994*, Appl. *1993*, 5pp, *Chem. Abstr,* 122(22), *1995,* 1268686h.

[62] Yongqiang H., Qi H., Zhenbao Y., Defeng Z. and Jinzong Y., (Dept. of Fine Chemicals, Dalian Univ. of Technology; Dialan, Peop. Rep. China 116012), Xiandia Huagong *1997*, 17(7), 22-24 (Ch), *Chem. Abstr,* 129(3), *1998,* 29385x.

[63] Rainer E., Eiko H. and Bernhard G.,(Henkel kGaA, Germany) PCT Int. Appl. WO 9845,307 (Cl.C07H15/04) *1998*, DE Appl. 19,714,255, *1997*; 13pp(Ger), *Chem. Abstr,* 129(21), *1998,* 277688g.

[64] Tae-Hee W., Dong-hyun C. and Kwang-ho P., (LG Chemical Ltd, S.Korea) PCT Int. Appl. WO 99 26,957 (Cl.C07H15/04), *1999*, KR Appl. 9,820,631, *1998*; 29pp (Eng), *Chem. Abstr,* 131(1), *1999,* 6871e.

[65] Xupeng G., Department of Chemical Engineering, Huadong Metallurgical Institute, Ma'anshan, Peop. Rep. China. Shanghai Huagong (*2000*), 25(18), 12-13, 26, CODEN: SHAHE2 ISSN: 1004-017X. Journal written in Chinese. CAN 134:282485 AN *2000:*750723 CAPLUS (Copyright *2004* ACS on SciFinder (R)

[66] Ya-dong Z., Deng-gao J., Zhi-wei L., Xiao-lei G. and En-cai H., College Chemical Eng., Zhengzhou Univ., Zhengzhou, Peop. Rep. China. Jingxi Huagong (*2001*), 18(9), 509-512. CODEN: JIHUFJ ISSN: 1003-5214. *Journal written in Chinese. CAN* 136:249264 AN *2001,*759787 CAPLUS (Copyright *2004* ACS on SciFinder (R)

[67] Yuqin J., Fukui X. and Desheng Z., (Research Inst. of Jilin Petro-Chemical Branch, China National Petroleum Co., Ltd., Peop. Rep. China). Faming Zhuanli Shenqing Gongkai Shuomingshu (*2001*), 5 pp, CODEN: CNXXEV CN 1315311 A 20011003 Patent written in Chinese. Application: CN *2000,*105522 20000330. CAN 137:385071 AN *2002*,942954 CAPLUS (Copyright *2004* ACS on SciFinder (R)

[68] Marie D. H.L., Hilde D. O.J., Horst G. and Harald R. W.,(Cerestar bolding B.V.) Eur. Pat. Appl. EP 334, 498 (Cl C07H15/04) *1989*, G.B. Appl. 88/6, 691, *1988*, 6 pp, *Chem. Abstr,* 112(16), *1990,* 141783a.

[69] Panntrarux C., Adachi C., Araki Y., Kimura Y. and Matsure R., *Enzyme and Microbial Technology,* 17(1), *1995.*

[70] Vic E., Thomson D. and Crout D., *Enzyme and Microbial Technology,* 20(8), *1997.*

[71] Jeon-Noel B., Philippe M. and Veronique D., De Baynast De Septfontaines, Fegis (Agro Industric Recherches Et Developments (A.R.D.); PCT Int. Appl. Wo9729, 115 (Cl C07H15/04), *1997*, FT Appl. 96/1, 542, *1996*; 54 pp, *Chem. Abstr,* 127(14), *1997,* 192193a.

[72] Markropoulou M., Chrastakopoulous P., Tsitsimpikou C., Kekos D., Kolisis F. N. and Macris B. J., *International Journal of Biological macromolecules,* 22,1 *1998.*

[73] Yan T.R., Department of Bioengineering; Liau J-Cf, Graduated school of chemical engineering, Tatung Inst. Of Technology, 40, Chung-Sherg North Road, 3rd Sec. Taipei Taiwan, China. *Biotechnology Letters*, 20(7), *1998.*

[74] Kobayashi T., Adachi S., Nakanish K. and Matsuno R., (Kyoto University), *Journal of Molecular Catalysis B: Enzymatic*, 11(1), *2000.*

[75] Kobayashi T., Adachi S., Nakanishi K. and Matsuno R., (Kyoto University), *Biotechnology Letters*, 22(23), *2000*.

[76] Kenji A., Yumi K., and Shizumi S., (Shin Nippon Pika KK, Japan) Jpn. Kokai Tokkyo Koho JP 08, 3A, 497 [96,319,497] (Cl. C11D1/68), 30 Dec. *1996*, Appl. 95/152,174, 26 May *1995*; 5 pp (Jpn), *Chem. Abstr*, 126(10), *1987,* 132904c.

[77] Randal H. M., Lander S., Michel W. and Mary G. K., (Staley Continental, Inc.) PCT Int. Appl. WO 8809,309 (Cl. C11D17/00), 01 Dec. *1988*, US Appl. 51,550 18 May *1987*; 25 pp, *Chem. Abstr* 110(20), 1989, 175551x.

[78] Richard B., Jochen H., Alfred O. and Dieter S., (BASF A.-G.) Ger. Offen. DE 3, 835, 199 (Cl. C07H15/04) 19 Apr. *1990*, Appl. 15 Oct. *1988*; 4 pp, *Chem. Abstr,* 113(12), *1990,* 99891c.

[79] Brigitte G., Alfred M., Andreas S. and Ingo W., (Henkel k.-G.a.A.) Ger. Offen DE 3, 827, 778 (Cl. C11D1/83), 22 Feb. *1990*, Appl. 16 Aug. *1988*, 5 pp, *Chem. Abstr,* 113(4), *1990,* 26015k.

[80] Alfred O., Stefan B., Richaid B., Alexender K., Ulrich G. and Johnannes P., (BASF A.-G.) Ger. Offen. DE 3,901,612 (Cl. C07H15/04), 26 Jul. *1990*, Appl. 20 Jan. *1989,* 10 pp, *Chem. Abstr,* 114(12), *1991,* 104809.

[81] Katsuhiko D., Kozo S. and Hiroyuki S.,(Kao Corp.) Eur. Pat. Appl. EP 388,810 (Cl. C11D1/66), 26 Sep. *1990*, JP Appl. 89/68, 698, 20 Mar. *1989*, 9 pp, *Chem. Abstr,* 114(12), *1991,* 104839u.

[82] Dieter B.,(Huels A.-G.) Eur. Pat. Appl. EP 45, 7965 (Cl. C11D10/04), 27 Nov. *1991*, DE Appl. 4, 016, 819, 25 May *1990*; 10 pp, *Chem. Abstr,* 116(8), *1992,* 62101f.

[83] Alfred O., Elisabeth K., Richard B. and Alexander K., (BASF A.-G.) Eur. Pat. Appl. EP 457,155 (Cl C07H15/04), 21 Nov. *1991*, DE Appl. 4,015,655, 16 May *1990*; 9 pp, *Chem. Abstr,* 116(14), *1992,* 131659h.

[84] Hans A. and Birgit M., (Henkel k.-G.a.A; Duesseldrof, Germany) *Tenside surfactants Deterg*, 28(6), *1991,* 413, *Chem. Abstr,* 116(8), *1992,* 620812.

[85] Georg M., Heinz W. M., Rolf P. and Juergen H., (Henkel k.-G.a.A.) PCT Int. Appl. WO 92,03,527 (Cl. C11D1/83), 05 March *1992*, DE Appl. 4,026,471, 22 Aug. *1990*; 17 pp, *Chem. Abstr,* 116(22), *1992,* 216817h.

[86] Manfred W., Thomas E., Hermann H. and Holger T., (Henkel k.-G.a.A.) Ger. Offen. DE 4,225,224 (Cl. B01F17/36), 03 Feb. *1994*, Appl. 30 July *1992*, 6 pp, *Chem. Abstr* 121(2), *1994,* 12286c.

[87] Eva K., Peter J., Birgit M. and Giuseppe C., (Henkel K. G. a. A.) Ger. Offen. DE 4,210,565 (Cl. C11D1/825) 07 Oct. *1993*, Appl. 30 March *1992*, 5 pp, *Chem. Abstr,* 121(2), *1994,* 12314k.

[88] Hermann H., Holger T., Hans K. U., Peter B., Dagmar S. and Barry S. A., (Henkel k.-G.a.A.) PCT Int. Appl. WO 93 23,512 (Cl. C11D1/83), 25 Nov. *1993*, US Appl. 884, 685, 18 May *1992,* 23 pp, *Chem. Abstr,* 120(18), *1994,* 220960z.

[89] Wolfgang P. and Christian N., (Henkel k.-G.a.A.) Ger. Offen DE 4, 210, 912 (Cl. C11D1/83), 07 Oct. *1993*, Appl. 02 Apr. *1992*, 6pp, *Chem. Abstr* ,120(10), *1994,* 110009p.

[90] Frank W., Wolfgang P. and Bert G., (Henkel k.-G.a.A.) Ger. Offen DE 4, 239, 390 (Cl. C07H15/04), 26 May *1994*, Appl. 24 Nov. *1992*, 5 pp, *Chem. Abstr,* 121(12), *1994,* 136643m.

[91] Lothar P., Martina K. and Wolfgang S., (Henkel k.-G.a.A.) Ger. Offen. DE 4,216,775 (Cl. C11D1/14), 25 Nov. *1993*, Appl. 21 May *1992*, 7 pp, *Chem. Abstr,* 120(18), *1994,* 220944x.

[92] Minglin G., Shenqing F. Z. and Shuomingshu G., CN 1,074,389 (Cl. 01F17/56), 21 July *1993*, Appl. 92,100,081, 15 Jan. *1992,* 6 pp, *Chem. Abstr,* 121(20), *1994,* 233503y.

[93] Jean Pierre B. and Bernard B., (Societe d'Exploitation de produits pour les Industries Chimiques (SEPPIC) PCI Int. Appl. WO 93,08,204 (Cl. C07H15/04), 29 Apr. *1993*, FR Appl. 91/12,841,17 Oct. *1991,* 23 pp, *Chem. Abstr,* 120(6), *1994,* 57220x.

[94] Manfred W., (Henkel k.-G.a.A.) Ger. Offen. DE 4,336,803 (Cl. C07H15/04), 04 May *1995*, Appl. 28 Oct. *1993,* 4 pp, *Chem. Abstr,* 123(18), *1995,* 232033z.

[95] Manfred W., (Henkel k.-G.a.A.) Ger. Offen DE 4,337,035 (Cl. C07H15/04), 04 May *1995*, Appl. 29 Oct. *1993,* 4 pp, *Chem. Abstr,* 123(22), *1995,* 290450u.

[96] Harald W. P., Paul S. A., Patricia B. E., Barry S. A., Michael C. W., Robert A. A., Patrick C. M., Robert S. and William K. G., et al. (Henkel kGaA) PCT Int. Appl. WO 9307,160 (Cl. C07H15/04), 15 Apr *1993*, US Appl. 774,430, 10 Oct *1991*; 85 pp, *Chem. Abstr* 120(4), *1994,* 33395j.

[97] Patrick C. M., Robert K. L., Michael G. W., James B. D. and Janet V. R., (Henkel kGaA) PCT Int. Appl. WO 94,02,494 (Cl. C07H1/00), 03 Feb *1994*, US Appl. 914,363, 15 Jul *1992,* 34pp, *Chem. Abstr,* 121(12), *1994,* 136637n.

[98] Paul S., Rainer E. and Herbert E., (Henkel kGaA) PCT Int. Appl. WO 9313,112 (Cl.C07H15/04), 08 Jul *1993*, DE Appl. 4.142,592, 21 Dec *1991*, 17pp, *Chem. Abstr,* 120(4), *1994,* 33397j.

[99] Anja K. and Nobert B., (Westfaelische Wilhelms Universitaet, 48149 Muenster, Germany). Camun. Jorn. Com. Esp. Deterg. *1995*, 26, 209, *Chem. Abstr,* 123(16), *1995,* 202942c.

[100] Juergen G. and Stefan S., (Huels Aktiengesell-schaft, Germany) Eur, Pat. Appl. EP 714,906 (Cl. C07H15/04), 5 Jun *1996*, DE Appl. 4,443,087, 3 Dec *1994,* 4 pp, *Chem. Abstr,* 125(8), *1996,* 89644h.

[101] Gerda G., Vwe T., Juergen G. and Stefan S., (Huels Aktiengesell-schaft, Germany) Ger. Offen. DE 4,432,623 (Cl. C11D1/66), 21 Mar *1996*, Appl. 4,432,623, 14 Sep *1994,* 3pp, *Chem. Abstr* 124(24), *1996,* 321079c.

[102] Karl S., Michael N., Manfred W. and Michael L., (Henkel k.-G.a.A., Germany) Ger. Offen. DE 4,436,761 (Cl. C07H15/04), 18 Apr *1996*, Appl. 4,436,761, 14 Oct *1994*, 5 pp, *Chem. Abstr* 125(4), *1996,* 36330w.

[103] Shuichi M., Kazuyasu I., Sadao Y., Kazuo K. and Isuyoshi U., (Fac. Sci. Technol., Koio Univ., Yokohama, Japan 223) JAOCS *1990*, 67(12), 996-1001 (Eng), *Chem. Abstr,* 114(8), *1991,* 64784p.

[104] Dieter B., (Huels A.-G.; Marl Germany) *Tenside, Surfactants, Deterg. 1991*, 28(6), 419-27 (Eng), *Chem. Abstr,* 116(8), *1992,* 62069b.

[105] Nickel D., Nitsch C., Kurzendorfer C-P., Rybinski W.,(Lab. Henkel kGaA, Duesseldrof, Germany). Prog Colloid Polym Sci *1992*, 89 (Trends Colloid Interface Sci. VI), 249-52 (Eng). *Chem. Abstr* 118(24), *1993,* 236437d.

[106] Aasa W., Emil M., Robert P. and Per C., (Dept. Chem., R. Inst. Technol., S-10044 Stockholm, Swed.). *J. dispersion Sci. Technol. 1994*, 15 (3), 273, *Chem. Abstr* 120(22), *1994,* 273555c.

[107] Susann W. and Norbert B., (D-48149 Muenster, Germany). GIT Fachz. Lab. *1995*, 39(5), 410, *Chem. Abstr,* 123(16), *1995,* 2029390x.

[108] Norbert B. and Susann W., (Anorganish- Chemishes Inst., D-48149 Muenster, Germany) *Tenside, Surfactants, Deterg. 1995*, 32(4), 336-9, *Chem. Abstr,* 123(16), *1995,* 202938f.

[109] Yeo-kyung Y. and Kyu-Suk C., (Coll. Eng., Hangang Univ. Seoul, 133-191 S. Korea). *Kongop Hwahak 1994*, 5(3), 431-6, *Chem. Abstr,* 123(22), *1995,* 290425q.

[110] Norbert B., Kruse A. and Susann W., (Westfaelische Wilhelms Universitaet, D-48149 Muenster, Germany). Agro-Food-Ind. Hi-Tech *1996*, 7(1), 6-8, *Chem. Abstr,* 125(8), *1996,* 89621y.

[111] Donald J. L., Kenneth M. B., Valenty V. and Vian (Staley. A.E. mfg. Co.) PCT Int. Appl. WO 8801,640 (Cl. C11D3/22), 10 Mar *1988*, US Appl. 904,363, 05 Sep *1986,* 36pp, *Chem. Abstr,* 109(16), *1988,* 131271f.

[112] Bernd F., Brigitte G., Karlheinz H., Martina K. and Robert P., (Henkel K. G. a. A.) Eur, Pat. Appl. EP 849, 906 (Cl. C11D1/83), 10 Jan *1990*, DE Appl. 3822997, 07 July *1998*; 10 pp, Chem Abstr, 112(20), *1990,* 181933f.

[113] Karlheinz H., Adolf A. and KarlHeinz S., (Henkel k.-G.a.A.) Eur. Pat. Appl. EP 396,044 (Cl B01D19/04), 07 Nov. *1990*, DE Appl. 3,914,906, 05 May *1989*; 5 pp, *Chem. Abstr,* 114(22), *1991,* 104813f.

[114] Dieter B., Rerdolf B. and Willi G., (Huels A.-G.) Eur. Pat. Appl. EP 474,915 (Cl. C11D10/04), 18 Mar *1992*, DE Appl. 4,029,035, 13 Sep *1990*, 10 pp, *Chem. Abstr,* 116(24), *1992,* 237830x.

[115] Hideji T., Yoshihiro I., Akiko I. and Fumio S., (Kao Corp) Jpn. Kokai Tokkyo Koho JP 05,239,493[93,239,493] (Cl. C11D1/86), 17 Sep *1993*, Appl. 92/43,368, 28 Feb *1992*, 8pp, *Chem. Abstr,* 120(2), *1994,* 10784u.

[116] Udo H., Rainer J. and Manfred W., (Henkel k.-G.a.A.) Ger. Offen. DE 4,311,159 (Cl. C11D3/22), 06 Oct. *1994*, Appl. 05 Apr *1993*, 6 pp, *Chem. Abstr*, 122(6), *1995,* 58907e.

[117] Eric L., (UK) *Spec. Chem.* 14(1), *1994*, 21-4, 26 (Eng). *Chem Abstr*, 122(8), *1995,* 84265p.

[118] Allen U. D. and Virginia L. L., (Henkel kGaA) PCT Int. Appl. WO 9405,753 (Cl.C11D1/62), 17 Mar *1994*, US Appl. 942,555, 09 Sep *1992*, 12pp, *Chem. Abstr,* 122(8), *1995,* 84318h.

[119] Knaut, J. and Kreinfeld G. (Henkel kGaA, Duesseldrof, Germany). *Riv. Ital. Sostanze Grasse* 71(6), *1994*, 307-11 (Eng). *Chem. Abstr,* 122(22), *1995,* 268655.

[120] Brigitte G., Wolfgang P., Karl S. and Walter S., (Henkel k.-G.a.A.) Ger. Offen DE 4,319,699 (Cl. C11D1/825), 22 Dec *1994*, Appl. 16 Jun *1993,* 8 pp, *Chem. Abstr,* 122(26), *1995,* 317479d.

[121] Ingegaerd J., (Berol Nobel AB) PCT Int. Appl. WO 9421,769 (Cl.C11D1/66), 29 Sep *1994*, SE Appl. 93/954, 23 M ar *1993,* 13pp, *Chem. Abstr,* 122(26), *1995,* 317469a.

[122] Brigitte G., Wolfgang P., Karl S. and Walter S., (Henkel k.-G.a.A.) Ger. Offen DE 4,319,710 (Cl. C11D1/825), 22 Dec *1994*, Appl. 16 Jun *1993*, 8 pp, *Chem. Abstr,* 122(26) , *1995,* 317478c.

[123] John F., (Emery Group, henkel Corp., Ambler, PA 19002 USA). Proc. World Conf. Lauric Oils: Sources, Process, Appl. *1994*, 127-32 (Eng). Edited by Applewhite, Thomsan H. AOCS: hampaign III. *Chem. Abst,*. 123(8), *1995,* 86600p.

[124] Alain M., Chantal A. and Bernard B., (SEPPIC, 75321 Paris, Fr.). Agro-Food-Ind. Hi-Tech *1995*, 6(2), 21-4, Chem. Abstr.123(16), *1995,* 202958n.

In: Heavy Metal Pollution
Editors: S. E. Brown and W. C. Welton

ISBN: 978-1-60456-899-8

Chapter 8

HEAVY METAL TOXICITY TO LEGUMES

Mohammad Saghir Khan, Almas Zaidi, Mohammad Oves and Parvaze Ahmad Wani

Department of Agricultural Microbiology, Faculty of Agricultural Sciences
Aligarh Muslim University, Aligarh-202002, India

ABSTRACT

Heavy metal pollution is one of the current most troublesome environmental problems due to the widespread use of metals for industrial and agricultural purposes. Heavy metals discharged often from the industrial operations and sewage sludge, cause a phenomenal increase in the extent of heavy metals and thus become a major global threat to the soil environment. Moreover, the introduction of pollutants to soil environment can often overwhelm the self cleaning capacity of recipient ecosystem and thus, result in their accumulation to even harmful level and consequently makes the soil unsuitable for sustainable agriculture. Legumes are used mainly for human consumption as well as for animal feeds and serve as a rich source of protein and contribute a substantial amount of N to the agronomic soils. They are grown in tropical and semi arid tropical countries. Once grown in the contaminated soils, the metals are taken up by plant organs to an undesired level, which in turn limits the growth and physiological processes, like the synthesis of macromolecules including protein and consequently affect the legume yields severely. However, the toxicity of heavy metals to legumes could be reduced/prevented through the application of phytoremediation technology that involves the use of plants to remove, destroy or sequester hazardous substances from the contaminated environment. Also, many plants possess efficient enzymatic and non-enzymatic antioxidant defense system that allows the scavenging of reactive oxygen species and provides protection to plants from oxidative damage. The antioxidant enzymes have been reported to be produced by plants under metal stress that in turn, reduces the toxicity of metal to plants. With these considerations, this chapter focuses mainly on the two aspects of metal in relation to legumes- (i) response of legumes to metal stressed environment and (ii) different strategies adopted by plants to protect itself from metal toxicity, which may help in assessing the ecological significance of metal stress.

1. INTRODUCTION

Heavy metals are chemical elements whose specific gravity is at least 5 times the specific gravity of water i.e. 1 at 4°C (39°F). Some well-known toxic metallic elements include arsenic, cadmium, iron, lead, and mercury etc. Heavy metals after discharge from industrial operation or from agricultural activities [Nriagu and Pacyna, 1988; Fernandes and Henriques, 1991; McGrath et al., 1995], persist in the top cultivated layer of agronomic soils [McGrath, 1987] . However, the amounts of metals accumulated in soils are dependent on the emission levels, the transport of the metal from the source to the accumulation site and the retention of the metal once it has reached the soil [Alloway, 1995]. Although some of these elements (essential metals) are required by organisms at low concentrations [Adriano, 2001] and play a role in different metabolic function. For example, zinc is the component of a variety of metalloenzymes or acts as cofactor for several enzymes (dehydrogenases, proteinases, peptidases, oxidase) [Hewitt, 1983] besides their role in the metabolism of carbohydrates, proteins, phosphate, auxins, in RNA and ribosome formation in plants [Mengel and Kirkby, 1982; Shier, 1994]. Similarly, copper at low concentration, contributes to several physiological processes such as photosynthesis, respiration, carbohydrate distribution, nitrogen, cell wall metabolism and seed production in plants [Kabata-Pendias and Pendias, 2001]. On the contrary, the non essential metals (e.g., cadmium) are not involved in any known biological processes and due to its higher mobility and bioavailabilty [Mac Laughin and Singh, 1999] in the rhizosphere, released from the the application of phosphate fertilizers [Zhu et al. 1999] or industrial and domestic sludges [Ibekwe et al. 1995], cadmium exerts a negative impact on both soil microbial communities [McGrath et al. 1988] and legumes [Wani et al., 2007a]. Generally, cadmium disturbs the (i) enzyme activities (ii) DNA-mediated transformation (iii) symbiosis between microbes and plants and [Wani et al. 2007b] (iv) increases plant predisposition to fungal invasion [Kabata-Pendias and Pendias, 2001].

In recent times, there has been increasing concern with heavy metal contamination because of their toxicity to plants including legumes. Toxicological effects of heavy metals on nodulation and N_2 fixation in legumes have been reported in lentil (*Lens esculenta*) [Wani et al. 2006] chickpea (*Cicer arietinum*) [Yadav Shukla, 1983; Wani et al. 2007a], greengram (*Vigna radiata* L. wilczek) [Wani et al. 2007b] and pea (*Pisum sativum*) [Wani et al. 2008]. The reduction in growth and grain yield of legumes, inoculated with or without their corresponding rhizobial symbionts, under metal stressed soils could be due to the toxic effects of metals which prevent the formation of nodules or the metal contamination may lead to the elimination of effective symbiont strains from the soil [Giller et.al. 1989]. In addition, there is evidence that suggests that the reduction in plant growth and nodule size, and nitrogenase activity in white clover (*Trifolium repens* L.) were due to cadmium, lead, and zinc, when plants were grown in soils highly contaminated with these metals [Rother et.al. 1983]. However, the evidence of metal toxicity on legumes is conflicting. For instance, symbiotic nitrogen fixation (acetylene reduction activity) was strongly affected by heavy metals in mine spoils or in sludge amended soil [Heckman et al. 1987] and also, pronounced metal toxicity on white clover was confirmed where nitrogen fixation was halved by increasing metal concentrations in sludge treated soil [Broos et al. 2005]. In contrast, Ibekwe et al. (1996) failed to detect any significant effect of metal contaminated sludge on nitrogen fixation in white clover. However, to combat stress from heavy metals, legumes, like other plants, have developed mechanisms, that include the formation and sequestration of heavy metals, conversion of toxic to non-toxic forms and direct efflux of metals out of the cell. Furthermore, certain enzymes (e.g. catalase, superoxide dismutase, oxygenase, glutathione reductase etc.) or phytochelatin produced under metal stress by the organs of legumes and released in to the metal contaminated rhizosphere soils, could help to detoxify the effect of heavy metals, and consequently, allow them to grow under metal contaminated soils.

2. Origin of the Metal Contamination in Soils

The presence of heavy metals in soils may be of natural or of anthropogenic origin. Natural sources include the transport of continental dusts, atmospheric emissions from volcanoes, and the weathering of metal-enriched rocks (Ernst, 1998). However, the major source of metal contamination of soil is the discharge and accumulation through anthropogenic activities. For instance, the consistent use of metal based pesticides and metal-enriched sewage sludges in agriculture, the emission from mines and smelters, combustion of fossil fuel, metallurgical industries and electronics (manufacture, use and disposal), adds a significant amount of heavy metals to soils (Alloway, 1995). The main problem with heavy metals in soils is that, unlike organic pollutants, they cannot be biodegraded and therefore, persist in the environment for long periods of times. Metal availability in soils is influenced by physico-chemical properties of the soils, such as cation exchange capacity (CEC), organic matter content, clay minerals and hydrous metal oxides, pH and buffering capacity, redox potential and extent of aeration, water content and temperature. Also, the biotic processes of plants, fungi and microbes affect the availability of metals in soil [Welp and Brummer, 1997; Wuertz and Mergeay, 1997; Dumestre et al. 1999]. Because of the high binding capacity for metallic micronutrients by soil particles, plants have evolved several strategies for increasing their soil bioavailability. These strategies include the production of metal-chelating compounds (phytosiderophores) which are synthesized in response to iron and possibly zinc deficiencies. In the rhizosphere, phytosiderophores chelate and mobilize iron, coper, zinc, and mangnese. Once chelated to phyto-siderophores, metal ions can be transported across the plasma membrane as a metal-phytosiderophore complex via specialized transporters. All these sources cause accumulation of metals and metalloids in the agricultural soils and pose a major threat to the soil fertility, and in turn, plant productivity, which ultimately may affect the safety of foods worldwide.

3. Metal Uptake, Translocation and Accumulation

There were wide variations of metal concentrations both among plant species and within different plant tissues of the same species and these fluctuations have been extensively documented (Shu et al. 2005; Remon et al. 2005]. However, the uptake of metals by the plant roots depends on- (i) diffusion of elements along the concentration gradient (ii) root interception, where soil volume is displaced by root volume due to extensive root growth and (iii) mass flow, transport from bulk soil solution along the water potential gradient. The apical region generally takes up metal, while some metals in plants can also be taken up by the entire root surface. Thereafter, metal is transported further into the cells (e.g., in the apoplast) and bind to cell wall substances. Part of the metal that enters apoplast is further transported through the plasma membrane into the cytoplasm, which concentrates nutrient elements, and help to exclude toxic elements. For instance, uptake and accumulation of chromium by *Medicago sativa,* and chickpea [Zlatareva et. al. 1999; Shanker, 2003], cadmium, chromium, nickel, zinc, and lead in roots and shoots of chickpea [Wani et al. 2007a], lead and zinc by lentil [Wani et al. 2006], and cadmium, chromium and copper in roots, shoots and grians of pea [Wani et. al. 2008] is reported. However, the phytoaccumulation of these metals was higher in roots compared with the shoots or grains of the test legumes. Moreover, the simultaneous application of metals reduced the uptake by plant organs. The variation in the uptake of metals by legumes could be due to the antagonistic effect of one metal on the other metal. A second possibility could be the interaction between metals at the root surface for plant uptake. A common model representing the mode of chromium entry in to plants is given in figure 1. Since there are some essential metals, at least the uptake of these ought to be regulated. Zinc is transported with zinc transporters, with a higher abundance in zinc accumulator species than in non-accumulator species (Lasat et al. 2000; Rout and Das, 2003). Zinc is also known to be actively transported as a free ion across the tonoplast and depends on the tolerance of the plant. Other metals (e.g. cadmium) easily enter the root through the cortical tissue and are translocated to the above ground tissues (Yang et al. 1998). As soon as cadmium enters the roots, it can reach the xylem through an apoplastic or symplastic pathway (Salt et al. 1995a), complexed by several ligands, such as organic acids and/or phytochelatins (Salt et al. 1995b). Normally, cadmium ions are retained in the roots and only very small amounts are transported to the shoots (Cataldo et al. 1983). It has been found that metal transporters with specificities for the respective metal cations exist (Clemens, 2001). However, these cations without specific metal transporters are likely to enter cells through cation transporters showing broad substrate specificity (Clemens, 2001).

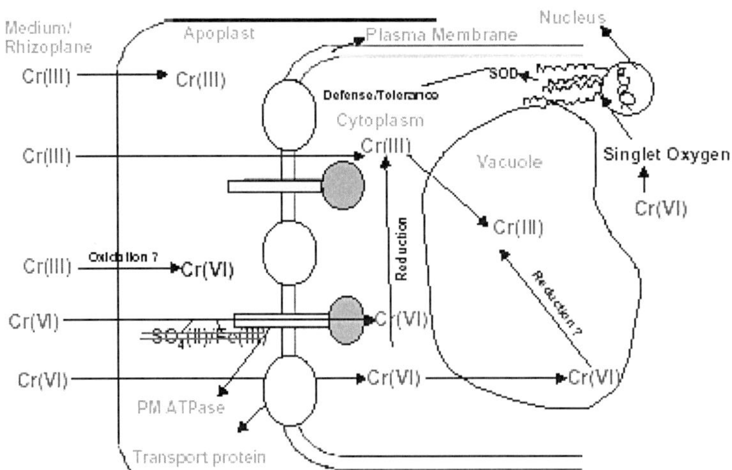

Figure 1. Hypothetical model of chromium transport and toxicity in plant roots [Shanker 2005].

Table 1. Composition of some the legume grains

Chemical Composition	Legumes						
	Green gram	Soy bean	Lentil	Black gram	Kidney bean	Cowpea	Bengal gram
Total carbohydrates (%)	62.6	20.9	5	59.6	59	60.3	63.5
Crude protein (%)	23.86	43.2	25.1	24	23-26	23.4	23
Fat (%)	1.15	19.5	0.7	1.4	0.7	1.8	5.3
Crude fibre (%)	5.27	3.7	0.7	0.9	4	-	
Ash (%)	3.32	-		-	4	-	
Food energy (calories)	347	432	343	3470	-	-	347.6
Digestibility (%)	79	-		-	-	-	
Biological coefficient (%)	72			-	-	-	
Moisture (%)	-	8.1	12.4	10.9	9.3	11	
Minerals	-	4.6	2.1	3.2	-	-	

Source: Rathore [2002]

4. PHYTOTOXICITY

Among the various consumable foods, legumes serve as a rich source of protein, though they also provide considerable amounts of carbohydrates (table 1) in the dietary system and are used mainly for human consumption as well as for animal feeds. Moreover, legumes add a substantial amount of nitrogen to the soils, which in turn is used up by plants to meet out their nitrogen demands. When grown in area enriched with metals, legumes suffer severely from metal toxicity. However, legumes respond differently to heavy metals in the soil whose toxicity varies with the plant genotypes, age and developmental stages of growth (Shaw and Rout, 2002; Shanker et al. 2005). Moreover, the toxicity of metals to legumes is also influenced by differences in metals solubility, absorbability, transport and chemical reactivity, route of exposure (leaf, shoot and root), mechanisms and mode of action, interactions with other metals, site of action (reproductive, respiratory, photosynthetic) and genetic composition of plants. Although some legumes are not affected by low concentrations of heavy metals [Wani et al. 2007a; Wani et al. 2008], elevated concentrations of such metals however, adversely affect the different metabolic activities of plants; leading thereby to a decrease in overall growth of plants [Wani et al. 2007b,c]. The toxicity symptoms observed in legume plants in the presence of excessive amount of heavy metals may be due to (i) changes in the permeability of the cell membrane (ii) reactions of sulphydryl (–SH) ADP or ATP and (iii) replacement of essential ions leading to an inhibition of activity or disruption of biomolecules including DNA and protein [Van Assche and Clijters, 1990; Das et al. 1997; Breen and Murphy, 1995] and (iv) alteration in photosynthetic pigments [Wani et al. 2007a; Bibi and Hussain, 2005]. Moreover, the prolonged exposure of soils to heavy metals has also shown a decrease in soil enzymes activity. Thus, in order to evaluate the meaningful physiological and biochemical effects of metal toxicity to legumes, it is essential to understand the composition and interaction of one metal with other metals. Moreover, the movement of metals including their absorption and translocation to different organs of legumes must be ascertained in order to establish their toxicity. However, the models designed to quantify the magnitude of relationship between exposure to metal ions and progressive yield losses are lacking [Taylor et al. 1991]. In the following sections, effect of heavy metals on the metabolic and physiological processes of different legumes will be discussed briefly.

4.1 Heavy Metals and Physiological Processes

4.1.1 Cell Wall and Plasma Membrane

Roots of the various plants including legumes are the first organ that is directly exposed to the metals in soils and hence, is the target of toxicity or provide tolerance to plants. The interaction of the metals with the cell wall has been reported [Ernst et al. 1990] but the binding properties and its role in the mechanism of metal tolerance have been controversial [Verkleij and Schat, 1990]. Most of the cell wall-associated heavy metals bind to polygalacturonic acids, to which the affinity of metal ions vary considerably [Ernst et al. 1992]. Though, a variety of metal detoxification mechanisms exist in plants, often the plasma membrane is the target [Chaoui et al. 1997]. The metal induce changes in membrane lipids both qualitatively and quantitatively in turn alter not only the membrane structure and function, but also the other cellular processes (figure 2). Some of these changes, however, provide protection to the plants against the toxicity of heavy metals. Such toxicity could result from various mechanisms including the oxidation and cross-linking of protein thiols, inhibition of key membrane proteins such as H^+-ATPase, or changes in the composition and fluidity of membrane lipids [Mehrag, 1993]. Among these, ATPase plays a significant role in the adaptation of plants to heavy-metal conditions [Dietz et al. 2001]. Among different metals affecting plasma membrane, toxic effect of chromium on the transport activities of plasma membrane was suggested by Zaccheo et al. [1982). The inhibition of ATPase activity is suggested to be due to the disruption of the membrane by free radical generated under metal stress. The decrease in ATPase activity in turn causes a decrease in proton extrusion which leads to a decrease in the transport activities of the root plasma membrane leading thereby to a reduction in the uptake of nutrient elements. Moreover, it is also likely that chromium may interfere with the mechanism controlling the intracellular pH; this possibility is supported by the fact that chromium could be reduced in the cells thereby utilizing the protons [Zaccheo et al. 1985]. Generally, the chromium stress can induce the following metabolic modification in plants (i) alteration in the production of pigments (e.g., chlorophyll) (ii) increased production of metabolites (e.g., glutathione, ascorbic acid) as a direct response to metal stress which may cause damage to the plants. Among the other metals, cadmium and copper have been found to adversely affect the lipid composition of membranes [Fodor et al. 1995; Quartacci et al. 2001]. Moreover, cadmium treatment has also been shown to reduce ATPase activity of the plasma membrane fraction of roots (Fodor et al. 1995).

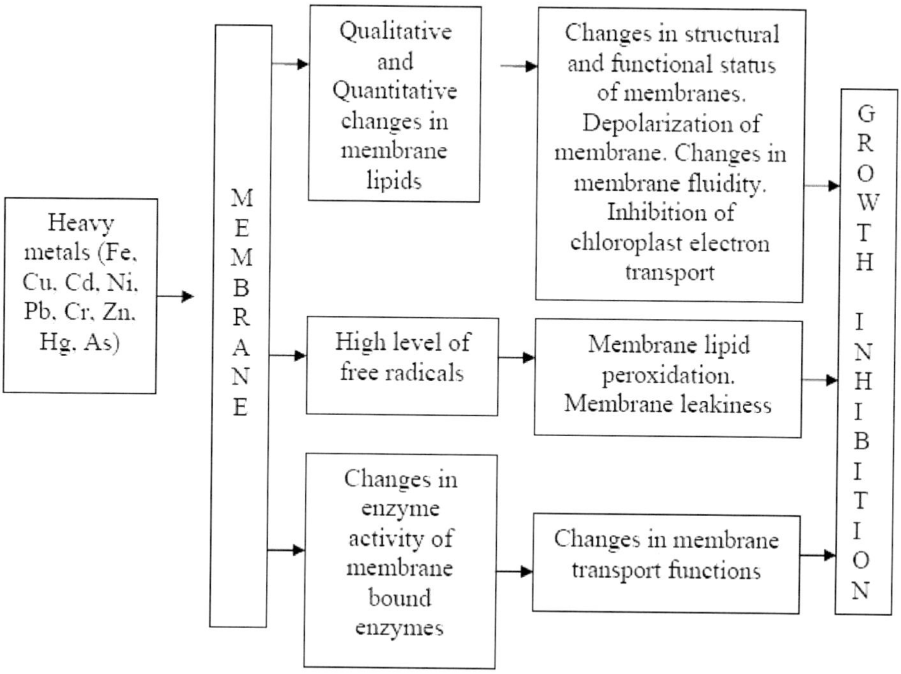

Figure 2. Heavy metal induced membrane changes play a major role in plant growth.

4.1.2. Lipid Peroxidation

In addition to the metal induced changes in fatty acid composition of membranes, membrane injury is also related often to an increased per-oxidation of membrane lipids, due to the action of highly toxic free radicals. Several metal ions cause peroxidation of lipids of both the plasma membrane and chloroplast membrane [Hernandez and Cooke, 1997]. As a result of this activity, the amount of reactive oxygen species (ROS) increases. Among metals, increased lipid peroxidation has been observed with cadmium [Chaoui et al. 1997], aluminum [Yamamoto et al. 1997], copper [Mazhoudi et al. 1997), zinc [Chaoui et al. 1997], nickel [Baccouch et al. 1998], chromium [Bagchi et al. 2001] and arsenate [Hartley-Whitaker et al. 2001a]. In general, the iron and copper compounds generates more free radicals and increases the peroxidation [Aust et al. 1985; Price and Hendry, 1991]. Thus, membrane dysfunction induced by metals could be due to changes in the membrane structure and peroxidation of membrane lipids [Cakmak and Horst, 1991]. For instance, Al causes lipid peroxidation by disorganizing the membrane structure [Jones and Kochian, 1997] while Al and Cu ions initiate the peroxidation reaction by generating free radicals [Weckx and Clijsters, 1996]. The increased lipid peroxidation is therefore, considered as an indicator of increased oxidative stress. Besides their role in altering the structure of membrane, the increased lipid peroxidation also changes membrane properties, such as fluidity and permeability and modulates the activities of membrane-bound ATPases [Shewfelt and Erickson, 1991]. Generally, peroxidation of membrane lipids is a chain reaction in which unsaturated fatty acids are converted stepwise into various small hydrocarbon fragments, such as malondialdehyde [Kappus, 1985]. The lipid peroxidation processes and the resulting substances in turn severely affect the functioning of the plasma membrane and finally lead to the death of the cells [Kappus, 1985].

4.1.3. Photosynthesis

When taken up by the plants, heavy metals can result in a wide variety of toxic effects including the effect on photosynthetic apparatus. It is generally believed that the toxic metals react with the photosynthetic apparatus at various levels of organization and architecture leading to (i) accumulation of metals in leaves (ii) metal interaction with cytosolic enzymes and organics (iii) alteration of the functions of chloroplast membrane and partitioning in leaf tissues like stomata, mesophyll and bundle sheath (iv) supra molecular level action particularly on photosystem I, photosystem II, membrane acyl liquids and carrier proteins in vascular tissues and (v) molecular level interactions, particularly with photosynthetic carbon reduction cycle enzymes, xanthophylls cycle and adenylates [Prasad, 1999]. The elevated concentration of metals are reported to significantly affect the photosynthetic pigments [Geiken et al. 1998; Franco et al. 1999; Zeid, 2001] and the photosynthetic process, like those involved in the reduction of carbon, when legumes are grown in heavy metal contaminated soils [Krupa and Baszynski 1995]. Similarly, excess concentrations of copper modified the ultrastructure of chloroplast in runner beans (*Phaseolus coccineus* L.) [Maksymiec et al. 1995] while reduction in phytosynthetic pigments (chlorophyll) following metal application in various legumes is reported [Bera et al. 1999; Mysliwa-Kurdziel and Stratka, 2002a]. The decrease in the chlorophyll a/b ratio [Shanker, 2003] following chromium application has been suggested due to the destabilization and degradation of the proteins of the peripheral part. The inactivation of enzymes involved in the chlorophyll biosynthetic pathway could thus contribute to the general reduction in chlorophyll content in most plants including legumes under heavy metal stress. However, the majority of reports on the impact of heavy metals on photosystem II activity have been observed for Cd^{2+} and Cu^{2+}, whereas Cd^{2+} affects both the PS II reaction center and the light harvesting complex (LHC) and cause an inefficient energy transfer from the LHC to the reaction center. Generally, enzymes of the photosynthetic carbon reduction (PCR) cycle are inhibited under heavy metal stress and the key steps of the Calvin cycle- (i) carboxylation, (ii) reduction and (iii) regeneration, have been found to be affected by heavy metals, the first step being the most sensitive one [Prasad, 1995a, 1998]. Among the metals, cadmium exerts its toxicity through membrane damage and inactivation of enzymes, possibly through reaction with sulfhydryl groups of proteins [Fuhrer, 1988] as also reported for Pb^{2+}, Cd^{2+}, Zn^{2+} and Cu^{2+} [Van Assche and Clijsters, 1990]. In some cases, heavy metal toxicity is however, reflected by an increase in the activity of these enzymes; for instance, malic enzyme, glucose-6-phosphate dehydrogenase and peroxidase in foliage [Van Assche and Clijsters, 1986]. Cadmium has been the most intensively studied inhibitor of dark reactions of photosynthesis. It was shown in isolated protoplasts treated with Cd^{2+} that the main target of this metal action was the reactions of the Calvin cycle and that activation of rubisco was not affected [Weigel, 1985a]. In contrast, Sheoran et al. [1990a] showed significant reduction of rubisco activity of pigeon pea (*Cajanus cajan* L.) plants, treated with Cd^{2+} at an early growth stage. However, in older plants, the activity of rubisco was not affected. They concluded that the reduction in photosynthesis was due to a decrease in chlorophyll content, effects on stomatal conductance and the electron transport system. In other *in vitro* study, Cd^{2+} treatment changed the structure of rubisco and resulted in dissociation of its small subunits [Stiborova, 1988].

4.2. Performance of Legumes under Heavy Metal Stress

4.2.1. Germination and Dry Matter Production

Seed germination is the first physiological process affected by heavy metals and the ability of a seed to germinate in a medium containing heavy metals would be indicative of its level of tolerance to metals [Peralta et al. 2001; Shanker et al. 2005]. In this context, a high level of hexavalent chromium (500 ppm) when applied to soil, reduced germination by 48% of bush bean (*Phaseolus vulgaris*) [Parr and Taylor, 1982] while 40 ppm of Cr (VI) reduced the germination of lucerne (*Medicago sativa* cv. alone) by 23% [Peralta et al. 2001]. The reduction in germination under heavy metals stress could be due to the depressive effect of these metals on the metabolic activity of roots and on the subsequent transport of sugars to the embryos. Also, protease activity is reported to be increased following metal treatment, which could contribute to the reduction in germination of metal-treated seeds [Zeid, 2001]. In a similar study, the germination of chickpea seeds was adversely affected by nickel (200 ppm) and cobalt (400 ppm) [Khan et al. 1996].

The toxicity of various metals to legume plants or their symbiotic partners (rhizobia) varies widely and often compound with the greatest heavy metal activities are the most damaging to both and legumes and nodule bacteria [Broos et al. 2005]. Heavy metals, therefore, affect both the viability of rhizobia [Chaudri et al. 2000] and the mechanisms involved in rhizobia- legume symbiosis [Giller et al. 1998; Wani et al. 2006]. The lower rates of metals when used either singly or in combination, in general, did not affect the growth and dry biomass possibly due to the release of organic acid along the root exudates or pH changes of rhizosphere, which regulate the uptake of metals and the metals involved [Jackson et.al. 1990; Prasad, 1999]. However, when metal concentrations become too high, the plant barrier looses its function, probably due to the toxic action of metals, and the uptake massively increase. Further, the growth retardation may be due to loss of cellular turgor resulting in decreased mitotic activities or inhibition of cell elongation [Prasad, 1999]. For instance, though zinc is a plant nutrient at low levels, at higher concentrations found in contaminated soils, zinc ions can be highly phytotoxic [Rout and Das, 2003]. In a study, 7.5mM of zinc extensively damaged the root cortical cells, caused major changes in the nucleus of the root tip cells and consequently, completely inhibited the elongation of pigeonpea (cv. ICPL 87) roots [Sresty and Madhava, 1999]. Similarly, the adverse effects of metals on plant height and shoot growth of legumes are reported [Rout et al. 1997; Chaudri et al. 2000]. For example, the shoot length of soybeans (*Glycine max*) was decreased by 12, 15 and 9%, for nickel, cadmium and chromium, respectively [Gupta et al. 2002] while reduction in root and shoot growth of azuki bean (*Vigna angularis*) and pea under cadmium [Bisessar et al. 1983], greengram under varying levels of chromate [Samantaray et al. 1999], chickpea and greengram under different metals [Wani et al.. 2007a,b; Vijayregan, 1994] and cowpea (*Vigna unguiculata*) under copper stress [Kopittke and Menzies, 2006] is reported. The reduction in the growth of roots following heavy metal application could be due to the inhibition of root cell division/root elongation or due to the extension of cell cycle of the roots. Further, under high metal concentrations, the reduction in root growth could also be due to the direct contact of seedlings roots with metal causing a collapse and subsequent inability of the roots to absorb water from the growing environment [Barcelo et al. 1986].

Toxic effects of heavy metals on extensiveness and proliferation of roots and their subsequent effects on shoot growth lead to suppression of dry matter production. However, the reports on the toxicity of metals on dry matter accumulation in legumes are conflicting. Interestingly, the most toxic metals, cadmium (at 12 mg/kg soil) and hexavalent chromium (at 68 mg/kg) when applied to sandy clay loam soils, increased the dry matter accumulation in pea while copper (1338 mg/kh soil) decreased the biomass production by 18% in pea plants [Wani et al. 2008]. The toxicity of copper could possibly be due to the inactivation of proteins, enzymes or genetic material of pea plants, through the generation of reactive oxygen intermediates. However, the tolerance of pea plants to cadmium and chromium could possibly be due to the formation of cadmium-phytochelatin or chromium-phytochelatin complexes, as reported for other plants [Griller al. 1987] but to the best of our information, no such complexes are reported for legumes. In other study, the effects of heavy metal toxicity on dry matter yield of field pea and fodder vetch (*Vicia* sp.) in contaminated soil is reported [Wang et al. 2002]. Moreover, when two metals are used in combination, the phytotoxicity could either increase (synergistic/additive effect) or decrease (antagonistic effect). The lesser toxicity following combined metal application could be due to the antagonistic effect of one metal over other metals. For example, zinc and cadmium have many physical and chemical similarities as they both belong to group II of the periodic table. They are usually found together in the ores and compete with each other for various ligands. Thus, interaction between zinc and cadmium in the biological system is likely to be similar. The fact that cadmium is a toxic heavy metal and zinc is an essential element which makes this association interesting as it raises the possibility that the toxic effects of cadmium may be preventable or tolerable by zinc [Chowdhury and Chandra, 1987].

4.2.2. Metals Affecting Symbiotic Traits

The formation of nodules on the root systems of legume plants through their host specific rhizobial partners is an important aspect of legume-*Rhizobium* symbiosis for making N available to legumes. The proper development and function of nodules and N_2 fixation however, appears to be hindered by the metal application [Zahran 1999; Wani et al. 2008]. For example, the nodulation in chickpea plants was reduced by the application of cadmium, zinc and copper while chromium showed an increase of 23 (67.5 mg/kg) to 54% (136 mg/kg) [Wani et al. 2007a]. Further, the dual metal combination of cadmium with chromium or cadmium with nickel has shown reduction in nodulation on chickpea plants while combination of chromium with lead stimulated the nodulation. Similarly, chromium significantly increased the number of nodules per plant in greengram while cadmium and copper reduced the number of nodules [Wani et al. 2007b]. In contrast to these findings, cadmium (24 mg/kg soil) and chromium (136 mg/kg soil) significantly increased the nodulation by 53 and 31%, respectively, in pea while copper (1338mg/kg soil) reduced the number of nodules by 25% [Wani et al. 2008]. There are several possibilities as to why these effects are seen- a) one or more of the metals present might have prevented the formation of effectively N_2 fixing nodules by effective *Rhizobium* strains present in the soil [Giller et al. 1993] or b) the metal application might have resulted in the elimination of effective *Rhizobium* strains from the soil [Broos et al. 2004] and as a result, indirectly affected the nitrogen fixation [Chaudri et al. 2000]. Giller et al. [1998] also concluded that clover rhizobia are far more sensitive to the toxic effects of heavy metals than are their host plant. The toxic effect of heavy metals on nitrogen fixation is, according to Giller et al. [1998], clearly due to the toxicity to rhizobia in the soil, which results in their gradual extinction. Further, the reduction in nodulation following metal application may be due to damage to root hairs and some effects associated with plant inducer and *Rhizobium* [Ibekwe et al. 1996, Martenson and Witter, 1990] or nodule structure. The damaged root system as a result of metal toxicity is supposed to be the reason for the lack of proper nodule formation. Apparently, normal bacteroids in these studies were found densely packed in central cells of the nodule, which was surrounded by an intact peribacteroid membrane and infection thread could be seen clearly in the meristematic region of the nodules. The symbiosis therefore, must fail to be effective due to incompatibility at a later stage of the process of infection and nodule development after bacteroid formation. In other study, the cadmium had greatest toxic effect on the nodulation when it was added to the soil either alone or in multiple metal combinations. Moreover, it is believed that metals not only decrease the size of the *Rhizobium* population but also the genetic diversity [Hirsch et al. 1993] including the spontaneous loss of symbiotic plasmids [Cassella et al. 1988], which would result in the loss of the ability of rhizobia to nodulate legumes, as the genes necessary for nodulation are present on this plasmid. In this context, the strains of rhizobia that survived in heavy metal contaminated soils had the least number of plasmid types and the increased number of plasmid bands per isolate [Lazkian et al. 2002]. Hence, it appeared that plasmids were involved in metal tolerance of rhizobia but the function of these plasmids is however, unclear. Similar evidence of toxicity of metals on nodulation in other legumes are reported [Chaudri et al. 1993]. Furthermore, an indicator of nodule activity is the presence of leghemoglobin (LHb), an iron containing protein similar to human hemoglobin, which binds to O_2. The LHb facilitates the O_2 diffusion throughout the interior of the nodule, while bacteroids require O_2 to maintain metabolic function and thus LHb in nodules affect the entire system of nitrogen fixation. It has been reported that the heavy metals adversely affect the synthesis of LHb, and reduce the nitrogenase activity [Skyujins et al. 1986; Wani et al. 2008] as well.

4.2.3. Seed Yield

Legumes when grown in soils contaminated with heavy metals have been shown to be adversely affected in terms of grain yield as well. However, there are conflicting reports on the effect of metals on grain production in legumes. For instance, increase in grain yield in lentil with chromium (68 mg/kg), nickel (290 mg/kg), and mixtures of nickel and lead is reported [Wani et al. 2006). In a similar study, the increase in grain yield in chickpea, pigeonpea, summer moong and lentil following waste water irrigation, having chromium, copper, nickel, lead and zinc, have been reported (Aziz et al. 1996). However, this finding cannot be compared with other results because the refinery waste water used as irrigation for these legume crops had the essential nutrients (g l^{-1}) : N 2.63; P 0.59; S 89; Ca 43.15; K 9; Mg 58.62, chlorides 127.4 and some of the micronutrients, that might have contributed to increase in growth of these legumes and consequently the grain yields. In contrast, cadmium (11.5 mg/kg) decreased the seed yield by 19% in chickpea inoculated with *Mesorhizobium* [Wani et al. 2007a] while at 24 mg/kg, it reduced the seed yield by 26% in greengram inoculated with *Bradyrhizobium* [Wani et al. 2007b]. Copper at 1338 mg/kg decreased the seed yield by 40% in greengram (Wani et al. 2007b) while the same concentration of copper decreased the seed yield by 15% in pea [Wani et al. 2008]. However, chromium (136 mg/kg) increased the seed yield in greengram by 74% [Wani et al. 2007b] while lead (195 mg/kg] increased the seed yield by 9% in chickpea [Wani et al. 2007a]. Interestingly, the average maximum increase of 13 and 8% has been reported to be observed for cadmium (24 mg/ kg) and chromium (136 mg/kg), respectively, in pea [Wani et al. 2008]. Though, chromium and lead significantly enhanced the grain yield but the grain yield decreased gradually with increase in their dosage. Increase in grain yield following chromium and lead could be due to the tolerance of these metals by rhizobial strain or due to the reduction of chromium by the rhizobial strain (Rajkumar et al. 2005) or the development of rhizobia resistant to these metals (Purchase and Miles, 2001). Furthermore, the mixtures of metals are reported to have additive and possibly even multiplicative effects. Generally, the composite application of heavy metals have shown a strong effect on growth, nodulation and seed yield of legumes and have been found far more superior than the single metal application suggesting synergism beyond simple additive effects. In this context, various interactions can occur when plants are exposed to unfavorable concentration of more than one heavy metals. Such combinations could some time exhibit additive, synergistic or antagonistic effect. The antagonistic relationship between metals may result from the competition between the metals for common sites on the surface of the cell with more efficient competitors preventing the uptake of other metal. In literature, both synergistic and antagonistic effects of metals and their consequent effect on grain yield of legumes have been reported. For instance, cadmium alone reduced the chickpea seed mass by 19 % which was synergistically enhanced by 21 and 24% with mixtures of cadmium and nickel and cadmium- chromium-nickel, respectively. The combination of heavy metals sometimes also shows an additive effect as well. For example, when chromium was used with lead, showed an additive effect and increased the seed yield of chickpea by 14% compared with 8.8% observed for each by chromium and lead alone [Wani et al. 2007a]. Among the antagonistic effect, cadmium has shown antagonistic effect on chromium [Wani et al. 2008] or zinc [Chaoui et al. 1997]. The reduction in seed yield following heavy metal application has in general been attributed to the effects of metals on the proliferation of roots and shoots (Ibekwe et al. 1996]. The reduction in roots and shoots then lead to the suppressive effect on dry matter production and consequently the seed yield [Bisessar et al. 1983]. Moreover, the heavy metals are known to affect not only the quantity of grains of various legumes, but they are also known to adversely affect the quality (protein) of seeds (Wani et al. 2007a,b; Wani et al. 2008].

5. Defense against Heavy Metals

5.1. Phytoremediation

Phytoremediation is defined as the use of plants to remove contaminants/pollutants from the environment [Cunningham et al. 1995; Salt et al. 1998]. However, the sensitivity or tolerance of plants towards metals is influenced greatly by plant species and genotypes. Broadly, plants can be grouped into three categories (i) excluders (ii) indicators and (iii) accumulators (figure 3). Among these, the plants belonging to excluder groups are sensitive to metals over a wide range of soil concentrations and survive through restriction mechanisms while indicators show poor control over metal uptake and transport processes and correspondingly respond to metal concentrations in soils. Grasses (e.g., sudangrass, bromegrass, fescue, etc.) and cereal crops (e.g. corn, soybean, wheat, oats, etc.) are included in the excluder and indicator groups, respectively. Plants in the accumulator group do not prevent metals from entering the roots and hence, have evolved specific mechanisms for detoxifying high concentrations of metal accumulated in the cells. Common plants included in this group are tobacco (*Nicotiana tabacum* L.), the mustard (*Brassica compestris*) and members of compositae families (e.g. lettuce, spinach, etc.). Among these accumulators, there are certain plants, which possess exceptionally high metal-accumulating capacity that allows them to survive and even thrive in heavily contaminated soils. These plants are generally called as hyper-accumulators. The term hyper-accumulator was introduced for the first time by Brooks et al. [1977] to describe plants which in their natural habitats were capable of accumulating more than 1000 mg Ni kg^{-1} dry weight of shoots. There are also higher plants capable of accumulating >100 mg Cd kg^{-1}, >1000 mg Cu, Ni, and Pb kg^{-1}, and >10 000 mg Zn kg^{-1} dry matter in their shoots when grown in heavy metal rich soils (Baker and Brooks, 1989; Baker et al. 1994a; Brooks, 1998; Baker et al. 2000). However, hyperaccumulators often accumulate only a specific heavy metals and have a low biomass production leading to long duration times before completion of remediation (Cunningham et al. 1995). The first hyper-accumulator plants belonged to the family *Brassicaceae* and *Fabaceae* and since then more than 400 plant species have been reported as hyper-accumulator plants and a considerable number of species show the capacity to accumulate two or more elements. Generally, hyper-accumulator plants accumulate one to three fold higher concentrations of metals than do the non-hyper-accumulator plants. Moreover, hyper-accumulators usually have a shoot to root metal concentration ratio of >1, whereas non-hyper-accumulator plants generally have higher metal concentrations in roots than in shoots [Baker et al. 1994b; Shen et al. 1997].

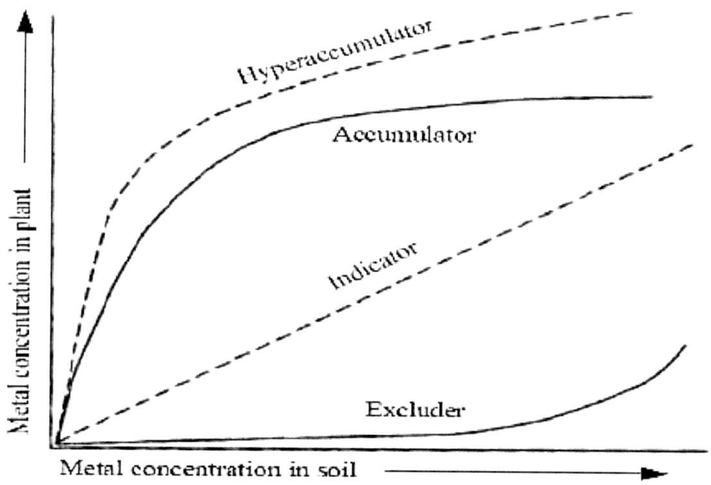

Figure 3. Strategies of metal uptake by plants depending on the concentration of metal in soil [adapted from Adriano, 2001].

The uptake and accumulation of metals in hyper-accumulator plants is a complex process that involves- (a) transport of metals across the plasma membrane of root cells (b) xylem loading and translocation and (c) detoxification and sequestration of metals both at the whole plant and cellular levels. However, mechanisms of metal accumulation in these plants are so far not fully explained but the accumulation ability of a plant is determined by the uptake and intracellular transportation capacity of plant (figure 4). The metal pollution of the soils may be corrected by using phyto-remediation technology, which is a cost-effective and plant-based approach and has shown considerable promise in field trials for environmental cleanup. Phyto-remediation technology can be applied to clean up the metal contaminated sites considering the contaminants, the site conditions, the level of clean-up required, and the plant genotype.

5.2. Mechanism Involved in Metal Tolerance by Plants

Various mechanisms involved in the detoxification of metals and thus, tolerance to metal stress includes- (i) binding of metal to cell wall and extracellular exudates (ii) reduced uptake or efflux pumping of metals by the plasma membrane (iii) chelation of metals in the cytosol by peptides such as phytochelatins (iv) repair of stress damaged proteins and (v) the compartmentation of metals in the vacuoles by tonoplast located transportes. The mechanism adopted by plants for the removal/detoxification of heavy metals from contaminated sites is briefly discussed in following section.

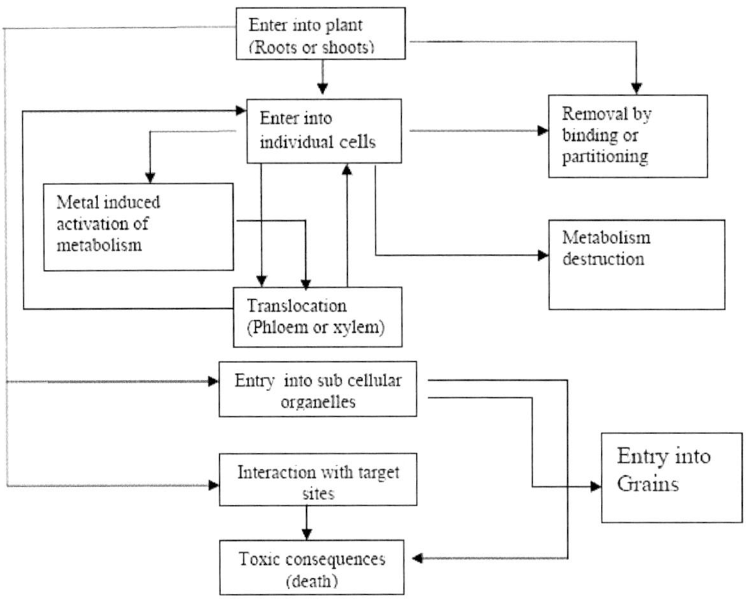

Figure 4. Flow chart showing the sequence of events from metal entry into a plant to death of the plant.

5.2.1. Phytostabilization

In phytostabilization process, certain plant species are used to immobilize metals in soil, through (i) absorption and accumulation of metals in roots by vacuole sequestration or cell wall binding (ii) complex formation by organic acids, root exudates etc. (iii) adsorption onto roots or precipitation within the root zone by complex formation and (iv) physical stabilization of metals on to soils. During this process, the mobility of contaminants is reduced and consequently, the migration of metals to the sites (groundwater or air) is prevented. This process generally helps to re-establish a vegetative cover at sites contaminated heavily with metal concentrations [Tordoff et al. 2000].

5.2.2 Phytofiltration

Phytofiltration involves the combination of rhizofiltration (plant roots) or blastofiltration (seedlings) to absorb or adsorb pollutants, mainly metals, from water and aqueous waste streams [Prasad and Freitas 2003]. The mechanisms involved in biosorption include complexation, ion exchange, micro precipitation, chemisorption, surface adsorption, and hydroxide condensation onto the bio-surface. However, either any one or combination of these processes can lead to the removal of metals from contaminated sites. During these events, plant roots or seedlings grown in aerated water absorb, precipitate and concentrate toxic metals from polluted effluents [Dushenkov and Kapulnik, 2000; Elless et al. 2005] and make the environment free of contamination.

5.2.3. Phytovolatilization

Among the various metals discharged from industrial operation, some metals (such as As, Hg, and Se) may exist as gaseous species in the environment. Recently, workers have emphasized that there could be the use of plants (either naturally-occurring or genetically-modified) possessing the ability to absorb elemental forms of these metals from the contaminated soil, so that they could biologically be converted in to gaseous forms within the plant itself, and consequently, are released into the atmosphere without adversely affecting microbial composition of soils or other plant species being grown in those areas. Such process is generally called as phytovolatilization. In a study, the volatilization of Se from plant tissues was suggested as means of its detoxification. The garlicky odour of plants that accumulate selenium is suggestive of release of volatile selenium compounds. Among the different phytoremediation technologies, this is the most controversial technologies since there is doubt about whether the volatilization of these elements into the atmosphere is desirable or safe (Watanabe, 1997), as reported for Hg and Se [Suszcynsky and Shann, 1995].

5.2.4. Phytoextraction

Phytoextraction, also called phytoaccumulation, refers to the uptake and translocation of metal contaminants in the soil by plant roots into above-ground organs (e.g., shoots, flower or grains) of the plants. For phytoextraction to be effective, plants must take up heavy metals from the soil, tolerate high plant or soil-heavy metal levels, and produce sufficient harvestable biomass [Chaney et.al. 1997; Pilon-Smit, 2005]. Phyto-remediation and phyto-extraction are used sometimes incorrectly as synonyms, but phytoremediation is a concept, whereas phytoextraction is a specific clean-up technology [Prasad and Freitas, 2003]. Phytoextraction involves different steps- (i) metal uptake by plants through apoplastic and symplastic pathways (ii) xylem loading of metals (iii) symplastic translocation to shoots (iv) biological conversion of metals in to gaseous forms and (v) release of metals in to the atmosphere in the gaseous forms. However, little is known about the molecular mechanisms of metal entry into root cells except for iron. However, once metal ions enter the roots, they can either be stored or exported to the shoot. Metal transport to the shoot primarily takes place through the xylem.

5.3. Enzymatic Detoxification of Metal

It is suggested that the toxicity of heavy metals, among other things, might result from the production of reactive oxygen species (ROS), such as superoxide, hydrogen radicals and hydrogen peroxide [Dietz et al. 1999, Schützendübel and Polle, 2002]. Plants growing on metal-contaminated sites therefore, need to develop some degree of tolerance to metal toxicity or ability to alleviate the toxicity of these metals in order to survive in contaminated sites. The suggested mechanisms for metal tolerance by plants include - (a) synthesis of organic compounds and its complexation with metals (b) compartmentalization of metals in certain cell compartments and (c) direct efflux of metals out the cell. However, once metal enters cell, induce the synthesis of certain proteins, such as, ferritins, metallothioneins, and phytochelatins that in turn, participate in excess metal storage and detoxification. When these systems are overloaded, oxidative stress defense mechanisms are activated.

5.3.1 Antioxidant Defenses

Exposure to heavy metals and other environmental factors can disrupt cellular homeostasis and enhance the production of several activated oxygen species (figure 5). Such reactive oxygen species includes superoxide (O_2^-), hydroxyl radicals (OH), oxygen singlet (O_2) and hydrogen peroxide (H_2O_2) and can be produced continuously by plants. These ROS in turn, affects various metabolic stages of plants including legumes (figure 6). For instance, nodules have a high capacity to produce these damaging chemicals because of the high rates of respiration, the strong reducing conditions required to reduce N_2, the tendency of leghaemoglobin to auto-oxidize and the likely ability of nitrogenase to directly reduce oxygen [Dalton, 1995]. However, plants possess very efficient enzymatic and non-enzymatic antioxidant defense systems that allows the scavenging of ROS and provide protection to plants from oxidative damage [Gratao et al. 2005]. Hydrogen peroxide (H_2O_2) and superoxide radical (O_2^-) are highly reactive and induce lipid peroxidation, thereby affecting the structural and permeability of cellular membrane [Dai et al. 1997]. Lipid peroxidation is associated with oxidative stress and contributes to the breakdown of the functional and the structural integrity of biological membranes, resulting in increases in the permability of the plasma membrane, which leads to K^+ ion and other solutes, amino acids oxidation and eventually may cause cell death [Gratao et al. 2005]. Plant species however, metabolizes activated oxygen species by invoking an increased activity of enzymatic antioxidants such as superoxide dismutase (SOD), catalase (CAT), peroxidase (POD) and glutathione reductase (GR) [Kondo and Kawashima, 2000; Isik et al. 2007]. Plants also possess large amounts of non-enzymatic antioxidants such, as proline, flavinoids, ascorbate (ASC) and carotenoids [Arora et al. 2002; Matysik et al. 2002] that aid the detoxification of the metals. In legume root nodules, the O_2 problem is dealt with by three mechanisms [Appleby, 1984; Dension, 1992]:- (i) an abundant amount of the O_2-binding protein leghaemoglobin to facilitate the flux of O_2 to symbiotic bacteria (*Rhizobium* or *Bradyrhizobium*), while maintaining an extremely low, nontoxic concentration of free O_2 (ii) a high rate of respiratory O_2 consumption and (iii) a variable diffusion barrier that controls the entry of O_2 into the central infected regions. The diffusion of O_2 into the nodule interior can be regulated by alterations in relative amounts of air, liquid or occluding glycoproteins within intercellular spaces [Van Cauwenberghe et al. 1993]. A major defense against activated O_2 in nodules is provided by ascorbate peroxidase, a hemoprotein that uses the reducing power of ascorbate to scavange H_2O_2. Although ascorbate peroxidase may be regarded as a nearly universal "housekeeper" in the cytosol and chloroplasts of plant cells, it is especially abundant in the cytosol of N_2-fixing root nodules, where it makes up almost 1% of the total soluble protein.

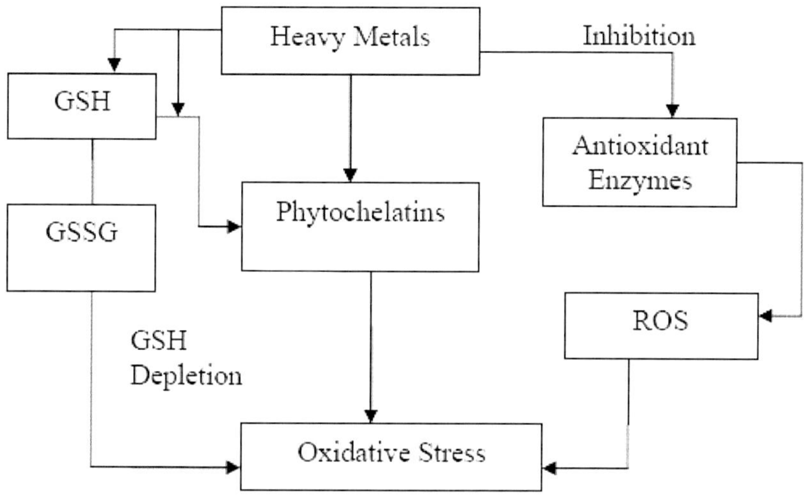

Figure 5. Heavy metal induced oxidative stress and related cellular processes.

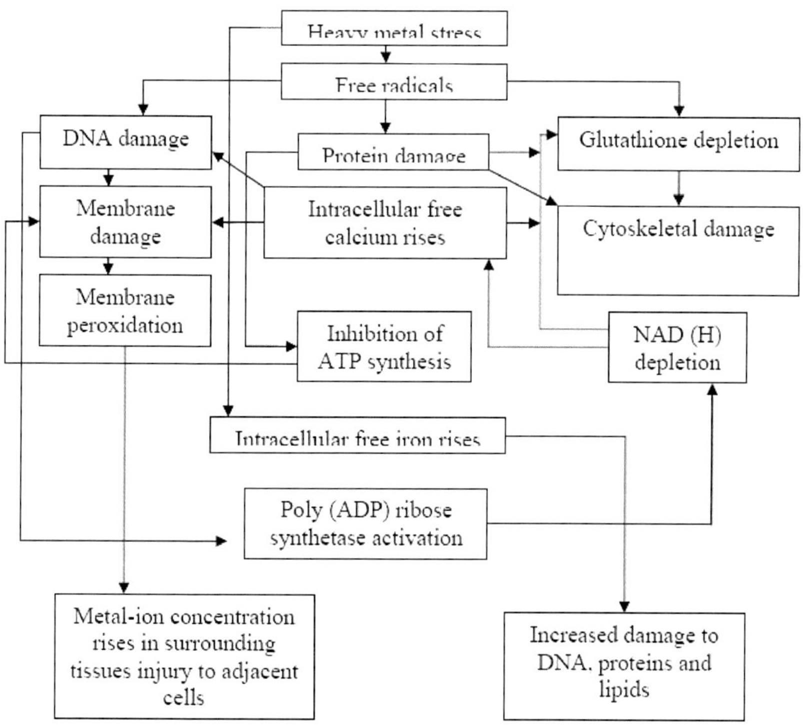

Figure 6. Cell components damaged by heavy metal generated free radicals.

Glutathione is an important antioxidant and metal chelator and has been shown to play an important role in cadmium detoxification in legume symbiont *Rhizobium leguminosarum* [Lima et al. 2006]. Two strains expressing different degrees of tolerance to cadmium stress were used and the influence of the addition of extracellular glutathione to the growth medium was determined. Oxidative stress indexes and ROS scavenging enzymes were evaluated, as well as changes in glutathione and oxidized glutathione levels. These findings suggested that $CdCl_2$ imposes oxidative stress in *R. leguminosarum*, which was characterized by an increase in oxidized glutathione formation and an induction of the activities of SOD, glutathione peroxidase (GPX), GR and CAT. Addition of reduced glutathione to the growth media showed a protective effect, especially in the sensitive strain. In a similar study, horsegram [*Macrotyloma uniflorum* (Lam.) verdc. Cv VZMI] and chickpea (cv annogiri) were exposed to different levels of lead stress and were observed for the extent of oxidative damage (lipid peroxidation), antioxidative response and the accumulation of lead in roots and shoots of both plants. Lead treated plants showed increased levels of lipid peroxidation (increased malondialdehyde content) coupled with the increase in the activities of SOD, CAT, POD, GR and glutathione S-transferase (GST) compared to control plants. A concentration dependent change in the activity of antioxidative enzymes was observed following lead treatment. For example, lead at 800 ppm increased the SOD, CAT and POD activities by 2-3 folds, GR activity by 3-5 fold and GST activity by 3-4 fold of roots and leaves of both plants. These findings suggest that Pb toxicity caused oxidative stress in legumes and the antioxidative enzymes could play a pivotal role against oxidative injury [Reddy et al, 2005]. Similarly, when phaseolus seedlings were treated with mercury (0.05 mM and 0.4 mM) and chromium (0.5 and 1.0 mM), inhibited the root and hypocotyls length but considerably increased the POD and oxidase. Ascorbate peroxidase (AP) is yet another key enzyme that scavenges potentially harmful H_2O_2 and prevents oxidative damage in plants, especially in N_2-fixing legume root nodules. For instance, the nodule endodermis of alfalfa (*Medicago sativa*) had elevated levels of AP protein, as well as the corresponding mRNA transcript and substrate (ascorbate). Enhanced AP protein levels were also found in cells immediately peripheral to the infected region of soybean, peas, clover (*Trifolium pretense*) and common bean nodules. The presence of high levels of antioxidants suggest that respiratory consumption of O_2 in the endodermis or nodule parenchyma may be an essential component of the O_2-diffusion barrier that regulates the entry of O_2 into the central region of nodules and ensures optimal functioning of nitrogenase [Dalton et al. 1998]. In other study, pea plants exposed to different concentrations of Cr (VI) for 7 days, affected total SOD activity of root mitochondria. At lower concentration, Cr (VI) increased the SOD activity by 29%, whereas the higher concentration of Cr (VI) significantly inhibited the SOD activity [Dixit et al. 2002]. Furthermore, a concentration dependent decrease in activity of catalase following chromium application was observed [Jain et al. 2000]. The increase in antioxidant enzymes activity observed might have been in direct response to the generation of superoxide radical by chromium induced blockage of the electron transport chain in the mitochondria. The increase recorded due to Cr (VI) indicated that Cr (VI) addition probably generates more singlet oxygen than Cr (III). The decrease in the activity of the enzyme as the concentration of the Cr increased might be because of the inhibitory effect of chromium ions on the enzyme system itself.

Similarly, the SOD and POD are two major protective enzymes reported for kidney beans. When plants grow in the polluted environment, changes of these two enzymes in plant organs can eliminate free radicals caused from pollution, and prevent the plants from injury. For example, addition of cadmium and zinc to the soils caused changes in SOD and POD activity in the plant organs suggesting that SOD and POD were stimulated to inhibit the free radicals. GSH is yet another low molecular mass thiol involved in the antioxidant defense through the ascorbate/GSH cycle. It provides protection to plant against abiotic and biotic stresses including heavy metal tolerance and xenobiotic detoxification [Foyer and Renmenber, 2000] .One of the characteristics of legumes is the presence of homoglutathione (γ-glutamylcysteine-_-alanine; hGSH), a homologue of GSH, which is present instead of or in addition to GSH. The concentration of GSH and hGSH has been reported more in the infected zones of both determinate and indeterminate nodules than in the other organs of the legume plants [Frendo et al. 1999; Matamoros et al. 1999]. Additionally, during the natural and induced senescence, thiol content and synthesis decline and the thiol pool becomes more oxidized [Evans et al.1999, Matamoros et al. 1999]. These results strongly suggest that GSH/hGSH play a role in the maintenance of the mature nodule. Moreover, GSH/hGSH are also known to play a role in the formation and the regulation of the nodule meristem, as they have shown a very important role in the functioning of root tip meristem [Vernoux et al. 2000] and in the regulation of the number of nodules per plant. Recently, the synthesis of glutathione reductase by the root and nodules of pea, grown in soils exposed to different regimes of nickel and [Wani et.al. 2007c]. In general, the glutathione reductase activity was found more in roots than nodules and protected the pea plants from the toxicity of both nickel and zinc. Similar reports on the effects of heavy metals on the glutathione reductase activity of other legumes [Reddy et al. 2005] and other detoxifying agents produced by rhizobia are reported [figueira et al. 2005].

5.3.2. Phytochelatin

As a result of increased uptake, metals interact with many cellular components, thereby interfering with the normal metabolic functions, causing cellular injuries and in extreme cases death of the plants through the generation of reactive oxygen intermediates. Although all reactive oxygen intermediates are more or less highly reactive and are toxic to living organisms, the ultimate damaging effect is, however, mainly by singlet oxygen ($^1O_2$) and hydroxyl radical (HO^*). The rapid and specific reaction of these radicals in turn damages all classes of bio-molecules [Breen and Murphy, 1995]. Oxidative stress due to the cadmium treatment has also been reported in pea leaves [McCarthy et al. 2001] while copper is known to interfere with oxidative enzymes in bean leaves [Shainberg et al. 2001]. However, legumes could be tolerant to the higher concentrations of metals probably due to the synthesis of phytochelatins (PC); a simple γ – glutamyl peptides [Grill et al. 1985]. The synthesis of phytochelatins is induced by most of the heavy metals including the multi – atomic anions [Maitani et al. 1996) in most of the higher plants [Gekeler et al. 1989] and the phytochelatin synthetase involved the synthesis of phytochelatins requires metals for its activation [Grill et al. 1989]. Since the phytochelatin synthetase activity has been detected largely in roots [Steffens, 1990], and root is the first organ exposed to the metal ions in the soil, the roots of the plants may provided an effective means of restricting the uptake of metals by forming a metal– phytochelatin complex. The phytochelatin– metal complexes has been reported mostly for cadmium but a few reports of phytochelatin forming complexes with copper is also available (Grill et al., 1987).

6. CONCLUSION

From these studies it is clear that the heavy metals invariably affect not only the growth and yield of various legumes but also the quality of grains of legumes. Although the regulatory criteria have been established with limits on environmental and human health, phytotoxic thresholds have not been listed for heavy metals. The metal concentration at which plants show phytotoxicity depend on a number of factors including soil type, legume genotypes and concentration and composition of metal combinations. Research is therefore, urgently required to examine the toxic effects of individual metal and their corresponding concentration on legumes grown under different agronomic zones of the world. However, to overcome the toxic effects of metals, bioremediation is being suggested as an alternative to chemical mode of detoxification/removal system. Among the various bioremediation approaches, hyper-accumulator plants have received considerable attention in recent times for their role in remediating the metal contaminated soils. The phyto-remediation approach, however, has certain limitation. For example, metal hyper-accumulating plants identified so far produce small biomass, and are not very adaptable to harsh environment. These limitations need to be addressed very seriously if the objective of this technology on commercial scale is to be realized. The low biomass production metal uptake efficiency can be increased by engineering common plants with hyper-accumulating genes. Furthermore, since major targets of phyto-remediation technology are likely to be areas highly contaminated with metals, extensive research programs on the basic and applied processes and problems affected by metals are therefore, needed before significant progress is to be made. Thus, a collective effort of researchers involving plant biologists, soil chemists, microbiologists, and environmental engineers becomes eminent for obtaining greater success in the field of phyto-remediation.

REFERENCES

Adriano, DC. Trace elements in terrestrial environments. Biogeochemistry, bioavailability and risks of metals. New York, NY: *Springer-Verlag*,; 2001.

Alloway, BJ. Introduction. Heavy Metals in Soils. *New York: Blackie Academic and Professional;* 1995; 3-9.

Arora, A; Sairam, RK; Srivastava, GC. Oxidative stress and antioxidant system in plants. Curr. Sci, (2002). 82, 1227-1238.

Appleby, CA. Leghemoglobin and *Rhizobium* respiration. *Ann. Rev. Plant Physiol*, (1984). 35, 443-478.

Aust, SD. Oxidation of persistant environmental pollutants by a white rot fungus. *Science,* (1985). 228, 1423-1436.

Aziz, O; Inam, A; Samiullah Siddiqi, AH. Long term effects of irrigation with petrochemical industry wastewater. *J. Environ. Sci. Health Part A*, (1996). 31, 2595-2620.

Baccouch, S; Chaoui, A; El Ferjani, E. Nickel-induced oxidative stress damage and antioxidant responses in Zea mays shoots. *Plant Physiol. Biochem*, (1998). 36, 689-694.

Bagchi, D; Bagchi, M; Stohs, SJ. Chromium (VI) induced oxidative stress, apoptotic cell death and modulation of p53 tumor suppressor gene. *Mol. Cellu. Biochem*, (2001). 222,

149-158.

Baker, AJM; Brooks, RR. Terrestrial higher plants which hyperaccumulate metallic elements—*a review of their distribution, ecology, and phytochemistry. Biorecovery*, (1989). 1, 81–126.

Baker, AJM; McGrath, SP; Reeves, RD, Smith, JAC. Metal hyper-accumulator plants: a review of the ecology and physiology of a biological resource for phyto-remediation of metal-polluted soils. In: Terry N, Baelos G, (eds). Phytoremediation of contaminated soil and water. Boca Raton, FL: Lewis Publishers; 2000; 85-107.

Baker, AJM; Reeves, RD, Hajar, ASM. Heavy metal accumulatoion and tolerance in British population of the metallophyte Thalaspi caerulescens J and C Presl (Brassicaceae). (1994a). New. Phytol, 127, 61-68.

Baker, AJM; McGrath, SP; Sidoli, CMD; Reeves, RD. The possibility of *in situ* heavy metal decontamination of polluted soils using crops of metal accumulating plants. *Resources, Conservation and Recycling,* (1994b). 11, 41-49.

Barcelo, J; Poschenrieder, C; Gunse, B. Water relations of chromium VI treated bush bean plants (*Phaseolus vulgaris* L cv contender) under both normal and water stress conditions. *J. Exp. Bot*, (1986). 37, 178-87.

Bera, AK; Kanta-Bokaria; AK; Bokaria, K. Effect of tannery effluent on seed germination, seedling growth and chloroplast pigment content in mungbean (*Vigna radiata* L.Wilczek*). Environ Ecol*, (1999). 17, 958- 61.

Bibi, M; Hussain, M. Effect of copper and lead on photosynthesis and plant pigments in black gram (*Vigna mungo* L*.). Bull. Environ. Contam. Toxicol*, (2005). 74, 1126–1133.

Bisessar, S; Rinne, RJ; Potter, JW. Effect of heavy metals and *Meloidogyne hapla* on celery grown on organic soil near nickel refinery. *Plant Dis*, (1983). 67, 11–14.

Breen, AP; Murphy, JA. Reaction of oxyl radicals with DNA. *Free Rad. Biol. Med.* (1995) 18, 1033-1077.

Brooks, RR; Lee, J; Reeves, RD; Jaffre, T. Detection of nickeliferous rocks by analysis of herbarium species of indicator plants. *J. Geochem. Explor*, (1977). 7, 49–57.

Brooks, RR. Plants that hyper-accumulate heavy metals Wallingford: *CAB International.* 1998. p. 384.

Broos, K; Uyttebroek, M; Mertens, J; Smolders, E. A survey of symbiotic nitrogen fixation by white clover grown on metal contaminated soils. *Soil Biol. Biochem,* (2004). 36, 633-640.

Broos, K; Beyens, H; Smolders, E. Survival of rhizobia in soil is sensitive to elevated zinc in the absence of the host plant. *Soil Biol. Biochem*, (2005). 37, 573-579.

Cakmak, I; Hrst, WJ. Effect of aluminium on net efflux of nitrate and potassium from root tips of soybean (glycine max L*.). j. Plant Physiol*, (1991). 130, 400-403.

Cassella S., Frassnetts S., Lupi F., Squartini A. Effect of cadmium, chromium and copper on symbiotic and free living *Rhizobium leguminosarum* bv. trifoli. FEMS Microbiol. lett, (1998). 49, 343-347.

Cataldo, DA; Garland, TR; Wildung, RE. Cadmium uptake kinetics in intact soybean plants. *Plant Physiology*, (1983). 73, 844-848.

Chaney, RL; Malik, M; Brown, YM; Brewer, SL; Angle, EP. Phytoremediation of soil metals. *Curr. Opinion Biotechnol*, (1997). 8, 279-283.

Chaudri, AM; McGrath, SP; Giller, KE; Rietz, E; Sauerbeck, DR. Enumeration of indigenous *Rhizobium leguminosarum* bv. trtfolii in soils previously treated with metal contaminated sewage sludge. *Soil Biol. Biochem*, (1993). 25, 301-309.

Chaudri, AM; Allain, CMG; Barbosa-Jefferson, VL; Nicholson, FA; Chambers BJ; McGrath, SP. A study of the impacts of Zn and Cu on two rhizobial species in soils of a long-term field experiment. *Plant Soil*, (2000). 221, 167-179.

Chaoui, A; Mazhoudi, S; Ghorbal, MH; Ferlani, EE. Cadmium and zinc induction of lipid peroxidation and effects on antioxidant enzyme activities in bean (*Phaseolus vulgaris* L.). *Plant Sci.* (1997). 127, 139-147.

Chowdhury, BA; Chandra, RK. Biological and health implications of toxic heavy metals and essential trace element interactions. *Progr. Food Nutr. Sci*, (1987). 11, 55-113.

Clemens, S. Molecular mechanism of plant metal tolerance and homeostasis. *Plant*, (2001). 212, 475-486.

Cunningham, SD; Berti, WR; Huang, JW. Phytoremediation of contaminated soils. *Trends in Biotechnol,* (1995). 13, 393–397.

Dai, Q; Yan, B; Huang, S; Liu, X; Peng, S; Miranda, MLL; Chavez ,AQ; Vergara, BS;
Olszyk, DM. Response of oxidative stress defense systems in rice (*Oryza sativa*) leaves with supplemental UV-B radiation. *Physiol. Plant*, (1997). 101, 301-308.

Dalton, DA. Antioxidant defenses of plants and fungi. In: Ahmad S (ed.). Oxidative stress and antioxidant defenses in biology. New York : *Chapman and Hall*; 1995; 298-355.

Dalton, DA; Joyner, SL; Becana, M; Iturbe-Ormaetxe, I; Chatfield, JM. Antioxidant defenses in the peripheral cell layers of legume root nodules. *Plant Physiol*, (1998). 116, 37-43.

Das, P; Samantaray, S; Rout, GR. Studies on cadmium toxicity in plants: a review. *Environ Pollut*, (1997). 98, 29-36.

Denison, FR. Mathematical modeling of oxygen diffusion and respiration in legume root nodules. *Plant Physiol*, (1992). 98, 901-907.

Dietz, KJ; Baier, M; Kramer, U. Free radicals and reactive oxygen species as mediators of heavy metal toxicity in plants. In: Prasad, MNV., Hagemeyer (eds.). Heavy metal stress in plants: from molecules to ecosystems. Berlin: *Springer-Verlag*; 1999; 73-97.

Dietz,K-J; Tavakoli, N; Kluge, C; Mimura, T; Sharma, SS; Harris, GC; Chardonnens AN; Golldack, D. Significance of the V type ATPase for the adaptation to stressful growth conditions and its regulation on the molecular and biochemical level. *J. Exp. Bot*, (2001). 52, 1969-1980.

Dixit, V; Pandey, V; Shyam, R. Chromium ions inactivate electron transport and enhance superoxide generation in vivo in pea (*Pisum sativum* L.cv. Azad) root mitochondria. *Plant Cell Environ*, (2002). 25, 687- 90.

Dumestre, A; Sauve, S; McBride, M; Baveye, P; Berthelin, J. Copper speciation and microbial activity in long-term contaminated soils. *Arch. Environ. Contam. Toxicol*, (1999). 36, 124–131.

Dushenkov, S; Kapulnik, Y. Phytofilitration of metals. In: Raskin, I; Ensley, BD. (eds.). Phytoremediation of toxic metals using plants to clean-up the environment. New York: Wiley; 2000; 89–106.

Elless, PM; Poynton, YC; Williams, AC; Doyle P.M., Lopez C.A., Sokkary A.D., et al. Pilot-scale demonstration of phytofiltration for drinking arsenic in New Mexico drinking water. Water Res, (2005). 39, 3863-3872.

Ernst, WHO; Verkleij, JAC; Schat, H. Evolutionary biology of metal resistance in Silene vulgaris. *Evol. Trends Plants*, (1990). 4, 45-51.

Ernst, WHO; Verkleij, JAC; Schat, H. Metal tolerance in plants. *Acta. Bot. Neerl*, (1992). 41, 229-248.

Ernst, WHO. The origin and ecology of contaminated, stabilized and nonpristine soils. In: Vangronsveld, J; Cunningham, SD. (eds.). Metal-contaminated soil. New York: *Springer*; 1998; pp. 17-29.

Evans, PJ; Gallesi, D; Mathieu, C; Hernandez, MJ; de Felipe, M;Halliwell, B. Oxidative stress occurs during soybean nodulesenescence. *Planta*, (1999). 208, 73–79.

Fernandes, JC; Henriques, FS. Biochemical, physiological, and structural effects of excess copper in plants. *Bot Rev*, (1991). 57, 246-273.

Figueira, EMAP; Lima, AIG; Pereira, SIA. Cadmium tolerance plasticity in Rhizobium leguminosarum bv. Viviae: glutathione as a detoxifying agent. *Can. J.Microbiol.* (2005). 51, 7-14.

Fodor, F; Boddi, B; Sarvari, E; Zaray, G; Cseh, E; Lang, F. Correlation of iron content, spectral forms of chlorophyll and chlorophyll-proteins in iron deficient cucumber (*Cucumis sativus*). *Physiol Plant*, (1995) 93, 750-756.

Foyer, CH; Rennenberg, H. Regulation of glutathione synthesis and its role in abiotic and biotic stress defence. In: Brunold, C. (ed.). Sulfur Nutrition and Sulfur Assimilation in Higher Plants. *Bern: Paul Haupt*; 2000; pp. 127–153.

Franco, E; Alessandrelli, S; Masojidek, J; margonelli, A; Giardi, Mt. modulation of D1 protein turnover under cadmium and heat stress monitored by [s-35] methionine incorporation. *Plant Sci*, (1999). 144, 53-61.

Frendo, P; Gallesi, D; Turnbull, R; Van de Sype, G; Hérouart, D; Puppo, A. Localisation of glutathione and homoglutathione in *Medicago truncatula* is correlated to a differential expression of genes involved in their synthesis. *Plant J*, (1999). 17, 215–219.

Fuhrer, J. Ethylene biosynthesis and cadmium toxicity in leaf tissue of beans (*Phaseolus vulgaris* L.). *Plant Physiol*, (1988). 70, 162-167.

Geiken, B; Masojidek, J; Rizutto, M; pompili, ML; giardi, MT. incorporation of [s-35] methionine in higher plants reveals that stimulation of the D1 reaction center II protein turnover accompanies tolerance to heavy metal stress. *Plant Cell Environ*, (1998). 21, 1265-1273.

Gekelr, W; Grill, WE; Winnacker, EL; Zenk, MH. Survey of the plant kingdom for the ability to bind heavy metals through phytochelatins. *Z Naturforsch*, (1989). 44 c, 361-369.

Giller, KE; McGrath, SP; Hirsch PR. Absence of nitrogen fixation in clover grown on soil subject to long-term contamination with heavy metals is due to survival of only ineffective Rhizobium. *Soil Biol. Biochem*, (1989). 21, 841–848.

Giller, KE; Nussbaum, R; Chaudri, A M; McGrath, SP. *Rhizobium meliloti* is less sensitive to heavy metal contamination in soil than *R. leguminosarum* bv. trtfolii or *R. loti*. Soil Biol. Biochem, (1993). 25, 273-278.

Giller, KE; Witter, E; McGrath, SP. Toxicity of heavy metals to microorganisms and microbial processes in agricultural soils: a review. *Soil Biol. Biochem*, (1998). 30, 1389-1414.

Gratao, PL; Polle, A; Lea, PJ; Azevedo R.A. Making the life of heavy metal stressed plants a little easier. *Func Plant Biol*, (2005). 32, 481–494.

Grill, E., Winnacker, E-L., Zenk, M.H Phytochelatins: the principal heavy-metal complexing peptides of higher plants. *Science,* (1985). 230, 674-676.

Grill, E; Winnacker, E-L; Zenk, MH. Phytochelatins, a class of heavy metal-binding peptides from plants, are functionally analogous to metallothioneins. *Proc. Natl. Acad.*

Sci. USA, (1987). 84, 439-443.

Grill, E; Loffler, S; Winnacker, E-L; Zenk, MH. Phytochelatins, the heavy-metal binding peptides of plants, are synthesized from glutathione by a specific glutamylcysteine dipeptidyl transpeptidase (phytochelatin synthase). *Proc. Natl. Acad. Sci. USA*, (1989). 86, 6838-6842.

Gupta, A; Meyer, JM; Goel, R. Development of heavy metal resistant mutants of phosphate solubilizing *Pseudomonas sp.* NBRI4014 and their characterization. *Curr. Microbiol.* (2002). 45, 323-327

Hartley-Whitaker, J; Ainsworth, G; Vooijs, R; Ten Bookum, W; Schat, H; Meharg, AA.

Phytochelatins are involved in differential arsenate tolerance in *Holcus lanatus*. Plant Physiol, (2001). 126, 299-306.

Heckman, JR; Angle, J S; Chaney, RL. Residual effects of sewage sludge on soybean. II. Accumulation of soil and symbiotically fixed nitrogen. *J. Environ. Qual*, (1987). 16, 118-124.

Hernández, LE; Cooke, DT. Modification of the root plasma membrane lipid composition of cadmium-treated Pisum sativum, *J. Exp. Bot*, (1997). 48, 1375-1381.

Hewitt, E.J. tals and micronutrients:uptake and utilization by plants. In: Robb, DA; Pierpoint, ES (eds.). London: academic Press; 1983; 277-300.

Hirsch, PR; Jones, MJ; McGrath, SP; Giller, KE. Heavy metals from past application of sewage sludge decrease the genetic diversity of *Rhizobium leguminosarum* biovar trifolii populations. *Soil Biol. Biochem*, (1993). 25, 1485-1490.

Ibekwe AM, Angle JS, Chaney RL, Van Berkum P (1996) Zinc and cadmium toxicity to alfalfa and its microsymbiont. *J Environ Qual* 25: 1032-1040.

Isik, K., Kayali, H.A.K., Sahin, N., Gundogdu, E.O., Tarhan, L Antioxidant response of a novel *Streptomyces* sp. M3004 isolated from legume rhizosphere to H_2O_2 and paraquat. *Process Biochem*, (2007). 42, 235-243.

Jackson ,PJ; Unkefer, PJ; Delhaize, E; Robinson, N.J. Mechanisms of trace metal tolerance in plants. In: Katterman, F (ed.), Environmental Injury to Plants. San Diego: Academic; 1990; 231-55.

Jain, R; Srivastava, S; Madan, VK; Jain, R. Influence of chromium on growth and cell division of sugarcane. Ind. *J. Plant Physiol*, (2000). 5, 228-31.

Jones, DL; Kochian, LV. Aluminum interaction with plasma membrane lipid and enzyme metal binding sites and its potential roles in Al cytotoxicity. FEBS Lett, (1997). 400, 51-57.

Kabata-Pendias, A; Pendias, H. Trace Elements in Soils and Plants. Boca Raton, Fla: CRC Press; 2001; 413.

Khan, MR; Khan, MW; Singh, K. Growth performance of chickpea under the influence of nickel and cobalt as soil pollutants. *J. Ind. Bot.Soc*, (1996). 75, 193-196.

Kappus, H. Lipid peroxidation: mechanisms, analysis enzymology and biological relevance. In: Sites, H (ed.) Oxidative stress. London: Academic Press; 1985; 273-310.

Kondo, N; Kawashima, M. Enhancement of the tolerance to oxidative stress in cucumber

(*Cucumis sativus* L.) seedlings by UV-B irridation: Possible involvement of phenolic compounds and antioxidative enzymes. *J. Plant Res,* (2000). 113, 311-317.

Kopittke1, PM; Menzies NW. Effect of Cu toxicity on growth of cowpea (Vigna unguiculata) *Plant and Soil*, (2006). 279, 287–296.

Krupa Z, Baszynski T. Some aspects of heavy metals toxicity towards photosynthetic apparatus—direct and indirect effects on light and dark reactions. *Acta Physiol Plant*, (1995). 17, 177– 190.

Lakzian, A; Murphy, P; Turner, A; Beynon, JL; Giller, KE. *Rhizobium leguminosarum* bv. viciae populations in soils with increasing heavy metal contamination: abundance, plasmid profiles, diversity and metal tolerance. *Soil Biol. Biochem*, (2002). 34, 519- 529.

Lasat, MM; Pence, NS; Garvin, DF; Ebbs, SD; Kochian, LV. Molecular physiology of zinc transport in the Zn hyperaccumulator Thlaspi caerulescens. *J. Exp. Bot*, (2000). 51, 71-79.

Lima, AIG; corticeiro, SC; Figueira, EMAP. Glutathione mediated cadmium sequestration in *Rhizobium leguminosarum.* Enz. Microb. Technol, (2006). 39, 763-769.

MacLaughin, MJ; Singh, BR. Cadmium in soils and plants. In: Prasad, MNV; Hagemeyer J (eds). Heavy metal stress in plants, from molecules to ecosystems. *Berlin: Spinger*; 1999; p. 1-7.

Maitani T., Kubota H., Sato K., Yamada T. The composition of metals bound to class II metallothionein (phytochelatin and its desglycyl peptide) induced by various metals in root cultures of *Rubia tinctorum*. Pl. Physiol, (1996). 110, 1145-1150.

Maksymiec, W; Bednara, J; Baszynski, T. Response of runner bean plants to excess copper as a function of plant growth stage: Effect on morphology and structure of primary leaves and their chloroplast ultrastructuure. *Photosynthetica,* (1995). 31, 427-435.

Martensson, AM; Witter, E. Influence of various soil amendments on nitrogen-fixing soil microorganisms in a long-term field experiment with special reference to sewage sludge. *Soil Biol. Biochem*, (1990). 22,977-982.

Matamoros, MA; Moran, JF; Iturbeormaetxe, I; Rubio, MC; Becana, M.Glutathione and homoglutathione synthesis in legume root nodules. Plant Physiol, (1999). 12,1 879–888.

Matysik, J; Alia, BB; Mohanty, P. Molecular mechanisms of quenching of reactive oxygen species by proline under stress in plants. *Curr. Sci*, (2002). 82, 525-532.

Mazhoudi,S; Chaoui, A; Ghorbal, MH; El Ferjani, E. Cadmium and zinc induction of lipid peroxidation and effects on antioxidant enzymes activities in bean (*Phaseolus vulgaris* L.). Plant Sci, (1997). 127, 139-47.

McCarthy, I; Romero-Puertas, MC; Palma, JM; Sandalio, LM; Corpas, FJ; Gomez, M; Del Rio, LA. Cadmium induces senescence systems in leaf peroxisomes of pea plants. *Plant Cell Environ,* (2001). 24, 1065-1073.

McGrath SP. Long term studies of metal transfers following application of sewage sludge. In: Coughtrey, JP ; Martin, MH ; Unsworth, MH. (Eds.). Pollutant transport and fate in ecosystems. Blackwell Scientific, Oxford: British Ecological Society Special Publication; 1987; No. 6.

McGrath, SP; Brookes, PC; Giller, KE. Effects of potentially toxic metals in soils derived from past applications of sewage sludge on nitrogen fixation by *Trifollium repens* L. Soil Biol Biochem, (1988). 20, 415- 24.

McGrath, SP. The uptake and translocation of tri- and hexavalent chromium and effects on the growth of oat in flowing nutrient solution and in soil. New Phytol, (1995). 92, 381–90.

Meharg, AA. The role of plasmalemma in metal tolerance in angiosperm. *Physiol. Plant,* (1993). 88, 191-198.

Mengel, K; Kirkby, EA. Principles of Plant Nutrition. International Potash Institute, Bern, (1982). 655.

Mysliwa-Kurdziel, B; Strzatka, K. Influences of metals on biosynthesis of photosynthetic pigments. In: Prasad, MNV; Strzatka, K. (eds.). Physiology and biochemistry of metal toxicity and tolerance in plants. *Dortrecht: Kluwer*; 2002a; pp. 201-227.

Nriagu, JO; Pacyna, JM; Quantitative assessment of worldwide contamination for air, water and soil by trace metals. *Nature*, (1988). 333, 134-139.

Parr, PD; Taylor, FG. Germination and growth effects of hexavalent chromium in Orocol TL (a corrosion inhibitor) on Phaseolus vulgaris. *Environ. Int*, (1982). 7,197-202.

Peralta, JR; Gardea, TJL; Tiemann, KJ; Gomez, E; Arteaga, S; Rascon, E. Uptake and effects of five heavy metals on seed germination and plant growth in alfalfa (*Medicago sativa*) L. Bull. Environ. *Contam. Toxicol*, (2001) 66, 727-34.

Pilon-Smits, EAH. Phytoremediation. *Ann. Rev. Plant Biol*, (2005). 56, 15-39.

Prasad, MN. Cadmium toxicity and tolerance in vascular plants. Environ. Exp. Bot. (1995a). 35, 525-545.

Prasad, MNV. Metal-biomolecule complexes in plants: occurrence, functions, applications. Analysis, (1998). 26-28.

Prasad, MNV; Strzayka, K. Impact of heavy metals on photosynthesis. Heavy Metal Stress in Plants: From Molecules to Ecosystems. *Springer*, Berlin, (1999). 117-128.

Prasad, MNV; Freitas, H. Metal hyperaccumulation in plants – Biodiversity prospecting for phytoremediation technology. *Elect. J. Biotechnol*, (2003). 6, 275-321.

Price, AH; Hendry, GAF. Iron catalyzed oxygen radical formation and its possible contribution to drug damage in nine native grasses and three cereals. *Plant Cell Environ*, (1991). 14, 477-484.

Purchase, D; Miles, RJ. Survival and nodulating ability of indigenous and inoculated *Rhizobium leguminosarum* trifolii in sterilized and unsterilized soil treated with sewage sludge. *Curr. Microbiol*, (2001). 42, 59-64.

Quartacci, MF; Cosi, E; Navari-Izzo, F. Lipid and NADPH-dependent superoxide production in plasma membrane vesicles from roots of wheat grown under copper deficiency or excess. *J. Exp. Bot*, (2001). 52, 77-84.

Rajkumar, M; Nagendran, R; Kui Jae, Lee; Wang Hyu, Lee. Characterization of a novel Cr (vi) reducing *Pseudomonas sp*. with plant growth promoting potential. *Curr Microbiol,* (2005) 50, 266-271.

Rathore, PS. Techniques and management of field crop production. Second edition. Jodhpur, India: *Agrobios* (India); 2002

Reddy, AM; Kumar, SG; Jyothsnakumari, G; Thimmanaik, S; Sudhakar, C. Lead induces changes in antioxidant metabolism of horsegram (*Macrotyloma uniflorum* (Lam.) verdc.) and bengalgram (*Cicer arietinum* L.). *Chemosphere*, (2005). 60, 97-104.

Remon, E; Bouchardon, JL; Cornier, B; Guy, B; Leclerc, JC; Faure, O. Soil characteristics, heavy metal availability and vegetation recovery at a former metallurgical landfill: Implications in risk assessment and site restoration. *Environ. Poll*, (2005). 137, 316-323.

Rother, J A; Millbank, J W; Thornton, I. Nitrogen fixation by white clover (*Trifolium repens*) in grasslands on soils contaminated with Cd, Pb, and Zn. J. Soil Sci, (1983). 34:127-136.

Rout, GR; Samantaray, S; Das, P. Differential chromium tolerance among eight mungbean cultivars grown in nutrient culture. *J. Plant Nutr*, (1997). 20, 473- 83.

Rout, GR; Das, P. Effect of metal toxicity on plant growth and metabolism: I. Zinc. Agronomie, (2003). 23, 3-11.

Salt, D.E., Prince, R.C., Pickering, I.J., Raskin, I. Mechanisms of Cd mobility and accumulation in Indian mustard. *Plant Physiol*, (1995a). 109, 427-433.

Salt, D.E., Blaylock, M., Kumar, NPBA., Dushenkov, V., Ensley, B.D., Chet, I., Raskin, I Phytoremediation: a novel strategy for the removal of toxic metals from the environment using plants. *Biotechnol*, (1995b). 13, 468-474.

Salt, DE; Smith, RD; Raskin,I. Phytoremediation. Ann. Rev. Plant Physiol. *Plant Mol. Biol,* (1998). 49, 643-668.

Samantaray S; Rout, GR; Das, P. Studies on differential tolerance of mungbean cultivars to metalliferous minewastes. *Agribiol. Res*, (1999). 52, 193-201.

Schutzendubel, A; Schwanz, P; Teichmann, T; Gross, K; Langenfeld- Heyser, R; Godbold, DL; Polle, A. Cadmium-induced changes in antioxidative systems, hydrogen per oxide content, and differentiation in Scot Pine roots, *Plant Physiol.* (2002). 127, 887- 898.

Shainberg, O; Rubin, B; Rabinowitch, HD; Telor, E. Loading beans with sub-lethal levels of copper enhances conditioning to oxidative stress. *J. Plant. Physiol*, (2001). 158, 1415-1421.

Shier, WT.metals as toxin in plants. *J. Toxicol-Toxin Rev*, (1994). 13, 205-216.

Shanker, AK. Physiological, biochemical and molecular aspects of chromium toxicity and tolerance in selected crops and tree species. Ph.D Thesis, (2003).Tamil Nadu Agricultural University, Coimbatore, India.

Shanker, AK; Carlos, C; Herminia, LT; Avudainayagamd, S. Chromium toxicity in plants. Environ. Internatl, (2005). 31 739– 753.

Shaw, BP; Rout, NP. Mercury and cadmium induced changes in the level of proline and the activity of proline biosynthesizing enzymes in *Phaseolus aureus* Roxb. and *Triticum aestivum* L. *Biol. Plant*, (2002). 45, 267-271.

Shen, ZG; Zhao, FJ; McGrath, SP. Uptake and transport of zinc in the hyper accumulator *Thlapsi caerulescens* and the non hyper accumulator *Thlapsi ochroleucum*. Plant Cell environ. Pollut, (1997). 20,898-906.

Sheoran, IS; Signal, HR; Singh, R. Effect of Cd and Ni on photosynthesis and the enzymes of the photosynthetic carbon reduction cycle in pigeonpea (*Cajans cajan* L.). *Photosyn Res,* (1990). 23, 345-351.

Shewfelt, RI; Erickson, MC. Role of lipid peroxidation in the mechanism of membrane associated disorders in edible plant tissue. Trends Food Sci. Technol, (1991). 2, 152-154.

Shu, WS; Ye, ZH; Zhang, ZQ; Lan, CY; Wong, MH. Natural colonization of plants on five lead/zinc mine tailings in southern China. *Restoration Ecology* (2005). 13, 49-60.

Skujins, J; Nohrstedt, H; Odens, S. Development of a sensitive biological method for determination of a low level toxic contamination in soil.1. Selection of nitrogenase activity. *Swed. J. Agric. Res*, (1986). 16, 113-118.

Steffens, JC. The heavy metal-binding peptides of plants. *Annu. Rev. Plant Physiol. Mol. Biol.* (1990). 41, 553-75.

Stiborova, M. Cd^{2+} ions affect quaternary structure ribulose-1, 5-biphosphate carboxylase from barley leaves. *Biochem. Physiol. Pflanzen*, (1988). 183, 371-378.

Strsty, TVS; Madhava Rao, KV. Ultrastructural alterations in response to zinc and nickel stress in the root cells of pigeonpea. *Environ. Exp. Bot*, (1999). 41, 3-13.

Suszcynsky, EM; Shann, JR. Phytotoxicity and accumulation of mercury subjected to different exposure routes. *Environ. Toxicol. Chem*, (1995). 14, 61- 67.

Taylor, R W; Williams, M L; Sistani, K R. Nitrogen fixation by soybean-*Bradyrhizobium* combinations under acidity, low P and high Al stresses. Plant Soil, (1991).131, 293-300.

Tordoff, GM; Baker, AJM; Willis, A J. Current approaches to the re-vegetation and reclamation of metalliferous mine wastes. *Chemosphere,* (2000). 41, 219-228.

Van Assche, F; Clijsters, H. Inhibition of photosynthesis in *Phaseolus vulgaris* L., by treatment with toxic concentration of zinc: effect on ribulose-1,5 bisphosphate carboxylase/oxygenase. *J. Plant Physiol*, (1986). 125, 355-360.

Van Assche, F; Clijsters, H. Effects of metals on enzyme activity in plants. *Plant Cell Environ.* (1990). 13, 195-206.

Van Cauwenbergh, R; Hendrix, P; Robberecht, H; Deelstra, HA. Daily dietary chromium intake in Belgium, using duplicate portion sampling. Z Lebensm Unters Forsch, (1993). 203, 203-206.

Verkleij, JAC; Schat, H. Mechanisms of metal tolerance in higher plants. In: Shaw, J. (ed..). Heavy metal tolerance in plants: Evolutionary aspects. Boca Raron: CRC Press; 1990; pp. 179-193.

Vernoux, T; Wilson, RC; Seeley, KA; Reichheld, JP; Muroy, S; Brown, S. The root meristemless1/cadmium sensitive*2* gene defines a glutathione-dependent pathway involved in initiation and maintenance of cell division during postembryonic root development. *Plant Cell*, (2000). 12 97–109.

Vijayarengan, P; Lakshmanachary, AS. Effects of water logging on growth and pigment content of blackgram (*Vigna mungo* (L.) Hepper). Geobios, Jodhpur Div. Environ. Sci., (1994). 21, 74-76.

Wang, M; Zhao, FJ; Meharg, AA; Raab, A; Feldmann, J; McGrath, SP. Mechanism of arsenic hyper accumulation in *Pteris vittata*: uptake kinetics, interaction with phosphate, and arsenic speciation. Plant Physiol, (2002). 130, 1552-1561.

Wani, PA; Khan, MS; Zaidi, A. An evaluation of the effects of heavy metals on the growth, seed yield and grain protein of lentil in pots. *Ann. Appl. Biol.* (Suppl. TAC) (2006). 27, 23-24.

Wani, PA; Khan, MS; Zaidi, A. Impact of heavy metal toxicity on plant growth, symbiosis, seed yield and nitrogen and metal uptake in chickpea. *Aus. J. Exp. Agric*, (2007a). 47, 712-720.

Wani, PA; Khan, MS; Zaidi, A. Cadmium, chromium and copper in greengram plants. Agron Sustain Dev, (2007b). 27, 145-153.

Wani, PA; Khan, MS; Zaidi, A. Effect of metal tolerant plant growth promoting Rhizobium on the performance of pea grown in metal amended soil. *Arch. Environ. Cotam. Toxicol*, (2007c). DOI; 10.1007/s00244-007-9097-y.

Wani, PA; Khan, MS; Zaidi, A. Effect of heavy metal toxicity on growth, symbiosis, seed yield and metal uptake in pea grown in metal amended soil. Bull. Environ. Contam Toxicol, (2008). DOI 10. 1007/s00128-008-9383-z

Watanabe, ME. Phyto-remediation on the brink of commercialization. *Environ. Sci. Technol*, (1997). 31, 182-186.

Weckx, JEJ; Clijsters, HMM. Oxidative damage and defense mechanisms in primary leaves of Phasolus vulgaris as a result of root assimilation of toxic amounts of copper. *Physiol. Plant,* (1996). 96, 506-512.

Weigel, HJ. Inhibition of photosynthetic reactions of isolated intact chloroplast by cadmium. *J. Plant Physiol*, (1985a). 119, 179-189.

Welp, G; Brummer, GW. Microbial toxicity of Cd and Hg in different soils related to total and water-soluble contents. *Ecotoxicol. Environ Saf*, (1997). 38, 200–204.

Wuertz, S; Mergeay, M. The impact of heavy metals on soil microbial communities and their activities. In: van Elsas, JD; Trevors, JT; Wellington, EMH (eds.). Modern Soil Microbiology. New York: Marcel Dekker,; 1997; 607–642.

Yadav, OP; Shukla, UC. Effect of zinc on nodulation and nitrogen fixation in chickpea (*Cicer arietinum* L.*). J. Agric. Sci,* (1983). 101, 559-563.

Yang, XE; Roemheld, V. Physiological and genetic aspect of micronutrient uptake by higher plants. In: Nielsen (ed.). Genetics and molecular biology of plant nutrition. Dordrecht, The Netherlands: Kluewer Academic Publishers; 1998; 151–86.

Yamamoto, Y; Hachiya, A; Matsumoto, H. Oxidative damage to membranes by a combination of aluminum and iron in suspension cultured tobacco cells. *Plant Cell Physiol,* (1997). 38, 1333-1339.

Zaccheo, P; Genevini PL; Cocucci, S. Chromium ions toxicity on the membrane transport mechanism in segments of maize seedling roots. *Plant Nutr*, (1982). 5, 1217- 27.

Zaccheo, P; Cocucci, M; Cocucci, S; Effects of Cr on proton extrusion, potassium uptake and transmembrane electric potential in maize root segments. *Plant Cell Environ*, (1985). 8,721-72 6.

Zahran, HH, *Rhizobium*-legume symbiosis and nitrogen fixation under severe conditions and in an arid climate. Microbiol. Mol. Biol. Rev, (1999). 63, 968–989.

Zeid, IM. Responses of *Phaseolus vulgaris* to chromium and cobalt treatments. *Biol. Plant,* (2001). 44, 111-115.

Zlatareva, E; Nikolov, N; Nikolaev, A. Uptake of Cr (III) by alfalfa depending on pH and the level of pollution of the soil. Pochvoznanie Agrokhimiya Ekologiya, (1999).34, 49- 53.

Zhu, YL; Pilon-Smits, AH; Jouarin, L; Terry, N. Over expression of glutathione synthase in Indian Mustard enhances cadmium accumulation and tolerance. *Plant Physiol*, (1999). 119, 73–80.

In: Heavy Metal Pollution
Editors: S. E. Brown and W. C. Welton
ISBN: 978-1-60456-899-8

Chapter 9

NATURAL AND SYNTHETIC MATERIALS FOR HEAVY METAL AND HUMIC ACID REMOVAL

Thanet Terdkiatburana, Shaobin Wang* and Moses O. Tadé
Department of Chemical Engineering, Curtin University of Technology,
GPO Box U1987, Perth, WA 6845, Australia

ABSTRACT

Heavy metals are significant contaminants in aqueous systems. All heavy metals can produce toxicity when ingested in sufficient quantities, but there are several important ones such as lead, mercury, copper, cadmium, arsenic, nickel and silver. These heavy metals are so pervasive and can produce toxicity at very low concentrations. Moreover, they may build up in biological systems and become a significant health hazard. Humic substances are macromolecules that naturally occur in all environments in which vegetation matters are present. Humic acid (HA) plays a fundamental role in many ecosystems since it interacts with toxic metal ions present in the system, resulting in a decrease in the bio-availability of such ions. Moreover, the availability of humic acid in water can react with other chemical compounds, such as chlorine to form trihalomethanes (including chloroform) and causes an increasing risk of cancer and may be linked to heart, lung, kidney, liver, and central nervous system damage.

Adsorption is approved as an effective and simple method for water and wastewater treatment. Many adsorbents then are developed for use in adsorption process such as montmorillonite, peat, activated carbon, zeolite, etc. In this research, humic acid and some heavy metals (Cu and Pb) were mainly selected for adsorption study. In the sorption experiment, several adsorbents such as synthesised MCM-22 zeolite (SZ), natural zeolite (NZ), powdered activated carbon (PAC), and fly ash (FA), were selected to examine the application for HA and heavy metal removal both in individual and simultaneous adsorption processes. The characteristics and interactions of the adsorbents with HA and heavy metals were systematically studied by batch laboratory experiments. The results show that the four adsorbents exhibit varying adsorption behaviour depending on the adsorbent characteristics. The synthesised zeolite MCM-22 presents higher adsorption for heavy metals and humic acid. The presence of HA will affect heavy metal adsorption on solids. Complexation of HA with metals will result in an increase of metal

* Email: Shaobin.wang@curtin.edu.au, Fax: +61 8 9266 2681

ions on adsorbent. However, the competitive adsorption can also occur which results in a decrease of HA and metal adsorption on solids.

1. INTRODUCTION

Heavy metals such as mercury (Hg), copper (Cu), cadmium (Cd), arsenic (As), chromium (Cr), thallium (Tl), and lead (Pb) exist in surface waters in colloidal, particulate, and dissolved phases. These metal ions can also be present in wastewater systems discharged from various industrial processes. In general, heavy metals generate their toxicity by forming complexes or ligands with organic compounds. These modified biological molecules lose their ability to function properly and result in the malfunction or death of the affected cells.

Natural organic matter (NOM) is part of all aquatic systems and mainly comprises humic substances (HS), which represent 60 to 80% of the total dissolved organic carbon (DOC). Humic substances give an undesired yellow colour to drinking water and also there is a horrible taste due to their ability to combine with heavy metals, pesticides, and herbicides, increasing these compound concentrations in water. In addition, humic substances will form toxic chemicals during water treatment, causing health problem to human beings. Humic substances can be divided into three main categories, humic acid (HA), fulvic acid (FA), and humin.

In water, the coexistence of humic acid and heavy metals introduces a strong complexation, which may reduce the toxicity level of simple hydrated ions and hydroxyl complexes. On the other hand, humic acid is also believed to play an important role in soil and water systems as it reduces the removal efficiency of heavy metals.

In the past decades, many techniques have been developed to remove heavy metals (Kurniawan et al., 2006, Mohammad and Najar, 1997, Rubio et al., 2002) and humic acid (Guibal et al., 2006, Guzzella et al., 2002, Uyguner et al., 2007, Zularisam et al., 2006) from water systems. Among these techniques, adsorption is usually a simple and cost-effective method and has been widely used. However, activated carbon as an adsorbent used in current practice is suffering drawbacks of high cost production and regeneration (Wang et al., 2005).

Many researchers have attempted to discover and develop adsorbents with high removal performance in humic acid and heavy metals individually and also humic acid-heavy metal complexation. They have found that there are some limitations to understand humic acid-heavy metal complexation because there are unrecognisable properties and reactions which depend on humic acid structure and binding formation. In this work, we will report an investigation of humic acid and heavy metal removal from aqueous solutions with various adsorbents. Adsorbents that were selected are synthesised zeolite MCM-22, natural zeolite, powder activated carbon and fly ash. The interaction between adsorbate and adsorbent in multi component system was systematically investigated. In addition, the effects of main control process parameters including initial concentration, solution pH, temperature, and contact time were also investigated.

2. EXPERIMENTAL

2.1. Adsorbents and Chemicals

Four adsorbents were employed for the adsorption experiment; synthesised zeolite MCM-22 (SZ), natural zeolite (NZ), powdered activated carbon (PAC) and fly ash (FA). MCM-22 was synthesised using an organic mixture solution containing hexamethyleneimine and caprolactam as a template. (Wang et al., 2006). The MCM-22 has a $SiO_2/Al_2O_3 = 30$. Natural zeolite was obtained from Zeolite Australia Ltd, which is mainly composed of clinoptilolite. Powder activated carbon was obtained from Calgon Company, USA. Fly ash was collected using electrostatic precipitators from Western Power, Australia. The chemical phases include quartz, mullite, with minor hematite and magnetite. Some of their physicochemical properties are presented in table 1. Other chemicals, humic acid, $Cu(NO_3)_2$, $Pb(NO_3)_2$, HNO_3, NaOH, were obtained commercially available from Aldrich or AJAX Chemicals and were used as received.

Table 1. Characteristics of SZ, NZ, PAC and FA

	S_{BET}	V_{total}	V_{micro}	V_{meso}		
Adsorbent	(m^2 g^{-1})	(cm^3 g^{-1})	(cm^3 g^{-1})	(cm^3 g^{-1})	Pore Size	pH
SZ	490	0.387	0.184	0.203	14 Å ×6 Å	5.7
NZ	16	0.040	0.001	0.038	7.9 Å ×3.5 Å	8.5
PAC	972	0.525	0.495	0.030	10.9 Å	8.7
FA	15.6	0.016	-	-	-	3.7

2.2. Adsorption Tests

2.2.1. Single Component Tests

The dynamic adsorption of HA and the effect of initial concentration on HA adsorption onto each adsorbent were first studied. In the experiments, the amounts of adsorbents at 250 mg/l were mixed with HA solutions varying from 10 to 100 mg/l. All solutions were set at 30 ^{o}C in the orbital shaker and kept at pH 5 adjusted using 0.1 M HNO_3 or NaOH. The concentration of humic acid was determined by spectroscopic method at 254 nm (Shimadzu spectrophotometer UV 120-02).

The effect of solution pH on adsorption of HA onto the adsorbents was also investigated. The amount of adsorbents, concentration of HA and solution temperature were fixed at 250 mg/l, 50 mg/l and 30 ^{o}C, respectively, while the solution pH was varied in a range of 3 to 7.

The effect of solution temperature was conducted in a temperature range of 30 to 60 ^{o}C and pH 5. The amount of adsorbents and concentration of HA were controlled at 250 mg/l and 50 mg/l, respectively.

Lead and copper adsorption experiments were conducted at the concentrations of metal ions varying from 10 to 50 mg/l. The pH of solutions was fixed at 5 by using 0.1 M HNO_3 and NaOH solutions. In the experiments, the amount of adsorbents was fixed at 250 mg/l. The orbital shaker was used to control the temperature of all solutions at 30 °C. The concentrations of heavy metal ions, Cu and Pb, were determined using an Atomic Absorption Spectrometer (AAS, Varian Spectra A-100). These two heavy metals will be investigated at 324.8 and 217 nm, respectively.

The effect of pH on the adsorption of heavy metal ions onto the adsorbents was also conducted. The amount of adsorbents, initial concentration of heavy metal ions and solution temperature were fixed at 250 mg/l, 50 mg/l and 30 °C, respectively, while the solution pH was varied from 3 to 6.

2.2.2. *Binary Adsorbate Tests*

The effect of heavy metal concentration was firstly studied. The concentration of heavy metals was varied from 10 to 30 mg/l. The HA concentration and the amount of adsorbents were fixed at 50 mg/l and 250 mg/l, respectively. The pH of solutions was controlled at pH 5 by using 0.1 M HNO_3 or NaOH and the temperature was controlled at 30 °C in the orbital shaker. The determination of the metal and HA concentrations was performed using the method as described above.

The effect of HA concentration was also conducted in which HA concentrations varied from 10 to 50 mg/l. The heavy metal concentration and amount of adsorbents were fixed at 10 mg/l and 250 mg/l, respectively. The solutions were set at 30 °C in the orbital shaker and adjusted to pH 5 using 0.1 M HNO_3 or NaOH.

3. Results and Discussion

3.1. Comparison of HA Adsorption on Four Adsorbents

The dynamic adsorption of humic acid on synthesised zeolite (SZ), natural zeolite (NZ), powder activated carbon (PAC), and fly ash (FA) are shown in figure 1. The adsorption of humic acid on the four adsorbents shows a fast increase in the first 20 h and then reaches the equilibrium after 50 h. There is no further rise in adsorption probably due to the limited active sites still available for adsorption of HA on these adsorbents. It is seen that the order of adsorption follows SZ > PAC > NZ > FA, which does not follow the order of surface area. Although PAC has the highest surface area while most of the pores available for adsorption are micropores, which can not accommodate large size molecules of HA. Starek et al. (1994) indicated that the size of HA molecules ranges between 1.2 and 4 nm. As shown in table 1, most of PAC pores are around 1.2 nm which is close to the smaller size of HA, hence, PAC exhibits a relative low adsorption. SZ presents much high mesoporous structure which will favour the adsorption of HA. In addition, the positive surface charge of SZ also promotes the adsorption of negative charged HA, making higher adsorption capacity of SZ. Even though NZ presents a lower surface area, the mesoporous structure with pore size around 4 nm makes it exhibiting high adsorption for HA, closer to the values of PAC and SZ. However, for FA, it shows the lowest adsorption capacity compared with other adsorbents. This might be because

FA has the lowest surface area and the pH value in this experiment might not be suitable for sorption of HA on FA.

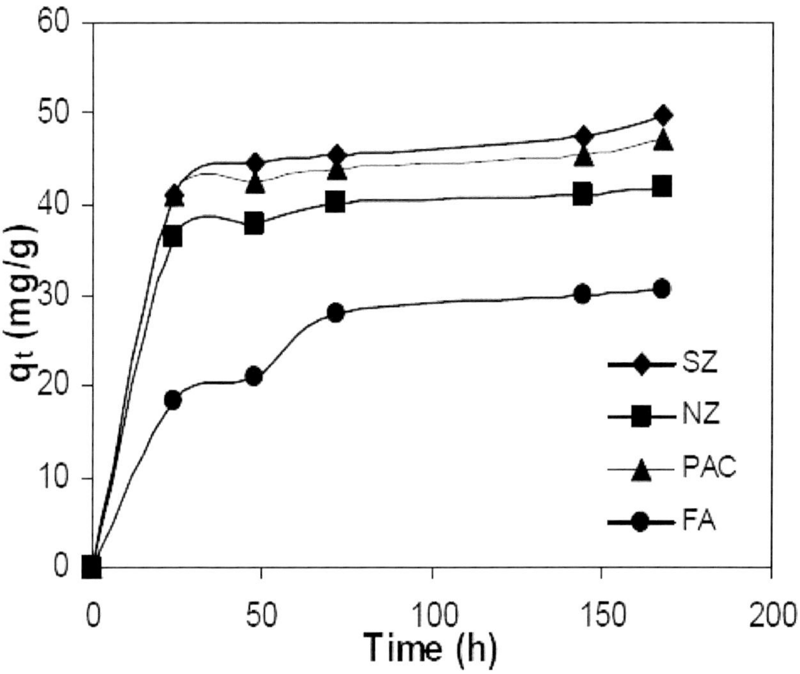

Figure 1. The dynamic adsorption of HA on adsorbents at $[HA]_0 = 50$ mg/l, 30 °C, pH 5.

In order to understand the kinetics of HA adsorption, two kinetic models were also applied to the experimental data. The pseudo first order kinetic model was developed by Lagergren in 1898.

$$\frac{d\,q_t}{d\,t} = k_1\,(q_e - q_t) \tag{1}$$

where q_e and q_t refer to the amount of samples adsorbed (mg/g) at equilibrium and time conducting, t (h), respectively, and k_1 is the equilibrium rate constant of the pseudo first order sorption (h^{-1}).

The pseudo second order model was proposed by Ho and McKay (1999). The equation that describes the pseudo second order is the following:

$$\frac{d\,q_t}{d\,t} = k_2\,(q_e - q_t)^2 \tag{2}$$

Integrating the above equation for the initial conditions $t = 0$, $q_0 = 0$ gives:

$$\frac{t}{q_t} = \frac{1}{k_2 q_e^2} + \frac{1}{q_e} t \qquad (3)$$

The second order rate constant, k_2, and the equilibrium adsorption capacity, q_e, were calculated from the intercept and slope of the plots of t/q_t versus t.

Figure 2 illustrates the fitting of the experiments using the first order kinetic model. Although the correlation coefficients for the samples are found to be higher than 0.98, the calculated equilibrium adsorption capacities do not agree well with the experimental values (table 2). This indicates that adsorption of humic acid onto SZ, NZ and PAC is not an ideal pseudo first order reaction. On the contrary, only humic acid adsorption onto FA seems to suite the pseudo first order model.

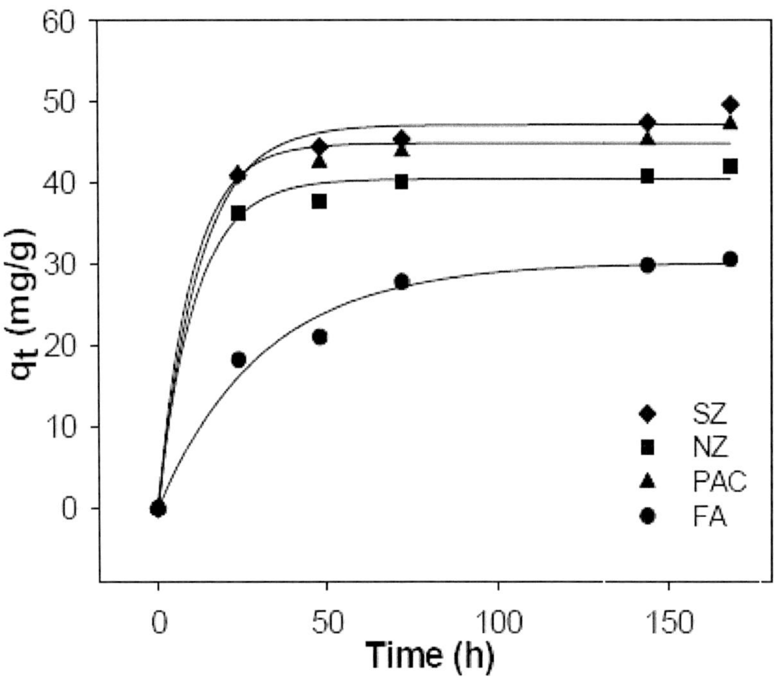

Figure 2. The first order kinetics model of HA adsorption on adsorbents.

The second order kinetics was also employed to fit the experimental data. If the pseudo second order kinetics is applicable, the plot of t/q_t versus t should show a linear relationship. The linear plots of t/q_t versus t show a good agreement of experimental data with the pseudo second order kinetic model for all the samples (figure 3).

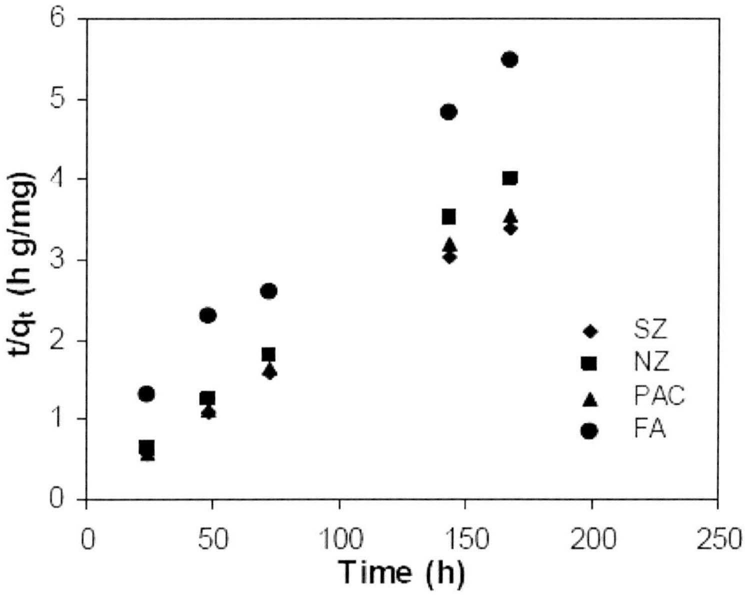

Figure 3. The second order kinetics model of HA adsorption on adsorbents.

Table 2. The kinetic parameters of HA adsorption on SZ, NZ, PAC and FA at [HA] = 10 to 100 mg/l, 30 °C, pH 5.

	[HA]	Pseudo 1st order			Pseudo 2nd order			Experiment
	(mg/l)	k_1 (h^{-1})	q_e (mg/g)	r^2	k_2 (g/mg h)	q_e (mg/g)	r^2	q_e (mg/g)
SZ	10	0.10	22.18	1.00	5.88	23.16	1.00	22.80
	30	0.08	44.29	1.00	9.45	46.55	1.00	45.37
	50	0.08	47.19	0.99	7.11	50.90	1.00	49.70
	70	0.07	55.17	0.99	6.27	60.80	0.99	59.50
	100	0.04	66.48	0.96	3.84	77.71	0.98	73.80
NZ	10	0.10	20.62	1.00	6.64	21.34	1.00	21.00
	30	0.11	36.17	1.00	8.31	38.23	1.00	37.73
	50	0.09	40.53	0.99	7.95	43.00	1.00	42.04
	70	0.09	51.29	1.00	11.42	53.98	1.00	52.91
	100	0.07	57.63	0.98	5.58	64.77	1.00	62.43
PAC	10	0.12	21.36	1.00	8.25	22.04	1.00	21.82
	30	0.11	44.33	1.00	17.55	45.61	1.00	45.09
	50	0.10	44.94	0.99	8.44	48.01	1.00	47.23
	70	0.08	52.49	0.98	5.78	58.67	1.00	56.27
	100	0.05	64.44	0.97	4.34	74.26	0.98	71.14
FA	10	0.04	19.23	1.00	1.08	21.75	1.00	19.38
	30	0.07	20.44	0.98	2.23	22.60	1.00	21.38
	50	0.03	30.24	0.98	1.43	35.11	0.99	30.62
	70	0.02	33.19	0.97	1.01	41.23	0.98	33.31
	100	0.05	33.18	0.98	2.72	37.14	1.00	34.53

The correlation coefficients (r^2) for the second order kinetics model are higher than 0.99. The second order rate constant, k_2, and the equilibrium adsorption capacity, q_e, were calculated from the intercept and slope of the plots of t/q_t versus t, which are summarised in table 2. The calculated q_e values agree very well with the experimental data. This suggests that the adsorption kinetics mainly followed the pseudo second order model. A similar result was reported by Lorenc-Grabowska and Gryglewicz (2005), who indicated that the adsorption of HA on activated carbon proceeds according to the pseudo second order model.

The effect of initial humic acid concentration on the equilibrium adsorption is shown in figure 4. As can be seen from the figure, when humic acid concentration increases, adsorption is also increased. SZ shows the highest adsorption capacity followed by PAC, NZ and FA. However, the initial concentration of HA shows less effect on the adsorption capacity of FA which differs from the other three adsorbents. Adsorption of humic acid generally increases, especially when the initial concentration is high. This behaviour may be attributed to a high driving force, which is the humic acid concentration difference between the liquid (i.e. sample solution) and the solid (i.e., adsorbents) phases, in cases of high initial concentration.

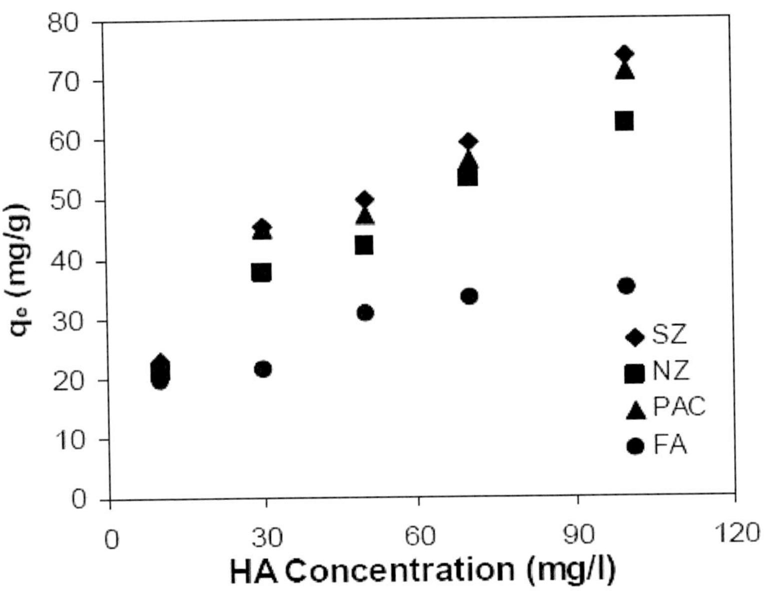

Figure 4. Adsorption of HA on adsorbents at various initial concentrations.

Similar results have been reported previously by Daifullah et al. (2004). They reported that the uptake of HA increased with increased initial concentrations of the solute up to 100 mg/l for a HA-carbon system. There was no further rise above this concentration and the equilibrium was obtained after 120 hours of contact between the solute and carbon W-75. This is probably due to the limited active sites available to adsorption of HA on the carbon surface. In the case of HA adsorption on the different types of adsorbents, Deng and Bai (2004) found that the adsorption of HA on aminated polyacrylonitrile fibers (APANFs) increased from approximately 3.0 to 12.2 mg/g when HA concentration was increased from 5 to 95 mg/l. Peng et al. (2005) used pillared bentonite (PB) to study the adsorption of HA and reported that the amount of HA adsorbed by PB increased with the increase in equilibrium concentration. Subsequently, the amount of humic acid adsorbed increased gradually. It remained constant when the equilibrium concentration exceeded approximately 460 mg/L and a high adsorption capacity of 537.0 mg/g was obtained.

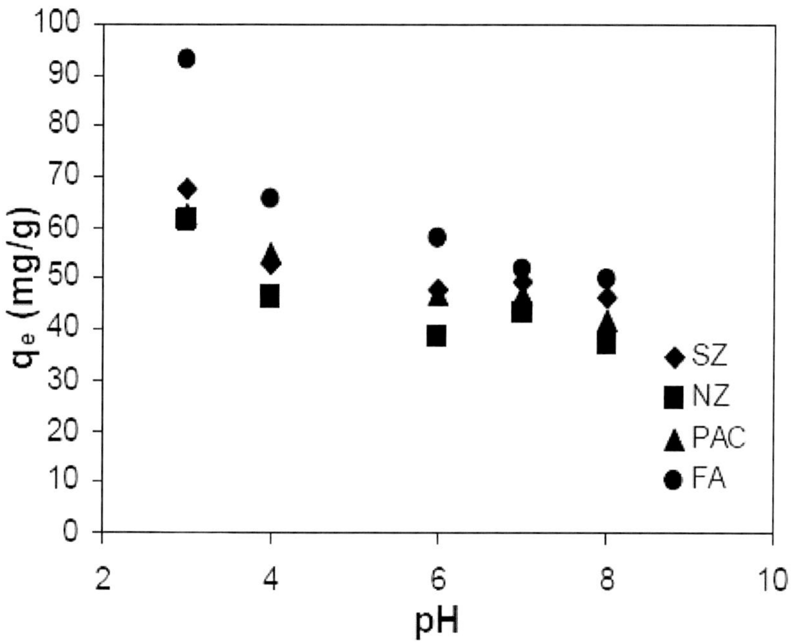

Figure 5. Effect of solution pH on HA adsorption on adsorbents at $[HA]_0$ = 50 mg/l, 30 °C.

It has been known that solution pH will significantly influence HA adsorption on solids. The effect of pH on HA adsorption is shown in figure 5. The results indicate that favourable adsorption of HA occurs at pH 3 to 4 for all adsorbents and an increase in pH will lead to a reduction of HA adsorption. However, most adsorption will not change significantly after pH > 6. When the solution pH is changed from 3 to 8, the adsorption of HA will decrease from 67.4 to 46.2 mg/g for SZ, 61.4 to 37.0 mg/g for NZ, 62.3 to 41.5 mg/g for PAC, and 93.0 to 50.0 mg/g for FA, respectively. FA shows the highest reduction in HA uptake followed by NZ, SZ and PAC.

In general, humic acid consists of many polymeric components and its overall adsorption behaviour should be dealt with in terms of multicomponent adsorption of its different fractions. Hence, this can be explained by considering HA solutions as mixtures of compounds with weakly acidic functional groups. At lower pH, more of these groups are in an uncharged state, hence more adsorbable. Therefore, it is expected that the less hydrophilic part of the humic polymer would bind hydrophobic compounds more effectively, and the binding constant would increase as pH is reduced.

According to zeolite, FA and other mineral properties, the main polyvalent cations responsible for the binding of humic acid to zeolites are Ca^{2+}, Mg^{2+}, Fe^{3+} and Al^{3+}. The divalent Ca^{2+} and Mg^{2+}ions do not form strong coordination complexes with organic molecules. In contrast, Fe^{3+} and Al^{3+} form strong coordination complexes with organic compounds. The polyvalent cations act as a bridge between two charged sites. Organic anions (HA), on the other hand, at low pH can be associated with the oxides and divalent ions of zeolites by simple reaction because of the increase of negative charge on the active sites of HA (i.e. carboxylic group and phenolic group). Therefore, anions from HA associating in this manner are readily connected by decreasing the pH.

For activated carbon, at the lower pH range (in acidic range), the surface of the adsorbent takes up the H^+ ions from the solution hence increasing the acidity of the sorbent surface. On the other hand, HA is predominantly negatively charged due to the abundance of carboxylic and phenolic groups and it has been noted that negatively charged groups on an adsorbed molecule HA will tend to associate with an equivalent number of positively charged groups on the carbon surface, forming an ion pair and thus become unavailable for adsorption process. In this concern, the carboxylic group starts to dissociate at pH values 4 to 6 and the phenolic group dissociates at relatively high pH values (Paajanen et al., 1997). This behaviour causes a gradient decrease in the adsorption of HA as pH increases due to the excess formation of ion-pairs. The remaining negatively charged groups on the HA are available and enhance the adsorption of HA by electrostatic attraction between the opposite charges.

It has been reported by several researchers (Han and Thompson, 2003; Bjelopavlic et al., 1999) that the adsorption of humic acid increases with decreasing pH. Chen and Wu (2004) also indicated that HA adsorption strongly depends on the solution pH and the adsorption capacity will decrease when the solution pH increases from 2.0 to 6.7. Only in a few cases (Katsumata et al., 2003), at very low pH, the adsorption capacity decreases again as a result of repulsion forces between the positive charge of the adsorbent surface and the positive charge of HA molecule.

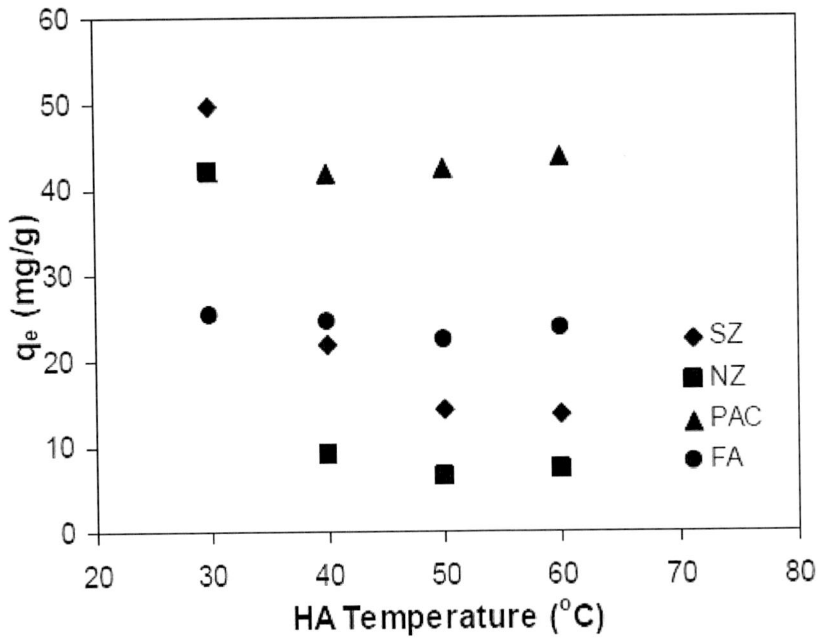

Figure 6. Effect of temperature on HA adsorption on adsorbents at $[HA]_0 = 50$ mg/l, pH 5.

The effect of adsorption temperature on HA adsorption on SZ, NZ, PAC and FA is presented in figure 6. It can be seen that temperature has an effect on adsorption depending on the types of adsorbents. For PAC and FA adsorbents, there is no significant change with an increasing solution temperature. However, the adsorption of SZ and NZ decreases with increased temperature. The SZ adsorption capacity drops dramatically from 49.7 to 22.0 mg/g and it will decrease from 42.0 to 9.2 mg/g for NZ, when the solution temperature is increased from 30 to 60 °C.

Due to the experimental results, the decrease in the amount of HA adsorbed at equilibrium with an increase in temperature may due to the blockage of some active sites on the adsorbent surface. According to HA properties, the increase of solution temperature could provide more mobility and quantity of HA molecules and causes molecules competition with other HA molecules, hence, resulting in reduction of the active site of both zeolites in use.

However, several researchers showed different results from this investigation. Daifullah et al. (2004) found that the uptake of humic acid on carbon W-75 increases as the temperature increases from 25 to 45 °C, indicating better adsorption at higher temperatures. Ghabbour et al. (2004) also reported that the adsorption of aqueous HA on kaolinite and monolayer HA adsorption is mostly endothermic and results in an entropy increase with increased temperature. Zhou et al. (1994) reported that the amounts of Aldrich HA and a water-derived HA sorbed by a kaolinite increased with increasing temperature in artificial seawater at pH 8, while those of water derived fulvic acid and hydrophilic fractions decreased with increasing temperature in the same medium.

The adsorption isotherm is important in understanding the adsorption capacity. In general, two adsorption isotherms, Langmuir and Freundlich models are widely used. The Langmuir model is shown below

$$\frac{q_e}{q_m} = \frac{K_L C_e}{1 + K_L C_e} \tag{4}$$

where q_e is the adsorption capacity of adsorbate in solid phase at equilibrium (mg/g), q_{m} is the maximum adsorption capacity (mg/g), C_e is the adsorbate equilibrium concentration (mg/l), and K_L is the Langmuir constant or adsorption equilibrium constant. The constants, q_m and K_L, are the characteristics of the Langmuir equation and can be determined from a linear regression approach for each set of experimental data.

$$\frac{1}{q_e} = \frac{1}{q_m} + \frac{1}{K_L q_m}\frac{1}{C_e} \tag{5}$$

The Freundlich sorption isotherm is one of the most widely used mathematical descriptions. It usually fits the experimental data over a wide range of concentrations. This isotherm gives an expression encompassing the surface heterogeneity and the exponential distribution of active sites and their energies. Unlike the Langmuir isotherm, the Freundlich isotherm does not provide information on monolayer adsorption. The equation has a form of:

$$q_e = K_F C_e^{1/n} \tag{6}$$

where K_F is an indicator of the adsorption capacity and $1/n$ is the adsorption intensity. In general, as the K_F value increases the adsorption capacity of adsorbent for a given adsorbate increases. Eq. (6) may be linearised by taking logarithms:

$$\log q_e = \log K_F - 1/n \log C_e \tag{7}$$

Values of K_F and n can be calculated from the intercepts and slopes of the plots. A value of n between 2 and 10 shows good adsorption. The numerical value of $1/n < 1$ indicates that adsorption capacity is only slightly suppressed at lower equilibrium concentrations.

The Langmuir equation was originally developed to describe individual chemical adsorbents and is applicable to physical adsorption (monolayer) within a low concentration range (Meghea et al., 1998). Although the Freundlich isotherm is an empirical equation, the parameter K_F can serve as a measure of the relative adsorptive capacity, and since $1/n$ is less than unity, the isotherm is characterised by a concave Freundlich isotherm, indicating that significant adsorption takes place at low concentrations, but the increase in the amount adsorbed with concentration becomes less significant at higher concentration (Teng and Hsieh, 1998).

Figure 7 present the linear forms of the Langmuir and Freundlich isothermal models of HA adsorption on four adsorbents, respectively. Values of the isotherm parameters calculated from the intercepts and slopes of the plots are listed in table 3. It can be seen that the adsorption follows the Langmuir and Freundlich isothermal models. The good agreement of the Langmuir plots with the experimental data suggests monolayer coverage of humic acids on the outer surface of the adsorbents. The monolayer adsorption capacity is as follows: SZ > PAC > NZ > FA while the Langmuir constant K_L is much similar for all adsorbents (table 3). The results suggest that humic acid is favourably adsorbed by SZ and PAC. Moreover, the values of the correlation coefficients (r^2) indicate that the Langmuir isotherm is better in fitting for the adsorption of humic acid on SZ.

Table 3. Adsorption isotherm parameters of HA on SZ, NZ, PAC and FA.

	pH	Langmuir isotherm			Freundlich isotherm		
Sample		q_m (mg/g)	K_L (l/mg)	r^2	K_F	n	r^2
SZ	5	84.09	0.04	0.99	1.54	2.04	0.98
NZ	5	67.83	0.04	0.98	1.47	2.16	0.99
PAC	5	81.17	0.04	0.99	1.50	2.06	0.96
FA	5	34.07	0.12	0.75	1.50	3.57	0.88

Many investigations have been conducted on HA adsorption on various solids. Compared with our data, Lin et al. (2005) studied the adsorption of HA on alginate gel (AG) beads and AG with activated carbon entrapped (AG–AC) beads. They found that the saturated adsorption amount of HA in AG was slightly higher than AC alone. The calculated adsorption capacity of AG-AC and AC using the Langmuir model was found to be 20.0 and 19.0 mg/g, respectively. Deng and Bai (2004) also examined the behaviour of HA. They used the Langmuir and Freundlich models to study the adsorption of HA on aminated polyacrylonitrile fibers (APANFs) and discovered that the experimental results can be well described by both isotherm models. The adsorption of HA on APANFs was found to be 16.22 mg/g. More results from literatures were reported in table 4. It can be seen that the maximum adsorption capacity of HA on SZ, NZ, PAC and FA by the Langmuir model (table 3) are found to be 84.1, 67.8, 81.2 and 34.1 mg/g, respectively. These results are higher than many adsorbents such as Carbon W-75, clay, and AG-AC, however, the maximum adsorption of HA on our adsorbents is still low compared to the adsorption of HA on chitosan-clay and chitosan bead which have maximum adsorption capacity more than 200 mg/g.

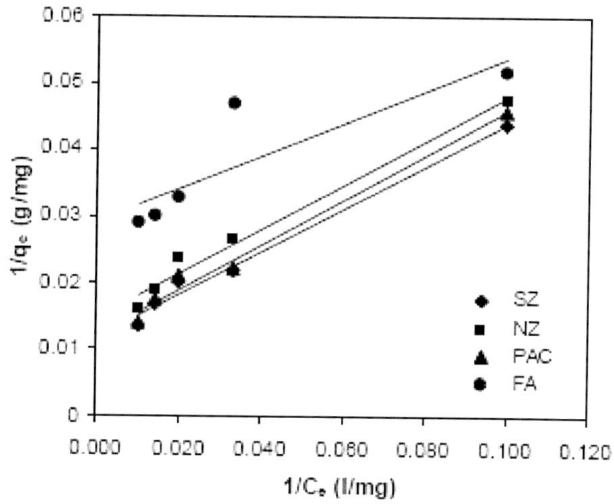

(A)

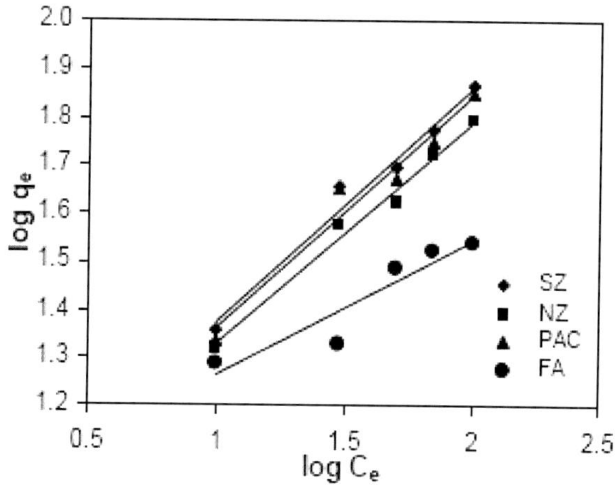

(B)

Figure 7. Adsorption isotherms of HA fitted with Langmuir and Freundlich models.

(A) Langmuir isotherm (B) Freundlich isotherm.

Table 4. Adsorption isotherm parameters of HA on various adsorbents

	Langmuir isotherm			Freundlich isotherm			References
	q_m (mg/g)	K_L (l/mg)	r^2	K_F	n	r^2	
Carbon W-75	45.40	0.50	0.99	32.30	11.36	0.99	Daifullah et al., 2004
Chitosan-PET	0.41	0.02	-	0.01	1.32	-	Zhang and Bai, 2003
Clay	28.20	0.02	-	1.60	2.04	-	Chang and Juang, 2004
Chitosan bead	262.00	0.03	-	1.90	1.56	-	
Chitosan-Clay	243.00	0.02	-	2.00	1.41	-	
AG-AC	19.00	0.10	1.00	-	-	-	Lin et al., 2005
PAC	-	-	-	1.83	0.79	-	Duan et al., 2003
APANFs	16.22	0.03	0.99	1.51	2.13	0.99	Deng and Bai, 2004

3.2. Heavy Metals (Cu and Pb) Adsorption

Figure 8 presents the dynamic adsorption of heavy metals (Cu and Pb) on SZ, NZ, PAC, and FA. The time required to reach equilibrium was found to be 50 hours in SZ NZ, PAC except for FA which is needed more than 120 hours to reach equilibrium. It can be seen that, when contact time increases, the adsorption of heavy metals also increases. Moreover, it shows that the order of adsorption follows SZ > FA > NZ > PAC for Cu, whereas for Pb adsorption, the order of equilibrium adsorption follows SZ > NZ > PAC. Many researchers reported the similar observation that the adsorption of heavy metals increases with contact time (Erol et al., 2005; Wang et al., 2006). Üçer et al. (2006) investigated the adsorption of heavy metals on tannic acid immobilised activated carbon and found that the adsorption of Cu(II), Cd(II), Zn(II), Mn(II) and Fe(III) ions increased with time. This is a similar trend to our results.

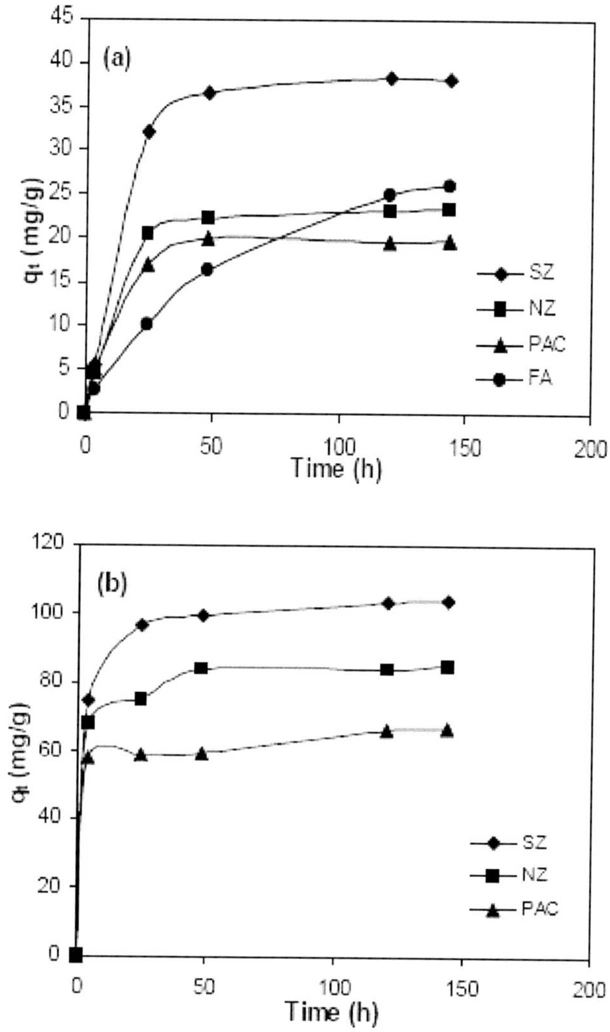

Figure 8. Adsorption of heavy metals on various adsorbents at $[Cu]_0$, $[Pb]_0$ = 10 mg/l, 30 °C, pH 5. (a) Cu (b) Pb.

Similarly, the kinetics of metal adsorption was also fitted by the two kinetic models. figures 9 and 10 illustrate the fitting of experiments using the first order kinetic and second order kinetic models in both heavy metals. The results from the kinetic calculation are summarised in table 5. As can be seen, the calculated q_e values agree very well with the experimental data. Moreover, the correlation coefficients (r^2) for both kinetic models are closer to 1.0. Hence, it was found that the adsorption of Cu on SZ, NZ, PAC and FA seems to fit with the pseudo first order model. However, the adsorption of Pb on SZ, NZ and PAC fit with the pseudo second order model. This suggests that the adsorption system follows the pseudo second order model in the Pb study. On the other hand, the adsorption system of Cu is more likely to match with the pseudo first order model. In addition, the order of equilibrium adsorption of Cu follows SZ > FA > NZ > PAC and the order of equilibrium adsorption of Pb follows SZ > NZ > PAC for both kinetic models. The increase of adsorption rate is Pb(II) > Cu(II).

Different results have been reported by several researchers. It was found that the adsorption kinetics of heavy metals on Moroccan stevensite was best described by the first-order model and the initial adsorption rates increased in the sequence Mn(II) > Pb(II) > Zn(II) > Cu(II) > Cd(II) (Benhammou et al., 2005). Sekar et al. (2004) reported that the kinetics of adsorption of Pb(II) on coconut shell-carbon followed the first-order kinetics. The removal of Zn(II), Pb(II) and Co(II) from solution using Jordanian natural silicate minerals was studied by Al-Degs et al. (2006). They discovered that kinetic data were accurately fitted to the pseudo first order kinetics.

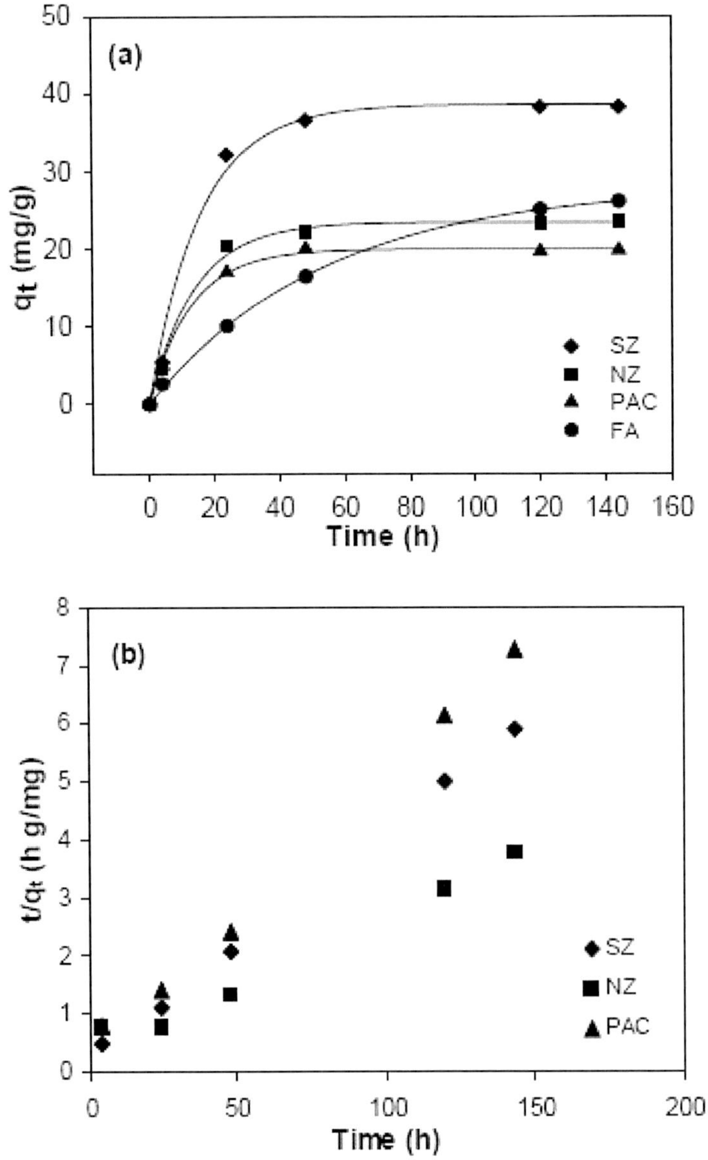

Figure 9. Comparison of kinetic models of Cu adsorption on adsorbents. (a) First order kinetics (b) Second order kinetics.

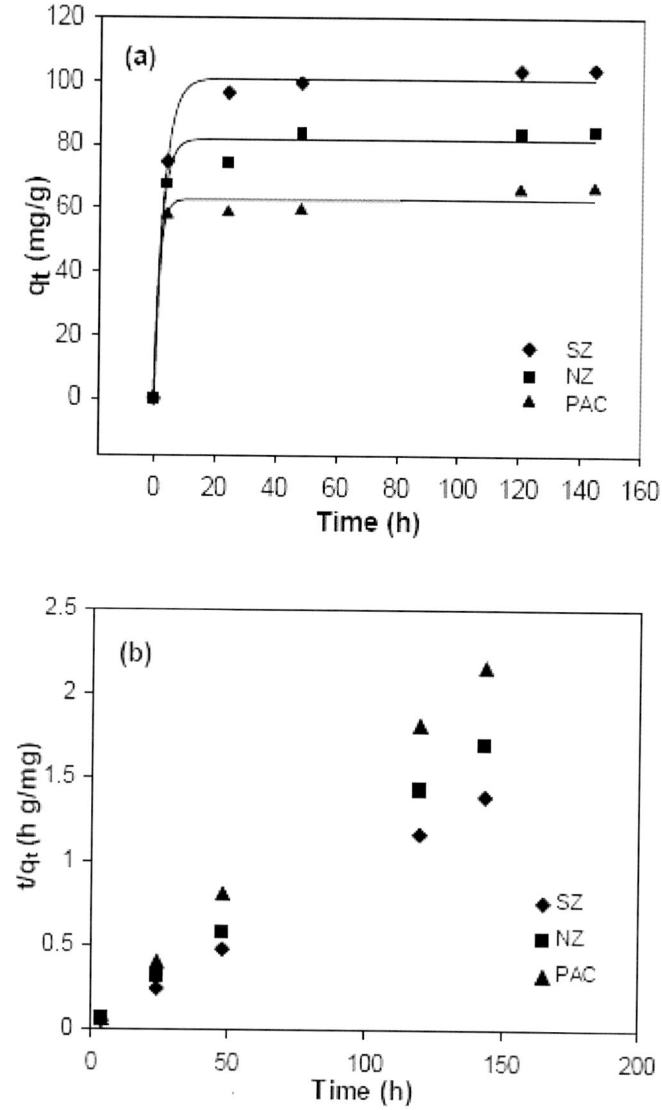

Figure 10. Comparison of kinetic models of Pb adsorption on adsorbents. (a) First order kinetics (b) Second order kinetics.

Table 5. The kinetics adsorption of heavy metals on SZ, NZ, PAC, and FA at 50 mg/l, 30 °C, pH 5.

		Pseudo 1st order			Pseudo 2nd order			Experiment
Metal	Adsorbent	k_1 (h^{-1})	q_e (mg/g)	r^2	k_2 (g/mg h)	q_e (mg/g)	r^2	q_e (mg/g)
Cu	SZ	0.06	38.58	0.99	8.09	39.67	1.00	38.22
	NZ	0.07	23.40	0.99	5.35	24.17	0.99	23.49
	PAC	0.08	19.94	1.00	0.14	20.15	1.00	19.77
	FA	0.02	27.99	1.00	0.63	36.62	1.00	26.06

Metal	Adsorbent	k_1 (h^{-1})	q_e (mg/g)	r^2	k_2 (g/mg h)	q_e (mg/g)	r^2	q_e (mg/g)
Pb	SZ	0.34	100.80	1.00	49.05	105.23	1.00	103.83
	NZ	0.44	81.72	0.99	46.02	85.44	1.00	84.51
	PAC	0.64	62.58	0.98	20.17	67.67	0.99	66.58
	FA	-	-	-	-	-	-	-

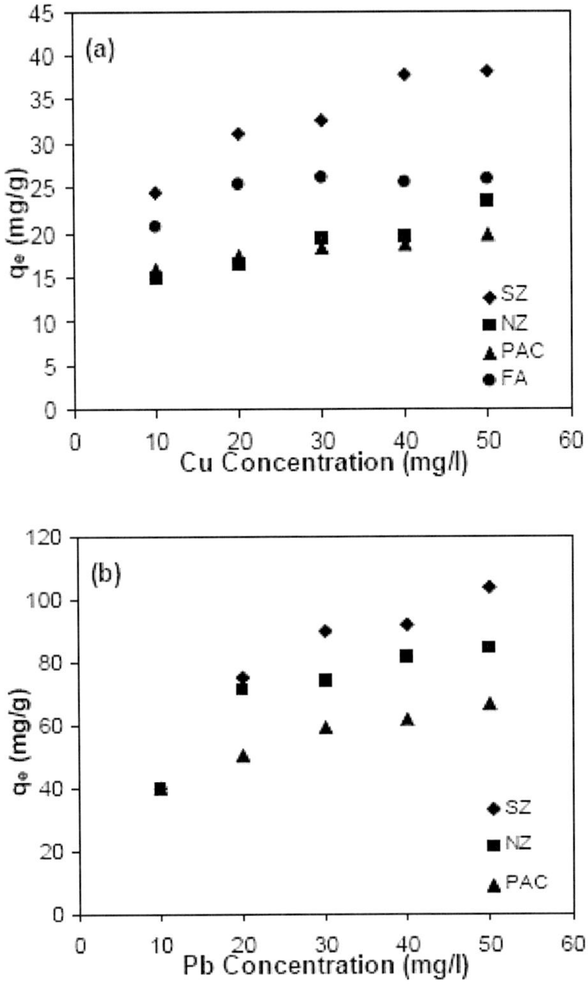

Figure 11. Adsorption of heavy metals on adsorbents at various concentrations. (a) Cu (b) Pb.

The effect of heavy metal concentration on the equilibrium adsorption is shown in figure 11. The data clearly show that SZ is more efficient than FA, NZ and PAC for Cu removal and SZ is also more efficient than NZ and PAC for Pb removal. The activated carbon presents the lowest adsorption in both heavy metals. Further investigations showed that the adsorption capacity of the adsorbents also increases continuously with the rise of initial heavy metal concentration up to 50 mg/l except for FA, which is steady when Cu concentration is higher than 20 mg/l. This is due to the fact that at lower concentration, the ratio of the initial moles of heavy metals to the available surface area is low and subsequently the fractional sorption becomes independent of initial concentration. However, at higher concentration, the moles of heavy metals present become higher compared to the available sites of sorption and hence the adsorption capacity of adsorbents is dependent upon the initial heavy metal concentration.

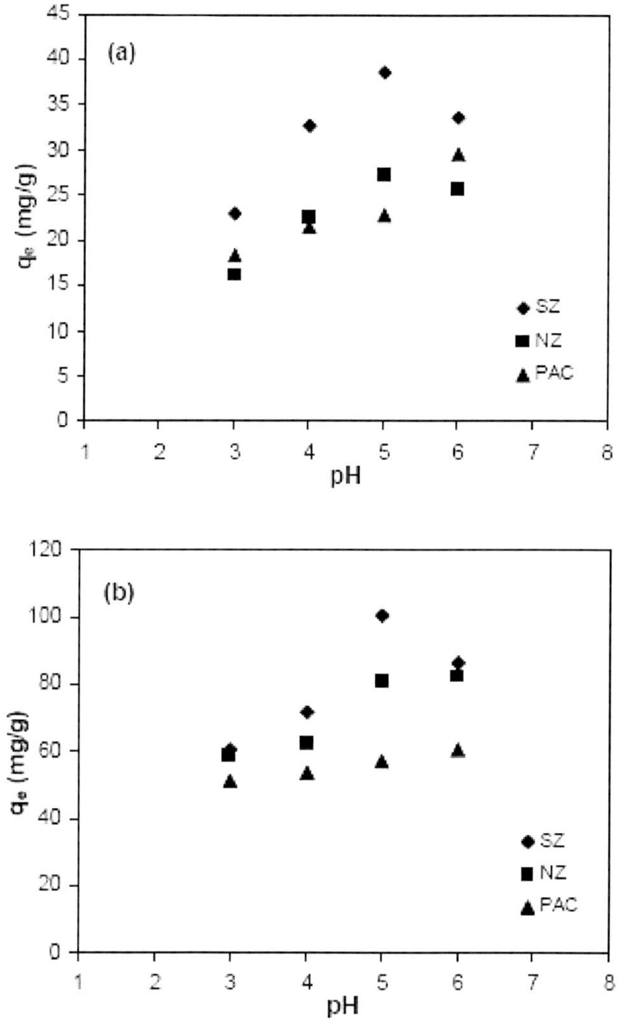

Figure 12. Effect of pH on heavy metal adsorption on adsorbents at $[Cu]_0$, $[Pb]_0$ = 50 mg/l, 30 °C. (a) Cu (b) Pb.

Compared with our observation, Sekar et al. (2004) revealed the similar results for the adsorption of Pb ions on activated carbon prepared from coconut shell. They found that the adsorption capacity was increased from 6.1 to 26.5 mg/g when they increased Pb concentration from 10 to 50 mg/l. Moreover, Vijayaraghavan et al. (2004) studied the removal of Cu by Marine Green Alga Ulva Reticulata. They also reported that when initial copper concentration increased from 250 to 1000 mg/l, the amount adsorbed increased from 36.71 to 70.28 mg/g at pH 5.5. Horsfall and Spiff (2004), on the other hand, reported a different trend from our results. They studied the adsorption of heavy metals onto Caladium bicolour (Wild Cocoyam) biomass and they found the removal of Pb and Cd was favoured at lower heavy metal concentrations.The heavy metal ions adsorption capacities of these adsorbents were determined under non-competitive conditions (i.e. single metal ions solutions) at different pH values of sample solutions. The adsorption results were given in figure 12. The results show the variation of adsorption capacity with pH for both Cu and Pb. It can be seen that higher pH results in higher adsorption. The adsorption of Cu(II) increases rapidly with increasing pH up to 38.6 mg/g at pH 5 for SZ, 27.2 mg/g at pH 5 for NZ, and 29.6 mg/g at pH 6 for PAC. The adsorption of Pb(II) also increases rapidly with increasing pH up to 100.5 mg/g at pH 5 for SZ, 82.4 mg/g at pH 6 for NZ, and 60.4 mg/g at pH 6 for PAC. In the pH range values, i.e. 3 to 6, the sorbent exhibits substantially higher adsorption capacity for Pb(II) ions than for Cu(II). The pH value has a significant impact on metals removal by zeolites and activated carbon since it can influence both the characteristics of the exchanging ions and adsorbents themselves.

The difference in adsorption behaviour of Cu(II) and Pb(II) can be explained by the different affinity of both heavy metal ions for the donor atoms in the ligand adsorbents The extent of complex formation varies with solution pH, the ionic composition, and metals.

Several investigators have observed a strong pH dependence on heavy metals removal by adsorbents. Kobya et al. (2005) reported that the adsorption of heavy metals onto the activated carbon prepared from apricot stone is strongly depending on solution pH. As the pH increases from 1 to 6, the adsorption capacity of activated carbon increases from 4.9 to 24.2 mg/g for Cu(II) and 6.7 to 22.8 mg/g for Pb(II), respectively. Their results showed maximum removal of 96.2% for Cu(II) at pH 4 and 99.9% for Pb(II) at pH 3 , respectively. The report by Jang et al. (2005) also supported that the initial removal rate of Pb(II), Cu(II) and Zn(II) by using mulch, increased with increased adsorbent doses and pH.

According to Horsfall and Spiff (2004) who studied the adsorption of Pb and Cd onto Caladium bicolour (Wild Cocoyam) biomass and reported a similar result that the pH for maximum adsorption of Pb was 7 and the sorption of Pb(II) increases when the solution pH increases. Chen et al. (1996) used a commercial granular activated carbon to study the adsorption of Cu. They demonstrated that the copper removal efficiency increased from 10 to 95% and the adsorption edge ranged from pH 2.3 to 7. Moreover, the removal efficiency of Pb(II), Cd(II), and Cu(II) by natural zeolites was almost over 99.5% in the range of pH 4 to 10, but the removal efficiency significantly decreased at pH values below 4 (Lee and Moon, 2001).

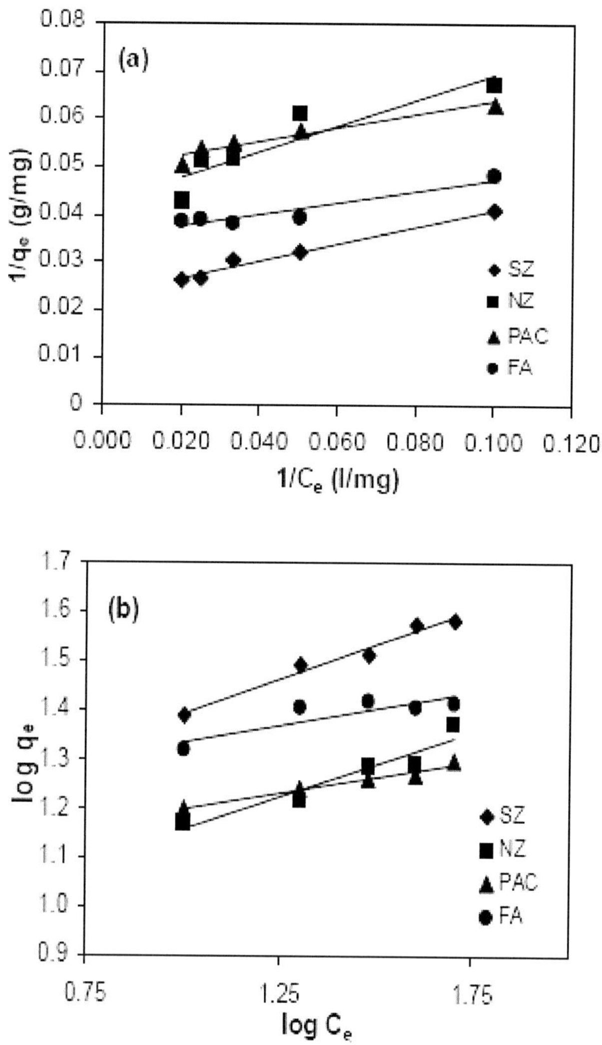

Figure 13. Adsorption isotherms of Cu. (a) Langmuir model (b) Freundlich model.

The Langmuir and Freundlich adsorption isotherms are also applied to Cu(II) and Pb(II) and the results are presented in figures 13 and 14. Correlation coefficients for the Langmuir and Freundlich equations indicate that both models adequately describe the adsorption of Cu and Pb, but the data are better fitted to the Langmuir isotherm. The monolayer adsorption capacity mainly follows SZ > FA > NZ > PAC. Furthermore, the adsorption capacity of Cu and Pb on these adsorbents follows the sequence Pb > Cu (table 6). Linear plots of log q_e versus log C_e show that adsorption of heavy metals on SZ, NZ, PAC, and FA also follows the Freundlich isotherm. Values of K_F and n calculated from the intercepts and slopes of the plots are listed in table 6. From the correlation coefficients, the Langmuir and Freundlich models have similar results and they can be used for the simulation of testing data.

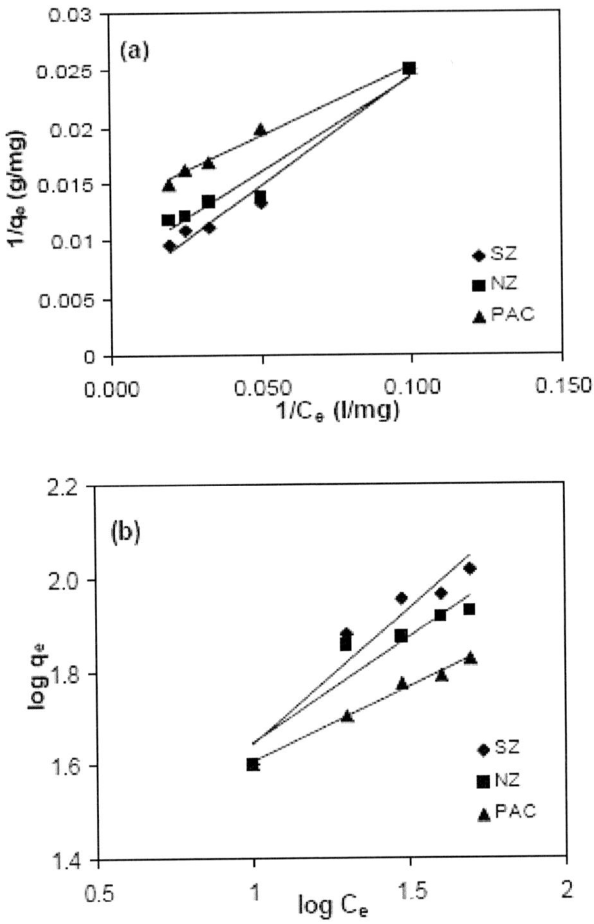

Figure 14. Adsorption isotherms of Pb. (a) Langmuir model (b) Freundlich model.

Several researchers reported a similar trend to our results. The adsorption of metal ions onto a natural zeolite has been investigated by Lee and Moon (2001). They suggested that the selectivity of the NaCl-treated natural zeolite and the chabazite for heavy metals follows the order Cs(I) >> Pb(II) > Cd(II) ≥ Cu(II). Jang et al. (2005) studied the removal of heavy metals by sorption on mulch and showed that under the same conditions (pH 6.0), the order of the initial removal rates was Pb(II) > Cu(II) > Zn(II). This order is possibly related to the differences in electronegativity and ionic radius of the atoms. The values of atomic radius and electronegativity of Pb(II) are higher than Cu(II) ions.

A value of n between 2 and 10 (table 6) shows good adsorption. The numerical value of $1/n < 1$ indicates that adsorption capacity is only slightly suppressed at lower equilibrium concentrations. This isotherm does not predict any saturation of the sorbent by the sorbate; thus infinite surface coverage is predicted mathematically, indicating multilayer adsorption on the surface (Hasany et al., 2002). The adsorption isotherm of other adsorbents which were studied for Cu and Pb is shown in table 7.

Table 6. Adsorption isotherm parameters of heavy metals on SZ, NZ, PAC and FA.

		Langmuir isotherm			Freundlich isotherm		
Metal	Adsorbent	q_m (mg/g)	K_L (l/mg)	r^2	K_F	n	r^2
Cu	SZ	43.52	0.13	0.97	2.59	3.56	0.97
	NZ	24.16	0.16	0.91	1.55	3.76	0.91
	PAC	19.71	0.40	0.99	2.34	7.67	0.98
	FA	28.59	0.28	0.90	2.34	7.45	0.75
Pb	SZ	190.74	0.03	0.98	2.36	1.75	0.93
	NZ	128.95	0.05	0.96	3.13	5.26	1.00
	PAC	76.81	0.11	0.99	3.90	3.16	0.99
	FA	-	-	-	-	-	-

Table 7. Adsorption isotherm parameters of heavy metals on various adsorbents

		Langmuir isotherm			Freundlich isotherm			References
Metal	Adsorbent	q_m (mg/g)	K_L (l/mg)	r^2	K_F	n	r^2	
Cu	Mulch	0.30	8.78	-	0.33	2.42	-	Jang et al., 2005
	MS	18.68	0.01	0.98	-	-	-	Benhammou et al., 2005
	TA-AC	2.73	4.76	0.99	1.91	3.13	0.91	Üçer et al., 2006
Pb	Biomass	54.80	8.41	0.96	1.70	1.11	0.68	Horsfall and Spiff, 2004
	Mulch	0.29	75.45	-	0.35	4.51	-	Jang et al., 2005
	MS	89.92	0.00	0.99	-	-	-	Benhammou et al., 2005

According to other literatures, Üçer et al. (2006) studied the adsorption of Cu(II), Cd(II), Zn(II), Mn(II) and Fe(III) ions by tannic acid immobilised activated carbon. They found that the data fitted better to the Langmuir isotherm model than the Freundlich model and the adsorption capacity of these heavy metals was found to be 2.7, 2.5, 1.8, 1.7 and 2.8 mg/g, respectively. Benhammou et al. (2005) examined the kinetics and isotherms of heavy metal adsorption on Moroccan stevensite and indicated that the maximum adsorption of Cu and Pb was found to be 18.7 mg/g and 89.9 mg/g, respectively, based on the Langmuir model. Compared to our results, the adsorption capacity of Cu(II) and Pb(II) on SZ, NZ, PAC and FA seems to be higher in most studies, except for Pb(II) adsorption on PAC, which is slightly lower than Moroccan stevensite.

3.3. Coadsorption of Heavy Metals and HA

Figures 15 and 16 present the dynamic coadsorption of HA and heavy metals (Cu^{2+} and Pb^{2+}) on SZ, NZ, PAC and FA. It can be seen that the simultaneous adsorption of HA and heavy metals reaches their adsorption equilibrium after 70 hours. However, in the case of HA adsorption on PAC and FA, there is a further increase in adsorption as it needs more than 168 hours to reach equilibrium. These results are different from the individual adsorption as described in the above sections. The adsorption of HA on SZ, NZ, PAC and FA decreases in the presence of heavy metals while the adsorption of Cu increases on most adsorbents in the presence of HA. Pb adsorption on these adsorbents, on the other hand, decreases in the presence of HA. Moreover Pb^{2+} ion is favoured to complex with HA more than Cu^{2+}, this might be because of the binding ability of HA towards heavy metals, which is known to depend on electronegativity and ionic radius of the atoms. Among the metal ions with the same charge, ions with a greater radius are able to effectively interact with the organic compounds, particularly humic materials (Rashid, 1970). Therefore, Pb(II) ions form more stable complexes with the functional groups of humic materials due to the larger ionic radius.

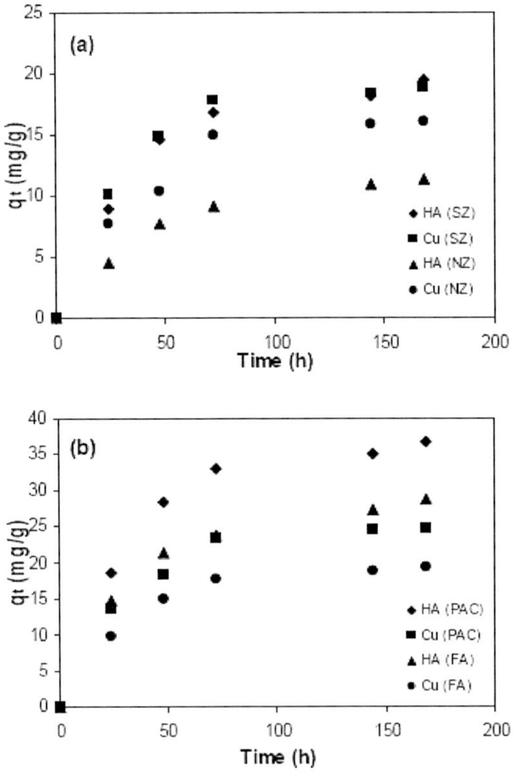

Figure 15. Coadsorption of HA and Cu at $[HA]_0$ = 50 mg/l, $[Cu]_0$ = 10 mg/l, pH 5.

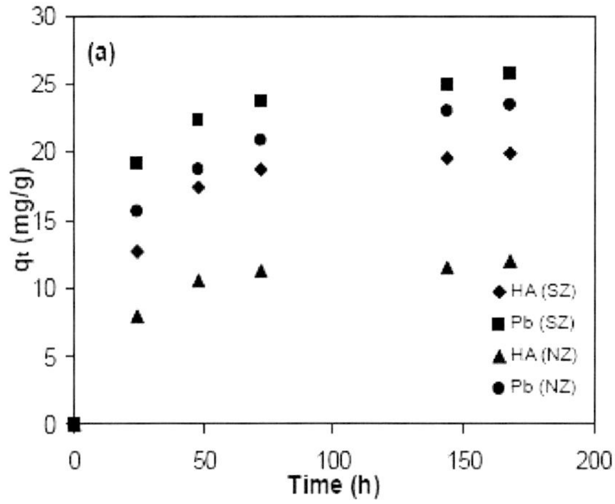

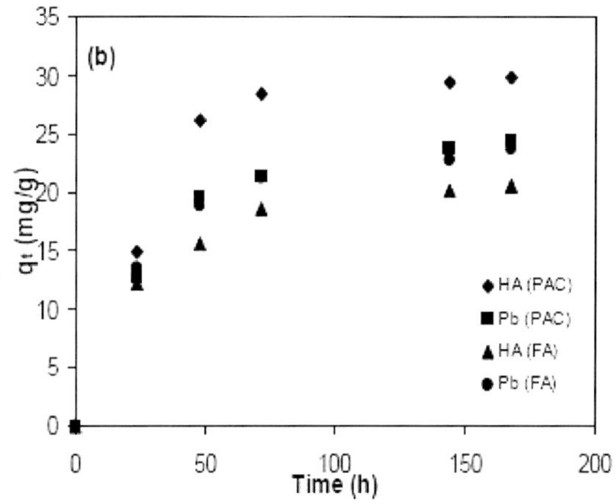

Figure 16. Coadsorption of HA and Pb at $[HA]_0 = 50$ mg/l, $[Pb]_0 = 10$ mg/l, pH 5.

The effect of initial HA concentration on heavy metal adsorption is presented in figure 17. As seen in figure 17a, the adsorption of Cu on SZ and FA decreases, however, the adsorption of Cu on PAC increases with increasing initial HA concentration, except for Cu adsorption on NZ which no significant effect is observed with increasing initial HA concentration.

Figure 17b shows the results of Pb adsorption on various adsorbents. It is observed that the adsorption of Pb is enormously decreased approximately 40 to 45% in most adsorbents in the presence of HA, except for FA. The adsorption of Pb on SZ and NZ is slightly decreased, whereas the adsorption of Pb on FA is increased with increasing HA concentration from 10 to 50 mg/l. Only in the PAC study, there is no significant effect when increasing HA concentration.

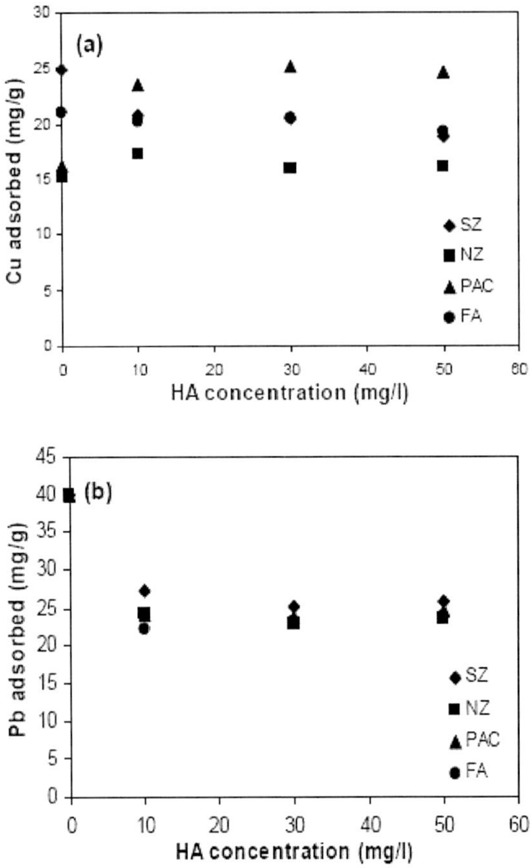

Figure 17. Effect of initial concentration of HA on heavy metals adsorption.

The effect of HA could mainly be explained by the reaction occurring at the surface of the adsorbent. The adsorption of HA on the surfaces of adsorbents may affect heavy metal adsorption throughout the following mechanisms: (1) HA binds to the same sites on adsorbent surface as heavy metal ions, (2) HA attaches to an adsorbent surface through a mechanism different from metal ions, and the total sites available for subsequent heavy metal adsorption remain the same, but a layer of HA adsorbed on the surface of an adsorbent may produce a barrier for heavy metals to connect to the available adsorption sites. Both mechanisms could provide extra space for heavy metals adsorption, and, as a result, it increases the adsorption of heavy metal ions (Yan and Bai, 2005).

However, the decrease of heavy metal adsorption on adsorbent in the presence of HA might be because the competitive adsorption by HA on the surface of solid adsorbent. According to our results, HA can be adsorbed by these adsorbents as well, which directly affect the adsorption rate of heavy metal in coadsorption. Therefore, this could lead to decreasing heavy metal adsorption as there is less space for heavy metal to bind with active sites of adsorbent. Moreover, the decrease of heavy metal adsorption on adsorbent in the presence of HA also supports that HA in this experiment seems not to provide an extra space for heavy metals adsorption as Yan and Bai (2005) mentioned previously.

The increase of Cu adsorption in the presence of HA on PAC, on the other hand, seems to be influenced by the complexation of HA. In this case, the small molecules of Cu(II) seem to prefer to be adsorbed with HA functional groups rather than the active sites of PAC. This might be because Cu molecules are less favoured to be adsorbed on carbon surfaces (Ford, 1981) and at this pH value (pH 5), HA is also favoured by activated carbon (Liu and Gonzalez, 1999), hence, Cu ions will combine with HA and allow HA to complex with PAC active sites and result in the increase of Cu adsorption as HA provides more active sites than activated carbon adsorbent. This means the complexation effect is more important for Cu adsorption on PAC than competitive adsorption.

However, the adsorption of Pb on PAC in the presence of HA indicates an opposite trend with Cu adsorption. The decrease in Pb adsorption might be because Pb ions (2.3) have higher electronegativity than Cu ions (1.9) which leads to a stronger adsorption with the PAC functional groups rather than combining with HA. Therefore, Pb binds with PAC and complex by HA in other sites (PAC-Pb-HA formation). This opinion is in agreement with Üçer et al. (2006).

The increase of heavy metal adsorption as an effect of HA concentration has been supported by Christl et al. (2005). They studied the effect of humic and fulvic concentration on Cu and Pb binding. They confirmed that the amount of Cu^{2+} and Pb^{2+} bound to fulvic and humic acid strongly increased with increasing free metal ion activity and increasing pH. Metal ion binding at constant pH and metal ion activity could be expected to be a function of the humic concentration because the molecules may form micellar-like structures at higher humic concentrations and that this could affect metal ion binding by making some functional groups less accessible. Also, from a thermodynamic point of view, bridge-like complexes might be expected to increase in relative importance with increasing humic concentration. Abate and Masini (2005) reported the presence of HA (HA-VM) increased Cd(II) adsorption, especially at pH 5.0, which is in agreement with literature data for other clay minerals (Arias et al., 2002). Furthermore, the presence of HA seems to reduce the competition between the heavy metal cation and H^+, providing strong surface complexing sites to bind Pb(II).

Our results are slightly different from Petrovic et al. (1999). They reported that sorption of Cu^{2+} ions on all the minerals studied rapidly increased as the concentration of dissolved HA increased. Furthermore, the amount of Pb^{2+} ions sorbed on sand particles slightly decreases while on kaolin the opposite effect is observed. The amount of lead sorbed onto kaolin increases for 15 to 20%. Then, the sorption of Zn^{2+} ions onto all minerals decreases as HA concentration was increased. Taylor and Theng (1995) studied the sorption of cadmium by complexes of kaolinite with HA. They found that an increase in the HA concentration and solution pH lead to the increase of cadmium adsorption by kaolinite.

On the other hand, for montmorillonite, Liu and Gonzalez (1999) verified that the adsorption of Cu(II), Cd(II) and Pb(II) was almost independent of the initial concentration of HA adsorbed on the clay mineral. For kaolinite, however, the adsorption capacity of metal cations increases with the increase of the amount HA adsorbed onto the mineral (Fu and Allen, 1992).

According to Alpatova et al. (2004), the increase of HA concentration did not influence the Cu adsorption but the pH value was more important in HA-heavy metals complexation. The sorption of cadmium by kaolinite increased with an increase in the HA concentration and solution pH (Taylor and Theng, 1995). Hankins et al. (2006) stated that the removal of Zn^{2+} ions increased slightly with initial concentration of HA similar to Pb^{2+} removal. Hence, the amount of bound metal ions increases with increasing concentration of HA, essentially as more binding sites become available in the solution for the heavy metals. Maguire et al. (1992), on the other hand, reported that the presence of HA in clay–Cs system depressed the sorption of Cs by clays and enhanced desorption of Cs.

The results of the effect of initial concentration of heavy metals on HA adsorption are shown in figure 18. As can be seen the adsorption of HA decreases when there is increased initial metal ion concentration. In HA-heavy metal complex, the presence of Cu^{2+} and Pb^{2+} ions will reduce the quantities of bound HA. For all the adsorbents, PAC presents less reduction in HA adsorption while NZ gives the highest reduction in HA adsorption.

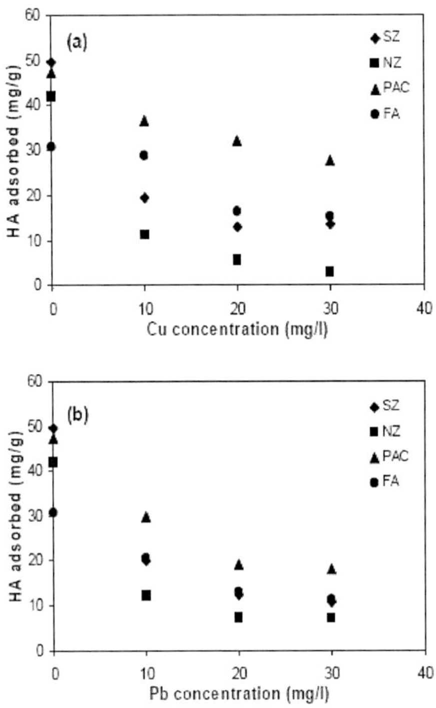

Figure 18. Effect of initial concentration of heavy metals on HA adsorption at $[HA]_0$ = 50 mg/l, pH 5, 30 °C.

The lower adsorption of HA in the presence of heavy metals on every adsorbent might occur because small molecules of heavy metal ions (Cu(II) and Pb(II) ions) seem to merge with the active sites of adsorbent surfaces easier than HA which has a very large molecular size. Moreover, this might be because the bulky molecules of HA could lead to the blockage of active sites on adsorbent surfaces during the adsorption process. Therefore, HA molecules are difficult to compete with heavy metals in order to bind with adsorbents. Consequently, complexation seems to have a significant effect on the adsorption of HA on adsorbents in the presence of heavy metals.

This opinion was supported by Katsumata et al. (2003). They used xonotlite to remove the HA-heavy metals complex and found that the presence of copper ions simply decreased HA adsorption. Tomaszewska et al. (2004) studied the removal of organic matter by coagulation enhanced with adsorption on PAC. They found that phenol occupies the active sites of PAC faster than HA and the accessibility of the carbon surface to HA is significantly reduced. Therefore, in the presence of HA the adsorption capacity of phenol or other small molecules does not change in the concentration range under investigation. The reason can be that phenol molecules are much smaller than HA molecules and therefore the accessibility of the carbon surface are much greater for phenol than for HA. These results are in agreement with those obtained by Cloirec et al. (1997) who found that the adsorption of a low molecular weight compound was not hindered by HA.

Hankins et al. (2006) also studied the removal of heavy metals bound to HA. They discovered that the initial concentration of heavy metal ions seems to have little (in this case even less) effect on removal efficiency. Regarding to binding with heavy metals, the highest removal efficiency of HA is still about 85%.

The complexation between HA-heavy metals was also explained by Vermeer et al. (1999). They suggested that this reaction might happen due to the interaction between the negatively charged humic acid and the positively charged metal oxide. The adsorption of metal ions to the metal oxide in the HA–oxide complex will be increased, whereas the adsorption of humic acid in the complex will be decreased.

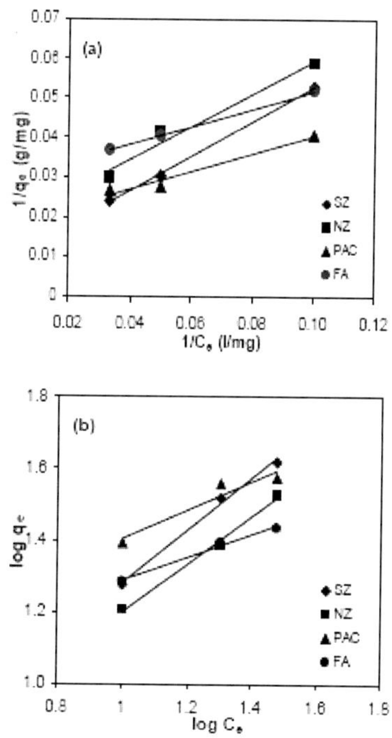

Figure 19. Adsorption isotherms of Cu on adsorbents in the presence of HA Langmuir model (b) Freundlich model.

For the binary adsorbate systems, adsorption isotherm was also obtained. Figures 19 and 20 show the linear form of the Langmuir and Freundlich isotherms of Cu and Pb adsorption in the presence of HA. As can be seen from the figures, the Langmuir and Freundlich models produce a good linear relationship, suggesting that our experiment data could be explained by both models. The adsorption capacities of Cu and Pb ions in the presence of HA seem to follow SZ > NZ > PAC > FA. The adsorption isotherm data are summarised and presented in table 8.

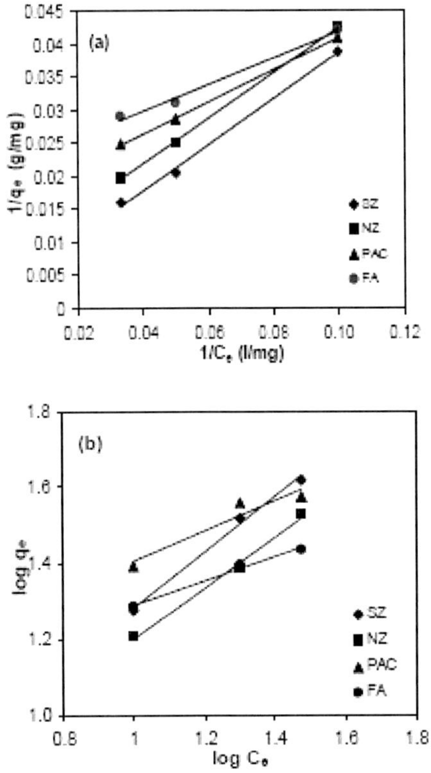

Figure 20. Adsorption isotherms of Pb on adsorbents in the presence of HA. (a) Langmuir model (b) Freundlich model.

Comparing the individual adsorption isotherm, the adsorption of Cu in the presence of HA increases from 43.5 to 110.2 mg/g for SZ, 24.2 to 55.9 mg/g for NZ, 19.7 to 54.9 mg/g for PAC, 28.6 to 34.3 mg/g for FA. The adsorption of Pb in the presence of HA increases from 190.7 to 254.5 mg/g for SZ, however, it decreases from 129.0 to 122.2 mg/g for NZ, and 76.8 to 60.31 mg/g for PAC, respectively.

Table 8. The parameters of metal adsorption isotherm of heavy metals-HA systems on SZ, NZ, PAC and FA

		Langmuir isotherm			Freundlich isotherm		
	Adsorbent	q_m (mg/g)	K_L (l/mg)	r^2	K_F	n	r^2
Cu-HA	SZ	110.24	0.02	1.00	2.36	1.38	0.99
	NZ	55.90	0.04	0.97	3.13	1.51	0.99

	Adsorbent	q_m (mg/g)	K_L (l/mg)	r^2	K_F	n	r^2
	PAC	54.89	0.08	0.97	3.90	2.51	0.91
	FA	34.25	0.13	1.00	3.90	3.17	0.99
Pb-HA	SZ	254.53	0.01	1.00	2.36	1.22	0.99
	NZ	122.26	0.02	1.00	3.13	1.42	0.99
	PAC	60.31	0.07	1.00	3.90	2.16	0.99
	FA	45.74	0.11	0.99	3.90	2.89	0.96

The adsorption isotherm of HA in the presence of heavy metals is also shown by the Langmuir and Freundlich isotherms (figures 21 and 22). It is seen that the adsorption follows the Langmuir and Freundlich models, except for NZ adsorption. However, the experiment results seem to be well described with the Langmuir model. The calculation values of the isotherm parameters are shown in table 9.

Table 9. The parameters of HA adsorption isotherms of HA-heavy metal systems on SZ, NZ, PAC and FA

		Langmuir isotherm			Freundlich isotherm		
	Adsorbent	q_m (mg/g)	K_L (l/mg)	r^2	K_F	n	r^2
HA-Cu	SZ	23.45	0.12	0.99	1.54	3.60	0.93
	NZ	18.65	0.07	0.73	1.47	2.82	0.46
	PAC	41.04	0.08	0.93	1.50	2.36	0.95
	FA	35.28	0.08	1.00	1.50	2.69	1.00
HA-Pb	SZ	50.90	0.01	1.00	1.54	1.37	0.98
	NZ	19.62	0.03	1.00	1.47	1.67	1.00
	PAC	77.82	0.01	0.99	1.50	1.39	0.96
	FA	40.90	0.02	1.00	1.50	1.53	0.98

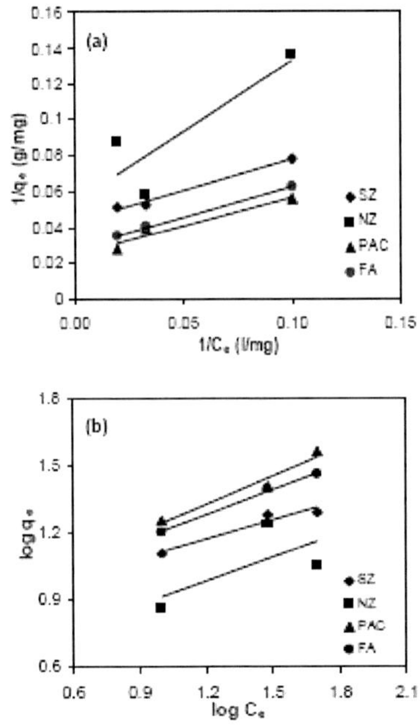

Figure 21. Adsorption isotherms of HA on adsorbents in the presence of Cu (a) Langmuir model (b) Freundlich model.

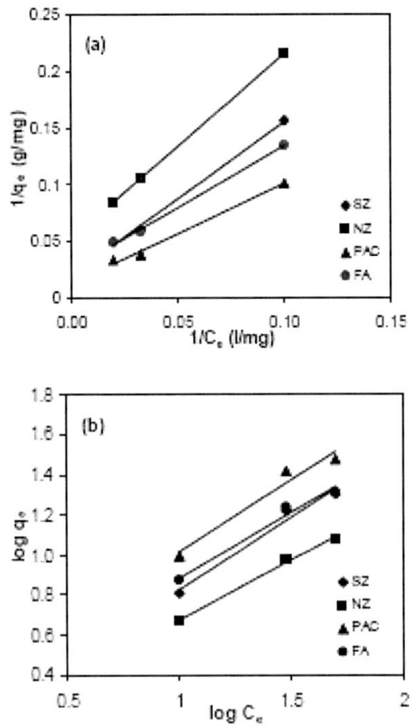

Figure 22. Adsorption isotherms of HA on adsorbents in the presence of Pb (a) Langmuir model (b) Freundlich model.

Compared with the individual adsorption isotherm, the adsorption of HA in the presence of Cu is decreased from 84.1 to 23.5 mg/g for SZ, 67.8 to 18.7 mg/g for NZ, 81.2 to 41.0 for PAC, but it is increased from 34.1 to 35.3 mg/g for FA. The adsorption of HA in the presence of Pb is decreased from 84.1 to 50.9 mg/g for SZ, 67.8 to 19.6 mg/g for NZ, and 81.2 to 77.8 mg/g for PAC, respectively.

4. CONCLUSION

1. Adsorbents, SZ, PAC, NZ, and FA, are all effective in HA and heavy metal adsorption. HA adsorption onto various adsorbents follows the order of SZ > PAC > NZ > FA. The adsorption of HA is favoured at a low pH and an increase in pH will lead to the reduction of HA adsorption. However, the adsorption will not change significantly after pH > 6. The adsorption capacity of HA on SZ and NZ decreases with increasing temperature but there is no significant effect on the adsorption of HA on PAC and FA. The equilibrium sorption capacities of HA on SZ, NZ, PAC and FA are 84.1, 67.8, 81.2 and 34.1 mg/g, respectively at 30 °C and pH 5.0.
2. For heavy metal removal, SZ is more efficient than FA, NZ and PAC for Cu removal. SZ is also more efficient than NZ and PAC for Pb removal. The ideal pH for heavy metal ions adsorption onto SZ, NZ, and PAC was found to be 5 and 6. Moreover, higher pH results in higher adsorption. The equilibrium sorption capacity of Cu and Pb ions on SZ, NZ, PAC and FA were 43.5, 24.2, 19.7, 28.6 and 190.7, 129.0, 76.8 mg/g, respectively at 30 °C and pH 5.
3. For coadsorption, when the initial HA concentration increases, Cu adsorption on SZ and FA decreases while the adsorption of Cu on PAC increases. In the adsorption of Cu on NZ, however, there is no significant effect. Pb adsorption is enormously decreased approximately 45% in most adsorbents in the presence of HA, except for the FA case. HA adsorption decreases with the increase in the initial heavy metal concentration. The adsorption of heavy metals on every adsorbent increases with increasing heavy metals concentration in the presence of HA. Throughout the presence of HA, SZ seems to be the best adsorbent to bind with Cu^{2+} and Pb^{2+} ions followed by PAC, NZ and FA. In the presence of heavy metals, PAC is the most effective adsorbent in order to remove HA through HA-heavy metal complexation, followed by FA, SZ and NZ, respectively.

REFERENCES

Abate, G. and Masini, J. C. (2005) Influence of pH, Ionic Strength and Humic Acid on Adsorption of Cd(II) and Pb(II) onto Vermiculite. *Colloids and Surfaces A: Physicochemical and Engineering Aspects*, 262(1-3): 33-39.

Al-Degs, Y.S., El-Barghouthi, M.I., Issa, A.A., Khraisheh, M.A., Walker, G.M. (2006) Sorption of Zn(II), Pb(II), and Co(II) using Natural Sorbents: Equilibrium and Kinetic Studies. *Water Research*, 40(14): 2645-2658.

Alpatova, A., Verbych, S., Bryl, M., Nigmatullin, R., Hilal, N. (2004) Ultrafiltration of Water Containing Natural Organic Matter: Heavy Metal Removing in the Hybrid Complexation-Ultrafiltration Process. *Separation Purification Technology*, 40: 155-162.

Arias, M., Barral, M.T., Mejuto, J.C. (2002) Enhancement of Copper and Cadmium Adsorption on Kaolin by the presence of Humic Acids. *Chemosphere*, 48: 1081–1088.

Benhammou, A., Yaacoubi, A., Nibou, L., Tanouti, B. (2005) Adsorption of Metal Ions onto Moroccan Stevensite: Kinetic and Isotherm Studies. *Journal of Colloid and Interface Science,* 282(2): 320-326.

Bjelopavlic, M., Newcombe, G., Hayes, R. (1998) Adsorption of NOM onto Activated Carbon: Effect of Surface Charge, Ionic Strength and Pore Volume Distribution. *Journal of Colloid and Interface Science*, 210: 271-280.

Chang, M.Y. and Juang, R.S. (2004) Adsorption of Tannic acid, Humic Acid, and Dyes from Water Using the Composite of Chitosan and Activated Clay. *Journal of Colloid and Interface Science,* 278: 18-25.

Chen, J.P., Yiacoumi, S., Blaydes, T.G. (1996) Equilibrium and Kinetic Study of Copper Adsorption by Activated Carbon. *Separation Technology*, 6: 133–146.

Chen, J. and Wu, S. (2004) Simultaneous Adsorption of Copper Ions and Humic Acid onto an Activated Carbon. *Journal of Colloid and Interface Science*, 280: 334-342.

Christl, I., Metzger, A., Heidmann, I., Kretzschmar, R. (2005) Effect of Humic and Fulvic Acid Concentrations and Ionic Strength on Copper and Lead Binding. *Environmental Science and Technology,* 39(14): 5319-5326.

Cloirec, P.L., Brasquet, C., Subrenat, E. (1997) Adsorption onto Fibrous Activated Carbon: Applications to Water Treatment. *Energy Fuels,* 11(2): 331-336.

Daifullah, A.A.M., Girgis, B.S., Gad, H.M.H. (2004) A Study of the Factors Affecting the Removal of Humic Acid by Activated Carbon Prepared from Biomass Material. *Colloids and Surfaces A: Physicochemical and Engineering Aspects*, 235(1-3): 1-10.

Deng, S. and Bai, R. (2004) Adsorption and Desorption of Humic Acid on Aminated Polyacrylonitrile Fibers. *Journal of Colloid and Interface Science*, 280(1): 36-43.

Duan, J., Wilson, F., Graham, N., Tay, J.H. (2003) Adsorption of Humic Acid by Powdered Activated Carbon in Saline Water Conditions. *Desalination*, 151: 53-66.

Erol, M., Küçükbayrak, S., Ersoy-Meriçboyu, A., Ulubaş, T. (2005) Removal of Cu^{2+} and Pb^{2+} in Aqueous Solutions by Fly Ash. *Energy Conversion and Management*, 46(7-8): 1319-1331.

Ford, D.L. (1981) Wastewater Characteristics and Treatment. in J.R., Perrich Activated Carbon Adsorption for Wastewater Treatment. CRC Press Inc., Boca Raton, Florida: 1-27.

Fu, G. and Allen, H.E. (1992) Cadmium Adsorption by Oxic Sediment. *Water Research*, 26(2): 225-233.

Ghabbour, E.A., Davies, G., Goodwillie, M.E., O'Donaughy, K., Smith, T.L. (2004) Thermodynamics of Peat-, Plant-, and Soil-Derived Humic Acid Sorption on Kaolinite. *Environmental Science and Technology*, 38(12): 3338 -3342.

Guibal, E., M. Van Vooren, B.A. Dempsey and J. Roussy, (2006) A Review of the Use of Chitosan for the Removal of Particulate and Dissolved Contaminants. *Separation Science and Technology 41*(11): 2487-2514.

Guzzella, L., D. Feretti and S. Monarca, (2002) Advanced Oxidation and Adsorption Technologies for Organic Micropollutant Removal from Lake Water Used as Drinking-Water Supply. *Water Research* 36(17), 4307-4318.

Han, N.Z. and M.L. Thompson, (2003) Impact of dissolved organic matter on copper mobility in aquifer material. *Journal of Environmental Quality* 32(5): 1829-1836.

Hankins, N.P., Lu, N., Hilal, N. (2006) Enhanced Removal of Heavy Metal Ions bound to Humic Acid by Polyelectrolyte Flocculation. *Separation and Purification Technology*, 51(1): 48-56.

Hasany, S.M., Saeed, M.M., Ahmed, M. (2002) Sorption and Thermodynamic Behaviour of Zinc(II)-Thiocyanate Complexes onto Polyurethane Foam from Acidic Solutions. *Journal of Radioanalytical Nuclear Chemistry*, 252: 477-484.

Ho, Y.S. and Mckay, G. (1999) Pseudo-Second Order Model for Sorption Processes. *Process Biochemistry*, 34(5): 451-465.

Horsfall, J.M. and Spiff, A.I. (2004) Studies on the Effect of pH on the Sorption of Pb^{2+} and Cd^{2+} Ions from Aqueous Solutions by Caladium Bicolour (Wild Cocoyam) Biomass. *Journal of Biotechnology*, 7(3): 313-323.

Jang, A., Seo, Y.W., Bishop, P.L. (2005) The Removal of Heavy Metals in Urban Runoff by Sorption on Mulch. *Environmental Pollution*, 133(1): 117-127.

Katsumata, H., Kaneco, S., Matsuno, R., Itoh, K., Masuyama, K., Suzuki, T., Funasaka, K., Ohta, K. (2003) Removal of Organic Polyelectrolytes and their Metal Complexes by Adsorption onto Xonotlite. *Chemosphere*, 52: 909-915.

Kobya, M., Demirbas, E., Senturk, E., Ince, M. (2005) Adsorption of Heavy Metal Ions from Aqueous Solutions by Activated Carbon Prepared from Apricot Stone. *Bioresource Technology*, 96(13): 1518-1521.

Kurniawan, T.A., G.Y.S. Chan, W.H. Lo and S. Babel, (2006) Physico-Chemical Treatment Techniques for Wastewater Laden with Heavy Metals. *Chemical Engineering Journal* 118(1-2), 83-98.

Lee, D.H. and Moon, H. (2001) Adsorption Equilibrium of Heavy Metals on Natural Zeolites. *Korean Journal of Chemical Engineering*, 18(2): 247–256.

Lin, Y.B., Fugetsu, B., Terui, N., Tanaka, S. (2005) Removal of Organic Compounds by Alginate Gel Beads with Entrapped Activated Carbon. *Journal of Hazardous Materials*, 120(1-3): 237-241.

Liu, A. and Gonzalez, R. (1999) Adsorption/Desorption in a System Consisting of Humic Acid, Heavy Metals, and Clay Minerals. *Journal of Colloid and Interface Science*, 218: 225-232

Lorenc-Grabowska, E. and Gryglewicz, G. (2005) Adsorption of Lignite-Derived Humic Acids on Coal-based Mesoporous Activated Carbon. *Journal of Colloid and Interface Science*, 284(2): 416-423.

Meghea, A., Rehner, H.H., Peleanu, I., Mihalache, R. (1998) Test-Fitting on Adsorption Isotherms of Organic Pollutants from Waste Water on Activated Carbon. *Journal of Radioanalytical Nuclear Chemistry*, 229: 105–110.

Mohammad, A. and M. Najar, (1997), Physico-Chemical Adsorption Treatments For Minimization Of Heavy Metal Contents In Water And Wastewaters. *Journal of Scientific and Industrial Research* 56(9): 523-539.

Paajanen, A., Lehto, J., Santapakka, T., Morneau, J.P. (1997) Sorption of Cobalt on Activated Carbons from Aqueous Solutions. *Separation Science and Technology*, 32(1–4): 813–826.

Peng, X., Luan, Z., Chen, F., Tian, B., Jia, Z. (2005) Adsorption of Humic Acid onto Pillared Bentonite. *Desalination*, 174(2): 135-143.

Petrovic, M., Kastelan-macan, M., Horvat, A.J.M. (1999) Interactive Sorption of Metal Ions and Humic Acids onto Mineral Particles. *Water Air Soil Pollution*, 111(1-4): 41-56.

Rashid, M.A. (1985) Geochemistry of Marine Humic Substances. Chapters 4 and 7, New York: Springer-Verlag.

Rubio, J., M.L. Souza and R.W. Smith, 2002, Overview of Flotation as a Wastewater Treatment Technique. *Minerals Engineering* 15(3), 139-155.

Sekar, M., Sakthi, V., Rengaraj, S. (2004) Kinetics and Equilibrium Adsorption Study of Lead(II) onto Activated Carbon Prepared from Coconut Shell. *Journal of Colloid and Interface Science,* 279(2); 307-313.

Starek, J., A. Zukal and J. Rathousky, 1994, Comparison of the Adsorption of Humic Acids from Aqueous-Solutions on Active-Carbon and Activated-Charcoal Cloths. *Carbon* 32(2), 207-211.

Taylor, M.D. and Theng, B.K.G. (1995) Sorption of Cadmium by Complexes of Kaolinite with Humic Acid. *Communications in Soil Science and Plant Analysis*, 26: 765-766.

Teng, H. and Hsieh, C.T. (1998) Influence of Surface Characteristics on Liquid Phase Adsorption of Phenol by Activated Carbons prepared from Bituminous Coal. *Industrial and Engineering Chemistry Research,* 37(8): 3618–3624.

Tomaszewska, M., Mozia, S., Morawski, A.W. (2004) Removal of Organic Matter by Coagulation Enhanced with Adsorption on PAC. *Desalination*, 161(1): 79-87.

Üçer, A., Uyanik, A., Aygün, Ş.F. (2006) Adsorption of Cu(II), Cd(II), Zn(II), Mn(II) and Fe(III) Ions by Tannic Acid Immobilised Activated Carbon. *Separation and Purification Technology,* 47(3): 113-118.

Uyguner, C.S., S.A. Suphandag, A. Kerc and M. Bekbolet, (2007), Evaluation of Adsorption and Coagulation Characteristics of Humic Acids Preceded by Alternative Advanced Oxidation Techniques. *Desalination* 210(1-3): 183-193.

Vermeer, A.W.P., Mcculloch, J.K., van Riemsdijk, W., Koopal, L.K. (1999) Metal Ion Adsorption to complexes of Humic Acid and Metal Oxides: Deviations from the Additivity Rule. *Environmental Science and Technology* 33: 3892–3897.

Vijayaraghavan, K., Jegan, J.R., Palanivelu, K., Velan, M. (2004) Copper Removal from Aqueous Solution by Marine Green Alga Ulva Reticulata. *Environmental Biotechnology*, 7(1): 61-71.

Yan, W.L. and Bai, R. (2005) Adsorption of Lead and Humic Acid on Chitosan Hydrogel Beads. *Water Research*, 39(4): 688-698.

Wang, S., Li, H.T., Xu, L.Y. (2006) Application of Zeolite MCM-22 for Basic Dye Removal from Wastewater. *Journal of Colloid and Interface Science*, 295: 71-78.

Wang, S.B., Z.H. Zhu, A. Coomes, F. Haghseresht and G.Q. Lu, (2005), The Physical and Surface Chemical Characteristics of Activated Carbons and the Adsorption of Methylene Blue from Wastewater. *Journal of Colloid and Interface Science* 284(2), 440-446.

Zhang, X. and Bai, R. (2003) Mechanisms and Kinetics of Humic Acid Adsorption onto Chitosan-Coated Granules. *Journal of Colloid and Interface Science*, 264: 30-38.

Zhou, J.L., Rowland, S., Fauzi, R., Mantoura, C., Braven, J. (1994) The Formation of Humic Coatings on Mineral Particles under Simulated Estuarine Conditions-A Mechanistic Study. *Water Research*, 28: 571-579.

Zularisam, A.W., A.F. Ismail and R. Salim, (2006) Behaviours of Natural Organic Matter in Membrane Filtration for Surface Water Treatment -A Review. *Desalination* 194(1-3), 211-231.

In: Heavy Metal Pollution
Editors: S. E. Brown and W. C. Welton

ISBN: 978-1-60456-899-8

Chapter 10

BINDING OF BIVALENT CATIONS BY CHARGED POLYSACCHARIDES FROM BACTERIAL EPS

Christian Mayer, Natascha Emmerichs, Guido Furth and Dirk Bergmann
University of Duisburg-Essen
FB Chemie
54 117 Essen, Germany

ABSTRACT

The binding of heavy metal cations in bacterial EPS is of significant ecological relevance: it may cause an undesired accumulation of heavy metals in naturally occurring biofilms, but may as well be used to remove heavy metals from polluted waste water in a biological approach. In an experimental study, the interaction between charged polysaccharides from bacterial EPS and bivalent cations was investigated by conductometry, viscosimetry, and nuclear magnetic resonance spectroscopy. It is shown that bacterial alginate, xanthan and hyaluronic acid bind various bivalent cations (Ca^{2+}, Mg^{2+}, Mn^{2+}, Fe^{2+}, Cu^{2+}, Zn^{2+}, Cd^{2+} and Pb^{2+}) in aqueous solutions. Molar relations between the bivalent cations and the disaccharide units in the resulting complexes were determined as well as their relative stabilities. Regarding the structure of the dissolved chain molecules, the addition of the cations has opposite effects on the polysaccharides: in case of alginate, the bivalent cations preferably form intermolecular cross-links, leading to the formation of a growing molecular network accompanied by the increase of the solution viscosity. In case of xanthan and hyaluronic acid, an increasing cation concentration is connected to a growing number of intramolecular links, reduced hydrodynamic radii and decreasing solution viscosities. Based on these data and results from NMR studies using a paramagnetic center, possible structures of the polysaccharide-cation complexes are proposed.

INTRODUCTION

Extracellular polymeric substances (EPS) are widely spread in ecological environments and are mostly connected to microbial growth. They principally consist of polysaccharides (Sutherland 2001 a, b), but may also contain proteins, nucleic acids or other biopolymers (Flemming 2001 and 2007). In presence of water, they swell and tend to form gels with high water content and, compared to the small amount of organic substance, surprisingly high mechanical strength. In their biological function, EPS represent the principal construction materials for microbial aggregates such as biofilms, flocs ("planktonic biofilms") and sludge (Flemming 2001). In these cases, the EPS matrix can be regarded as the “house of cells” (Flemming 2007). This given, the capability of EPS to bind heavy metals in a polluted water system becomes a crucial concern for environmental protection. If a heavy metal ion is selectively bound by an EPS constituent, it is likely to be accumulated in flocs and biofilms and hence is very likely to enter the food chain with all known undesired consequences. On the other hand, this phenomenon offers a technical approach for an efficient removal of this pollutant from industrial waste water as long as the separated flocs and biofilms are separated and taken care of properly. The adsorption of heavy metals to algal (Apiratikul 2006, Pavasant 2006, Papageorgiou 2006) or bacterial EPS (Philippis 2007) have been proposed as possible applications of this technique.

With a large degree of diffusive exchange between the EPS and the aqueous environment, the EPS matrix forms a separating phase between the micro-organism and the surrounding water. All constituents present in the water phase at a concentration c_W will occur in the EPS matrix at a concentration c_{EPS} given by a thermodynamic distribution equilibrium $K = c_{EPS} / c_W$. Naturally, this equilibrium constant is strongly influenced by any tendency of the EPS constituents to irreversibly or reversibly bind the cations, e.g. by the formation of complexes.

As the most important structural component, polysaccharides may also play the most important role in binding cations (Son 2004). In this respect, polysaccharides which carry negative charges are the primary candidates due to the electrostatic interactions which are likely to occur between the polymer and the cations. These interactions are expected to be especially strong in case of bivalent cations. This is only partially due to the higher charge itself: a more important reason for a stronger binding of bivalent cations is their capability to form inter- or intramolecular bridges between two negatively charged positions on the polysaccharide chain. This not only changes the structure of the polysaccharide in aqueous solution, it also leads to the formation of very stable chelate-like complexes and hence to a very strong cation binding. A well known example is the binding of calcium ions by alginate, a negatively charged polysaccharide often found in algal or bacterial EPS. Together with Ca^{2+}-ions, algal alginate forms a very stable gel where the ions are embedded in an egg-box-like structure (Bryce 1974, Ertesvag 1998). Lately, it was proposed that other structures of similar stability may compete with the ideal egg-box conformation (Li 2007). In many cases, the stability of the calcium-alginate complex is the most important contribution to the mechanical strength of a biofilm.

However, calcium is not necessarily the only metal ion which is bound by anionic polysaccharides. In aquatic environments, naturally occurring cations such as Mn^{2+} or Mg^{2+} may compete with calcium, even if they form less stable and complexes with less specific binding sites (Lattner 2003, Emmerichs 2004). Of further importance are possible interactions between anionic polysaccharides and heavy metal cations which are expected in industrial waste waters. In aqueous solution, the majority of ecologically relevant heavy metals occur as bivalent cations, such like Pb^{2+}, Cd^{2+}, Hg^{2+}, or Cu^{2+}. Hence, it is very likely that heavy metals ions are capable of replacing the calcium ion in the polysaccharide complex. That means that every biofilm or plankton which contains anionic polysaccharides in its EPS matrix is a potential trap for heavy metals in aqueous solution. This given, these systems may accumulate these pollutants with all desired or undesired consequences mentioned above.

In the following, we want to present a study which is dedicated to the elucidation of the binding mechanism between bivalent metal cations and charged polysaccharides of bacterial origin. The consequences of the complex formation on the overall cation mobility are studied by means of conductometric titration which also yields data on the stoichiometry and the intensity of the binding. The molecular structure of the cation-polysaccharide is detected by nuclear magnetic resonance using the effect of a paramagnetic bivalent cation (Mn^{2+}). Finally, cation-induced intermolecular interactions and changes of the conformation of the dissolved polysaccharide molecules are observed using solution viscosimetry.

BACTERIAL ALGINATE

Probably the most prominent case of cation binding by polysaccharides is the interaction between calcium ions and alginate. Chemically, alginate represents a non-branched anionic copolymer from the two monosaccharides D-mannuronate (M) and L-guluronate (G) (Ertesvåg 1998) (figure 1). The guluronate units are statistically distributed among a majority of mannuronate units and typically contribute to 30% of the chain (Schürks 2002).

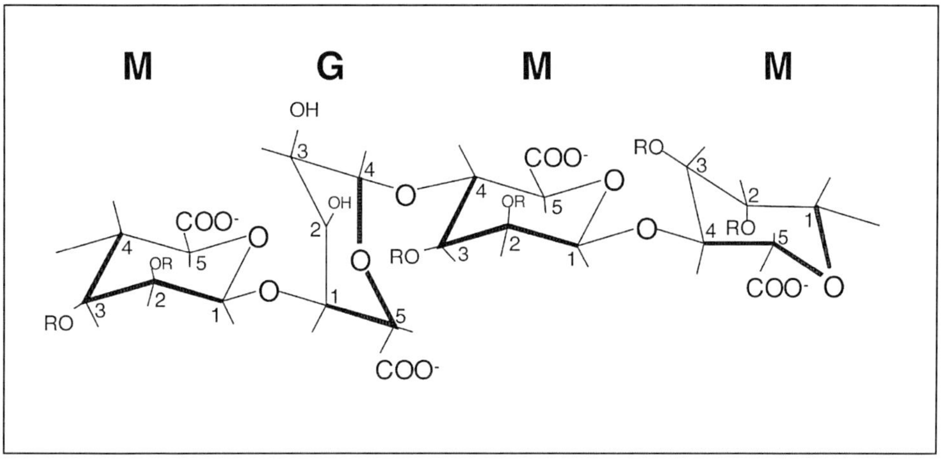

Figure 1. Characteristic fraction of an alginate chain of bacterial origin (M: mannuronate, G: guluronate residues). The carbon atoms C-2 and C-3 of the mannuronate residues are partially acetylated (R = -H or R = $-COCH_3$). All C-5 carbon atoms carry a carboxylate group which may be partially protonated. The G-units are statistically distributed among a majority of M-units and typically contribute to 30% of the chain.

It is well known as an exopolysaccharide of algal origin, but may as well be produced by bacteria such as *Pseudomonas aeruginosa*, *Pseodomonas medocina*, and *Azotobacter vinelandii* (Christensen 1990, Sabra 2001). In contrast to algal alginate, bacterial alginate is partially acetylated at the C-2 and the C-3 carbons of the mannuronic acid units which give the chain a somewhat more hydrophobic behavior. In addition, the alginates of bacterial origin miss the blocks of guluronic acid $-(G)_n$-which are generally seen as the origin of the formation of the egg box structure together with calcium ions (Bryce 1974, Ertesvag 1998). Nevertheless, bacterial alginate shows specific interaction with calcium ions. The binding of the cations is directly observed in a conductometric titration experiment (figure 2).

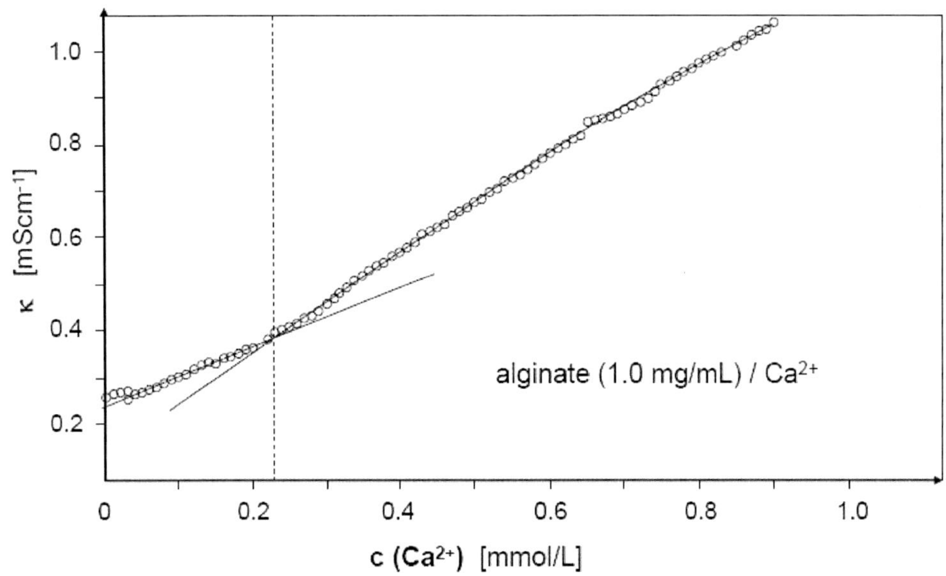

Figure 2. Conductometric titration of a partially acetylated sodium alginate solution (1.0 mg/mL) with Ca^{2+}-ions. The specific conductivity κ is plotted as a function of the Ca^{2+}-concentration. The dotted line indicates the equivalence point between the active binding sites of the alginate and the calcium cations which is observed at a cation concentration near 2.3 mmol/L. On both sides of the equivalence concentration, the plot is approximated by polynomial fits (solid lines).

The resulting plot of the specific conductivity vs. the cation concentration is easily interpreted: in absence of calcium ions (c (Ca^{2+}) = 0 mmol/L), the starting value of the specific conductivity results from the presence of the sodium counter ions of the alginate. With addition of calcium chloride solution, the conductivity increases at a reduced rate, as a significant quantity of the calcium ions are immobilized at the polysaccharide chains. Only after all of the active binding sites are occupied, the increase of the specific conductivity follows the expected dependence according to Kohlrausch's law. The intersection between both parts of the conductivity plot marks the equivalence point where the number of cations corresponds to the number of binding sites. For the partially acetylated alginate typical for bacterial EPS, it is found that it takes approximately twenty monomers (or ten repetitive units as shown in figure 1) to bind a single cation. In contrast, algal alginate shows an equivalence of only five monomers to a single cation (Siew 2005).

How is the cation immobilized by bacterial alginate? The structure of the resulting alginate-cation complex has been elucidated using solid-state carbon nuclear magnetic resonance (^{13}C-NMR) spectroscopy on ^{13}C-labelled alginate in presence of calcium and magnesium ions (Lattner 2003). The solid state NMR spectra indicate that the cation interactions primarily affect the local mobility of guluronate units, even though the G-residues contribute only around one third of the overall chain and G-blocks are missing completely (Schürks 2002). Additional NMR studies using manganese cations as paramagnetic centers lead to the conclusion that the preferred binding site for bivalent cations is represented by pockets formed in between adjacent guluronate and mannuronate monomers as depicted in figure 3 (Emmerichs 2004).

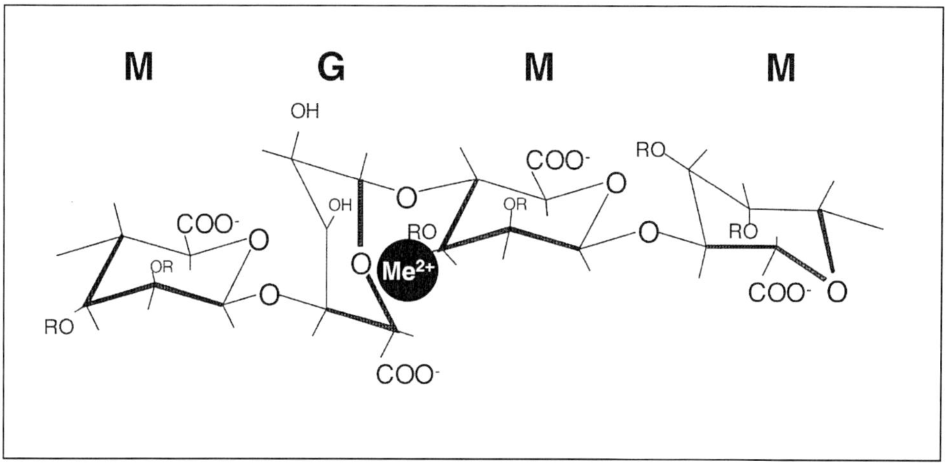

Figure 3. Proposed cation binding site in bacterial alginate formed by a G-M-pair within the polysaccharide chain (Emmerichs 2004).

This idea has been supported by the observation of a specific cation binding capacity of a strictly alternating mannuronate – guluronate (MG) polymer (Donati 2005). Similar to the more prominent egg-box structure for algal alginate, this type of cation binding can lead to the cation-induced gelation of bacterial alginate, even though the resulting gel is much weaker (Körstgens 2001, Wloka 2004). The early steps of cation-induced gelation can be observed in a viscometric titration experiment (figure 4).

In case of algal alginate (full symbols in figure 4), the presence of calcium ions leads to a rapid increase of the solution viscosity caused by the formation of intermolecular cross-links between the dissolved macromolecules. In contrast, the partially acetylated alginate typical for the bacterial variety shows a much weaker increase and even exhibits a shallow minimum at low calcium concentrations. On the other hand, it was found that the acetylated variety shows a higher ability to form gels with Fe^{3+}-ions (Lee 1996). Manganese ions are bound by both types of alginate, but generally have little effect on the solution viscosities of both alginate varieties.

Generally, bacterial alginate has a much weaker tendency to bind bivalent cations than algal alginate. In addition, the gelation caused by calcium ions leads to a lower degree of cross-linking and therefore to a weaker gel structure. Nevertheless, bacterial alginate is capable of binding different bivalent metal ions even if, on average, twenty monomer units are needed for a single cation. Hence, it is expected that the biofilms formed by bacteria such as *Pseudomonas aeruginosa*, *Pseodomonas medocina*, and *Azotobacter vinelandii* which produce alginate as an EPS constituent will immobilize and accumulate bivalent cations from their aqueous environment.

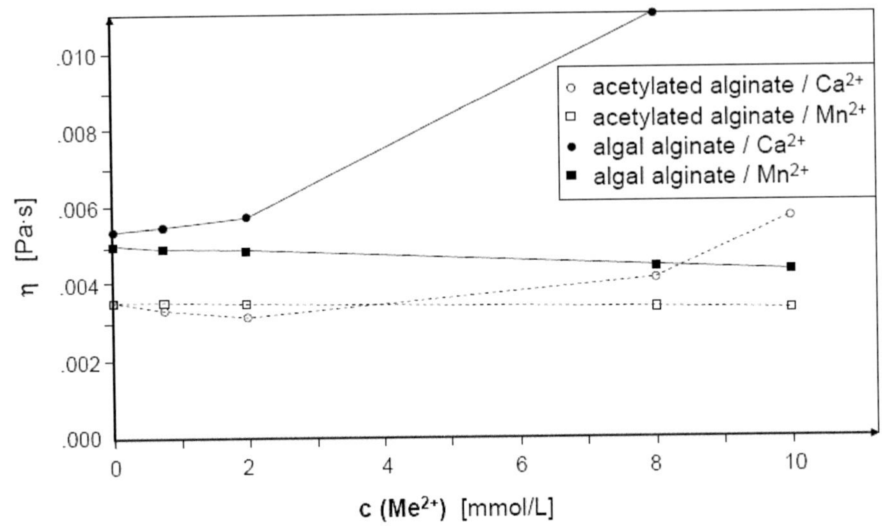

Figure 4. Comparison of the apparent viscosities of solutions of acetylated alginate (open symbols) and original algal alginate (full symbols) in presence of the bivalent cations of calcium (circles) and manganese (squares). Connecting lines (full lines: algal alginate, broken lines: acetylated alginate) are added to improve the readability of the data.

Xanthan

The bacterial polysaccharide xanthan is produced by organisms such as *Xanthomonas campestris, Pseudomonas aeruginosa or Azetobacter vinelandii* under aerobic conditions. It represents a branched anionic copolymer with a relatively complicated composition (figure 5): the main chain consists of glucose (Glc) molecules connected by β-1,4 glycosidic links and is similar to cellulose. Every second glucose unit (Glc') carries a side chain which is composed of β-D-mannose (Man), β-1,4-D-glucuronic acid (GlcA) and α-1,2-D-mannose (Man'). A variable percentage of the terminal mannose residues (Man') carry a pyruvic acid (Pyr) unit, 60% to 70% of the internal mannose residues (Man) are acetylated (Ac) (Shatwell 1990). Similarly to cellulose, xanthan is stable over a wide pH range. It is generally assumed that the conformation of natural xanthan consists of right-handed single and double helices with a twist level of 47 Å. The molecular weight of natural xanthan lies in the range of 10^6 – 10^7 g/mol (Lopez 2001).

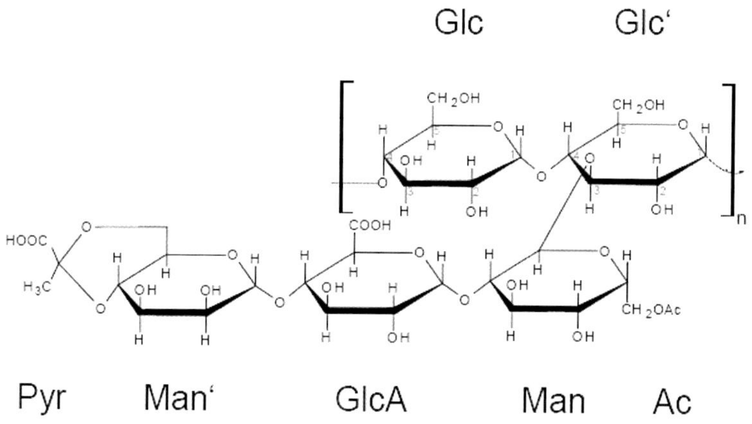

Figure 5. Repetitive unit of the branched polysaccharide chain of a xanthan molecule. The main chain is formed by pairs of glucose units (Glc and Glc', top). Every Glc'-unit carries a side chain of three monosaccharides: a mannose unit (Man) which is partially acetylated (Ac), a glucuronic acid unit (GlcA) and a terminal mannose unit (Man') which may be linked to a pyruvate residue (Pyr) (Bergmann 2008).

On average, the repetitive unit of five monosaccharide units carries up to two negative charges. Hence, it is expected that xanthan will bind more cations than the same amount of a neutral polysaccharide and that the polysaccharide-cation interactions will be stronger. In fact, in comparison with an uncharged polysaccharide such as dextran, the cation binding by xanthan is quite significant. This again is directly observed by a conductometric titration experiment. Figure 6 shows a comparison of two plots obtained on dextran and xanthan solutions in a conductometric titration with cadmium ions. The dextran solution shows no significant trace of cation binding. Essentially, the titration curve exhibits the same shape as for water; the dependence of the specific conductivity is well described by Kohlrausch's law (figure 6, full symbols). In contrast, the titration plot of xanthan indicates that the increase of the conductivity connected to the increasing concentration of mobile cadmium cations is distinctly reduced by the presence of this polysaccharide (figure 6, open symbols). Only above a concentration of 1.0 mmol/L, the addition of Cd^{2+} leads to the expected increase of the specific conductivity as observed in case of dextran. Due to the initial presence of the counter ions of the xanthan, the starting conductivity is significantly higher than in case of dextran. All bivalent cations tested so far show a similar behavior in the conductometric plots (Bergmann 2008).

If the same experiment is repeated for an aqueous solution in absence of the polysaccharide, but in presence of a similar quantity of counter ions (such that the starting value for κ is similar to the original plot), a reference curve is generated which allows one to quantify the cation binding effect (dotted line in figure 6). The difference $\Delta\kappa$ obtained at the equivalence point characterizes the stability of the polysaccharide-cation complex and may be used to compare the binding efficiency of various polysaccharides for a given cation (Furth 2008). These conductivity plots have been applied to determine the equivalence point and the binding efficiency of xanthan with respect to different biologically and ecologically relevant cations (Bergmann 2008). The result is summarized in table 1.

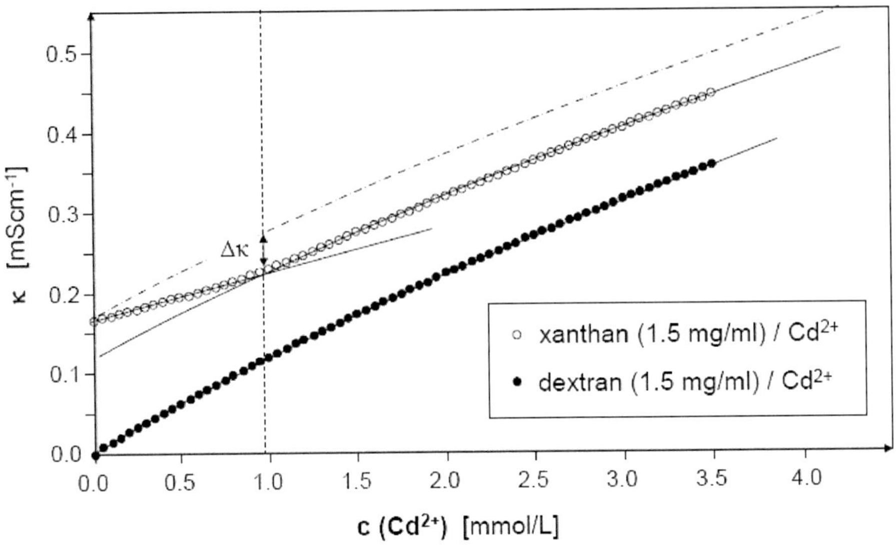

Figure 6. Conductometric titrations of dextran and xanthan solutions (each 1.5 mg/mL) with Cd^{2+}-ions. The specific conductivity κ is plotted as a function of the Cd^{2+}-concentration. The vertical broken line indicates the equivalence point between the active binding sites of xanthan and the cadmium cations which is observed at a cation concentration near 1.0 mmol/L. The two fractions of the plot for xanthan (open symbols) and the full plot for dextran (full symbols) are approximated by polynomial fits (solid lines). The dotted line marks the conductivity plot in absence of the polysaccharide, but with a similar quantity of counter ions to yield the same starting value for c (Cd^{2+}) = 0 mmol/L. The difference Δκ between the respective specific conductivities at the equivalence point represents a characteristic value for the cation binding strength.

Table 1. Molar relation between cations and disaccharide units of the main chain (n_{cat}/n_{dis}) and binding induced loss of specific conductivity Δκ for various bivalent cations in presence of xanthan. Each value of n_{cat}/n_{dis} from the conductometric titration has been derived by averaging over different xanthan concentrations (Bergmann 2008). The corresponding results for n_{cat}/n_{dis} from the viscometric titration and those for Δκ have been obtained on a xanthan concentration of 1.5 mg/mL.

cation (in interaction with xanthan)	n_{cat}/n_{dis} via conductometric titration	n_{cat}/n_{dis} via viscometric titration	Δκ via conductometric titration
Ca^{2+}	1 / 2.06	1 / 1.9	0.078 mS/cm
Mg^{2+}	1 / 2.11	1 / 2.0	0.045 mS/cm
Mn^{2+}	1 / 1.86	1 / 2.0	0.032 mS/cm
Zn^{2+}	1 / 2.06	1 / 2.0	0.082 mS/cm
Cu^{2+}	1 / 2.07	1 / 1.8	0.059 mS/cm
Fe^{2+}	1 / 2.05	1 / 1.9	0.058 mS/cm
Cd^{2+}	1 / 1.81	1 / 1.9	0.093 mS/cm
Pb^{2+}	1 / 1.90	1 / 2.0	0.130 mS/cm

With the large intersection angle between the two fractions of the conductometric plots for xanthan, the precise localization of the intersection point is difficult. Hence, the potential error in the detection of the equivalence point is as large as ±10%. However, all data for n_{cat}/n_{dis} listed in table 1 are in the vicinity of 1 / 2.00 within the expected error margin. This given, two repetitive units such as depicted in figure 5 are necessary to bind a single bivalent cation (Bergmann 2008). The comparison of the binding efficiency as indicated by the reduction of the specific conductivity Δκ shows that the bivalent ions of heavy metals such as cadmium or lead generally bind stronger to xanthan than those of light metals such as magnesium or calcium.

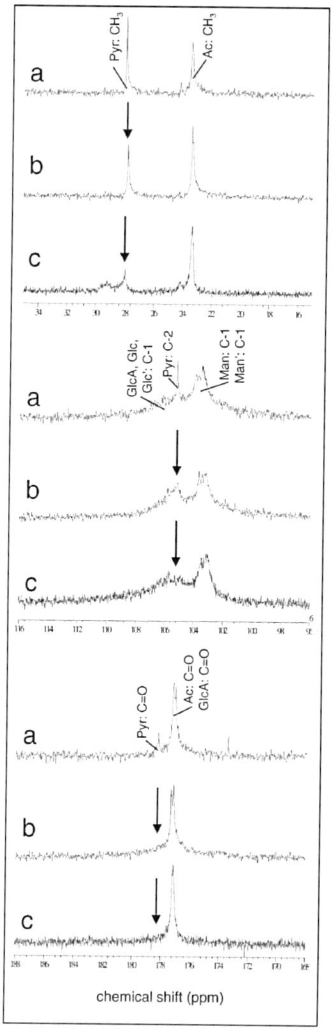

Figure 7. Fractions of a ^{13}C spectrum of xanthan under the influence of increasing manganese concentration (a: 0 mmol/L, b: 0.3 mmol/L, c: 1.2 mmol/L). All three carbon signals of the pyruvate residue (Pyr) are subject to strong signal widening and amplitude loss on addition of Mn^{2+} (arrows): the methyl group (top), the C-2 position (center) and the carbonyl group (bottom). The signals of all other groups, notably the ones of the acetyl group (Ac) and the carbonyl group of the glucuronic acid (GlcA), remain largely unchanged (Bergmann 2008).

Again, the structure of the polysaccharide-cation complex is studied by nuclear magnetic resonance using the manganese cation as a paramagnetic center. The signals in the ^{13}C-spectrum of the xanthan solution can be assigned to individual carbon positions of the xanthan chain (Horton 1985, Gamini 1991, Bergmann 2008).

Due to the dipolar interaction between the Mn^{2+}-nucleus and the observed carbon nuclei, it is possible to estimate the distance of individual carbon positions to the metallic center.

On addition of manganese cations, the signals of the carbon atoms in the direct vicinity of the binding site suffer increasing widening connected to a corresponding loss of signal amplitude. Three fractions of the full carbon spectrum under the influence of increasing manganese concentration (a: 0 mmol/L, b: 0.3 mmol/L, c: 1.2 mmol/L) are shown in figure 7.

Obviously, all changes are restricted to the three carbon signals of the pyruvate group. All other signals, notably the ones for carbon atoms of the acetyl group and the carbonyl group of the glucuronic acid residue, remain largely unchanged. This means that the cations are exclusively bound by the terminal ends of the side chains which carry a pyruvate residue. Even though the glucuronic acid may adopt a negative charge as well as the pyruvate group, no cation interaction with the latter is observed. The intensity of the methyl signal of the pyruvate residue (figure 7, top) shows that, in the given case, practically all side chains are equipped with a pyruvate group. Together with the information about the stoichiometry of the xanthan-cation complex, one may hence postulate that the metal is "pinned" between the terminal ends of two side chains as depicted in figure 8 (Bergmann 2008). This is in contradiction to an earlier model postulated by Tako et al. which includes an active role of the carboxylate group of the glucuronic acid (Tako 1987). However, the present NMR data are quite clearly excluding this alternative.

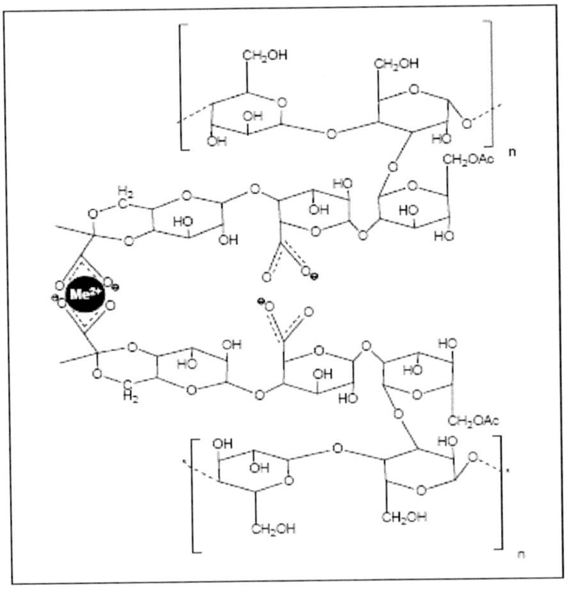

Figure 8. Proposed binding site for bivalent cations formed by two repetitive units of xanthan. The carboxylate groups of the glucuronic acid groups (center) are obviously not involved in the specific cation binding. Viscometric titration data indicate that the two repetitive units belong to the same main chain of a single xanthan molecule (Bergmann 2008).

So far, the two side chains may or may not belong to the same main chain. In order to answer this question, it is helpful to analyze the variation of the apparent viscosity of a xanthan solution with increasing cation content (figure 9).

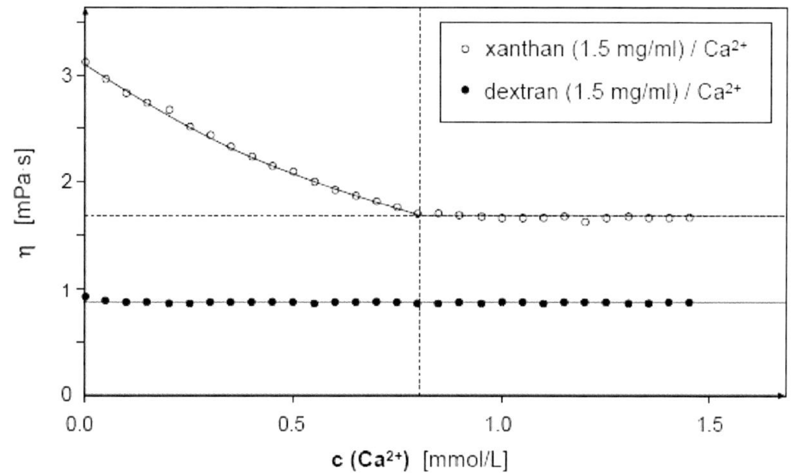

Figure 9. Apparent viscosities of solutions of xanthan (open symbols) in dependence on the calcium concentration (Bergmann 2008). Data for dextran (full symbols) are shown for comparison. The vertical broken line indicates the equivalence point between the cations and the xanthan binding sites as observed by the viscometric titration. Solid lines represent polynomial fits to the data points.

While dextran shows no dependence of the apparent viscosity on the calcium concentration, the plot for xanthan indicates a strong interaction with calcium. Up to a concentration of 0.08 mmol/L (corresponding to approximately two repetitive units of the xanthan per calcium ion), the solution viscosity decreases with increasing Ca^{2+}-concentration. For calcium concentrations above 0.08 mmol/L, the value remains constant just as for dextran. Even at higher calcium concentrations, no increase of the solution viscosity is observed. Similar plots are obtained for other bivalent cations. As for the conductometric titration, the equivalence points defined by the intersection between the two parts of the plots are determined for each cation and listed in table 1.

From this observation we conclude that the hydrodynamic radius of the dissolved xanthan molecules decreases with the binding of bivalent cations. Together with the finding that a single cation is bound by two repetitive units of the xanthan, this justifies the assumption that the connection which is formed by the bivalent cation is intramolecular and leads to an increased coiling of the macromolecule. Hence, the structure depicted in figure 8 should typically refer to two repetitive units of a single xanthan chain. In contrast to bacterial alginate, xanthan shows no tendency to develop intermolecular interactions which would lead to cross-linking and, eventually, to the formation of a gel state.

HYALURONIC ACID

Another representative for a bacterial polysaccharide, hyaluronic acid (or hyaluronate as the deprotonated form), can be regarded as an unbranched anionic copolymer. It is formed by organisms such as *Bacillus subtilis*, *Streptococcus zooepidemicus*, *Streptococcus equisimilis* or *Streptococcus equi*. Moreover, it occurs in animal and human tissue, e.g. in the synovial fluid of the joints or in the vitreous body of the human eye. Chemically, hyaluronate is a linear copolymer formed by two strictly alternating monomers: 2-*N*-acetyl-2-deoxy-D-glucopyranose (adGluP) and D-glucopyranoside (GluPA) (figure 10). In presence of sodium or potassium ions, hyaluronate forms left-handed 4-fold helices, whereas calcium ions induce the formation of left-handed 3-fold helices (Sheehan 1983).

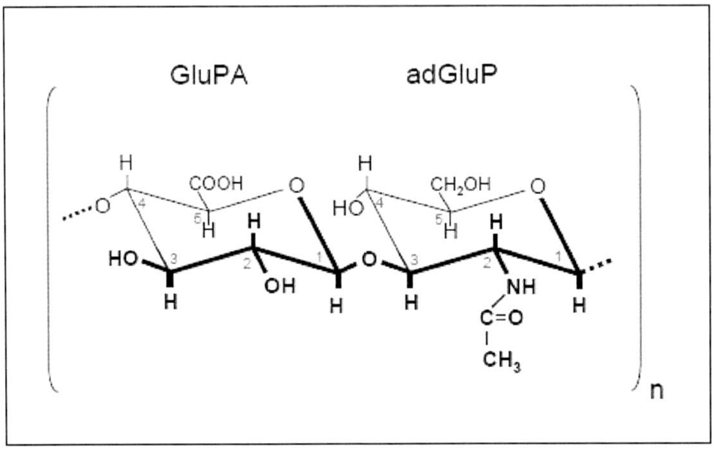

Figure 10. Repetitive unit of the unbranched polysaccharide chain of a hyaluronic acid molecule. The linear chain is formed by two alternating monomers: 2-*N*-acetyl-2-deoxy-D-glucopyranose (adGluP, right) and D-glucopyranoside (GluPA, left).

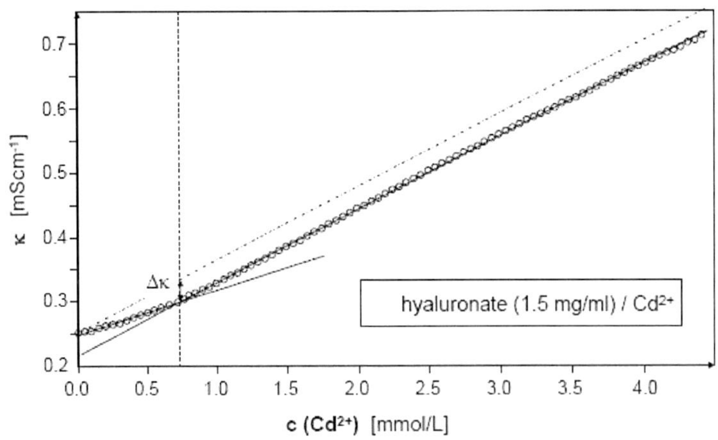

Figure 11. Conductometric titrations of a hyaluronate solution (1.5 mg/mL) with Cd^{2+}-ions. The specific conductivity κ is plotted as a function of the Cd^{2+}-concentration. The vertical broken line indicates the equivalence point between the active binding sites of xanthan and the cadmium cations which is observed at a cation concentration near 0.7 mmol/L. The two fractions of the plot are approximated by polynomial fits (solid lines). The dotted line marks the conductivity plot in absence of the polysaccharide, but with a similar quantity of counter ions to yield the same starting value for c = 0 mmol/L. The difference Δκ between the respective specific conductivities at the equivalence point represents a characteristic value for the cation binding strength.

The structure of the hyaluronate chain resembles the one of alginate. Regarding the expected binding capacity for cations, the main difference can be seen in its potential electrostatic capacity: with just a single negative charge per two hexose segments in a repetitive unit, hyaluronate is expected to bind less bivalent cations than alginate. However, conductometric plots on hyaluronate solutions (an example for a titration with cadmium ions is shown in figure 11) indicate that hyaluronate offers more binding sites for cations than bacterial alginate

Table 2. Molar relation between cations and disaccharide units of the main chain (n_{cat}/n_{dis}) and binding induced loss of specific conductivity Δκ for various bivalent cations in presence of hyaluronate. Each value of n_{cat}/n_{dis} from the conductometric titration has been derived by averaging over different hyaluronate concentrations. The corresponding results for n_{cat}/n_{dis} from the viscometric titration and those for Δκ have been obtained on a hyaluronate concentration of 1.5 mg/mL.

cation (in interaction with hyaluronate)	n_{cat}/n_{dis} via conducto-metric titration	n_{cat}/n_{dis} via viscometric titration	Δκ via conducto-metric titration
Ca^{2+}	1 / 5.00	1 / 4.4	0.074 mS/cm
Mg^{2+}	1 / 5.09	1 / 5.0	0.040 mS/cm
Mn^{2+}	1 / 5.02	1 / 5.0	0.039 mS/cm
Zn^{2+}	1 / 4.62	1 / 5.0	0.075 mS/cm
Cu^{2+}	1 / 5.26	1 / 5.0	0.075 mS/cm
Fe^{2+}	1 / 4.68	1 / 4.4	0.064 mS/cm
Cd^{2+}	1 / 4.78	1 / 4.7	0.084 mS/cm
Pb^{2+}	1 / 4.97	1 / 5.0	0.102 mS/cm

Regarding the mass-specific cation binding, hyaluronate is approximately as efficient as xanthan: in both cases, 1.5 mg polysaccharide bind approximately 0.7 to 0.9 mmol of bivalent cations. In relation to the disaccharide units of the main chain, it is found that, on average, five of these units are needed to immobilize a single cation (table 2) as opposed to two in case of xanthan (Furth 2008). The polysaccharide-induced decrease of the specific conductivity (Δκ) caused by hyaluronate is in the same range as for xanthan and shows the same general pattern regarding the dependence of the atomic mass of the metal: heavy metals seem to bind stronger than light metals (table 2).

Again, nuclear magnetic resonance spectroscopy is applied to study the molecular nature of the binding site. Figure 12 shows ^{13}C-NMR spectra on aqueous solutions of hyaluronate in presence of different concentrations of manganese ions. The signals are assigned to the individual carbon positions according to Sicińska et al. (Sicińska1993). The strongest effect induced by the presence of the paramagnetic center is observed for the signal of the carbonyl position of the acetyl group in the D-glucopyranoside (GluPA: C=O). A somewhat weaker amplitude loss is detected for the carboxylate group of the 2-*N*-acetyl-2-deoxy-D-glucopyranose (adGluP: C=O) and for the methyl residue of the acetyl group of the D-glucopyranoside (GluPA: CH_3). Finally, some characteristic shifts are observed for the signals of some ring carbon positions of both monosaccharides in the range between 70 – 80 ppm.

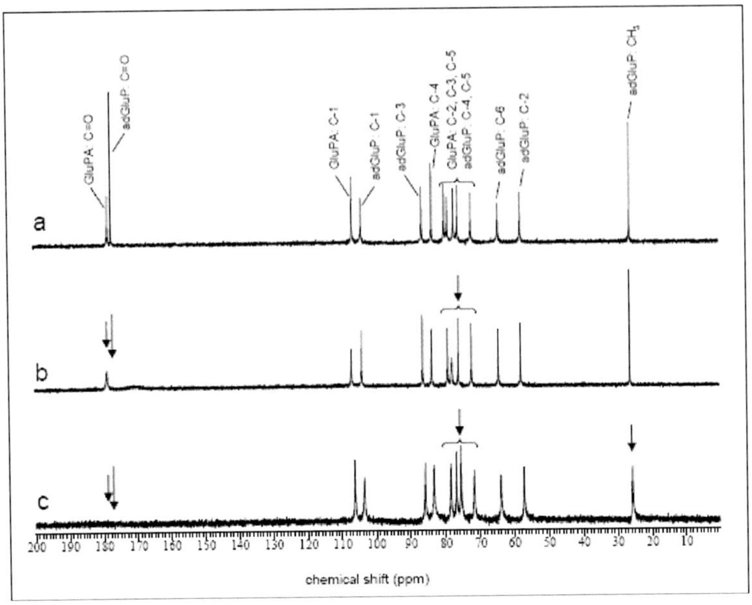

Figure 12. Set of ^{13}C spectra of hyaluronate under the influence of increasing manganese concentration (a: 0 mmol/L, b: 0.3 mmol/L, c: 1.2 mmol/L). The carbon signals of the carbonyl group of GluPA and of the acetyl carbonyl group in adGluP as well as the methyl signal of the acetyl group in adGluP are subject to strong signal widening and amplitude loss on addition of Mn^{2+} (arrows left and right). In addition, characteristic paramagnetic shifts are observed for several ring positions of both monosaccharide units (central arrow) (Furth 2008). All signal assignments derive from Sicińska et al. (Sicińska1993).

These observations are in accordance with a chelate-like structure proposed by Furth et al. (Furth 2008) which is depicted in figure 13. Herein, the metallic center is bound between two disaccharide (GluPA-adGluP) units. The acetyl groups of both adGluP monomers as well as the carboxylate groups of both GluPA monomers represent the primary binding partners to the cation.

This structure would involve two disaccharide units for the binding of a single cation. However, based on the data from the conductometric titration which led to a stoichiometric relation of five disaccharide units per cation, one must expect three more disaccharide units in the saturated polysaccharide-metal complex. Furthermore, the viscometric titration data (an example is shown in figure 14) indicate that the hydrodynamic radius of the dissolved molecules is reduced by the binding of cations. Hence, all cation bindings seem to result in intramolecular links, no molecular network is being formed by the bridging between pairs of disaccharide units. On increasing cation concentration, the dissolved molecules adopt an increasingly spherical shape as could be detected by a complete loss of the shear rate dependence of the solution viscosity (Furth 2008).

Altogether, these findings result in the following model: two disaccharide units (GluPA-adGluP) bind a single cation via their acetyl and carboxylate groups as indicated in figure 13. Both disaccharide units are part of a single molecular chain and are, on average, connected by three more disaccharide residues such that five disaccharides are involved altogether in a single complex. Similar to xanthan, hyaluronate is not capable of forming three-dimensional networks with bivalent cations such as calcium, not even at concentrations near the saturation limit.

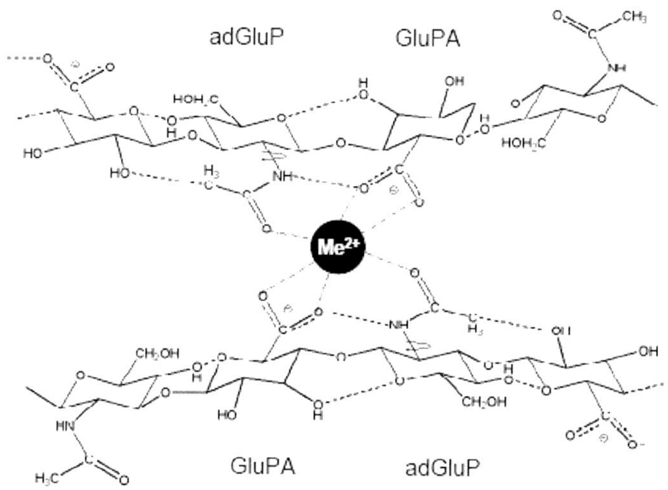

Figure 13. Proposed binding site for bivalent cations formed by two repetitive units of hyaluronate. The cation is bound between two carboxylate groups and two acetyl groups as indicated by the NMR data. Viscometric titration data indicate that the two repetitive units belong to the same main chain of a single hyaluronate molecule (Furth 2008). Based on the stoichiometric relation of five disaccharide units per single cation, one may assume that, on average, three disaccharide units are needed to connect the two disaccharides which bind the cation.

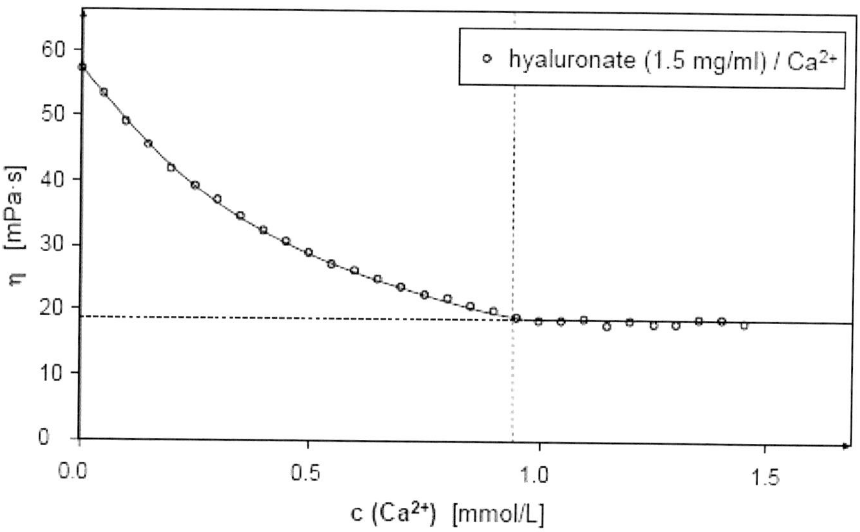

Figure 14. Apparent viscosity of a solution of hyaluronate in dependence on the calcium concentration (Furth 2008). The solid line represents a polynomial fit to the data points. The vertical broken line indicates the equivalence point between the cations and the hyaluronate binding sites as observed by the viscometric titration. Stoichiometric relations obtained from the position of the equivalence point for various bivalent cations are listed in table 2.

Conclusion

Polysaccharides from bacterial EPS may bind and accumulate bivalent cations. Especially the charged varieties such as bacterial alginate, xanthan and hyaluronic acid form relatively stable chelate-like complexes. In case of alginate, the formation of the complex can lead to intermolecular cross-linking and eventually to the formation of a gel-like solid. In case of xanthan and hyaluronate, the binding of the cation involves two disaccharide positions on the same polysaccharide chain and hence results in an intramolecular link. This leads to a stronger coiling of the molecule and consequently to a decreased hydrodynamic radius. Hence, alginate solutions have a strong tendency to increase their viscosity at higher cation concentrations, while the opposite is observed for xanthan and hyaluronate.

As expected, the strength of the bivalent cation binding varies with the type of the polysaccharide and the nature of the cation. From the conductometric titration, the complex stability can be quantified by the polysaccharide-induced reduction of the specific conductivity at the equivalence point $\Delta\kappa$. Due to the similarities in their general behavior, a direct comparison is possible between xanthan and hyaluronate. Generally, it is found that both polysaccharides tend to bind heavy metals such as lead and cadmium stronger than light metals such as magnesium. However, a light metal such as calcium can be found among the strongly binding species, whereas copper is among the ions with weaker binding. The variation of the complex stability is most likely linked to the individual ionic radius which is generally, but not always, larger for heavy metals. A correlation between the ionic radius of the bivalent cations and the corresponding value for $\Delta\kappa$ is plotted in figure 15.

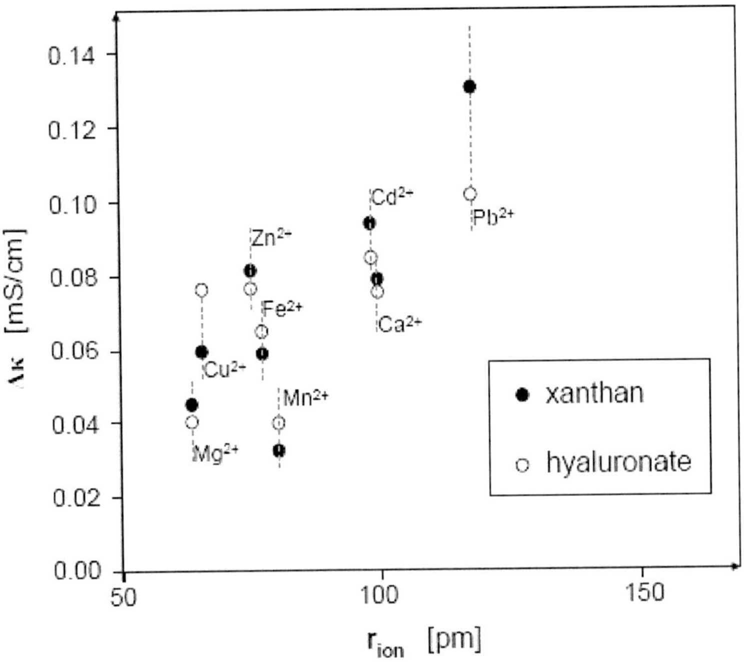

Figure 15. Values $\Delta\kappa$ for the binding induced loss of conductivity at the equivalence point plotted against the hydrodynamic radii r_{ion} of various bivalent cations. Full symbols refer to binding by xanthan, open symbols to binding by hyaluronate.

It shows a general trend that xanthan as well as hyaluronate have a stronger affinity towards bivalent ions with a larger radius such as cadmium or lead. It is very likely that the same rule holds for corresponding polysaccharides in bacterial EPS. Therefore, it must be expected that lighter metals with smaller ionic radii such as magnesium may be replaced by lead or cadmium in the EPS of a bacterial biofilm in a natural environment.

References

Apiratikul, R; Pavasant, P. Sorption isotherm model for binary component sorption of copper, cadmium, and lead ions using dried green macroalga, *Caulerpa lentillifera. Chemical Engineering Journal*, 2006 119, 135 – 145.

Bergmann, D; Furth, G; Mayer, C. Binding of biyalent cations by xanthan in aqueous solution. *International Journal of Biological Macromolecules*, 2008, in press.

Bryce, TA; McKinnon, AA; Morris, ER; Rees, DA ; Thom, D. Chain conformations in the sol-gel transitions for polysaccharide systems, and their characterization by spectroscopic methods. *Faraday Discussions of the Chemical Society*, 1974 57, 221- 229.

Christensen, BE; Characklis, WG. Physical and chemical properties of biofilms. In: Characklis, WG; Marshall, KC. *Biofilms*. New York: Wiley; 1990; 93 – 131.

Donati, I; Holtan, S; Mørch, YA; Borgogna, M; Dentini, M; Skjåk-Bræk, G. New hypothesis on the role of alternating sequences in calcium-alginate gels. *Biomacromolecules*, 2005 6, 1031 – 1040.

Emmerichs, N; Wingender, J; Flemming, HC; Mayer, C. *International Journal of Biological Macromolecules*, 2004 34, 73 – 79.

Ertesvag, H; Valla, S. Biosynthesis and applications of alginates. *Polymer Degradation and Stability*, 1998 59, 85 – 91.

Flemming, HC; Wingender J. Relevance of microbial extracellular polymeric substances (EPS) - Part I: Structural and ecological aspects. *Water Science and Technology*, 2001 43, 1 – 8.

Flemming, HC; Neu, TR; Wozniak, DJ. The EPS Matrix: the "house of biofilm cells". *Journal of Bacteriology*, 2007 189, 7945 – 7947.

Furth, G; Knierim, R; Buss, V; Mayer, C. Binding of bivalent cations by hyaluronate in aqueous solution. *International Journal of Biological Macromolecules*. 2008 42, 33 – 40.

Gamini, A; de Bleijser, J; Leyte, JC. Physico-chemical properties of aqueous solutions of xanthan: an n.m.r. study. *Carbohydrate Research*, 1991 220, 33 – 47.

Horton, D; Mols, O; Walaszek, Z; Wernau, WC. Structural and biosynthetic studies on xanthan by ^{13}C-n.m.r. spectroscopy. *Carbohydrate Research*, 1985 141, 340 – 346.

Körstgens, V; Flemming, HC; Wingender, J; Borchard, W. Influence of calcium ions on the mechanical properties of a model biofilm of mucoid *Pseudomonas aeruginosa. Water Science and Technology* 2001 43, 49 – 57.

Lattner, D; Flemming, HC; Mayer, C. ^{13}C-NMR study of the interaction of bacterial alginate with bivalent cations. *International Journal of Biological Macromolecules*, 2003 33, 81 – 88.

Lee, JW; Ashby, RD; Day, DF. Role of acetylation on metal induced precipitation of alginates. *Carbohydrate Polymers,* 1996 29, 337 – 345.

Li, L; Fang, Y; Vreeker, R; Appelqvist, I. Reexamining the egg-box model in calcium-alginate gels with X-ray diffraction. *Biomacromolecules*, 2007 8, 464 – 468.

Lopez, M; Ramos-Cormenzana, A. The effect of olive mill wastewaters variability on xanthan production. *Journal of Applied Microbiology*, 2001 90, 829 – 835.

Papageorgiou, SK; Katsaros, FK; Kouvelos, EP; Nolan, JW; Le Deit, H; Kanellopoulos, NK. Heavy metal sorption by calcium alginate beads from *Laminaria digitata*. *Journal of Hazardous Materials*, 2006 137, 1765 – 1772.

Pavasant, P; Apiratikul, R; Sungkhum, V; Suthiparinyanont, P; Wattanachira, S; Marhaba, T. Biosorption of Cu^{2+}, Cd^{2+}, Pb^{2+}, and Zn^{2+} using dried marine green macroalga *Caulerpa lentillifera*. *Bioresource Technology*, 2006 97, 2321 – 2329.

Philippis, R; Paperi, R; Sili, C. Heavy metal sorption by released polysaccharides and whole cultures of two exopolysaccharide-producing cyanobacteria. *Biodegradation*, 2007 18, 181 – 187.

Sabra, W; Zeng, AP; Deckwer, WD. Bacterial alginate: physiology, product quality and process aspects. *Applied Microbiology and Biotechnology*, 2001 56, 315 – 325.

Schürks, N; Wingender, J; Flemming, HC; Mayer, C. Monomer composition and sequence of alginates from *Pseudomonas aeruginosa*. *International Journal of Biological Macromolecules*, 2002 30, 105 – 111.

Shatwell, KP; Sutherland, IW; Ross-Murphy, SB. Influence of acetyl and pyruvate substitutions on the solution properties of xanthan polysaccharide. *International Journal of Biological Macromolecules*, 1990 12, 71 – 78.

Sheehan, JK; Atkins, EDT. X-ray fibre diffraction study of conformational changes in hyaluronate induced in the presence of sodium, potassium and calcium cations. *International Journal of Biological Macromolecules*, 1983 5, 215 – 221.

Sicińska, W; Adams, B; Lerner, L. A detailed ^{1}H and ^{13}C NMR study of a repeating disaccharide of hyaluronan: the effects of temperature and counter ion type. *Carbohydrate Research* 1993 242, 29 – 51.

Siew, CK; Williams, PA; Young, NWG. New insights into the mechanism of gelation of alginate and pectin: Charge annihilation and reversal mechanism. *Biomacromolecules*, 2005 6, 963 – 969.

Son, BC; Park, K; Song, SH; Yoo, YJ. Selective biosorption of mixed heavy metal ions using polysaccharides. *Korean Journal of Chemical Engineering*, 2004 21, 1168 – 1172.

Sutherland, IW. Exopolysaccharides in biofilms, flocs and related structures. *Water Science and Technology*, 2001 43, 77 – 86.

Sutherland, IW. Biofilm exopolysaccharides: a strong and sticky framework. *Microbiology*, 2001 147, 3 – 9.

Tako, M; Nakamura, S. Rheological Properties of Ca Salt of Xanthan in Aqueous Media. *Agricultural and Biological Chemistry*, 1987 51, 2919 – 2923.

Wloka, M; Rehage, H; Flemming, HC; Wingender, J. Rheological properties of viscoelastic biofilm extracellular polymeric substances and comparison to the behavior of calcium alginate gels. *Colloid and Polymer Science,* 2004 282, 1067 -1076.

In: Heavy Metal Pollution
Editors: S. E. Brown and W. C. Welton

ISBN: 978-1-60456-899-8

Chapter 11

PHYTOREMEDIATION OF HEAVY METAL CONTAMINATED MEDIA

Teresa J. Cutright and Harry van Keulen
Department of Civil Engineering, The University of Akron
Department of Biological, Geological and Environmental Sciences,
Cleveland State University

ABSTRACT

Phytoremediation is a promising method for the treatment of media contaminated with toxic heavy metals. The extraction efficiencies of other technologies are often ineffective due to the low metal concentrations present. Certain plants, such as *Thlaspi caerulescens*, *Brassica juncea*, and *Helianthus annuus* are classified as hyperaccumulators since they are able to uptake much higher than average levels of metal ions (by a factor of 10-100) without exhibiting symptoms of toxicity. However, how such ability might be acquired is poorly understood. One possibility is that when plants are exposed to low levels of toxic metal, over time they adapt to increase their capacity to sequester metals in their leaves or other tissues by inducing metal binding proteins. This chapter will review phytoremediation efficiency of hydroponic and soil based plants and the effect of certain metal chelators. Efficiency will be based on the total metal uptake, selectivity, and sequestration location of the heavy metals. Preliminary data on the isolation and characterization of proteins used to bind and transport the specific metals will also be discussed. The identification of the proteins and possible enzymes involved in the root-to-shoot transport or regulation of metal sequestration should provide researchers with novel targets for engineering metal accumulation.

INTRODUCTION

Phytoremediation uses living plants for the in situ treatment of contaminants by extraction, accumulation and sequestration of toxins in plant tissues or release of enzymes and exudates to stimulate microbial activity as a means of detoxification (Nichols et al., 1997). All plants have the potential to absorb a wide variety of metals from the soil. For the most part, plants tend to only absorb those metals that are essential for their survival and growth. Occasionally a plant will uptake and sequester a toxic metal that is an analogue (i.e., resembles) an essential element (Ghosh and Singh, 2005). In certain instances the plant can tolerate, uptake, and translocate elevated concentrations of heavy metals that would otherwise be toxic to any other known organism. Such plants are termed "hyperaccumulators" and include species whose leaves contain >100 mg/kg Cd, >1000 mg/kg Ni and Cu, or >10000 mg/kg Zn and Mn (Brown, 1995). Hyperaccumulation is a function of 1) increase in metal transporters at the root membrane; 2) release and readsorption of natural chelators; and 3) change in rhizosphere that enables metal mobilization (Gove et al., 2002). Plants that can hyperaccumulate heavy metals are used for phytoextraction. For instance, there are more than 350 known Ni hyperaccumulators, 3/4 belong to the *Brassicaceae* family (Robinson et al., 2003). A partial listing of known hyperaccumulators is provided in table 1.

Table 1. Partial listing of known hyperaccumulators

Cultivar	Metal	Reference
Alyssum bertolonii	Ni	Li et al., 2003; McGrath et al 2001
	Cd	Doran 2002; Keller et al., 003
Alyssum lesbiacum	Ni	Wycisk et al., 2004
Arabidopsis halleri	Zn	MacNair 2002; Swartz et al, 2003
	Cd	Zhao et al. 2003
Arabidopsis thaliana	Cd	Lee and Korban, 2002
	Ni	Wycisk et al., 2004
Berkheya codii	Ni	Alkorta et al., 2004; Robinson et al., 2003
Brassica campestris	Cd	Ghosh and Singh, 2005
Brassica juncea	Cd	Kunito et al., 2001; Heiss et al., 2003
	Cr	Mei et al., 2002
	Ni	Robinson et al., 2003
Haumaniastrum robertii	Co	Alkorta et al., 2004; Reeves, 2003
Helianthus annuus	Cr	Chen and Cutright, 2001; Mei et al, 2002
	Cd	Gallego et al., 2005; Lesage et al., 2005; Turgut et al., 2005
	Ni	Chen and Cutright, 2001
	Pb	Solhi et al., 2005
Iberis intermedia	Th	Alkorta et al., 2004
Pityrogramma calomelanus	As	Francesconi et al., 2002
Petris vittata	As	Poynton et al., 2004; Zhang et al., 2002
Thlaspi caerulescens	Zn	Ferot et al, 2003; Reeves et al., 2001; Schwartz et al., 2003
	Cd	Zhao et al., 2003; Keller et al., 2003; Lasat, 2002; Mizuno et al, 2003

The efficiency of phytoextraction as a treatment method is dependent on the cultivar used, heavy metal availability, efficient translocation from root to shoot, and metal sequestration (Cho et al., 2003; Keller et al., 2003; Robinson et al., 2003). Currently it is believed that most cultivars can achieve hyperaccumulation with only one heavy metal (Gisbert et al., 2003). The only documented exceptions to the 'one-metal rule' are *Thlaspi caerulescens* (Cd, Zn) (Zhao et al., 2003) and *Helianthus annuus* (Cr, Cd) (Chen and Cutright, 2001; Turgut et al, 2005). The ability to hyperaccumulate more than one metal is directly dependent on the specific strain of the cultivar. Furthermore, most hyperaccumulators such as *Thlaspi* and *Alyssum* have restricted root systems (Keller et al., 2003), which can limit uptake unless the metals are in close proximity to the roots.

Plants have evolved several strategies for increasing the bioavailability of metals in soil. The first strategy is the plant's ability to produce metal-chelating compounds such as phytochelatins, metallothioneins, and phytosiderophores, to mobilize metal compounds from soil (Heiss et al., 2003; Rauser, 2003). The second approach involves the solubilization of metals by exuding protons or low molecular weight organic acids from roots to acidify the rhizosphere soil (McGrath et al., 2001).

As shown in figure 1, hydrated metal ions and/or metal-chelate complexes first bind to the exterior root wall. Once bound, the metals transfer across the root wall, entering the xylem. Here either the same or a different protein molecule is used to transport the heavy metal from the root to the shoot. Various transporters regulate intracellular distribution through the plant, where the type of transporter depends on the metal present. Once at the appropriate section (leaf vacuoles, flower, etc.), the metal is sequestered in the tissue via a different set of proteins.

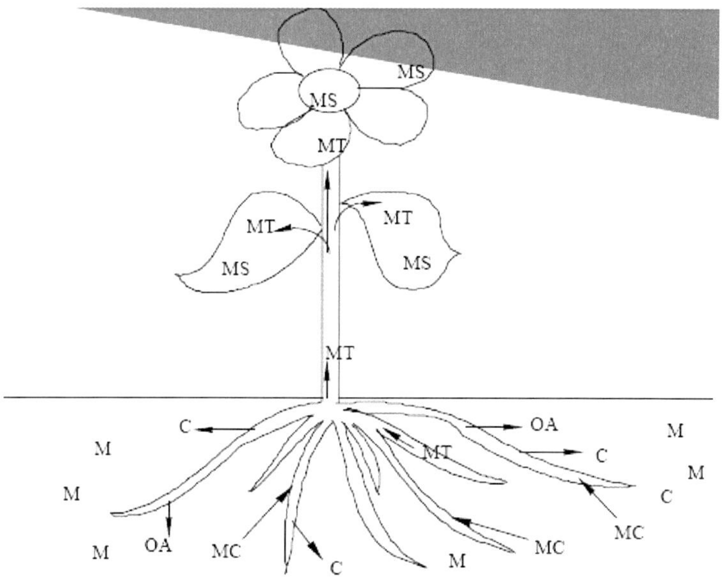

Figure 1. Simplified schematic of phytoextraction process. M – metal ion; C – plant based chelator; OA- organic acid; MC – metal chelate complex; MT- metal-protein complex for transporting heavy metal ion; MS – sequestration of heavy metal.

There are two classes of protein families associated with metal complexation that facilitate metal homeostatis and tolerance in plant systems – phytochelatins (PC) and metallothioneins (MT). Phytochelatins are the product of the enzymatic reaction of phytochelatins synthase (PCS) with glutathione (GSH) and peptides are produced that are not gene encoded and have the sequence: Glu -(Cys)n-Gly (n: 2-8) (Grill et al. 1989, Pilon-Smith 2005, Yang et al. 2005). PCS is a unique cysteine transpeptidase (Rea et al. 2004). Over-expression of PCS in *Arabidopsis* has resulted in plants that became more As tolerant but at the same time more sensitive to Cd (Li et al. 2004), indicating that PCs, while known to be chelators, are just one of the mechanisms involved in metal binding and other unknown phenomena may play a role in metal accumulation. PCS is constitutively present in the cytoplasm of plant cells (Stolt et al., 2003). It is activated by binding the GSH-metal chelate (Cazale and Clemens, 2001). GSH has also been found in the roots, yet phytochelatins are not always generated. A lower phytochelatin concentration could be indicative of the transport of reduced sulfur (from GSH) from the leaves to roots is needed to sustain the higher phytochelatins needed for plants grown in severely contaminated soils (de Knecht et al., 1995). How the GSH conjugates accumulate within the plant, as well as the structure and synthesis of subsequent phytochelatins have been thoroughly reviewed by Beck et al. (2003) and Rauser (2003).

Previously it was thought that only phytochelatins were involved in plant based complexation and sequestration. Recent research has shown that metallothioneins are also involved. Metallothioneins are low molecular weight, gene encoded, proteins rich in cysteine. Metallothioneins have been identified in animals and plants and can bind a variety of metals through a cysteine mercaptide bond (Cobbett and Goldsbrough 2002). A key role for metallothioneins in animals is their ability to detoxify transition metals such as Cd (Honda et al., 2005). Although this is a well-established role in most vertebrates, it was not until recently that metallothioneins were associated with plants. Their lack of identification was associated with the sampling techniques that exposed samples to air. Metallothioneins are very unstable in the presence of oxygen and would therefore degrade before the analysis could be completed. However Lee et al. (2004) has reported that the Cd bound in the leaf vacuoles of tabacco was due to the chelation complex with the cysteine fraction of metallothionein. By using a Cd sensitive yeast strain Lee et al. (2004) have shown by transforming them with a cDNA library obtained from *Arabidopsis* that some cells became Cd tolerant. These either contained a PCS clone or two different metallothionein clones. It was shown that these *Arabidopsis* MT genes enhanced resistance to Cd when they were expressed in *Vicia faba* guard cells. Similarly, MT genes cloned from *Typha latifolia* conferred increased resistance to Cu and Cd when introduced into *Arabidopsis* (Zeng et al. 2004).

As stated earlier, synthetic chelators can be added to enhance the effectiveness of phytoremediation by increasing the removable metal concentrations. Recent studies have found that the addition of certain synthetic chelators may interfere with or terminate the complexation reaction needed for the uptake/sequestration if phytochelatin (PC) synthase is the enzyme responsible for binding (Lesage et al., 2005; Maier et al, 2003). For example Sagner et al. (1998) found that phytochelatins were not able to complex Ni when glutathione (a key intermediate for PC synthase) was also present, implying that glutathione can compete with PC for metal binding. Instead, citric acid was able to sequester Ni although the extent of complexation was only 50% (Sagner et al., 1998).

In addition to binding to the transporter enzyme, chelators can interfere with phytoremediation efficiency by mobilizing a metal concentration that is toxic to the plant or by leaching metals to the groundwater. The first phase of experiments presented here, investigated the best chelator-cultivar combination for a silty-clay loam soil from Ohio contaminated with cadmium, chromium and nickel. The total metal uptake and selectivity was evaluated for three different cultivars of *H. annuus* (Dwarf Sunspot, Sundance, and Teddy Bear), nine different strains of *Brassica juncea*, *T. caerulescens* and *Alyssum montanum* in the absence and presence of different chelators. The chelators evaluated were EDTA, HEDTA, and citric acid at dosage levels from 0.1 mg/kg to 3 mg/kg. The best chelator-cultivar combination was then used in the hydroponic experiments to identify the plant proteins involved in heavy metal binding at the root, root-shoot transfer and sequestration. The hydroponic approach was used to decrease mass transfer limitations (prevent sorption to soil); enable thorough contact between chelator, metals and proteins; and to ensure that all of the root system could be collected.

PRELIMINARY SOIL BASED EXPERIMENTS TO DETERMINE CULTIVAR-CHELATOR COMBINATION

Preliminary soil based experiments were conducted with nine different strains of *B. juncea* (Indian mustard from the USDA Plant Production Station in Iowa), three strains of *H. annuus* (dwarf sunflowers, Jung Seeds), *A. montanum* (North Central Regional Plant Introduction Station, Iowa), and *T. caerulescens* (USDA, NY) at metal concentrations ranging from 5 to 250 mg/kg per metal. Several of the *B. juncea,* the *Alyssum* and *Thlaspi* strains can not grow in Northeast Ohio during the winter season making them less useful for practical phytoremediation in this area. Furthermore, although *B. juncea* is a proven hyperaccumulator for Ni, when multiple metals were present, *H. annuus* cultivars exhibited a statistically higher uptake rate. *Thlaspi* did achieve hyperaccumulator status, but in terms of total metal uptake it was not as effective due to is significantly smaller biomass (figure 2). Therefore, subsequent studies were conducted exclusively with *H. annuus*.

Figure 2. Examples of different biomass size for four week old seedlings transferred to contaminated soil (a) Sundance species of *H. annuus* and (b) *T. caerulescens.*

The first set of experiments with the dwarf sunspot strain of *H. annuus* grown in soil evaluated the uptake of the mixed metal contamination in concentrations of 30-270 mg/kg (10 mg/kg to 90 mg/kg per metal) without the use of chelators. As the metal concentration increased, the biomass decreased, a typical sign of toxicity. The best uptake results were achieved with the 210 mg/kg experiment (70 mg/kg each metal). For this experiment the total uptake of Cd in the roots, leaf and stems was 300 mg/kg, 125 mg/kg and 120 mg/kg, respectively. Ni also showed the highest accumulation in the roots (210 mg/kg) followed by the leaves (75 mg/kg) and stems (50 mg/kg). Cr appeared to be the most difficult to translocate by the plant, resulting in not more than 5 mg/kg accumulation in the stems and leaves. When the total metal concentration was 250 mg/kg, all plants died within two days.

Although chelators may increase the efficiency by increasing the mobile metal fraction, it may also pose toxicity by mobilizing too high of a metal concentration. Adding chelators at high doses increased the mobile metal fraction by more than 40%. As shown in table 2, as the EDTA concentration increased, toxicity also increased as exhibited by the decrease in dwarf sunspot biomass. Other researchers have reported that chelators can actually inhibit the metal uptake (Bennett et al., 1998). This corroborates the finding that as the chelator source and concentration changed, the selectivity (preferred metal to uptake and translocation through the plant) changed as well (table 2).

Table 2. Metal uptake rates, total tissue levels and selectivity for *H. annuus* (dwarf sunspot) grown in silty-clay loam soil contaminated 30 mg/kg of each metal and amended with different chelator dosages

Chelator	Dose (g/kg)	Uptake (mg/g)	Biomass (g)	Total Metal (mg)	Selectivity
Control	0	0.116	8.604	1.000	Cd>Ni>>Cr
EDTA	0.1	0.545	1.337	0.729	Cd>Cr>>Ni
	0.3	0.516	1.602	0.342	Cr>Cd>>Ni
	0.5	0.506	1.265	0.400	Ni>Cd>>>Cr
	1.0	0.873	2.425	0.360	Cd≈Ni>>>Cr
HEDTA	0.5	0.063	1.596	0.100	Cd>>Cr≈Ni
	1.0	0.045	1.564	0.07	Cd>Cr≈Ni
Citric Acid	1.0	0.359	1.249	0.449	Cr>Cd>>Ni
	3.0	0.315	1.238	0.390	Cr>Cd>>Ni

Several researchers have attributed metal accumulation to the activity of PC-synthase (Chen et al., 2003; Stolt et al., 2003); however, these studies have only been conducted in the presence of one metal. The exact mechanism(s) responsible for the change in selectivity and translocation or inhibition by the presence of chelators is not well known. Furthermore, recent studies have further shown that chelators may interfere with or terminate the complexation reaction needed for the uptake/sequestration (Maier et al, 2003).

As expected, the change in selectivity and total metal uptake were also different when the EDTA dosage was held constant and the cultivar source changed. *T. caerulescens* did achieve hyperaccumulator status for both EDTA dosages with the same selectivity (Cr>>>Ni>>Cd). However, its total metal uptake (~0.2 mg) was significantly less ($p<0.05$) than that obtained with the different *H. annuus* strains (table 3). This was due to the substantially smaller biomass. As such, one of the factors for determining in situ phytoremediation efficiency is the amount of biomass that can be utilized for metal sequestration.

Table 3. Total metal tissue levels and selectivity for different culitvars grown in silty-clay loam soil contaminated 30 mg/kg of Cd, Cr, and Ni and amended with either 0.1 or 0.3 g/Kg EDTA

Cultivar	EDTA dose (g/Kg)	Biomass (g)	Total metal (mg)	Selectivity
H. annuus (dwarf sunpot)	0.1	1.34	0.73	Cd>Cr>>Ni
	0.3	0.43	0.34	Cr>Cd>>Ni
H. annuus (sundance)	0.1	0.96	0.53	Cr>>Ni>Cd
	0.3	1.53	0.71	Cr>>Ni>Cd
H. annuus (teddy bear)	0.1	1.18	0.71	Cr>Cd>>Ni
	0.3	0.63	0.57	Cd>Cr>>Ni
T. caerulescens	0.1	0.24	0.21	Cr>>>Ni>>Cd
	0.3	0.17	0.26	Cr>>>Ni>>Cd

The change in selectivity was also present for different strains of the same cultivar family. For instance, the Dwarf Sunspot had the biggest biomass (1.34 g) and metal uptake (0.73) for the 0.1 g/kg EDTA dosage ($p<0.05$) with a selectivity of Cd>Cr>>Ni. At the same EDTA level, the Teddy Bear cultivar had a similar uptake, however the selectivity was Cr>Cd>>Ni. The change in uptake selectivity between Cr and Cd could be due to the PC-Cd complexes in the root cells delaying the radial Cd transport into the xylem (Stolt et al., 2003). Conversely, the Sundance cultivar showed the same selectivity regardless of the EDTA dosage, with the highest uptake and biomass with the 0.3 g/kg EDTA level (table 3). The difference in selectivity and uptake within the same family corroborates the hypothesis that the proteins needed for metal binding, transport, and sequestration could be cultivar specific (Gallego et al., 2005; Stolt et al., 2003).

To address the question how proteins might be involved in metal accumulation it was decided to identify and characterize such metal binding proteins. This induction is hypothesized to be detectable at the protein and the mRNA level. However, some proteins might be induced while their corresponding mRNAs might not be easily identified and vice versa, some mRNAs might be induced, while the proteins they code for might be difficult to detect. Therefore, a dual approach was used: identification of induced proteins at the protein and mRNA level.

Experimental Methods

Heavy Metal Source and Levels

The first set of experiments evaluated exposure to individual metals (Cd, Cr, or Ni) at a concentration of 30 mg/L. The next set of experiments determined the uptake and selectivity of a metal solution contained 30 mg Cd^{+2} (as $CdCl_2{\cdot}2.5H_2O$), 30 mg Cr^{+3} ($CrCl_3{\cdot}6\ H_2O$) and 30 mg Ni^{+2} ($NiSO_4{\cdot}6H_2O$) for a total metal concentration of 90 mg/kg soil. The combined scenario was evaluated both in the presence and absence of EDTA.

Cultivar Source and Seedling Preparation

Two different strains of the cultivar *H. annuus* were used: Sundance and Teddy Bear sunflower. The cultivars were grown hydroponically in a greenhouse with supplemental lighting to maintain a 15-hour photo-period. Each type of seedling was initially sown in Rockwool, a non-reactive, non-absorbent support system and sustained with a standard hydroponic nutrient solution (Cropking Corp., Seville OH). At the end of four weeks, half of the plants were collected to be used as controls. The remaining plants were exposed to heavy metals. Thus, after four weeks of growth, the plants were then subjected to 4 L of contaminated water operated at a complete recycle rate of 6 gallon per hour. Two different experiments were conducted – one with four days of exposure and one with 16 days of exposure to heavy metals. After the appropriate exposure time, the plants were harvested and the biomass sectioned into root, leaf, and stems and weighed. Half of the collected tissue was immediately used (i.e., not dried) to determine the location and quantity of free and metal-complexed proteins. The remaining tissue was dried followed by an acid digestion to determine the heavy metal uptake and selectivity.

Heavy Metal Determination in Plant Biomass

After drying, the tissue sections were weighed and milled with mortar and pestle. The crushed tissue sections were then digested as follows: 1.0 g of milled plant matter was soaked in 20 ml of concentrated nitric acid (HNO_3) for 6 hours. The mixture was boiled to 50% of its original volume. Then, 4 ml of perchloric acid ($HClO_4$) was added and the mixture refluxed for 90 min. Finally, the solution was diluted with distilled water to 25 ml of total volume and analyzed with flame atomic adsorption spectroscopy (Buck 200 AA). Standard solutions of 2, 5, 10, 25, and 50 mg/L of each metal were used to generate calibration curves to convert absorbance readings to concentrations. The detection limit obtainable with the extraction procedure and AA analysis was 0.2 mg/L for each metal. Blind samples (samples with concentrations known by the laboratory manager and not the AA operator) were also analyzed for quality control purposes.

Isolation, Purification, and Characterization of Metal-Binding Proteins

Plant material was ground to a fine powder in a mortar in the presence of liquid nitrogen and stored at -80 °C until use. Proteins were extracted from 100 – 200 mg of leave material using a method described by Heiss et al. (1999) with modifications. In brief, the leaf powder was added to 1 ml extraction buffer (50 mM Tris.HCL pH 7.5, 2 mM EDTA, 0.5 % polyvinylpolypyrrolidone, 1 mM PMSF, 0.1% Triton X-100, 0.1% NP-40, 0.5% 2-mercaptoethanol), with an equal volume of glassbeads (0.5 mm) and incubated in a mini-beadbeater two times 30 s each and then centrifuged for 20 min at 15,000 x g at 4 °C. Supernatants were collected and the protein concentrations measured using the Bradford method with human serum albumin as the reference protein solution. Protein extracts were size fractionated by standard one-dimensional polyacrylamide gel electrophoresis under denaturing conditions (PAGE-SDS) to determine protein profiles of both the control and the metal-exposed plants (leaves). A comparison between the control and metal-exposed extracts indicated that protein induction may be possible (figure 3). Two proteins are clearly present in As-treated plants, which are not present in Cr treated ones.

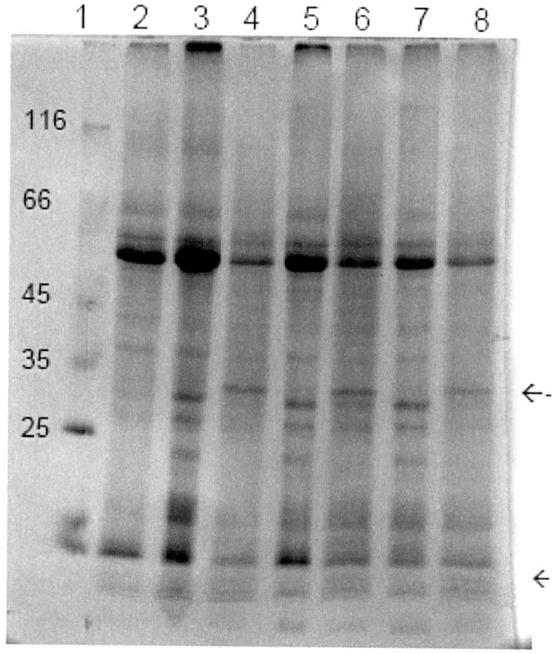

Lane 1.Molecular weight markers in kDa
Lane 2. Control leaves, supernatant
Lane 3. Control leaves, pellet
Lane 4. All metals (Cd, Ni, Cr & As) treated plants
Lane 5. Control leaves total lysate
Lane 6. All metals (Cd, Ni, Cr & As) treated plants
Lane 7. Cr treated plants
Lane 8. As treated plants

Figure 3. PAGE-SDS of leave protein samples.

The second approach that was used was to identify metallothioneins (MT) in leaf material by measuring mRNA levels. MTs are widely distributed low molecular weight proteins, which are cysteine rich and bind heavy metals. To this end primers were designed to amplify MT mRNA by reverse transcription followed by the polymerase chain reaction (RT-PCR). Total RNA was isolated from leaves of control plants and those exposed to Cd and Ni. RNA was then reverse transcribed into cDNA using a MT specific primer. A semi-quantitative RT-PCR was used. First cDNA was synthesized using Moloney Murine Leukemia Virus Reverse Transcriptase with a metallothionein specific primer and the same amounts of RNA (2.5 μg) from each of the samples. A rRNA (18S rRNA) specific primer was used as a control of rRNA loading. A PCR reaction was then conducted on part of the obtained cDNA using the appropriate primers, for 25 cycles and 40 cycles, respectively.

The identity of the amplified DNA was confirmed by isolating, cloning and sequencing of the amplified DNA. The sequence showed that it belonged to the type 2 metallothioneins.

Statistical Analysis of Experimental Data

Due to the complexity associated with the interaction between the individual heavy metals, plant biomass, proteins and EDTA, a statistical analysis program will be used to analyze the experimental data. Both a fixed-effects factorial and randomized block design will be used for the analysis to determine if there is any statistical significance for interaction factors of individual, between-sample, and within-sample. An α of 0.05 was used, therefore $p>0.05$ are indicative of insignificant results and $p<0.5$ indicate the results are statistically significant.

Results and Discussion

Exposure to Individual Heavy Metals for One Week

Figure 4 contains the results of six experiments that exposed individual metal solutions to either Sundance or Teddy Bear cultivars for one week. Both cultivars exhibited the highest metal concentration in the roots. This was corroborated with translocation factors (TF) contained in table 4 for each individual metal experiment being significantly less than one. The only exception to this is when Ni was used with Teddy Bear sunflowers it showed a slightly higher concentration in the stems ($p < 0.05$).

Table 4. Translocation factors for Sundance and Teddy bear cultivars exposed to individual (30 mg/L) and mixed metal (90 mg/L) solutions without EDTA

Cultivar	Treatment	Duration	Cd	Cr	Ni
Sundance	Individual	7 days	0.64	0.64	0.19
	Mixed	7 days	0.32	1.70	0.93
	Mixed	16 days	1.45	0.14	2.22
Teddy Bear	Individual	7 days	0.13	0.37	0.35
	Mixed	7 days	0.10	0.18	0.24
	Mixed	16 days	0.12	0.50	0.46

The TF is defined as the metal ion concentration in the shoots (stems and leaves) divided by the metal ion concentration in the roots. Therefore the lower the TF, the more pronounced is the difficulty of transporting the metal ion(s) from the root to shoot. TFs substantially lower than one can eliminate a plant for use in soil systems since for in-situ applications, not all of the roots may be harvested. Thus if the majority of the toxic metal is retained in the root system, it will be reintroduced into the ecosystem after the aerial portion is harvested and the root matter decays. It is also important to note that TFs maybe misleading in that by dividing concentration values the amount of biomass available for sequestration is masked. Critical evaluations of phytoremediation must therefore also include the total mass of each contaminant in the above ground biomass.

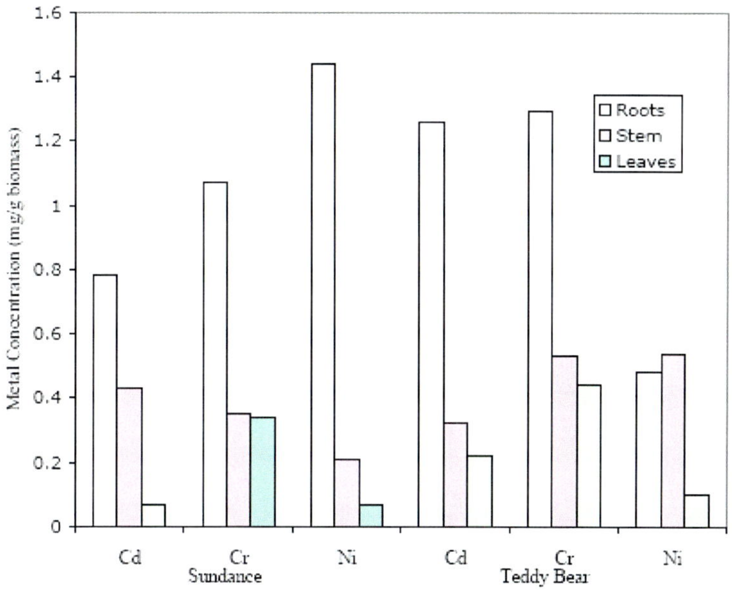

Figure 4. Metal uptake results for individual experiments with 30 mg/L of Cd, Cr, or Ni exposed to Sundance and Teddy Bear cultivars.

The difference in metal uptake was attributed to the induction of proteins by the plant for transporting the metal ion. As stated in the introduction the ability and rate of protein induction will be dependent on the individual cultivar. The PCR results demonstrate that there was a difference for the Sundance protein induction depending on the metal present. An rRNA (18S rRNA) specific primer was used (lanes 4-6, figure 5) as a control on RNA loading. A PCR reaction was then conducted on part of the obtained cDNA using the appropriate primers, for 25 cycles and 40 cycles, respectively. After 25 cycles no bands were observed in the control and Ni treated samples (lanes 1 and 3, left side), but a band of the expected size for the metallothionein (230 nucleotides) was observed in the Cd treated samples (lane 3). The intensity of this band increased after 40 cycles (lane 3, right side). In addition, a band became visible for the Ni treated sample but not for the control samples. These results indicated that the mRNA for metallothionein was induced in both Cd and Ni treated plants, but much stronger in the Cd treated ones. The rRNA control showed that in all cases similar amounts of RNA were used and the control without Reverse Transcriptase showed that the PCR reaction did not amplify genomic DNA of which trace amounts could have contaminated the RNA samples (lanes 7-9, figure 5).

In comparing the two cultivars, Teddy Bear achieved a higher metal uptake ($p < 0.05$) within one week than the Sundance for Cd (0.43 mg versus 0.3 mg). Conversely, Sundance accumulated more Ni (0.32 mg, versus 0.22 mg with Teddy Bear) ($p < 0.05$). The cultivars were able to uptake the same amount of Cr (~0.55 mg, $p>0.05$). When the cultivars were subjected to solution containing 30 mg/L of each metal, the Sundance cultivar depicted significantly better uptake ($p < 0.05$). This was based on the Sundance plants achieving a total metal uptake of 3.55 mg within one week whereas the Teddy Bears only had 2.15 mg. However both cultivars had the same selectivity Cr>Ni>Cd.

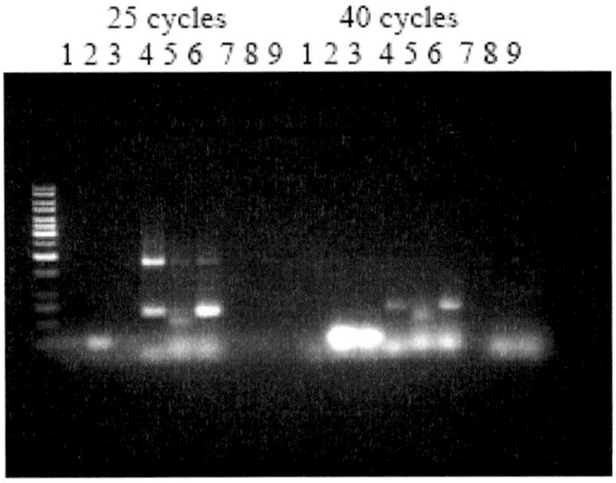

Lane 1. Control not treated using Metallothionein primers
Lane 2. Cd treated, Metallothionein primers
Lane 3. Ni treated, Metallothionein primers
Lane 4. Control, rRNA primers
Lane 5. Cd treated, rRNA primers
Lane 6. Ni treated, rRNA primers
Lane 7. Control, no reverse transcriptase, Metallothionein primers
Lane 8. Cd treated, no reverse transcriptase, Metallothionein primers
Lane 9. Ni treated, no reverse transcriptase, Metallothionein primers

Figure 5. Semi-quantitative RT-PCR of sunflower RNA.

Exposure to Solution Containing All Three Heavy Metals

Prolonging exposure from one week to 16 days changed TF and selectivity for the three metals. After one week the selectivity with Sundance was Cr>>Ni>Cd and changed to Cr>Ni>>Cd by the sixteenth day. The most significant difference was with Cr whose TF went form 1.7 (7day) to 0.04 (16 days) indicating that most of Cr in roots. The more pronounced diminished translocation was attributed to the conversion of Cr(VI) to Cr(III) (Zayed and Terry, 2003). Although the Cd and Ni translocation with Sundance sunflowers was in same order of magnitude regardless of exposure duration, the actual TF values were slightly higher for the 16 day treatment thereby indicating that the additional time enabled more of the metal to be translocated from roots to shoot. This was not the case with the Teddy Bear sunflowers where the TF only increased slightly (table 4), yielding a selectivity of Ni>Cr>>Cd.

The ability of the Sundance cultivar to perform better than Teddy Bear was corroborated by the total metal uptake and concentration (figure 6).

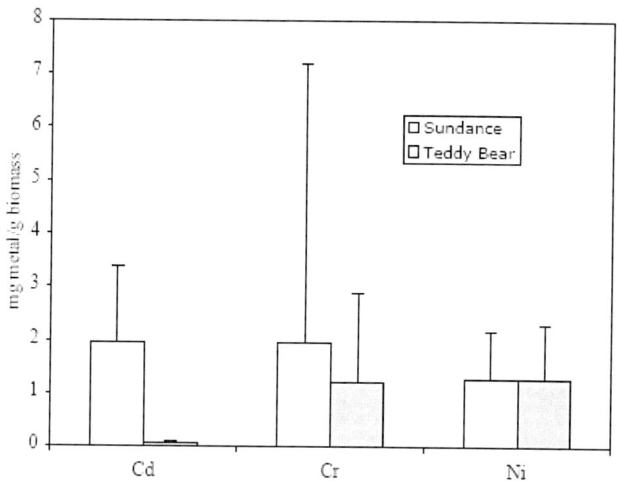

Figure 6. Metal uptake results for mixed experiments with 30 mg/L each of Cd, Cr, or Ni exposed to Sundance and Teddy Bear cultivars for 16 days.

After 16 days, the Sundance cultivar had accumulated 1.1 mg Ni within one week and increased to 4.14 mg ($p < 0.05$) after 16 days (table 5) while the total Cr content was 5.76 mg. Conversely, Teddy Bear only has a smaller biomass that resulted in a Cd, Cr, Ni metal accumulation of 0.57, 3, and 3.59 mg, respectively.

Table 5. Total metal uptake for cultivars exposed to mixed solution of Cd, Cr, and Ni (30 mg/l each) for 16 days in the presence and absence of EDTA

Cultivar	EDTA	Biomass (g)	Cd (mg)	Cr (mg)	Ni (mg)
Sundance	No	7.74	0.66	5.76	4.14
	Yes	2.18	2.15	2.36	2.07
Teddy bear	No	4.76	0.57	3.00	3.59

There have been conflicting results reported for Cd. Some researchers have found Cd to decrease essential enzyme activities required for growth in *H. annuus* (Chen et al., 2003) while others have shown exposure to Cd to increase both Cd tolerance in the plant and phytochelatin levels (Yurekli and Kucuckbay, 2003). Even with the conflicting reports, the ability to uptake Cd rather than Ni, regardless of whether present as an individual contaminant or as a part of a mixture, appears to be dependent on protein induction. As shown earlier in figure 5, the mRNA for metallothionein was induced in both Cd and Ni treated plants (as sole contaminants). However the induction was much stronger in the Cd treated ones. Tsuji et al. (2003) found Cd to be the strongest inducer for PC-synthase. The inability of the extended treatment time to increase the Cd uptake was attributed to the presence of multiple metals, particularly Cr. Cr(VI) inhibits the induction of metal-responsive transcription factor I needed by Cd for translocation via MT (Majumder et al., 2003).

Impact of EDTA on Total Metal Accumulation, Translocation and Selectivity

As stated earlier one of the factors for effective phytoremediation is that the heavy metal must be bioavailable to the roots (close proximity and not sorbed to soil). In some situations the naturally produced phytochelatins cannot overcome the mass transfer limitations. For instance, Rauser (2003) documented that phytochelatins did not bind all Cd adjacent to the roots. The extent of binding was a function of metal-phytochelatin exposure time, metal concentration and the exact type of phytochelatins. Although this will be less of a problem for hydroponic systems, researchers have shown that adding chelators to soil to enhance metal mobility can be detrimental. The potential of EDTA to impede translocation by terminating complexation with plant-based chelates was investigated by conducting the hydroponic mixed metal contaminant experiment with Sundance sunflowers in the presence of EDTA. Teddy Bear cultivars were not investigated since they had exhibited significantly smaller biomass in the treatments without EDTA.

EDTA may have stunted the plant growth by binding too high of a metal concentration at the roots. As shown previously in table 5, the Sundance biomass was reduced from 7.74 g to 2.18 g in the presence of EDTA. Madrid et al (2003) also found EDTA posed toxicity by decreasing dry root weight and shoot weights by one-third and one-half respectively. It has also been documented to be better at overcoming limitations associated with metal to root surface transfer than enhancing the barrier of root to shoot translocation (Lombi et al., 2001). This was attributed to EDTA can interfering with initial complexation, particularly for Cr, prior to translocation.

Figure 7 shows the metal concentrations within each biomass section measured in the presence and absence of EDTA. Although EDTA resulted in a diminished biomass, limiting the amount of metal accumulation (figure 7a), it did assist with the translocation of Cd. Without EDTA the uptake selectivity was Cr>Ni>>Cd. When EDTA was added the selectivity was changed to Cr>Cd>Ni. The total metal content when EDTA was present was 2.36 mg Cr, 2.15 mg Cd, and 2.07 mg Ni.

The reduced metal uptake was attributed to both a decreased biomass (less area for sequestration) and EDTA-metal complexation interfering with binding to the biomass. Researchers have found that Ni is preferentially stored in the vacuoles. In order for this to happen, complexation with histidine and other low molecular weight acids must occur to detoxify the Ni (Robinson et al., 2003). The higher the free histdine pool, the higher the Ni tolerance and subsequent loading into the xylem (Wycisk et al., 2004). It is hypothesized that once Ni had bound to EDTA it was not able to complex with histdine, thereby limiting translocation. Complexation with EDTA appeared to have no effect on Cd transport. Boominathan and Doran (2003) also found the ability to complex with natural, plant based organic acids to be more important for Ni than Cd accumulation within *T. caerulescens* and *A. bertolini* than Cd.

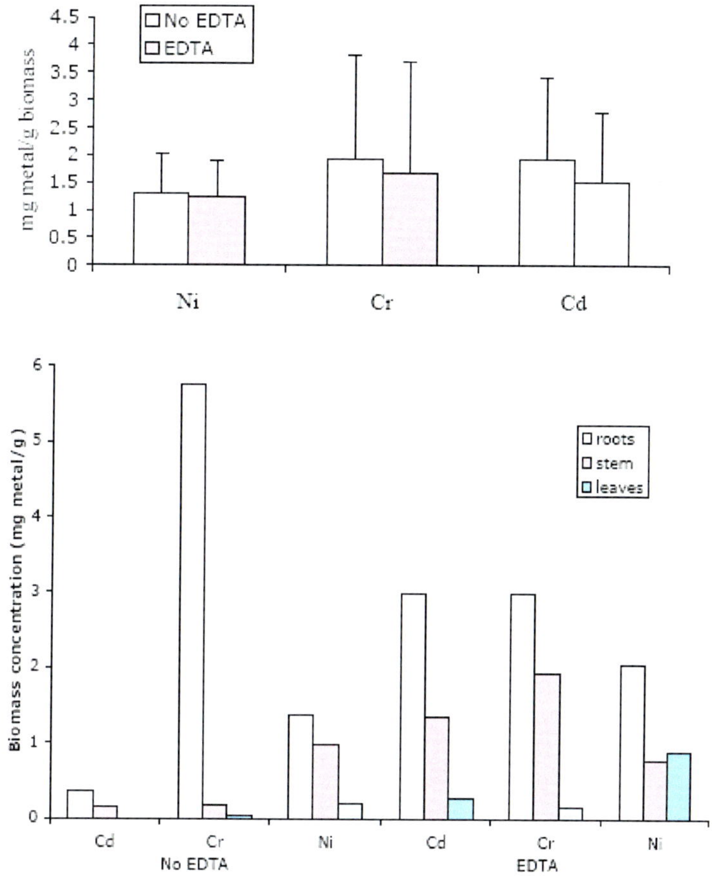

Figure 7. Metal accumulation in Sundance in the presence and absence of EDTA (a) Total metal concentration and (b) metal concentration in individual biomass sections.

Conclusion

In order for phytoremediation to be an effective treatment method, the plants to be used should be able to grow in the climate where the contaminated site is located. For instance, *Alyssum montanum* is a known hyperaccumulator but it cannot be cultivated in northeast Ohio. Furthermore, the plant should have a relatively rapid growth, large shoot mass and root zone, achieve a high TF, and be able to hyperaccumulate more than one metal. These factors provide several advantages. A rapid growth would enable multiple plantings per season, thereby increasing the total metal removal. A large aerial biomass provides more tissue for sequestering the extracted metal. As stated earlier, experiments with *Thlaspi caerulescens* showed that it did achieve hyperaccumulator status in terms of the metal concentration. However due to the small biomass, *Thlaspi* only accumulated a total of 0.2 mg, which was substantially less than that achieved with the *Helianthus annuus* cultivars. This was also due in part to the small root zone. A larger root zone would increase the area that could be treated. Regardless of the root span, a high TF is critical for soil based systems. This would ensure that the majority of the metals were transferred to the shoots, thereby providing removal from the site upon harvesting. In addition, the ability to hyperaccumulate or at least tolerate, multiple metals is critical since most contaminated sites contain at least four heavy metals.

In order to increase the treatment efficiency in contaminated soils, chelators have been implemented to make metals more available to the plant. The 'optimal' chelator source and level would be that which increases metal availability without posing toxicity or inhibiting uptake and translocation within the plant. The hydroponic experiments were used to determine if EDTA would interfere with the complexation involved with the uptake and translocation through Sundance and Teddy bear sunflowers. Without EDTA, the Sundance and Teddy bear cultivars achieved a selectivity of Cr (5.76 mg)>Ni (4.14 mg)>> Cd (0.66 mg) and Ni (5.39 mg)>Cr (3 mg)>>Cd (0.57 mg), respectively. The addition of EDTA stunted the Teddy Bear biomass too much for it to be effective. For the Sundance cultivar, the uptake selectivity was changed to Cr>Cd>Ni. The total metal accumulated was decreased (2.36 mg Cr, 2.15 mg Cd, and 2.07 mg Ni) due to the substantial reduction in biomass caused by the toxic effects of metal accumulation. It was also hypothesized that the metal-EDTA complex may have interfered with translocation. Leaves obtained from plants that were not treated to EDTA had indicated that the mRNA for metallothionein was induced when they were grown in the presence of Cd or Ni , but the induction was much stronger in the Cd treated ones. Since less Ni was translocated in the presence of EDTA, it may have inhibited the induction of proteins needed for Ni binding. In conclusion, we have started to investigate the potential of dwarf sunflowers to be used in the phytoremediation of contaminated sites in northeast Ohio and have made a beginning in our goal to unravel the molecular mechanism of metal accumulation.

Acknowledgements

The authors would like to thank several individuals whose assistance made this work possible:

Dr. Kochian of the USDA laboratory (Ithaca NY) for the donation of *Thlaspi caerulescens* seeds.

North Central Regional Plant Introduction Station in Iowa for the donation of the *B. juncea* strains and Laura Marek for the *Alyssum montanum.*

Dr. Randy Mitchell, Dept Biology University of Akron for use of greenhouse space to grow the cultivars.

Andrea Rennick and Mary January for their assistance conducting the hydroponic experiments.

Nomenclature

AA	atomic adsorption spectroscopy
EDTA	ethylene diamine tetracetic acid
GSH	glutathione
HEDTA	N-(2-hydroxyethyl)-ethylenediaminetriacetic acid
MRNA	messenger RNA

MT	metallothionein
PC	phytochelatins
PCR	polymerase chain reaction
PCS	phytochelatin synthase
RT-PCR	reverse transcription followed by the polymerase chain reaction
TF	translocation factor
USDA	United States Department of Agriculture

References

Alkorta, I., Hernandez-Allica, J., Beccerril, J.M., Amezaga, I., Albizu, I., Garbisu, C., Recent findings on the phytoremediation of soils contaminated with environmentally toxic heavy metals and metalloids such as Zn, Cd, Pb, and As. *Re. Environ. Sci. Biotechnol.* 3: 71-90, 2004.

Beck, A., Lendzian, K., Oven, M., Christmann, A., Grill, E., Phytochelatin synthase catalyzes key step in turnover of glutathione conjugates. *Phytochemistry* 62: 423-431, 2003.

Bennett, F.A., Tyler, E.K., Brooks, R.R., Gregg, P.E., Stewart, R.B., Fertilization of hyperaccumulators to enhance their potential for phytoremediation and phytomining, in Brooks, R.R. (ed), Plants that Hyperaccumulate Metals, *CAB International*, Walligford, 1998.

Boominathan, R., Doran, P.M., Organic acid complexation, heavy metal distribution and the effect of ATPase inhibition in hairy roots of hyperaccumulator plant species, *J. Biotechnol.* 101: 131-146, 2003.

Brown, K.S., The Green Clean: The Emerging Field of Phytoremediation Takes Root. *Bioscience,* 45(9): 579-582, 1995.

Cazale, A.C., Clemens, S., *Arabidopsis thaliana* expresses a second functional phytochelatins synthase. FEBS Letters 507: 215-219, 2001.

Chen, H., Cutright, T., EDTA and HEDTA effects on Cd, Cr, and Ni uptake by *Helianthus annuus*. *Chemosphere* 45, 21-28, 2001.

Chen, Y.X., He, Y.F., Luo, Y.L., Yu, Y.L., Lin, Q., Wong, M.H., Physiological mechanisms of plant roots exposed to cadmium, *Chemosphere* 50: 789-793, 2003.

Cho, M., Chardonnens, A.N., Dietz, K.J., Differential heavy metal tolerance of *Arabidopsis halleri* and *Arabidopsis thaliana*: a thin leaf slice test, New Phytol. 158: 287-293, 2003.

Cobbett, C., Goldsbrough, P., Phytochelatins and metallothioneins: roles in heavy metal detoxification and homeostasis. *Annual Reviews Plant Biology* 53: 150-158, 2002.

de Knecht, J.A., van Baren, N., Ten Bookum, W.M., Sang, H., Koevoets, P., Schat, H., Verkleij, J., Synthesis and degradation of phytochelatins in cadmium-sensitive and cadmium tolerant *Silene vulgaris*. *Plant Science* 106: 9-18, 1995.

Doran, P.M., Properties and applications of hairy root cultures, *Plant Biotechol.* Transgenic Plants, 143-161, 2002.

Frerot, H., Petit, C., Lefebvre, C., Gruber, W., Collin, C., Escarre, J., Zinc and cadmium accumulation in controlled crosses between metallicolous and nonmetallicolous populations of *Thlaspi caerulescens* (Brassicaceae), *New Phytol.* 157: 643-648, 2003.

Francesconi, K., Visoottiviseth, P., Sridokchon, W., Goessler, W., Arsenic species in an arsenic hyperaccumulating fern, *Pityrogramma calomelanos:* a potential phytoremediator of arsenic-contaminated soils. *Sci. Tot Environ* 284: 27-55, 2002.

Gallego, S.M., Kogan, M.J., Azpilicueta, C.E., Pena, C., Tomaro, M. Glutathione-mediated antioxidativec mechanisms in sunflower *(Helianthus annuus L.*) cells in response to cadmium stress. *Plant Growth Regulation* 46(3): 267-276, 2005.

Ghosh, M., Singh, S.P., A comparative study of cadmium phytoextraction by accumulator and weed species. *Environ Poll* 133: 365-371, 2005.

Gisbert, C., Ros, R., De Haro, A., Walker, D., Bernal, M., Serrano, R., Navarro-Avino, J., A plant genetically modified that accumulates Pb is especially promising for phytoremediation, *Biochem. Biophys. Res. Commun*. 303: 440-445, 2003.

Gove, B., Hutchinson, J.J., Young, S.D., Craigon, J., McGrath, S.P., Uptake of metals by plants sharing a rhizosphere with the hyperaccumulator *Thlaspi caerulescens, Int. J. Phytorem*. 4(4): 267-281, 2002.

Grill, E, Loeffler, S, Winnacker, E Zenk, M, Phytochelatins, the heavy-metal-binding peptides of plants, are synthesized from glutathione by a specific γ-glutamyl cysteine dipeptidyl transpeptidase (*phytochelatin synthase*). *PNAS. U.S.A.,* 86: 6838-6842, 1989.

Heiss, S., Wachter, A., Bogs, J., Cobett, C., Rausch, T., Phytochelatin synthase (PCS) protein is induced in *Brassica juncea* leaves after prolonged Cd exposure*, J. Experimental Botany* 54(389): 1833-1839. 2003.

Honda, R.T., Araujo, R.M., Horta, B.B., Val, A.L., Demasi, M., One-step purification of metallothioneins extracted from two different sources. *Journal Chromatography* B 820: 205-210, 2005.

Keller, C., Hammer, D., Kayesr, A., Richner, W., Brodbeck, M., Sennhauser, M., Root development and heavy metal phytoextraction efficiency: comparison of different plant species in the field, Plant Soil 249(1): 67-81, 2003.

Kunito, T., Saeki, K., Nagaka. K., Oyaizu, H., Matsumoto, S., Characterization of copper resistant bacterial community in rhizosphere of highly copper-contaminated soil, *Eur. J. Soil Biol* 37: 95-102, 2001.

Lasat, M.M., Phytoextractions of toxic metals: a review of biological mechanisms, *J. Environ. Qual.* 31(1): 109-120, 2002.

Lee, S., Korban, S.S., Transcriptional regulation of *Arabidopsis thaliana* phytochelatin synthase (AtPCS1) by cadmium during early stages of plant development, Planta 215(4): 689-693, 2002.

Lee, J., Shim, D., Wong, WY., Hwang, I., Lee, Y., Arabidopsis metallothioneins 2a and 3 enhanced resistance to cadmium when expressed in Vicia faba guard cells. *Plant Molecular Biology* 54: 805-815, 2004.

Lesage, E, Meers, E, Lamsal, S, Hopgood, M, Vervaeke, P, Tack, FMG, Verlook MG, Enhanced phytoextraction: II Effect of EDTA and citric acid on heavy metal uptake by *Helianthus annuus* from a calcareous soil*, Int. J. Phytorem* .7:143-152, 2005.

Li, Y.M., Chaney, R., Brewer, E., Roseberg, R., Angle, J., Baker, A., Reeves, R., Nelkin, J., Development of a technology for commercial phytoextraction of nickel: economic and technical considerations, *Plant Soil* 249(1): 107-115, 2003.

Li, Y., Dhankher, O.P., Carreira, L., Lee, D., Chen, A., Schroeder, J.I., Balish, R.S., Meagher, R.B., Overexpression of phytochelatin synthase in *Arabidopsis* leads to enhanced arsenic tolerance and cadmium hypersensitivity. *Plant Cell Physiology* 45: 1787-1797, 2004.

Lombi, E., Zhao, F.J., Dunham, SJ., McGrath, S.P., Phytoremediation of heavy metal contaminated soils: natural hyperaccumulation versus chemically enhanced phytoextraction. *J. Environ. Qual.* 30: 1919-1926, 2001.

MacNair, M.R., Within and between population genetic variation for zinc accumulation in *Arabidopsis halleri*, New Phytol. 155: 59-66, 2002.

Madrid, F., Liphadzi, M.S., Kirkham, M.B., Heavy metal displacement in chelate-irrigated soil during phytoremediation, *J. Hydrology*, 272: 107-119, 2003.

Maier, T., Yu, C., Kullertz, G., Clemens, S., Localization and functional characterization of metal binding sites in phytochelatin synthase, Planta 218: 300-308, 2003.

Majumder, S., Ghoshal, K., Summers, D., Bai, D., Jharna, J., Samson, T. Chromium(VI) down-regulates heavy metal induced metallothioneins gene transcription by modifying transactivation potential of key transcription factor, metal-responsive transcription factor 1. *J. Biological Chemistry* 278(8): 26216-26226, 2003.

McGrath, S.P., Zhao, F.J., Lombi, E., Plant and rhizosphere processes involved in phytoremediation of metal contaminated soils, Plant Soil 232: 207-214, 2001.

Mei, B., Puryear,J.D., Newton, R.D., Assessment of Cr tolerance and accumulation in selected plant species, *Plant Soil*. 247: 223-231, 2002.

Mizano, T., Sonoda, T., Khorie, K., Senook K., Tanaka, A., Mizuno, N., Obata, H., Cloning and characterization of phytochelatin synthase from a nickel hyperaccumulator *Thlaspi japonicum* and its expression on yeast, *Soil Sci. Plant Nutr.* 249(2): 285-290, 2003.

Nichols, T.D., Wolf, D.C., Rogers, H.B., Beyrouty, C.A., Reynolds, C.M., Rhizosphere microbial populations in contaminated soils. Water Air Soil Pollution, 95: 165-178, 1997.

Pilon-Smith, E, Phytoremediation. *Annu. Rev. Plant Biol.* 56:15-39, 2005

Poynton, C.Y., Huang, J.W., Blaylock, M.J., Kochian, L.V., Elless, M.P., Mechanisms of arsenic hyperaccumulation in *Pteris species*: root As influx and translocation. Planta 219: 1080-1088, 2004.

Rauser, W.E., Structure and function of metal chelators produced by plants: the case for organic acids, amino acids, phytin and metallothioneins. *Cell Biochem. Biophys.* 31: 19-48, 2003.

Rea, P.A., Vatamaniuk, O.K., Rigden, D.J., Weeds, worms and more. Papain's long lost cousin, phytochelatins synthase. *Plant Physiology* 136: 2463-2474, 2004.

Reeves, R.D., Tropical hyperaccumulators of metals and their potential for phytoextraction, Plant Soil, 249(1): 57-65, 2003.

Reeves, R.D., Schwartz, C., Morel, J.L., Edmonson, J., Distribution and metal accumulating behavior of *Thlaspi caerulescens* and associated metallophytes in France. *International J. Phytoremediation* 3(2): 145-172, 2001.

Robinson, B.H., Lombi, E., Zhao, F.J., McGrath, S.P., Uptake and distribution of nickel and other metals in the hyperaccumulator *Berkheya coddiias*, New Phytol, 158: 279-283, 2003.

Sagner, S., Kneer, R., Wanner, G., Cosson, JP, Deus-Neumann, B., Zenk, M.H. Hyperaccumulation, complexation, and distribution of nickel in *Sebertia acuminata,* Phytochem. 47(3): 339-347, 1998

Schwartz, C., Echevarria, G., Morel, J.L., Phytoextraction of cadmium with *Thlaspi caerulescens*, Plant Soil 249(1): 27-35, 2003.

Solhi, M., Shareatmandari, H., Hajabbasi, M., Lead and zinc extraction potential of two common crop plants: *Helianthus annuus* and *Brassica napus*. Water Air Soil Pollution 167(12-4): 59-71, 2005.

Stolt, J.P., Sneller, F.E.C., Bryngelsson, T, Lundborg, T., Schat, H. Phytochelatin and cadmium accumulation in wheat. *Environ. Experimental Botany* 49: 21-28, 2003.

Tsuji, N., Hirayanagi, N., Iwabe, O., Namba, T., Tagawa, M., Miyamoto, S., Miyasaka, H., Takagi, M., Hirata, K., Miyamoto, K. Regulation of phytochelatins synthesis by zinc and cadmium in marine green alga, Dunaliella tertiolecta. *Phytochem* 62: 453-459, 2003.

Turgut, C., Pepe, M.K, Cutright, T.J. The effect of EDTA on *Helianthus annuus* uptake, selectivity, and translocation of heavy metals when grown in Ohio, New Mexico and Colombia soils. *Chemosphere,* 58(8): 1087-1095, 2005.

Wycisk, E., Kim, E.J., Schroeder, J.I., Kramer, U. Enhancing the first enzymatic step in the histidine biosynthesis pathway increases free histidine pool and nickel tolerance in Arabidopsis thaliana. FEBS Lett 578: 128-134, 2004.

Yang, X, Feng, Y, He, Z, Stoffella, P.J. Molecular mechanisms of heavy metal hyperaccumulation and phytoremediation. J. Trace Elements in Med. and Biol. 18: 339-353, 2005.

Yurekli, F., Kucukbay, Z., Synthesis of phytochelatins in *Helianthuus annuus* is enhanced by cadmium nitrate, *Acta Botanica Coratica* 62(1): 21-25, 2003.

Zayed, A.M., Terry, N., Chromium in the environment: factors affecting biological remediation. Plant Soil 249(1): 139-156, 2003.

Zhang, W., Cai, Y., Tu, C., Ma, L.Q., Arsenic speciation and distribution in arsenic hyperaccumulating plant, Sci. Total Environ 300: 167-177, 2002.

Zhao, F.J., Lombi, E., McGrathm S.P., Assessing the potential for zinc and cadmium phytoremediation with the hyperaccumulator *Thlaspi caerulescens*, Plant Soil 249(1): 37-43, 2003.

Zeng, Y-W., Tam, N., Wong, Y.S., Cloning and characterization of type 2 metallothionein-like gene from a wetland plant, *Typha latifolia*. Plant Science 167: 869-877, 2004.

In: Heavy Metal Pollution
Editors: S. E. Brown and W. C. Welton

ISBN: 978-1-60456-899-8

Chapter 12

INTERFERENCES OF Zn^{2+} AND Pb^{2+} ON Cd^{2+} BIOSORPTION ONTO ALGAE *GELIDIUM*, ALGAL WASTE AND A COMPOSITE MATERIAL

Vítor J. P. Vilar, Cidália M. S. Botelho and Rui A. R. Boaventura*
LSRE-Laboratory of Separation and Reaction Engineering
Departamento de Engenharia Química, Faculdade de Engenharia da Universidade do Porto, Rua Dr. Roberto Frias, 4200-465 Porto, Portugal

ABSTRACT

Biosorption of Pb(II), Cd(II) and Zn(II) from binary metal solutions onto algal derived biomass, was investigated in a batch system at pH 5.3 and 20°C. Discrete and continuous equilibrium models were able to predict equilibrium results at different equilibrium concentrations and pH values. The following sequence was obtained for the maximum uptake capacity and equilibrium affinity constants of metal ions to the active sites: Pb(II) > Cd(II) > Zn(II). The "iso-concentrations" cuts from the three dimensional sorption isotherms, show that the uptake capacity of Cd(II) and Zn(II) decreased drastically in the presence of Pb(II). On the other hand, for the Cd(II)/Zn(II) system it was concluded that Zn(II) uptake capacity is more sensitive to the presence of Cd(II) than the opposite. From discrete model the Langmuir affinity parameter, K_L, can be determined as a function of pH and co-cation concentration. The K_L values for Cd(II) and Zn(II) decreased drastically in the presence of Pb(II), showing a higher affinity of the binding sites to Pb(II).

Keywords: Algae *Gelidium*, Algal Waste, Composite Material, Metals, Modelling

* Tel.: +351 225081683; Fax.: +351 225081674; bventura@fe.up.pt; vilar@fe.up.pt; cbotelho@fe.up.pt

1. INTRODUCTION

Biosorption of metal ions is an alternative technology to remove these hazardous, persistent and synergetic pollutants from aqueous solutions by using low-cost biomass. A huge market exists for cheap biosorbents, as electroplating, metal finishing operations, mining, and other processing operations, as smelters, tanneries, batteries, etc. All together, release more than a thousand tons of metal ions to the world waters. So, the potential application for a low cost biosorption process appears to be enormous, compared to the high costs associated with the ion exchange process [1]. To evaluate the performance of the biosorbents, different metal mixtures must be tested before the application to a real effluent. The waste materials from the zinc and cadmium plating processes are mainly: oils and greases, acids, cyanides, zinc and cadmium compounds and lead from zinc plating rinses and dumped zinc plating baths [2].

Chong and Volesky [3], studied the biosorption of Cd(II)/Zn(II) in *Ascophylum nodosum* crosslinked with formaldehyde, and concluded that the uptake of Zn decreased drastically when Cd were present, using Langmuir-type models to describe the multi-component equilibrium. Saeeda et. al. [4], studied the removal and recovery of Pb(II) from single and multimetal (Cd, Cu, Ni, Zn) solutions by crop milling waste (black gram husk). The selectivity order of the biosorbent was Pb > Cd > Zn > Cu > Ni. The biosorbent removed Pb from binary and ternary solutions with Cd, Cu, Ni and Zn, with the same efficiencies as in one-metal system. The presence of Pb in binary and ternary solutions also did not significantly affect the sorption of other metal ions.

In this work, the interference of zinc and lead on cadmium biosorption onto algae *Gelidium*, algal waste and composite material has been studied. A discrete and continuous model considering, respectively, homogeneous and heterogeneous distributions of carboxylic groups [5], as binding sites, has been used to describe the equilibrium of three binary mixtures (Cd/Zn, Cd/Pb and Zn/Pb).

2. MATERIALS AND METHODS

2.1. Preparation of Biosorbents

An algal waste from the agar extraction industry and the same residue granulated by Polyacrylonitrile-PAN was used in this study as well as algae *Gelidium*, which is the raw material for agar extraction. The characteristics of the materials and their preparation were presented in previous works [6, 7].

2.2. Preparation of Metal Solutions

Pb(II), Cd(II) and Zn(II) solutions were prepared by dissolving a weighed quantity of anhydrous lead (II) chloride (Sigma-Aldrich, 98%), anhydrous cadmium (II) chloride (Sigma-Aldrich, 99%) and anhydrous zinc (II) chloride (Merck, 98%) in distilled water. pH was controlled by adding 0.01 M of HCl or 0.01 M NaOH solutions.

2.3. Sorption Equilibrium Studies

The experiments were conducted in duplicate, using 100 ml Erlenmeyer flasks, at constant pH (5.3) and temperature (T = 20°C). Metal solutions containing combinations of either Cd/Zn, Cd/Pb or Zn/Pb metal ions were in the concentration range of 0 to 300 mg l^{-1} of each one. A given weight of adsorbent was suspended in 100 ml of a two-metal solution and stirred at 100 rpm; the pH was adjusted and controlled by adding NaOH and HCl diluted solutions; the temperature was maintained constant by using a HOTTECOLD thermostatic refrigerator. Once equilibrium was reached, one hour later, samples were taken out and centrifuged (Eppendorf Centrifuge 5410) and the supernatants analysed for metals ions.

2.4. Analytical Procedure

The concentration of residual Pb(II), Cd(II) and Zn(II) solutions was determined by atomic absorption spectrometry (GBC 932 Plus Atomic Absorption Spectrometer). The amount of each metal adsorbed per gram of biosorbent was calculated as:

$$q = \frac{V\left(C_i - C_f\right)}{W} \quad (1)$$

where q is the metal uptake (mmol metal g^{-1} of the biosorbent), C_i and C_f the initial and final metal concentrations in the solution (mmol metal l^{-1}), V the volume of solution (l), and W the dry weight of biosorbent (g).

3. Results and Discussion

3.1. Equilibrium Models

For each binary system an isotherm was performed at pH 5.3 and 20°C for the same initial concentrations of both metal ions (mg l^{-1}). In order to quantify the effect of the co-cation in the primary metal biosorption, the experimental data were fitted to two equilibrium models:

Discrete equilibrium model [8]:

$$q_T = \frac{Q_{max}\left(K_{M_1}C_{M_1} + K_{M_2}C_{M_2}\right)}{1 + K_H C_H + K_{M_1}C_{M_1} + K_{M_2}C_{M_2}} \quad (2)$$

where q_T is the total metal uptake (mmol g^{-1}), Q_{max} is the maximum amount of binding sites (carboxylic groups) (mmol g^{-1}), K_H and K_{M_i}, C_H and C_{M_i} are apparent equilibrium binding constants and equilibrium concentrations for protons and divalent metal ions M_i, respectively.

Continuous equilibrium model [9]:

$$\theta_{T,X} = \frac{\left(K'_X C_X\right)^{n_X}}{\sum_X \left(K'_X C_X\right)^{n_X}} \times \frac{\left[\sum_X \left(K'_X C_X\right)^{n_X}\right]^p}{1+\left[\sum_X \left(K'_X C_X\right)^{n_X}\right]^p} \tag{3}$$

where K'_X is the average value of the affinity constants distribution for specie X, and n_X reflects the overall non-ideality which can be due to lateral interactions and/or stoichiometry effects . The effect of the intrinsic heterogeneity of the carboxylic group (p) can be isolated from the non-ideal behaviour of species X (n_X): $m_X = n_X \times p$. The total bound amount, q_X, is given by:

$$q_X = Q_{max}\left(\theta_{T,X}\right)\left(\frac{n_X}{n_H}\right) \tag{4}$$

where the factor n_X / n_H is added to give thermodynamic consistency and has some other advantages [10].

Mathematical models were fitted to experimental data using the Solver (Quasi-Newton algoritm). The obtained parameters were calculated using the matrix inverse approach [11] and statistical parameters are presented in table 1.

To fit the continuous model to experimental data, Q_{max}, pK'_H and m_H parameters were pre-determined from potentiometric titrations. The p values obtained from the single metal adsorption experiment were used. The model gives the pK'_M and n_M values for each metal ion in each binary system (table 2).

The same parameters were also determined using all the experimental data (single-metal and two-metal systems) for each biosorbent. In this case, the model gives only a value of pK'_M and n_M for each metal/biosorbent (table 3).

The statistical parameters in table 1 and 2 shows that both models describe well the experimental data. The model performance was compared using the F-test for 95% confidence level, and no significant statistical difference exists between the two models when considering all binary systems. So, as the continuous model was developed considering an heterogeneous distribution of binding sites, it allows to obtain a better interpretation of the experimental results and gives the distribution of the binding constants, the rate of ion exchange, the heterogeneity of the biosorbents active sites and the competition between metal ions and protons to the binding sites. On the other hand, the discrete model, as a simple equation, can be used in the mathematical mass transfer modelling in batch and continuous systems.

When $n_M / n_H < 1$, then the maximum metal ion binding is less than the total site density, as defined by the protons biosorption. This could reflect some degree of multi-dentism. Table 3 presents n_M / n_H values greater than unity, which could reflect some degree of cooperativity [9].

Table 1. Adjustable parameters for the multi-component discrete equilibrium model (Eq. 2) (value ± standard error)

Biosorbent	System	Metal	Discrete model			R^2	S_R^2 (mmol $g^{-1})^2$
			Q_{max} (mmol g^{-1})	pK_M	pK_H		
Gelidium	Cd^{2+} / Zn^{2+}	Cd^{2+}	0.26 ± 0.01	3.2 ± 0.1	4.2 ± 0.1	0.919	3.8×10^{-4}
		Zn^{2+}		3.0 ± 0.1			
	Pb^{2+} / Cd^{2+}	Pb^{2+}	0.28 ± 0.01	3.98 ± 0.05	4.05 ± 0.05	0.926	4.3×10^{-4}
		Cd^{2+}		3.16 ± 0.04			
	Pb^{2+} / Zn^{2+}	Pb^{2+}	0.29 ± 0.01	3.89 ± 0.05	3.88 ± 0.06	0.915	3.9×10^{-4}
		Zn^{2+}		2.89 ± 0.04			
Algal waste	Cd^{2+} / Zn^{2+}	Cd^{2+}	0.19 ± 0.01	3.1 ± 0.1	4.4 ± 0.1	0.914	2.0×10^{-4}
		Zn^{2+}		3.0 ± 0.1			
	Pb^{2+} / Cd^{2+}	Pb^{2+}	0.20 ± 0.01	3.99 ± 0.06	4.35 ± 0.06	0.911	2.8×10^{-4}
		Cd^{2+}		3.12 ± 0.04			
	Pb^{2+} / Zn^{2+}	Pb^{2+}	0.20 ± 0.01	3.97 ± 0.06	4.26 ± 0.08	0.900	2.5×10^{-4}
		Zn^{2+}		2.84 ± 0.05			
Composite material	Cd^{2+} / Zn^{2+}	Cd^{2+}	0.097 ± 0.004	3.6 ± 0.1	4.3 ± 0.1	0.862	1.1×10^{-4}
		Zn^{2+}		3.6 ± 0.1			
	Pb^{2+} / Cd^{2+}	Pb^{2+}	0.098 ± 0.002	4.50 ± 0.05	4.58 ± 0.05	0.924	6.0×10^{-5}
		Cd^{2+}		3.42 ± 0.04			
	Pb^{2+} / Zn^{2+}	Pb^{2+}	0.098 ± 0.002	4.50 ± 0.06	4.58 ± 0.07	0.919	5.4×10^{-5}
		Zn^{2+}		3.42 ± 0.05			

Table 2. Adjustable parameters for the multi-component continuous equilibrium model (Eq. 4), using only the experimental data points for the binary system (value ± standard error)

Biosorbent	System	Metal	Continuous model		R^2	S_R^2 (mmol g^{-1})^2
			pK'_M	n_M		
Gelidium	Cd^{2+} / Zn^{2+}	Cd^{2+}	3.0 ± 0.1	0.55 ± 0.03	0.956	2.0×10^{-4}
		Zn^{2+}	2.7 ± 0.1	0.58 ± 0.04		
	Pb^{2+} / Cd^{2+}	Pb^{2+}	3.8 ± 0.1	0.64 ± 0.03	0.924	4.5×10^{-4}
		Cd^{2+}	3.1 ± 0.1	0.50 ± 0.03		
	Pb^{2+} / Zn^{2+}	Pb^{2+}	3.8 ± 0.1	0.64 ± 0.03	0.919	5.4×10^{-4}
		Zn^{2+}	2.9 ± 0.1	0.46 ± 0.05		
Algal waste	Cd^{2+} / Zn^{2+}	Cd^{2+}	3.1 ± 0.1	0.69 ± 0.05	0.957	1.1×10^{-4}
		Zn^{2+}	3.0 ± 0.1	0.64 ± 0.05		
	Pb^{2+} / Cd^{2+}	Pb^{2+}	4.1 ± 0.1	0.67 ± 0.04	0.896	3.3×10^{-4}
		Cd^{2+}	3.2 ± 0.1	0.59 ± 0.04		
	Pb^{2+} / Zn^{2+}	Pb^{2+}	4.2 ± 0.1	0.65 ± 0.05	0.886	4.0×10^{-4}
		Zn^{2+}	2.9 ± 0.1	0.63 ± 0.08		
Composite material	Cd^{2+} / Zn^{2+}	Cd^{2+}	3.0 ± 0.1	0.69 ± 0.07	0.900	8.5×10^{-5}
		Zn^{2+}	2.9 ± 0.1	0.78 ± 0.09		
	Pb^{2+} / Cd^{2+}	Pb^{2+}	4.3 ± 0.1	0.69 ± 0.04	0.920	6.5×10^{-5}
		Cd^{2+}	3.0 ± 0.1	0.63 ± 0.04		
	Pb^{2+} / Zn^{2+}	Pb^{2+}	4.1 ± 0.1	0.73 ± 0.04	0.938	4.3×10^{-5}
		Zn^{2+}	3.0 ± 0.1	0.65 ± 0.05		

The ratio n_H / n_M has a particularly strong influence on the exchange ratio (r_{ex}) since it is linearly related to it. r_{ex} can change over the range $0 < r_{ex} < n_H / n_M$, the greatest value being found for small p (high heterogeneity), low metal concentration and low pH. This is true when most of the sites are occupied by protons or other ions as potassium, sodium, calcium and magnesium and the binding reaction approaches a "normal" ion exchange reaction. Under these conditions, the ratio n_H / n_M represents some kind of reaction stoichiometry [9]. Most n_H / n_M values presented in table 3 are close to the unity suggesting a 1:1 stoichiometric exchange.

The homogeneity of the binding sites follows the order: algae *Gelidium* > algal waste > composite material, as indicated by m_M values. Higher m_H values give a narrower distribution of the binding constants and, consequently more homogeneous biosorbent active sites.

Table 3. Adjustable parameters for the multi-component continuous equilibrium model (Eq. 4), using all the experimental data points for all metal ions (value ± standard error)

Biosorbent	Metal	Continuous model					
		pK'_M	n_M	$\frac{n_M}{n_H}$	m_M	R^2	S_R^2 (mmol g^{-1})^2
Gelidium	Pb^{2+}	3.7 ± 0.1	0.64 ± 0.03	1.4	0.57	0.944	5.4×10^{-4}
	Cd^{2+}	2.9 ± 0.1	0.53 ± 0.05	1.1	0.48		
	Zn^{2+}	2.7 ± 0.3	0.53 ± 0.09	1.1	0.48		
Algal waste	Pb^{2+}	4.2 ± 0.1	0.66 ± 0.04	1.2	0.44	0.928	2.9×10^{-4}
	Cd^{2+}	3.2 ± 0.1	0.63 ± 0.05	1.1	0.42		
	Zn^{2+}	2.9 ± 0.3	0.70 ± 0.09	1.3	0.46		
Composite material	Pb^{2+}	4.0 ± 0.1	0.75 ± 0.08	1.0	0.33	0.885	2.0×10^{-4}
	Cd^{2+}	2.9 ± 0.1	0.7 ± 0.1	0.9	0.30		
	Zn^{2+}	2.8 ± 0.3	0.8 ± 0.1	1.1	0.35		

3.2. Interferences of Co-Cations on Metal Ion Biosorption

The most appropriate, and correct, way of representing the biosorption equilibrium of a two-metal system is by constructing a three-dimensional (3-D) sorption isotherm plot whereby the metal uptake is plotted as a function of the two metals equilibrium concentrations in solution for constant pH. In this case, the metal uptake can be either the uptake of one of the metals or the total uptake of both metals. In this paper cadmium and zinc uptakes are presented as a function of two metal concentrations in solution (figure 1).

The surfaces were constructed for the continuous model and table 2 parameters using MATLAB. The plotted points refer to single and binary systems for the respective metal at pH 5.3 and 20°C. It can be observed that, Cd and Zn uptake is reduced in the presence of lead ions, due to competition for the binding sites. Competition between Cd and Zn is found to be softer than with lead ions. In figures 1 (a_1 and a_2), (b_1 and b_2) and (c_1 and c_2), the lower slope of the surface suggests higher inhibition of Cd or Zn uptake in presence of Pb.

While 3-D isotherm surfaces represent the summary of the two-metal equilibrium results, the selected cuts through 3-D diagrams presented in figures 2 (a), (b) and (c) for Cd/Zn, better reveal the quantitative trends observed in the two-metal sorption systems.

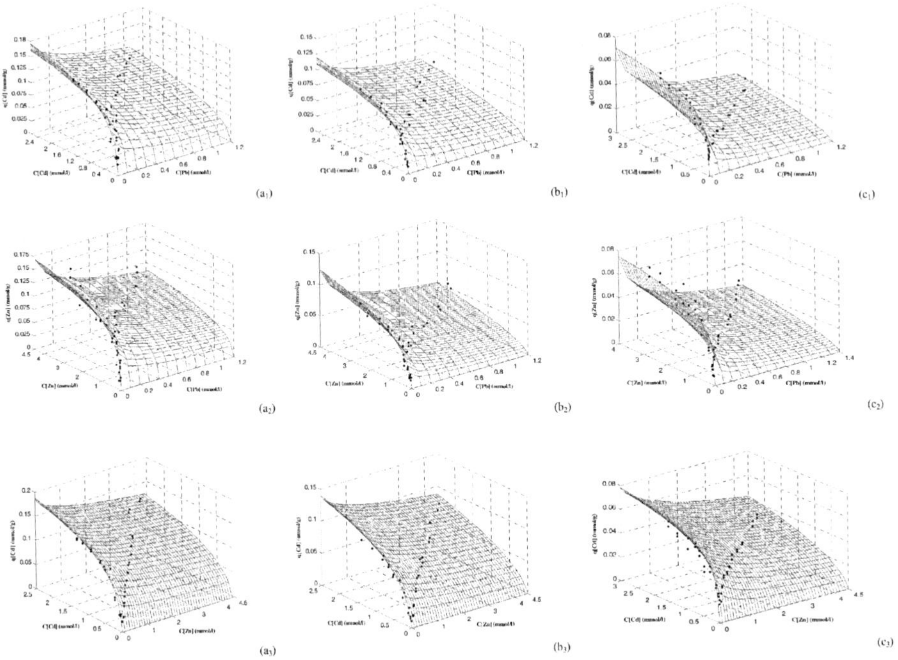

Figure 1. Two-metal sorption isotherm surfaces at pH = 5.3 and T = 20°C. (a_i) Algae *Gelidium*, (b_i) algal waste and (c_i) composite material: i = 1 – Pb/Cd; i = 2 – Pb/Zn; i = 3 – Cd/Zn.

These cuts are called "iso-concentrations", which means, cuts of the biosorption isotherm surface of the binary system M_1/M_2, where the total metal uptake vs. the equilibrium concentration of one metal is represented, while the equilibrium concentration of the other metal ion remains constant. For a given C_{Zn}, as C_{Cd} increases, the uptake capacity of cadmium also increases for the three biosorbents (figures 2 (a), (b) and (c)). However, this increase becomes gradually smaller at higher zinc concentrations, due to the saturation of the active sites. The total uptake of metals increases when increasing C_{M_1} and keeping C_{M_2} constant. For the same equilibrium concentration of cadmium and zinc (0.5 mmol/l or 2.5 mmol/l for each metal ion), about 55% of the total metal uptake is due to cadmium biosorption. From "iso-concentrations" plots it can be concluded that the equilibrium concentration of zinc needs to be slightly higher than cadmium to obtain the same proportion of metal binding. So, the three biosorbents have a slight preference for Cd biosorption over Zn.

The "iso-concentrations" for the binary systems Pb/Cd and Pb/Zn were presented in a previous work [8]. In those systems a lead uptake around 85% was obtained for the same equilibrium concentrations of both metals in solution and for the three biosorbents.

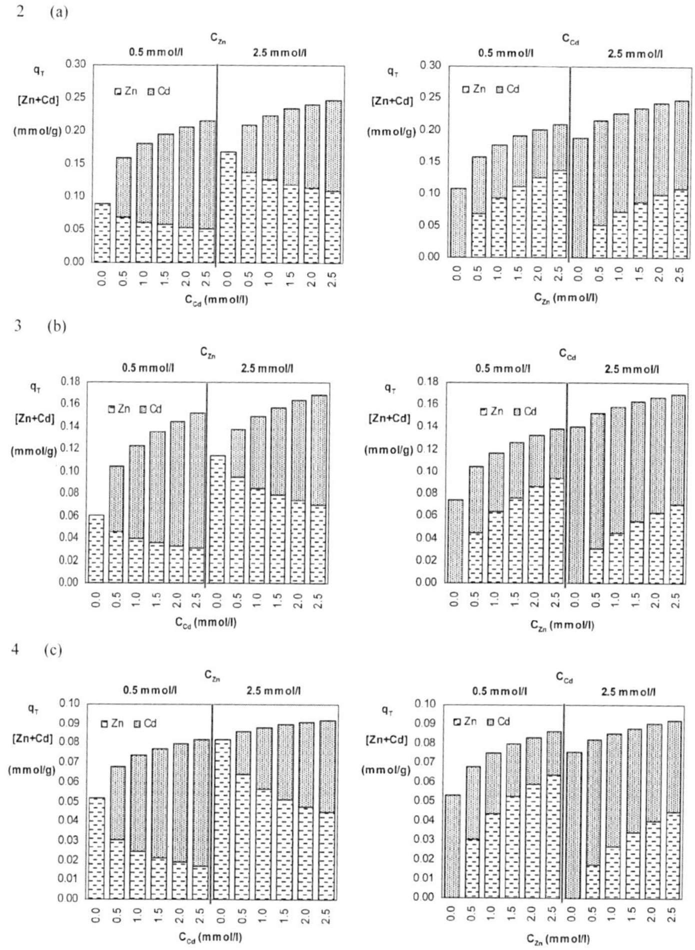

Figure 2. Iso-concentration cuts of the Cd/Zn sorption isotherm surface. (a) Algae *Gelidium*, (b) algal waste and (c) composite material.

In order to better evaluate the effect of the presence of the second metal (co-cation) (M_2) on the uptake capacity of the primary metal (M_1), the uptake percentage of cadmium and zinc ions was represented as a function of the concentration of the co-cation for pH 5.3 (see figures 3 (a_i), (b_i) and (c_i)). The curves resulting from the cuts, are well described by the Langmuir model [8], where K_L represents the apparent affinity of the metal ion towards the binding sites. K_L changes with the equilibrium concentration and the primary metal, co-cation and proton affinity constants. At pH 5.3, it is possible to obtain values of K_L, for the primary metal (cadmium or zinc) as a function of the co-cation concentration (figures 4 (a_i), (b_i) and (c_i)). For this discussion, two concentrations were arbitrary chosen to represent "low" (0.1 – Pb or 0.5 – Cd and Zn mmol/l) and "high" (1.0 – Pb or 2.5 – Cd and Zn mmol/l) concentrations. The binding of Cd is more sensitive to the presence of Pb than Zn. Langmuir K_L values for Cd decreased from 1.5, 1.1 and 2.1 to 0.26, 0.20 and 0.15, respectively for algae, algal waste and composite material, in the presence of 0.5 mmol/l of Pb, but the decrease was smaller in the presence of high Zn concentrations (2.5 mmol/l). For high Cd concentrations, a decrease of about 65% is observed for Cd binding in the presence of 0.5 mmol/l of Pb. In the presence of the same Zn concentration the decrease is only around 22%.

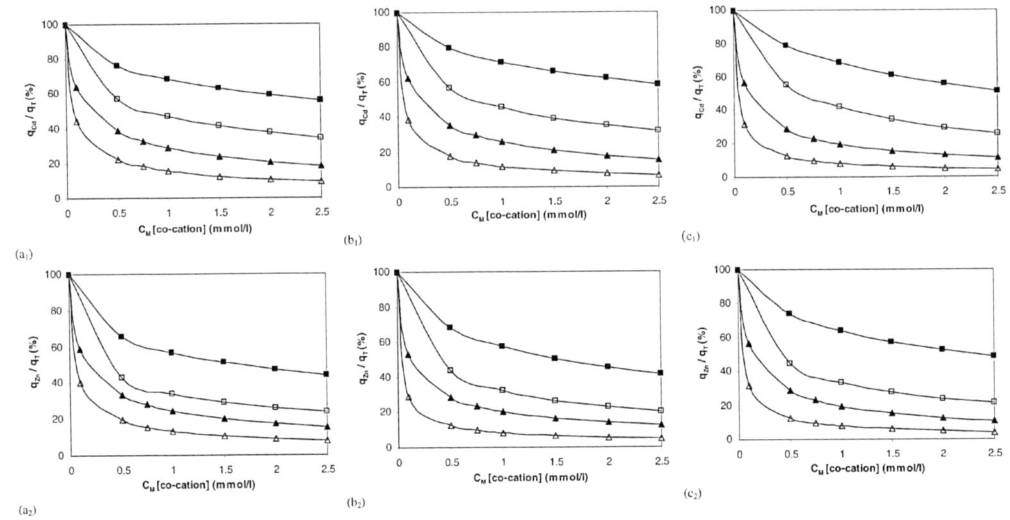

Figure 3. Co-cation (M_2) effect on the primary metal (M_1) uptake. Curves:(△) C_{M_1} = 0.5 mmol/l, M_2: Pb; (▲) C_{M_1} = 2.5 mmol/l, M_2: Pb; (□) C_{M_1} = 0.5 mmol/l, M_2: Cd or Zn; (■) C_{M_1} = 2.5 mmol/l, M_2: Cd or Zn. (a_i) Algae *Gelidium*, (b_i) algal waste and (c_i) composite material: i = 1 – Cd, i = 2 – Zn.

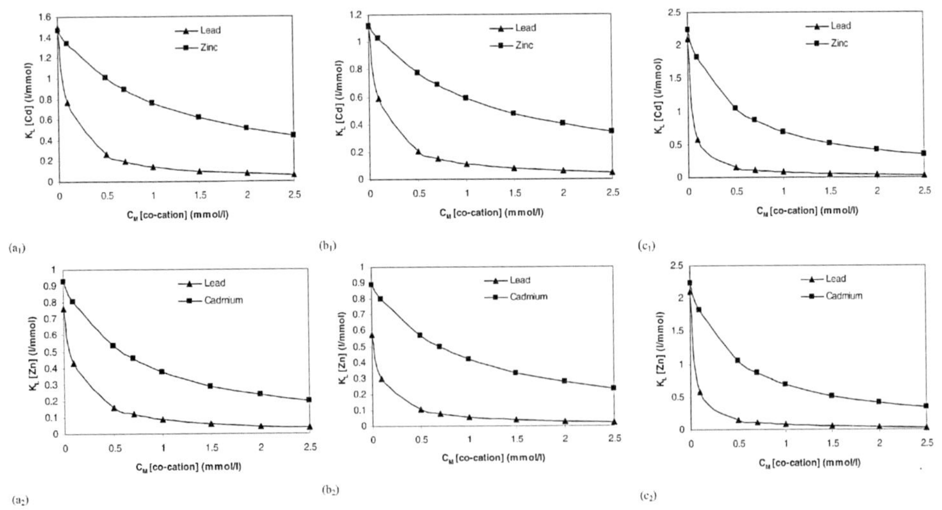

Figure 4. Co-cation effect on the Langmuir K_L value of the primary metal. (a_i) Algae *Gelidium*, (b_i) algal waste and (c_i) composite material: i = 1 – Cd, i = 2 – Zn.

As concluded for cadmium ions, Zn uptake is much more sensitive to the presence of Pb than of Cd. For high zinc concentration, a decrease of about 70% on Zn uptake is observed in the presence of 0.5 mmol/l of Pb. However, in the presence of Cd, the uptake capacity of Zn decreases around 30%. The Langmuir constant, K_L, decreases drastically (≈ 90%) for Zn in the presence of Pb. But in the presence of an higher Cd concentration (2.5 mmol/l), the decrease is around 70%.

Considering the results obtained for the three systems, we can conclude that the biosorbents exhibited a metal preference in the order: Pb > Cd ≈ Zn. This order is in agreement with the affinity order, given by $K'_M = 10^{-pK'_M}$ values (table 3). Similar sorption characteristics have been reported by Puranik and Paknikar [12] using *Streptoverticillium cinnamoneum* and *Penicillium chrysogenum* biomass in biosorption studies using binary systems. In the binary systems Pb/Cd, Pb/Zn, Pb/Cu, Pb/Ni, Pb/Co, Cu/Zn, Cd/Zn, Ni/Zn and Co/Zn, *Streptoverticillium cinnamoneum* and *Penicillium chrysogenum* exhibited preferential sorption orders: Pb > Zn ≈ Cu > Cd > Ni > Co and Pb > Cu > Zn > Ni > Co, respectively. Kim [13] showed that Cd did not affect Pb removal using crab shell biomass.

These observations can be viewed in a wider context: Zn is classified as predominantly "hard" ion, especially in aqueous solution [14]. Pb and Cd share characteristics of "hard" and "soft" ions, being defined as borderline ions. The "hard" ions tends to form strong bonds with highly electronegative donors, which are difficult to polarize (hard donors – O or F), while soft cations form more stable complexes with soft donors (N, S, P, As) [14]. Soft ions participate in more covalent bonding where the free energy is mostly enthalpic, and hard ions participate in more ionic bonding [15]. The surface groups of the biosorbents used in this work have been characterized as, predominantly, carboxylic and hydroxyl groups [5]. Because experiments were performed at pH values below 7, carboxylic groups (donor – O) are the only free active sites to bind metal ions.

Some characteristics of the three studied metal ions are listed in table 4.

It can be seen that, according to the (z^2 / r_{hyb}) criterion, the strength of the ionic bonds is Pb > Cd > Zn, but the differences among the metals ions are small. Using the ΔX_m, or the $1 - exp\left(-\Delta X_m^2/4\right)$ criteria, one can conclude that the binding of Zn is more electrostatic than the binding of Cd or Pb. Pb binding is highly covalent when compared with the other two metals. Nieboer and McBryde [16] introduced the parameter $X_m^2\left(r_{cryst} + 0.85\right)$ as a measure for the strength of covalent bonding, where 0.85 stands for the contribution of N or O donors to the bond distance. "Hard" ions are characterized by $X_m^2\left(r_{cryst} + 0.85\right) < \approx 4.2$, and "soft" ions by $X_m^2\left(r_{cryst} + 0.85\right) > 7$. The $X_m^2\left(r_{cryst} + 0.85\right)$ criterion confirms that the relative contribution of covalent bonding follows the order Pb > Cd > Zn. This can explain the weaker Zn binding as reflected by the parameter for the total binding strength, ξ.

Table 4. Binding strength characterization parameters

	H	Na	K	Ca	Mg	Zn	Cd	Cu	Pb
z (charge)	1	1	1	2	2	2	2	2	2
r_{cryst} [a] (Å)	---	1.02	1.51	1.00	0.72	0.74	0.95	0.73	1.19
r_{hyb} [b] (Å)	2.82	3.58	3.31	4.12	4.28	4.30	4.26	4.19	4.01
X_m [c]	2.20	0.93	0.82	1.00	1.31	1.65	1.69	1.90	1.8
z^2 / r_{cryst} [d1] (1/Å)	---	0.98	0.66	4.00	5.56	5.41	4.21	5.48	3.36
z^2 / r_{hyb} [d2] (1/Å)	0.35	0.28	0.30	0.97	0.93	0.93	0.94	0.95	1.00
$X_m^2(r_{cryst} + 0.85)$ [e] (Å)	---	1.51	1.51	1.85	2.26	4.07	5.20	6.32	6.61
ΔX_m [f]	1.3	2.6	2.7	2.5	2.3	1.9	1.8	1.5	1.7
$1 - exp(-\Delta X_m^2/4)$ [g]	0.34	0.82	0.84	0.79	0.73	0.59	0.56	0.43	0.51
$\xi^h = z^2/r_{hyb} / 1 - exp(-\Delta X_m^2/4)$ (1/Å)	1.03	0.34	0.36	1.23	1.27	1.56	1.69	2.22	1.94

[a] Shannon crystal radii [17]
[b] Nightingale hydrated ion radii ([18]
[c] Pauling electronegativity [19]
[d] Parameter for hydration ([1]) or ionic bonding ([2]) strength [20]
[e] Parameter for covalent bond character (0.85 is an appropriate constant assumed to reflect the radius of O and N donor atoms) [16]
[f] Parameter for ionic bond character (Electronegativity of the metal relatively to the oxygen) [19]
[g] Fraction of ionic bond character [21]
[h] Parameter for total binding strength

4. CONCLUSIONS

From the binary systems Cd/Zn, Pb/Cd and Pb/Zn it can be concluded that Pb is preferentially bound to all the studied biosorbents, when compared with Cd and Zn. The effect of Pb as co-cation, on Cd and Zn biosorption is high. These inhibitory effects on metal adsorption can be attributed to the characteristic class behaviour of each cation as suggested by Pearson [14]. According to the ΔX_m or the $1 - exp(-\Delta X_m^2/4)$ criteria, it can be concluded that Zn binding is more electrostatic than Cd and Pb. The application of the $X_m^2(r_{cryst} + 0.85)$ criterion [16], confirms that the relative contribution of covalent bonding follows the order Pb > Cd > Zn.

ACKNOWLEDGEMENTS

Financial support for this work was in part provided by national research grant FCT/POCTI/AMB/57616/2004 and by LSRE financing by FEDER/POCI/2010, for which the authors are thankful. V. Vilar's acknowledges his Ph.D. scholarship by FCT (SFRH/BD/7054/2001).

REFERENCES

[1] Volesky B. Sorption and Biosorption. Quebec, BV Sorbex, Inc., 2003.

[2] ESCWA. Environmentally Sound Technologies in the Metal Finishing Industry. In New York, Economic and Social Commission for Western Asia, 2001.

[3] Chong KH, Volesky B. Description of Two-Metal Biosorption Equilibria by Langmuir-Type Models. *Biotechnology and Bioengineering* 47: 451-460, 1995.

[4] Saeeda A, Iqbala M, Akhtarb MW. Removal and Recovery of Lead(II) from Single and Multimetal (Cd, Cu, Ni, Zn) Solutions by Crop Milling Waste (Black Gram Husk). *Journal of Hazardous Materials* B117: 65-73, 2005.

[5] Vilar VJP. Remoção de iões metálicos em solução aquosa por resíduos da indústria de extracção do agar. Ph.D. Thesis, Faculty of Engineering University of Porto, Porto, 2006.

[6] Vilar VJP, Botelho CMS, Boaventura RAR. Equilibrium and kinetic modelling of Cd(II) biosorption by algae *Gelidium* and agar extraction algal waste. *Water Research* 40(2): 291-302, 2006.

[7] Vilar VJP, Sebesta F, Botelho CMS, Boaventura RAR. Equilibrium and kinetic modelling of Pb^{2+} biosorption by granulated agar extraction algal waste. Process *Biochemistry* 40(10): 3276-3284, 2005.

[8] Vilar VJP, Botelho CMS, Boaventura RAR. Effect of Cu(II), Cd(II) and Zn(II) on Pb(II) biosorption by algae *Gelidium*-derived materials. *Journal of Hazardous Materials* 154: 711-720, 2007.

[9] Kinniburgh DG, Riemsdijk WHv, Koopal LK, Borkovec M, Benedetti MF, Avena MJ. Ion Binding to Natural Organic Matter: Competition, Heterogeneity, Stoichiometry and Thermodynamic Consistency. Colloids and Surfaces A: *Physicochemical and Engineering Aspects* 151: 147-166, 1999.

[10] Koopal LK, Saito T, Pinheiro JP, Riemsdijk WHV. Ion binding to natural organic matter: General considerations and the NICA-Donnan model. Colloids and Surfaces A: *Physicochemical and Engineering Aspects* 265: 40-54, 2005.

[11] Chapra SC, Canale RP. Numerical Methods for Engineers. McGraw-Hill, 1998.

[12] Puranik PR, Paknikar KM. Influence of Co-Cations on Biosorption of Lead and Zinc-a Comparative Evaluation in Binary and Multimetal Systems. *Bioresource Technology* 70: 269-276, 1999.

[13] Kim DS. The Removal by Crab Shell of Mixed Heavy Metal ions in Aqueous Solution. *Bioresource Technology* 87: 355-357, 2003.

[14] Pearson RG. Hard and Soft Acids and Bases. *Journal of American Chemical Society* 85: 3533-3539, 1963.

[15] Buffle J. Complexation Reactions in Aquatic Systems: An Analytical Approach. Chichester, *Ellis Horwood* Ltd., 1988.

[16] Nieboer E, McBryde WAE. Free-Energy Relationship in Coordination Chemistry. III. A Comprehensive Index to Complex Stability. *Canadian Journal Chemistry* 51: 2512-2524, 1973.

[17] Evans HT. Ionic Radii in Crystals. Boca Raton, In: Lide, D.R. (Ed.), CRC Handbook of *Chemistry and Physics,* CRC press, 1993.

[18] Marcus Y, Kertes AS. Ion Echange and Solvent Extraction of Metal Complexes. London, Wiley Interscience, *John Wiley and Sons*, 1969.

[19] Dean JA. Lange´s Handbook of Chemistry. New York, McGraw-Hill, 1999.

[20] Phillips CSG, Williams RJP. Inorganic Chemistry. Principles and Non-Metals. New York, Oxford University Press, 1965.

[21] Pauling L. Nature of the Chemical Bond. Ithaca, NY, Cornell University Press, 1967.

In: Heavy Metal Pollution
Editors: S. E. Brown and W. C. Welton

ISBN: 978-1-60456-899-8

Chapter 13

HEAVY METALS AND MOTOR NEURON DISORDERS

Julien Praline and Philippe Corcia
Neurology Department, CHRU de Tours, France and INSERM U930, Tours, France

ABSTRACT

Motor neuron disorders are progressive neurodegenerative diseases involving the motor neurons of at least one of the following areas: cerebral cortex, brainstem and spinal cord. Amyotrophic lateral sclerosis is the most common motor neuron disorder of adult onset. Even in cases of a familial inheritance of the pathologic trait that suggests a genetic pattern, the pathogenesis of this affection often remains unclear. Based on epidemiological surveys and similitudes with toxic disorders, environmental factors should be considered.

Heavy metals have been suspected owing to their proved neurotoxicity. Some heavy metals are key constituents of metalloproteins and their accumulation leads to systemic disease with neurologic symptoms. Such heavy metals might enter the nervous system by different pathways such as inhalation, digestion, skin absorption, and retrograde axonal transport.

ALS cases related to heavy metals intoxication and professional exposure have been reported. Some epidemiological surveys showed that occupational exposure to heavy metals such as mercury or lead could be considered as a risk factor for the development of motor neuron disorder. This relationship between heavy metals intoxication and motor neuron disease was also underscored through neuropathological examination that showed accumulation of metals within cortical and spinal motor neurons. Experimental studies based on autometallography analysis demonstrated such accumulation in mice or rats after heavy metal exposure. Animal studies or experiments also demonstrated the pathological role of metals in the motor neuron homeostasis.

INTRODUCTION

Motor neuron disorders encompass progressive neurodegenerative diseases involving the motor neurons of at least one of the following areas: cerebral cortex, brainstem and spinal cord. ALS is the most common motor neuron disorder of adult onset. This disease affects both upper and lower motor neurons of the three areas. Incidence of ALS in the general population ranges about 1.5 per 100,000 inhabitants. Usually sporadic, familial cases are encountered and 10% of them are associated with mutations in Superoxide dismutase (SOD1) gene (Rosen et al., 1993).

Other motor neuron disorders affect either upper or lower motor neurons in one or several defined areas. These differential diagnoses of ALS may be genetically determined (such as spino-bulbar muscular atrophy, childhood-onset spinal muscular atrophy or hereditary spastic paraplegias) but in others (Hirayama's disease, primary lateral sclerosis for example), the cause remains unknown.

We focus on involvement of heavy metals in ALS owing to this topic was predominantly studied in this disease and few data are available in other motor neuron disorders.

SUSPECTED CAUSES OF ALS

To date, even in cases of a familial inheritance of the pathologic trait that suggests a genetic pattern, the cause of ALS often remains unknown. The aetiology of ALS is likely to be multifactorial, a complex interplay between genetic and environmental factors (Cluskey and Ramsden, 2001). Some susceptibility genes have been identified in sporadic ALS. Studies on apolipoprotein E4 genotype as risk factor provide contradictory results (Li et al., 2004). The copy number of SMN1 gene seems to be the most solid susceptibility factor (Corcia et al., 2006). Others association studies revealed HFE and angiogenin genotype as risk factors (Wu et al., 2007, Goodall et al., 2005).

Based on epidemiological surveys and similitudes with toxic disorders, environmental factors should be considered. As that observed in poliomyelitis, persistent viral infection might cause sporadic ALS. Detection of enterovirus RNA in the spinal cords of patients with ALS has not been proven (Berger et al., 2000; Walker et al., 2001) but Anderson et al. create animal model of motor neuron disease by an infection with two viruses (Anderson et al. 1995). Furthermore, motor neuron disease has also been reported in a small number of patients infected with the human immunodeficiency virus (HIV) (Moulignier et al., 2001) and has been reversed by anti-viral therapy in a few of them (Mc Gowan et al., 2001). Toxic environmental factors have been suspected considering reports of clusters and conjugal cases of ALS (Corcia et al., 2003). Guam disease is the most representative example of such a cluster. The parkinsonism-dementia-ALS complex of Guam seemed to result of interaction between toxic and genetic factors. Toxicity comes from the direct or indirect consumption of cycad nuts (Cox and Sacks, Neurology 2002) and the frequency of mutant polymorphisms of tau gene and CYP2D6 gene are higher in the Chamorro population than in Caucasians (Poorkaj et al., 2001; Chen et al., 1996). Moreover, cycasin, a cycad-derived toxin increases the expression of tau RNAm (Esclaire et al., 1999). Other toxic factors that induce motor neuron disorder have been described, such as lathyrism due to guinea pea consumption and Konzo linked to intoxication by cyanides (Howlett et al., 1990; Ludolph and Spencer, 1996).

Although the cause remains unknown, several pathophysiological processes leading to motor neuron death have been identified in ALS. Oxidative stress, protein aggregation, mitochondrial dysfunction, excitotoxicity, and impaired axonal transport should be associated to involve a programmed cell death pathway resembling apoptosis (Shaw, 2005).

Role of Metals in the Metabolism

Some metals are key constituents of metalloproteins such as haemoglobin, ceruleoplasmin and ferritin. Zinc, iron and copper are the metals which are bound to proteins via sulphydryl groups of amino-acids to form metalloenzymes. Thus, body stores and concentrations of metals in cells and fluids are regulated. One of these metalloenzymes must be considered. Superoxide dismutase 1 is a metalloenzyme that contains an essential copper atom at the active site which is alternately reduced and oxidised by superoxide. The presence of zinc is thought to stabilise the protein structure. This enzyme catalyzes the conversion of toxic superoxide radicals to hydrogen peroxide and oxygen and also has pro-oxidant activities, including peroxidation, the generation of hydroxyl radicals, and the nitration of tyrosine. Mutations in the SOD1 gene affect approximately 20% of patients with familial ALS (Rosen et al., 1993). Studies on animal models have led to the conclusion that mutant SOD1 protein exerts its pathological effects through a toxic gain of function rather than a loss of function. The gain of function is caused by modifications concerning both enzymatic capacities and protein structure. Structural changes could increase copper exposure leading to the generation of free radicals and peroxynitrite. Decreased zinc affinity should explain functional abnormalities (Crow et al, 1997 ; Estevez et al, 1999). These changes lead to several pathophysiological processes such as oxidative stress, mitochondrial dysfunction, excitotoxicity, protein aggregation, and inflammation (Shaw, 2005).

Neurological Manifestations of Heavy Metals Poisoning

Metals with physiological function can provide neurotoxic effects. For example, Wilson disease is characterized by copper accumulation in the liver and nervous system. Parkinsonism and psychiatric disturbances are the most common symptoms in adults (Waggoner et al., 1999). Increased iron concentration in the body leads to hemochromatosis. Accumulation of iron commonly induces liver, pancreas and cardiac disturbances. Dementia, ataxia and parkinsonism have been described (Nielsen et al., 1995).

Others metals without known physiological function but with described neurotoxicity are mercury and lead. Among all metals implicated in environmental or occupational intoxication, mercury appears as one of the most neurotoxic. Neurological clinical features suggestive of mercury intoxication comprise peripheral and central nervous system signs as erethism, tremor, peripheral neuropathy or cortical blindness (Clarkson et al., 2003).

Chronic exposure to lead induces peripheral neuropathy in a context of systemic illness accompanied by abdominal pain, constipation and microcytic anemia (Windebank, 1993).

The upper limbs are involved to a greater extent than the lower limbs. Distal weakness usually asymmetric, often begin in the finger extensor. Interstitial nephropathy may coexist.

Heavy Metals Exposure and Motor Neuron Disorder

Heavy metals have been suspected for long time in the aetiology of ALS. The role of exposure to lead in ALS was first suspected by Aran in 1850 (Aran, 1850). Few other lead-related ALS cases were reported (Livesley and Sissons, 1968; Currier and Haerer, 1968; Boothby et al., 1974; Petkau et al., 1974; Couratier et al., 1998; Oh et al., 2007). The role of professional lead exposure has been studied. Risk of ALS was associated with occupational exposure to lead with a dose response for lifetime days of exposure (Chancellor et al., 1993; Kamel et al., 2002). However, other studies showed conflicting results (Armon et al., 1991; Gresham et al., 1986; Strickland et al., 1996; McGuire et al., 1997).

ALS cases related to mercury intoxication (Brown, 1954; Kantarjian, 1961; Barber, 1978; Adams et al., 1983; Schwarz et al., 1996; Praline et al., 2007) and professional exposure have been reported (Currier and Haerer, 1968). This relationship between mercury intoxication and motor neuron disease was also underscored through neuropathological examination that showed accumulation of mercury within cortical and spinal motor neurons in a patient after a suicide attempt with mercury injection (Pamphlett and Waley 1996 b).

A case of ALS associated with exposure to cadmium has been reported (Bar-Sela et al., 2001). Cadmium is suspected to decrease brain SOD1 level and activity and enhance excitotoxicity of glutamate and oxidative stress (Huang et al., 2006J Cell Biochemistry).

Although ALS cases associated with aluminium exposure have not been proven, two ALS cases with increased CNS aluminium concentration were reported (Yasui et al., 1991 Neurotoxicology). Other studies have looked at metals levels in blood of patients with ALS but the results were inconsistent (House et al., 1978; Khare et al., 1990; Pamphlett et al., 2001; Vinceti et al., 1997).

Experimental Data on Toxicity of Metals For Motor Neurons

Some animal's experiments support the hypothesis of toxin as cause of ALS. Indeed, metal toxins enter motor neurons selectively (Arvidson, 1989) and motor neurons commonly contain metals in their cytoplasm (Pamphlett and Waley, 1998). Suspected on cases reports, the relationship between mercury intoxication and motor neuron disease was also underscored through neuropathological examination that showed accumulation of mercury within cortical and spinal motor neurons in a patient after a suicide attempt with mercury injection (Pamphlett and Waley, 1996 b). Experimental studies based on autometallography analysis demonstrated an accumulation of metallic mercury in motor neurons in mice or rats after blood (Su et al., 1998) or peritoneal (Pamphlett and Waley, 1996 a) injection of mercury or after exposure to vapours of mercury (Pamphlett and Coote, 1998). Another experience showed that mercury is transported retrogradely from muscular nerves terminals to spinal and brainstem motor neurons in rats or mice following intramuscular injection of inorganic mercury (Arvidson, 1992; Schionning, 1993). To date, the exact mechanism of mercury neurotoxicity remains unknown. Nevertheless, it has been demonstrated that mercury reacts and depletes free sulfhydryl groups and causes a reduction in the activity of SOD (Stohs and Bagchi, 1995). Thus the presence of mercury in motor neurons may lead to the formation of an oxidative stress which is implicated in the pathophysiology of ALS (Shaw, 2005). The tubulin which is a kinesin protein of microtubule is disintegrated after mercury exposure in an experimental study and this phenomenon leads to neurofibrillary aggregates (Leong et al., 2001). In mice, acute mercury vapour exposure led to atrophy of large myelinated motor axons and irregularity in axon shape which may be caused by damage of cytoskeletal components (Stankovic, 2006). Abnormal axonal transport is another suspected mechanism implicated in pathophysiology of ALS. Indeed, accumulation and abnormal assembly of neurofilaments are common pathological hallmarks of ALS (Shaw PJ, 2005).

Exposure to manganese can cause neurotoxicity with a parkinsonian syndrome (Perl and Olanow, 2007). The manganese concentration in spinal cords from autopsied cases with amyotrophic lateral sclerosis (ALS) was increased and more prominent in the anterior horn and lateral column than in the posterior column (Miyata et al., 1983; Kihira et al., 1990). After chronic low-dose intracisternal administration of aluminium, rabbits developed clinical and pathological features of ALS (Triarhou et al., 1985 ; Strong and Garruto, 1991 ; Wakayama et al., 1996). The injection of aluminium into the subepineurial space of the sciatic nerve induces morphological changes and degeneration of the spinal motor neurons (Kihira et al., 1995).

Lead can induce neurotoxicity through several mechanisms. Lead can be substituted by calcium in many intracellular reactions causing damage to mitochondria and neural tissues (Adonaylo and Oteiza, 1999). It should increase glutamate excitotoxicity (Salvolainen et al., 1998). As with mercury, retrograde axonal transport of lead from muscle to spinal cord is demonstrated in rodents (Baruah et al., 1981; Pamphlett and Bayliss, 1992). Furthermore, lead impairs the slow axonal transport in rats (Yokohama and Araki, 1992).

Conclusion

The relationship between heavy metals and motor neuron disorders is based on clinical reports and experimental data. However, their pathological role cannot explain the motor neuron damage entirely. Heavy metals may step in these diseases as susceptibility environmental factors.

References

Adams CA, Ziegler DK, Lin JT. Mercury intoxication simulating amyotrophic lateral sclerosis. *JAMA* 1983; 250: 642-643.

Adonaylo VN, Oteiza PI. Lead intoxication : antioxidant defenses and oxidative damage in rat brain. *Toxicology* 1999; 135: 77-85.

Anderson GW, Palmer GA, Rowland RR, Even C, Plagemann PG. Infection of central nervous system cells by ecotropic murine leukemia virus in C58 and AKR mice and in in utero-infected CE/J mice predisposes mice to paralytic infection by lactate dehydrogenase-elevating virus. *J. Virol.* 1995; 69(1): 308-319.

Aran FA. Recherches sur une maladie non encore décrite du système musculaire. *Arch. Gen. Med.* 1850 ; 24 : 1-35.

Armon C, Kurland LT, Daube JR, O'Brien PC. Epidemiologic correlates of sporadic amyotrophic lateral sclerosis. *Neurology* 1991; 41: 1077-1084.

Arvidson B. Inorganic mercury is transported from muscular nerve terminals to spinal and brainstem motoneurons. *Muscle Nerve* 1992; 15(10): 1089-1094.

Arvidson B. Retrograde axonal transport of metals. J Trace Elem Exp Med 1989; 2: 343-347.

Barber TE. Inorganic mercury intoxication reminiscent of amyotrophic lateral sclerosis. *J. Occup. Med.* 1978; 20: 667-669.

Bar-Sela S, Reingold S, Richter ED. Amyotrophic lateral sclerosis in a battery-factory worker exposed to cadmium. *Int. J. Occup. Environ. Health* 2001; 7(2):109-112.

Baruah JK, Rasool CG, Bradley WG, Munsat TL. Retrograde axonal transport of lead in rat sciatic nerve. *Neurology* 1981; 31(5): 612-616.

Berger JR, Espinosa P, Kissel J. Brachial Amyotrophic Diplegia in a patient with human immunodeficiency virus infection. *Arch. Neurol.* 2005; 62: 817-823.

Boothby JA, DeJesus PV, Rowland LP. Reversible forms of motor neuron disease: lead "neuritis". *Arch. Neurol.* 1974; 31: 18-23.

Brown IA. Chronic mercurialism. A cause of the clinical syndrome of amyotrophic lateral sclerosis. *Arch. Neurol. Psychol.* 1954; 72: 674-681.

Chancellor AM, Slattery JM, Fraser H, Warlow CP. Risk factors for motor neuron disease: a case-control study based on patients from the Scottish Motor Neuron Disease Register. *J. Neurol. Neurosurg. Psychiatry* 1993; 56(11): 1200-1206.

Chen X, Xia Y, Gresham LS, Molgaard CA, Thomas RG, Galasko D et al. ApoE and CYP2D6 polymorphism with and without parkinsonism-dementia complex in the people of Chamorro, guam. *Neurology* 1996; 47(3): 779-784.

Clarkson TW, Magos L, Myers GJ. The toxicology of mercury-Current exposures and clinical manifestations. *N. Engl. J. Med.* 2003; 349: 1731-1737.

Cluskey S, Ramsden DB. Mechanisms of neurodegeneration in amyotrophic lateral sclerosis.*Mol. Pathol.* 2001; 54(6): 386-392.

Corcia P, Camu W, Halimi JM, Vourc'h P, Antar C, Vedrine S, et al. SMN1 gene, but not SMN2, is a risk factor for sporadic ALS. *Neurology.* 2006; 67: 1147-1150.

Corcia P, Jafari HF, Mazyad H, Giraud P, Lardillier D, Clavelou P, Pouget J, Camu W. A clustering of Conjugal Amyotrophic Lateral Sclerosis in Southern France. *Arch. Neurol.* 2003; 60: 553-557.

Couratier P, Bernet Nernardy P, Truong T, et al. Lead intoxication and amyotrophic lateral sclerosis. *Rev. Neurol.* (Paris) 1998; 154: 345-347.

Cox PA, Sacks OW. Cycad neurotoxins, consumption of flying foxes, and ALS-PDC disease in Guam. *Neurology* 2002; 58(6): 956-959.

Crow JP, Sampson JB, Zhuang Y, Thompson JA, Beckman JS. Decreased zinc affinity of amyotrophic lateral sclerosis-associated superoxide dismutase mutants leads to enhanced catalysis of tyrosine nitration by peroxynitrite. *J. Neurochem.* 1997; 69: 1936-1944.

Currier RD, Haerer AF. Amyotrophic lateral sclerosis and metallic toxins. *Arch. Environ. Health* 1968; 17: 712-719.

Esclaire F, Kisby G, Spencer P, Milne J, Lesort M, Hugon J. The Guam cycad toxin methylazoxymethanol damages neuronal DNA and modulates tau mRNA expression and excitotoxicity. *Exp. Neurol.* 1999; 155(1): 11-21.

Estevez AG, Crow JP, Sampson JB, Reiter C, Zhuang Y, Richardson GJ, et al. Induction of nitric oxide-dependent apoptosis in motor neurons by zinc-deficient superoxide dismutase. *Science.* 1999; 286: 2498-2500.

Goodall EF, Greenway MJ, van Marion I, Carroll CB, Hardiman O, Morrison KE. Association of the H63D polymorphism in the hemochromatosis gene with sporadic ALS. *Neurology* 2005; 65: 934-937.

Gresham LS, Molgaard CA, Golbeck AL, Smith R. Amyotrophic lateral sclerosis and occupational heavy metal exposure: a case-control study. *Neuroepidemiology* 1986; 5: 29-38.

House AO, Abbott RJ, Davidson DLW, Ferguson IT, Lenman JAR. Response to penicillamine of lead concentrations in CSF and blood in patients with motor neurone disease. *Br. Med. J.* 1978; 2: 1684.

Howlett WP, Brubaker GR, Mlingi N, Rosling H. Konzo, an epidemic upper motor neuron disease studied in Tanzania. *Brain* 1990; 113 (Pt 1): 223-235.

Huang YH, Shih CM, Huang CJ, Lin CM, Chou CM, Tsai ML, Liu TP, Chiu JF, Chen CT. Effects of cadmium on structure and enzymatic activity of Cu,Zn-SOD and oxidative status in neural cells. *J. Cell Biochem.* 2006; 98(3): 577-589.

Kamel F, Umbach DM, Munsat TL, Shefner JM, Hu H, Sandler DP. Lead exposure and amyotrophic lateral sclerosis. *Epidemiology* 2002; 13(3): 311-319.

Kantarjian AD. A syndrome clinically resembling amyotrophic lateral sclerosis following chronic mercurialism. *Neurology* 1961; 11: 639-644.

Khare SS, Ehmann WD, Kasarskis EJ, Markesbery WR. Trace element imbalances in amyotrophic lateral sclerosis. *Neurotoxicology* 1990; 11(3): 521-532.

Kihira T, Mukoyama M, Ando K, Yase Y, Yasui M. Determination of manganese concentrations in the spinal cords from amyotrophic lateral sclerosis patients by inductively coupled plasma emission spectroscopy. *J. Neurol. Sci.* 1990; 98(2-3): 251-258.

Kihira T, Yoshida S, Komoto J, Wakayama I, Yase Y. Aluminum-induced model of motor neuron degeneration: subperineurial injection of aluminum in rabbits. *Neurotoxicology* 1995; 16(3): 413-424.

Leong CC, Syed NI, Lorscheider FL. Retrograde degeneration of neurite membrane structural integrity of nerve growth cones following in vitro exposure to mercury. *Neuroreport.* 2001; 12(4): 733-737.

Li YJ, Pericak-Vance MA, Haines JL,·Siddique N, McKenna-Yasek D, Hung WY, et al. Apolipoprotein E is associated with age at onset of amyotrophic lateral sclerosis. *Neurogenetics* 2004; 5: 209-213.

Livesley B, Sissons CE. Chronic lead intoxication mimicking motor neuron disease (letter) *BMJ* 1968; 4: 387-388.

Ludolph AC, Spencer PS. Toxic models of upper motor neuron disease. *J. Neurol. Sci.* 1996; 139 Suppl: 53-59.

MacGowan DJ, Scelsa SN, Waldron M. An ALS-like syndrome with new HIV infection and complete response to antiretroviral therapy. *Neurology.* 2001; 57: 1094-1097.

Mc Guire V, Longstreth WT, Nelson LM, Koepsell TD, Checkoway H, Morgan MS, Van Belle G. Occupational exposures and amyotrophic lateral sclerosis. A population-based case-control study. *Am. J. Epidemiol.* 1997; 145(12) :1076-1088.

Miyata S, Nakamura S, Nagata H, Kameyama M. Increased manganese level in spinal cords of amyotrophic lateral sclerosis determined by radiochemical neutron activation analysis. *J. Neurol. Sci.* 1983; 61(2): 283-293.

Moulignier A, Moulonguet A, Pialoux G, Rozenbaum W. Reversible ALS-like disorder in HIV infection. *Neurology*. 2001; 57: 995-1001.

Nielsen JE, Jensen LN, Krabbe K. Hereditary haemochromatosis: a case of iron accumulation in the basal ganglia associated with a parkinsonian syndrome. *J. Neurol. Neurosurg. Psychiatry* 1995; 59(3):318-321.

Oh S, Kim EA, Lee S, Kim M, Kang S. A case of amyorophic lateral sclerosis in electronic parts manufacturing worker exposed to lead. *NeuroToxicology* 2007; 28: 324-327.

Pamphlett R, Bayliss A. Lead uptake in motor axons. Muscle Nerve 1992; 5(5): 620-625.

Pamphlett R, Coote P. Entry of low doses of mercury vapor into the nervous system. *Neurotoxicology* 1998; 19(1): 39-47.

Pamphlett R, McQuilty R, Zarkos K. Blood levels of toxic and essential metals in motor neuron disease. *Neurotoxicology* 2001; 22(3): 401-410.

Pamphlett R, Waley P. Mercury in human spinal motor neurons. Acta Neuropathol (Berl) 1998; 96(5): 515-519.

Pamphlett R, Waley P. Motor neuron uptake of low dose inorganic mercury. *J. Neurol. Sci* 1996; 135(1): 63-67. a

Pamphlett R, Waley P. Uptake of inorganic mercury by the human brain. *Acta Neuropathol.*(Berl) 1996; 92(5): 525-527. b

Perl DP, Olanow CW. The neuropathology of manganese-induced Parkinsonism. *J. Neuropathol. Exp. Neurol.* 2007; 66(8): 675-682.

Petkau A, Sawatzky A, Hillier CR, Hoogstraten J. Lead content of neuromuscular tissue in amyotrophic lateral sclerosis: case report and other considerations. *Br. J. Int. Med.* 1974; 31: 275-287.

Poorkaj P, Tsuang D, Wijsman E, Steinbart E, Garruto RM, Craig UK et al. TAU as a susceptibility gene for amyotropic lateral sclerosis-parkinsonism dementia complex of Guam. *Arch Neurol.* 2001; 58(11): 1871-8.

Praline J, Guennoc AM, Limousin N, Hallak H, de Toffol B, Corcia P. ALS and mercury intoxication: a relationship? *Clin. Neurol. Neurosurg.* 2007 ; 109(10):880-883.

Rosen DR, Siddique T, Rouleau G, Brown RH. Mutations in Cu/Zn superoxide dismutase gene are associated with familial amyotrophic lateral sclerosis. *Nature* 1993; 364: 362.

Salvolainen KM, Loikkanen J, Eerikainen S, Naaral J. Interactions of excitatory neurotransmitters and xenobiotics in excitotoxicity and oxidative stress: glutamate and lead. *Toxicol Lett* 1998; 102-103: 363-367.

Schionning JD. Retrograde axonal transport of mercury in rat sciatic nerve. Toxicol Appl *Pharmacol* 1993; 121(1): 43-49.

Schwarz S, Husstedt I, Bertram HP, Kuchelmeister K. Amyotrophic lateral sclerosis after accidental injection of mercury. *J. Neurol. Neurosurg. Psychiatry* 1996; 60(6): 698.

Shaw PJ. Molecular and cellular pathways of neurodegeneration in motor neurone disease. *J. Neurol. Neurosurg .Psychiatry* 2005; 76: 1046-1057.

Stankovic R. Atrophy of large myelinated motor axons and declining muscle grip strength following mercury vapor inhalation in mice. *Inhal. Toxicol.* 2006 ; 18(1): 57-69.

Stohs SJ, Bagchi D. Oxidative mechanisms in the toxicity of metal ions. *Free Radic Biol. Med.* 1995; 18(2):321-336.

Strickland D, Smith SA, Dolliff G et al. Amyotrophic lateral sclerosis and occupational history: a pilot case-control study. *Arch. Neurol.* 1996; 53: 730-733.

Strong MJ, Garruto RM. Chronic aluminum-induced motor neuron degeneration: clinical, neuropathological and molecular biological aspects. *Can. J. Neurol. Sci.* 1991 ; 18(3 Suppl):428-431.

Su M, Wakabayashi K, Kakita A, Ikuta F, Takahashi H. Selective involvement of large motor neurons in the spinal cord of rats treated with methylmercury. *J. Neurol.Sci. 1998*; 156(1): 12-7.

Triarhou LC, Norton J, Bugiani O, Ghetti B. Ventral root axonopathy and its relation to the neurofibrillary degeneration of lower motor neurons in aluminum-induced encephalomyelopathy. *Neuropathol. Appl. Neurobiol.*1985; 11(6): 407-430.

Vinceti M, Guidetti D, Bergomi M, Caselgrandi E, Vivoli R, Olmi M, Rinaldi L, Rovesti S, Solime F. Lead, cadmium, and selenium in the blood of patients with sporadic amyotrophic lateral sclerosis. *Ital. J. Neurol. Sci.* 1997; 18(2): 87-92.

Waggoner DJ, Bartnikas TB, Gitlin JD. The role of copper in neurodegenerative disease. *Neurobiol. Dis.* 1999; 6(4): 221-230.

Wakayama I, Nerurkar VR, Strong MJ, Garruto RM. Comparative study of chronic aluminum-induced neurofilamentous aggregates with intracytoplasmic inclusions of amyotrophic lateral sclerosis. *Acta Neuropathol (Berl)* 1996; 92(6): 545-554.

Walker MP, Schlaberg R, Hays AP, Bowser R, Lipkin WI. Absence of echovirus sequences in brain and spinal cord of amyotrophic lateral sclerosis patients. *Ann. Neurol.* 2001; 49: 249-53.

Windebank AJ. Metal neuropathy. In: Dyck PJ, Thomas PK, Griffin JW, Low PA, Poduslo JF. Peripheral neuropathy, 3rd ed. Philadelphia: WB Saudners Company 1993: 1549-1570.

Wu D, Yu W, Kishikawa H, Folkerth RD, Iafrate AJ, Shen Y, Xin W, Sims K, Hu GF. Angiogenin loss-of-function mutations in amyotrophic lateral sclerosis. *Ann. Neurol.* 2007; 62(6): 609-617.

Yasui M, Yase Y, Ota K, Mukoyama M, Adachi K. High aluminum deposition in the central nervous system of patients with amyotrophic lateral sclerosis from the Kii Peninsula, Japan: two case reports. *Neurotoxicology* 1991; 12(2): 277-283.

Yokoyama K, Araki S. Assessment of slow axonal transport in lead-exposed rats. Environ Res 1992; 59(2): 440-446.

In: Heavy Metal Pollution
Editors: S. E. Brown and W. C. Welton

ISBN: 978-1-60456-899-8

Chapter 14

MERCURY AND BISMUTH FILM MICROELECTRODES IN DETERMINATION OF TRACE HEAVY METALS BY ANODIC STRIPPING VOLTAMMETRY

V. Rehacek and I. Hotovy
Department of Microelectronics, Faculty of Electrical Engineering and Information Technology, Slovak University of Technology, Ilkovicova 3, 812 19 Bratislava, Slovak Republic

ABSTRACT

Bismuth electrodes are very promising in the field of electrochemical analysis. Bismuth film electrodes (BiFEs) can be formed on the same substrates as mercury, which is used most frequently for simultaneous determination of heavy metals (such as Cu(II), Pb(II), Cd(II), Zn(II)) by anodic stripping voltammetry (ASV), and have a potential to replace toxic mercury. The most important results with bismuth film and mercury hemispherical microelectrodes that we reached in this field on different substrates are summarized in this work. High conductive (0.008 - 0.024 Ωcm) p- or n-type silicon substrates have been investigated as a support for mercury plating. The influence of many parameters on mercury plating has been studied. The stripping current response of trace heavy metals (Cu(II), Pb(II) and Cd(II) in concentration 3.3×10^{-8} mol/L of each) has been demonstrated by square wave (SW) ASV. Mercury-plated as well as bismuth-plated nitrogen-doped diamond-like carbon (DLC) microelectrode arrays were investigated for Cu(II), Pb(II) and Cd(II) determination by differential pulse (DP) ASV or SW ASV. The current responses of all metals were linear with concentration in the range from 1.6×10^{-8} to 8.3×10^{-8} mol/L (in the case of mercury) and 2×10^{-8} to 1.2×10^{-7} mol/L (in the case of bismuth). The stripping current responses of the bare nitrogen-doped DLC microelectrode array reached under the same conditions were much lower in comparison with the mercury or bismuth plated arrays. Interesting results have been obtained on a single pencil-lead graphite electrode with a bismuth film. Detection limits 2.4×10^{-9} mol/L for Pb(II), 2.9×10^{-9} mol/L for Cd(II) and 1.2×10^{-8} mol/L for Zn(II) were estimated in this case. Concerning the design of electrodes, we used a single disc graphite electrode (0.5 mm in diameter) or microelectrode arrays consisting of 50 625 microdiscs with 2 μm in diameter and interelectrode distances of 20 μm.

1. INTRODUCTION

General trends for environmentally friendly analytical methods and alternative materials have continuously been sought. From this point of view bismuth film electrodes are very promising in the field of electrochemical analysis. Bismuth has a potential to replace toxic mercury used most frequently for determination of heavy metals (such as Cu, Cd, Pb, Zn) by anodic stripping voltammetry (ASV) [1, 2]. The attractive properties of bismuth films include simple preparation, high sensitivity, excellent peak resolution, insensitivity to dissolved oxygen in aqueous solutions, negligible toxicity and ability to operate in highly alkaline media [3-5].

Anodic stripping voltammetry is a very attractive electroanalytical technique for determination of trace heavy metals. The method is characterized by remarkable sensitivity and extremely low detection limits (< ppb). Other advantages such as the ability to determine simultaneously several metals with high selectivity, rapidity, simplicity and low cost of the analyses are important. The high sensitivity and low detection limits are mainly given by the first step involving preconcentration of the analyte species in accumulating metal (bismuth or mercury), which is carried out under a fixed reduction potential for a predetermined time. In the second, stripping step, the accumulated species are stripped from bismuth or mercury into the solution by reoxidation. The stripping current peaks are proportional to the concentration of the species accumulated in bismuth or mercury and, hence, to the concentration in the analyte (figure 1).

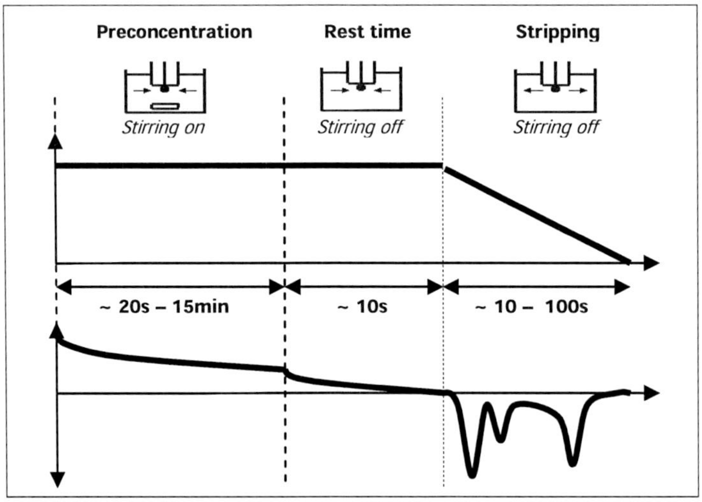

Figure 1. Principle of anodic stripping voltammetry.

Toxic metals can simultaneously be determined for concentration levels down to sub-ppb. The stripping step can involve a variety of potential waveforms. Pulse techniques such as differential pulse voltammetry (DPV) and mainly square wave voltammetry (SWV) are preferred rather than linear-potential scan techniques. In addition, the stripping analysis by SWV eliminates the time-consuming deaeration step because the stripping step is less sensitive to irreversible processes at high square wave frequency, such as the reduction of oxygen which is dissolved in aqueous solutions [6, 7]. Thanks to this technique, the oxygen interference with analytical signal is strongly reduced.

There are two methods which are the most frequently used. The first method, *ex-situ* plating, involves electroplating the accumulating metal before transffering the electrode into the analyzed solution [3, 5]. In the second method, *in-situ* plating, bismuth or mercury ions are added into the analyzed solution and target metals are co-deposited with bismuth or mercury on the electrode. The latter procedure is usually recommended because it gives better ASV performance in terms of sensitivity, selectivity and reproducibility. Another advantage of *in-situ* technique is the new surface generated at each analysis [8]. On the other hand, *in-situ* procedure cannot be employed when additions of accumulating metal into the analyte cause contamination, precipitation or changes in the speciation of the sample [9]. In the case of *ex-situ* procedure with mercury the stability of preplated mercury can be, in comparison with bismuth, strongly influenced by experimental and storing conditions or during the transfer. Mercury microelectrodes have to be stored in a deoxygenated solution at a negative potential to prevent oxidation of mercury which causes a decrease of their performance in time [9, 10].

In general, the performance of anodic stripping voltammetry with accumulating metal film is strongly affected by the supporting electrode material. An ideal electrode should possess low ohmic resistance, chemical and electrochemical inertness, good mechanical properties, high hydrogen and oxygen overvoltage (wide potential window), low background current, ease of reproduction of the electrode surface. The choice of the electrode support that bismuth or mercury is deposited onto is crucial. Bismuth can be plated on the same substrates as mercury. Glassy carbon [11-14], carbon paste [11, 15], carbon fibre [3, 16, 17], impregnated graphite [11], pencil-lead [18], gold [5, 16], platinum [5] and others [4, 5] have been reported for bismuth.

In the case of mercury film formation, metal film electrodes such as Ag, Au or Pt are not suitable because of their dissolution by mercury and formation of intermetallic compounds with the analyzed metals. Ir electrodes were found to be a useful substrate for mercury electroplating because of its good wettability, chemical and mechanical resistivity and low solubility in mercury [2, 19-21].

Diamond-like carbon (DLC) is a promising material for application in electroanalyses [22-25]. Analytical performance of the bismuth film electrode on DLC is influenced by the properties of DLC. DLC thin films are characterized by excellent mechanical properties (high hardness, high elastic modulus), chemical inertness to any acids, alkaline solutions or organic solvents [23]. Generally, high ohmic resistivity of DLC can be decreased by using doping elements such as nitrogen during the fabrication of the DLC. When we compare doped DLC electrodes with traditionally used metal film electrodes such as Au or Pt, the DLCs are not sensitive to surface oxidation or reduction and no chemical interaction occures with metal deposits. As well, the overvoltage of DLC to hydrogen is higher than that for Au or Pt.

In this work, results of heavy metal determination by anodic stripping voltammetry with mercury and bismuth microelectrode arrays are summarized. In this field traditionaly used mercury could be replaced by attractive bismuth. Different supports, highly conductive silicon [26] and nitrogen-doped DLC thin films on silicon [27] were used for *in-situ* determination with mercury hemispherical microelectrodes. In the case of bismuth, a single pencil-lead graphite electrode [28] and nitrogen-doped DLC films on silicon [28, 29] were used as a support. A survey of many parameters which influence the performance of ASV is given.

2. EXPERIMENTAL

The silicon microelectrode arrays as a support for mercury electroplating were fabricated very simply by the standard silicon technology (figure 2). The array consists of 50 625 silicon disc-shaped microelectrodes with disc diameter of 2 μm each and interelectrode spacing 20 μm was patterned by photolithography on highly conductive (0.008 - 0.024 Ωcm) p- or n-type silicon substrates.

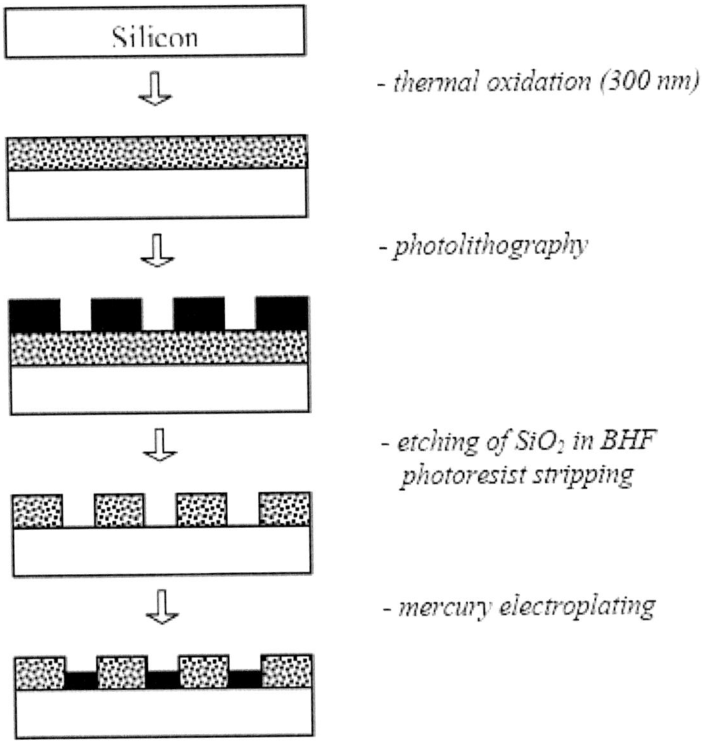

Figure 2. Fabrication process of the silicon based mercury microelectrode array.

The nitrogen-doped DLC microelectrode arrays were fabricated by vacuum pulse sputtering of graphite on highly conductive silicon substrates. The areas which act as working electrodes (the array of microdiscs with the same geometry as mentioned above) were defined photolithographically using a positive photoresist. A cross-sectional view of the microelectrode array is shown in figure 3.

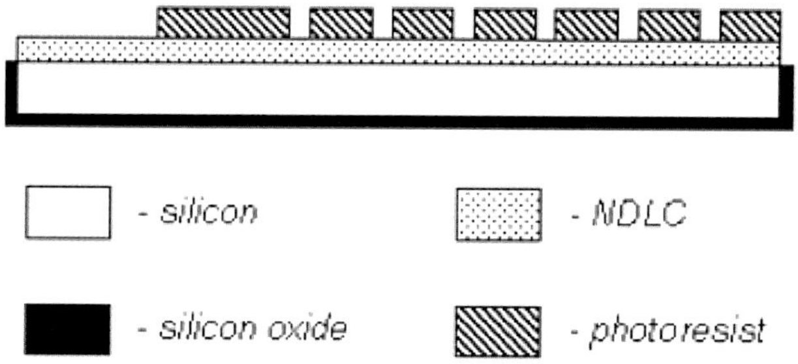

Figure 3. Cross-sectional view of the microelectrode array.

The pencil-lead graphite disc electrode was prepared from a pencil-lead rod (HB, 0.5 mm in diameter) which was inserted into a polyethylene tube and fixed by epoxy glue. The tip of this electrode (figure 4) was mechanically polished. Before use, the disc electrode was rinsed only with deionized water and dried.

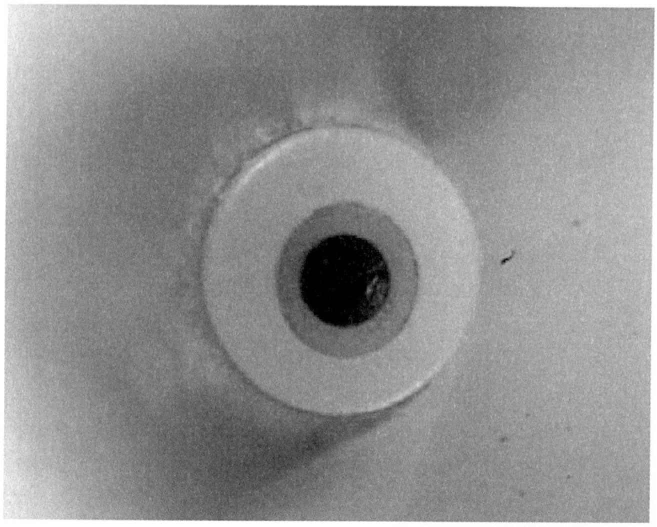

Figure 4. Single disc graphite electrode in the middle (Ø = 0.5 mm).

All the stock solutions were prepared from analytical grade chemicals ($Hg(NO_3)_2 \times 0.5$ H_2O – >99.0 %, $Bi(NO_3)_3 \times 5$ H_2O – 99.999 %, $Pb(NO_3)_2$ – 99.0 %, $Cd(NO_3)_2 \times 4$ H_2O – 99.0 %, $Zn(NO_3)_2 \times 6$ H_2O – 99.0 %, $Cu(SO_4) \times 5$ H_2O - >99.0 %, CH_3COOH – 99,8 %, CH_3COONa – 99.0 %, KNO_3 – 99.0 %) in 18 MΩcm deionized water. A three-electrode arrangement was used in all experiments. As a reference, a home-made Ag/AgCl/agar Cl^- (3 mol/L)/agar NO^{3-} (1 mol/L) electrode and, as a counter electrode, a platinum wire were used. Voltammetric experiments were performed with an electrochemical sensor interface (PalmSens, Palm Instruments BV) in combination with a magnetic stirrer and personal computer. Electrochemical properties of the graphite electrodes and of the nitrogen-doped DLC electrode arrays were investigated by cyclic voltammetry (CV). SWV or DPV were used in the measurement of the stripping voltammograms. The composition of nitrogen-doped DLC films was analyzed by Raman spectroscopy.

3. Results and Discussion

3.1. Microelectrode and Microelectrode Array

There are several definitions of a microelectrode (also the term ultramicroelectrode is used). One of them, the so-called geometrical definition, identifies a microelectrode as any electrode with at least one dimension not greater than 25 μm [30]. Other definitions [31-33] give a different value (20 μm) of the critical electrode dimension. From the point of view of microelectrode array behavior, the more convenient is an operational definition of the microelectrode as any electrode whose characteristic dimension is, under the given experimental conditions, comparable to or smaller than the diffusion layer thickness. Under these conditions, a steady-state or a pseudo steady-state is reached [34].

A microelectrode offers a number of attractive features in the field of electroanalyses over conventional macroelectrodes. The major advantageous feature of a microelectrode is its radial diffusion, which allows steady-state conditions to be reached rapidly, in comparison with linear diffusion of a macroelectrode. It is known that the steady-state limiting current of a single microdisc electrode is proportional to the radius of the disc and is governed by equation [35] (1):

$$I_{limS} = 4nFDCr \qquad (1)$$

where n is the number of electrons involved in the electrochemical reaction, F is Faraday`s constant, D is the diffusion coefficient of the reactant, C is the bulk reactant concentration, and r is the electrode radius.

For the microelectrode behavior a high mass transport is specific showing a sigmoidal cyclic voltammogram with a limiting current. Due to a reduced charging current and a small RC time constant, the voltammetric experiments can be performed at high scan rates. Extremely small currents and small ohmic drop make possible to measure in high resistance solutions without a basic electrolyte as well as without a third auxiliary electrode. Thanks to the reduced charging current, the ratio of faradaic-to-charging current (the smaller the electrode surface, the higher the measured signal-to-noise ratio) is improved and thereby the sensitivity can be higher. If the dimensions of the microelectrode are extremely small, the current response can be independent of the convection of the solution. On the other hand, the main drawback of a single microelectrode is its extremely small current which is difficult to measure.

The most important benefit in using a microelectrode array is to enhance the measured currents considerably and, at the same time, to maintain the advantageous features of a single microelectrode. For an inlaid microdisc array, the steady-state limiting current is proportional to the both the microdisc radius and the number of microdisc electrodes [36] (2):

$$I_{limA} = 4mnFDCr \qquad (2)$$

where *m* is the number of microdisc electrodes, *n* is the number of electrons involved in the electrochemical reaction, *F* is Faraday`s constant, *D* is the diffusion coefficient of the reactant, *C* is the bulk reactant concentration, and *r* is the electrode radius.

In comparison with a single microelectrode, the microelectrode array has to accomplish certain geometrical requirements in order to get a collective current response while maintaining the anventageous features of single microelectrodes. In the case of a microdisc array a minimum interelectrode distance has to be kept.

In general, if the interelectrode distance is too small, the array behaves at voltammetric measurements as a single macroelectrode due to the so-called shielding effect caused by overlapping of the diffusion layers of adjacent discs. Therefore, this distance should be at least 10 times larger than the microdisc diameter [2, 36]. If this requirement is satisfied, each microdisc behaves independently and the array provides characteristic advantages of a microelectrode.

Single microelectrodes of different shapes, such as disc, ring, band, hemisphere, sphere, etc. are often prepared by sealing conductive fibers or foils [37]. On the other hand, single microelectrodes and microelectrode arrays (microdisc array, interdigitated band array) are usually fabricated using microlithography [38], or screen-printing [39].

If the fabrication of array by microlithography is based on the use of a thin insulating layer (polyimide, Si_3N_4) with an array of openings on a conductive electrode layer, the thickness of the insulation layer is also important. An effect of recessed microdisc on the electroanalytical response was reported [36, 40]. In comparison with an inlaid microelectrode array, the steady-state current was reduced by ~10% for a microdisc array with disc radius of 5 μm and a 0.8 μm thick silicon nitride layer [40].

A method commonly used for characterization of microelectrodes and microelectrode arrays is cyclic voltammetry. The cyclic voltammetric behavior of microelectrode arrays is more complicated in comparison with single microelectrodes. The same array can exhibit behavior of micro- or macroelectrodes. It depends on the scan rates of voltammetric measurement [41]. At very high scan rates the peak-shaped voltammograms indicate linear diffusion within individual microelectrodes, but steady-state behavior is not reached. At very slow scan rates a similar non-steady-state peak-shaped voltammograms were observed, but linear diffusion is restored due to the overlap of the diffusion layers of individual microelectrodes. While at very high scan rates the current is proportional to the area of individual microelectrodes, at very slow scan rates it is proportional to the total area of the array [42]. In this case the set of microelectrodes behaves as a single macroelectrode. For intermediate scan rates, the radial diffusion is dominant and thereby voltammograms with a steady-state behavior are recorded. In the range of scan rates from 2 to 500 mV/s a sigmoidal response of cyclic voltammograms was found with a steady-state current in the case of lower scan rates (2 – 10 mV/s), and a peak-shape for higher ones (>100 mV/s) for an array with disc radius of 20 μm [43].

The behavior of microdisc arrays with disc diameter of 2 μm at a different interelectrode distance (2 μm and 20 μm) (figure 5) was characterized by cyclic voltammetry in 10 mmol/l $K_3Fe(CN)_6$/0.1 mol/L KNO_3 solution in a potential range from –1.2 V to +0.6 V vs. Ag/AgCl/Cl^- and scan rates from 10 to 200 mV/s.

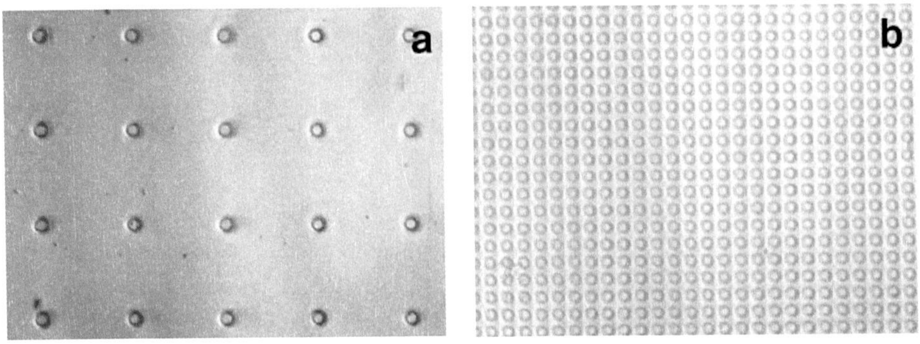

Figure 5. Microelectrode array with disc diameter of 2 μm each and interelectrode distance of 20 μm (a) and 2 μm (b).

According to microelectrode theory, a sigmoidal response was obtained with the microelectrode array (figure 6) for only a array with interelectrode diameter of 20 μm, independently of the scan rate in the given range. Compared with a packed microelectrode array, peak-shaped responses were observed with the microelectrode array with interelectrode distance of 2 μm (figure 7) at all scan rates due to the overlap of the diffusion layers of adjacent microdiscs.

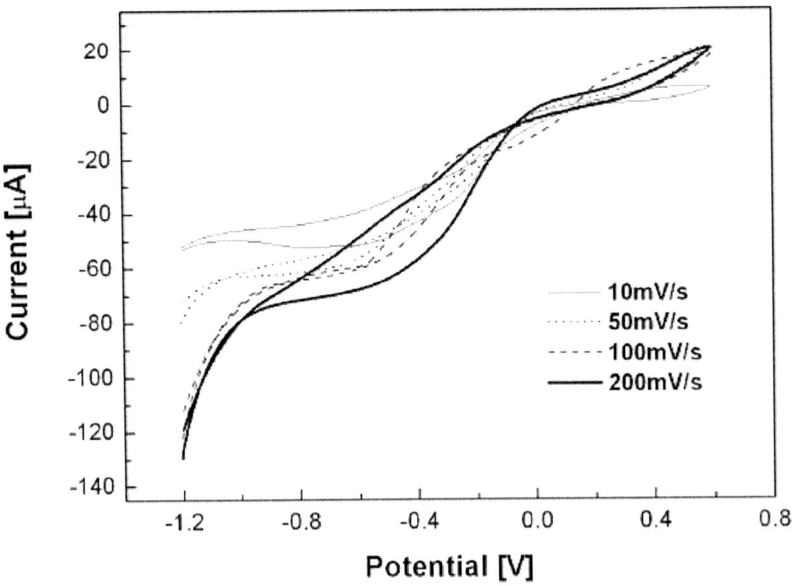

Figure 6. Cyclic voltammograms of gold microelectrode array in 0.1 mol/L acetate buffer (pH 4.5), interelectrode distance of 20 μm.

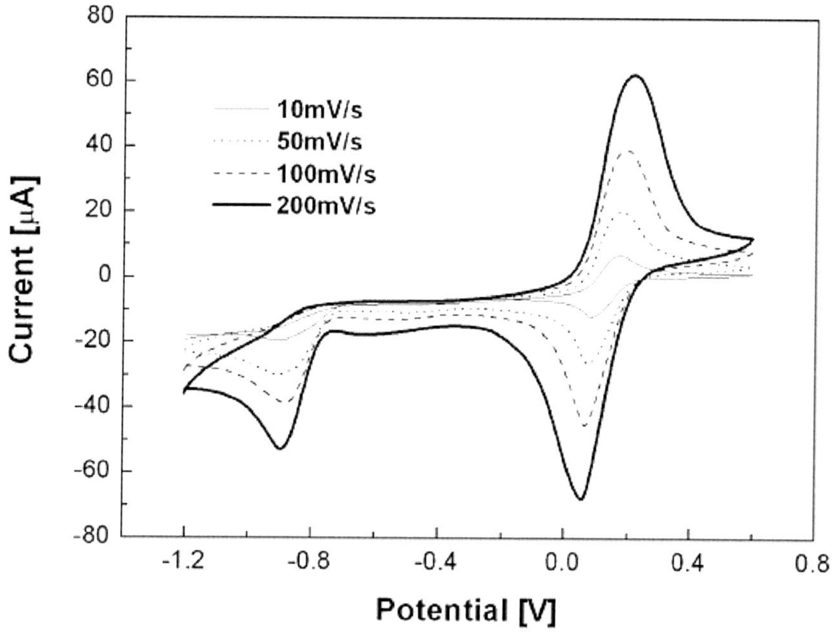

Figure 7. Cyclic voltammograms of gold microelectrode array in 0.1 mol/L acetate buffer (pH 4.5), interelectrode distance of 2 μm.

3.2. Diamond-Like Carbon Electrodes

Diamond-like carbon is a promising environment-friendly material for application in electroanalyses, because of its excellent mechanical (high hardness, high elastic modulus), chemical (inertness to any acids, alkaline solutions or organic solvents) and electrochemical properties (high hydrogen overvoltage, low background current) [22, 23]. DLC electrodes show a similar potential windows with respect to electrochemical properties like boron-doped diamond (BDD) electrodes used more frequently in electroanalyses [44-48]. DLC films can be deposited at room temperature, while BDD films have to be fabricated at high substrate temperature of about 800 °C or above [22].

DLC films can be prepared by CVD processes from the gas phase, by ion source techniques, pulse laser deposition, ion beam deposition, microwave plasma-assisted deposition, electrodeposition from organic liquids or by sputtering [48, 49].

Generally, DLC films are characterized by high ohmic resistance. During fabrication, the electrical properties can be controlled by changing the ratio between sp^2 and sp^3 bonded carbons, by the concentration of bonded hydrogen or by doping with nitrogen or different metallic elements [23]. Nitrogen atoms incorporated during layer sputtering influence the bonding formation (ratio between sp^2 and sp^3 bonded carbon) in the diamond-like carbon structure. Usually, an excess of N atoms will lead to graphitization of DLC and, hence, the conductance of nitrogen-doped DLC will be increased [22]. For successful utilization of nitrogen-doped DLC in electrochemical applications in aqueous solutions, nitrogen-doped DLCs with conductance as high as possible are required, of course.

When we compare nitrogen-doped DLC electrodes with traditionally used metal film electrodes such as Au or Pt, the DLCs are not sensitive to surface oxidation or reduction and no chemical interaction occures with metal deposits. Thereby, the long-term stability and reproducibility of the nitrogen-doped DLCs at multiple operation in electroanalyses is non-comparably higher. As well, the overvoltage of nitrogen-doped DLC to hydrogen is higher than that for Au or Pt. A wide electrochemical window of doped nitrogen-doped DLC (approximately –1 to 2.5 V vs. Ag/AgCl reference electrode [23], -1.6 to 1.6 V vs. SCE [50]) has been reported.

Nitrogen-doped DLC films prepared by vacuum pulse sputtering of graphite under various parameters on highly conductive (0.008-0.024 Ωcm) silicon substrates were used as a support for deposition of mercury or bismuth films. Selected parameters of nitrogen-doped DLC film preparation are summarized in table 1. Sample DLC_3 was deposited for time twice longer than samples DLC_1 and DLC_2.

Table 1. Selected parameters of preparation of nitrogen-doped DLC films

Sample	Flow rate of Ar [sccm]	Flow rate of N_2 [sccm]	Film thickness [nm]	Note
DLC_1	60	70	297	without Ti sublayer
DLC_2	70	60	405	with Ti sublayer
DLC_3	70	60	790	with Ti sublayer

Composition of the films was analyzed by Raman spectroscopy. In figures 8a – 10a, typical Raman spectra are shown.

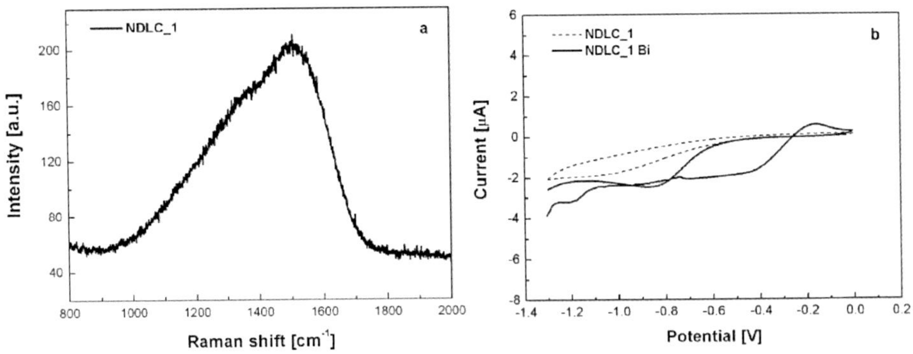

Figure 8. Raman spectra (a) and I-V curves (b) of nitrogen-doped DLC_1. I-V curves were recorded at scan rate of 50 mV/s in 0.1 mol/L acetate buffer (pH 4.5) without and with bismuth film.

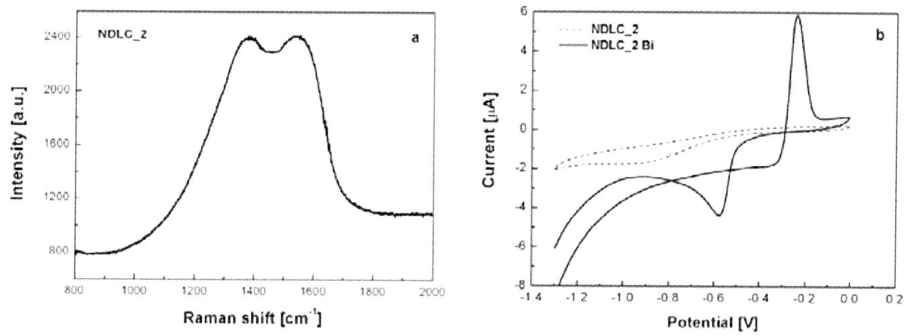

Figure 9. Raman spectra (a) and I-V curves (b) of nitrogen-doped DLC_2. I-V curves were recorded at scan rate of 50 mV/s in 0.1 mol/L acetate buffer (pH 4.5) without and with bismuth film.

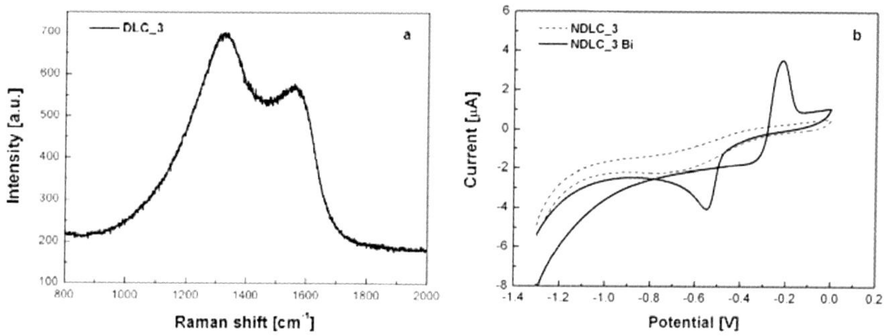

Figure 10. Raman spectra (a) and I-V curves (b) of nitrogen-doped DLC_3. I-V curves were recorded at scan rate of 50 mV/s in 0.1 mol/L acetate buffer (pH 4.5) without and with bismuth film.

In most cases, two intensive peaks were observed in the nitrogen-doped DLC spectra, referring to the so-called G-line at 1520 cm^{-1} and D-line at 1350 cm^{-1}. It is know that single crystal graphite displays only a single peak (G) and disordered or fine graphite crystallites show an additional disordered peak (D) [50]. The position, intensity and a width of the G and D peak are influenced by nitrogen partial pressure at sputter deposition. The conductance of the nitrogen-doped DLC films dependes mainly on nitrogen partial pressure during the film growth.

Electrochemical properties of the nitrogen-doped DLC film macroelectrodes were investigated by cyclic voltammetry in 0.1 mol/L acetate buffer solution (pH 4.5) in a potential range from –1.3 to 0 V at a scan rate of 50 mV/s (dashed lines in figures 8b – 10b). Remarkably different cyclic voltammograms were recorded after addition of 1×10^{-5} mol/L Bi(III) into the electrolyte, when the nitrogen-doped DLCs were preplated with a bismuth film for 60 s at –1.3 V (solid lines in figures 8b – 10b). The peak at approximatelly –215 mV comes from oxidation reaction of the bismuth film. Although, the area of each electrode as well as preplating conditions were the same, none or different peak heights from bismuth were observed. We assume that differences in voltammograms were caused by different conductances of nitrogen-doped DLCs.

For analyses of metals by ASV a low background current of electrodes is very important in a broad potential range so as to achieve the highest signal-to-noise ratio and, thereby, the highest sensitivity. In our case, no sharp increase of the reduction current due to hydrogen overvoltage was observed in the range of potentials from 0 to –1.2 V.

3.3. Silicon Support for Mercury Microelectrode Arrays

In anodic stripping voltammetry with mercury electrodes it is very important to select a supporting electrode material with solubility in mercury as low as possible [19]. The solubility of silicon in mercury is 0.001% [19], which is lower than for frequently used metals in this field such as Au, Ag or Pt, but higher than for Ir. For optimum electroplating of mercury, an electrode material with a great potential window is required. In figure 11, cyclic voltammograms are shown in acetate buffer (pH 4) recorded for different p- and n-type silicon wafers. Intensive hydrogen gas evolution started at approximately –1 V for Si 9218 (n-type) and –1.2 V for Si 9105 (n-type) and Si 9610 (p-type).

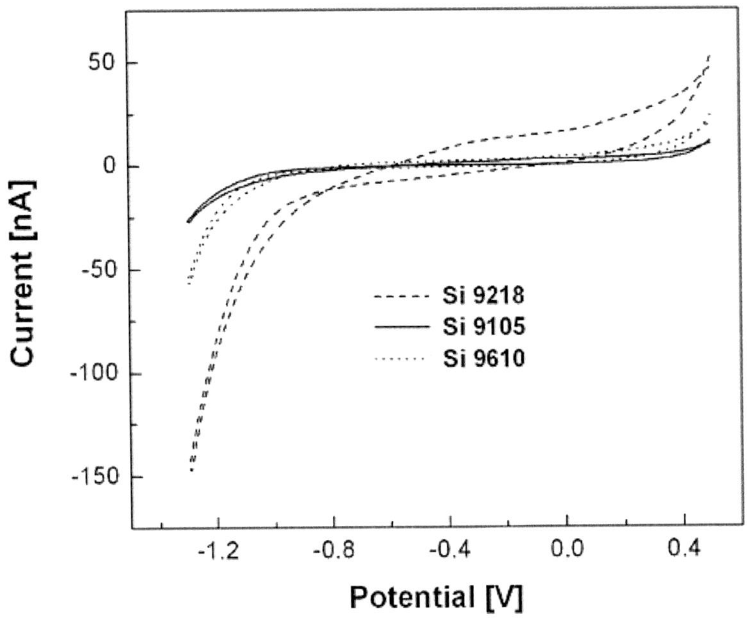

Figure 11. I-V curves of different silicon microelectrode arrays at scan rate of 50 mV/s in 0.1 mol/L acetate buffer.

The quality of a mercury plated microelectrode greatly influences the sensitivity of anodic stripping voltammetry. A lot of parameters have an influence on the formation of the mercury deposit including plating solution, plating potential and time, etc. In our preliminary experiments, plating solutions consisting of 0.1 mol/L acetate buffer (pH 4) with different contents of Hg(II) (1×10^{-4}, 1×10^{-3} and 1×10^{-2} mol/L) were used. We obtained the best results in the solution with concentration of Hg(II) 1×10^{-3} mol/L. Almost no mercury was deposited in the solution with 1×10^{-4} mol/L concentration.

As far as the plating potential is concerned, a typical value of –1.3 V vs. Ag/AgCl/Cl^- for deposition of mercury on metal or carbon type supports it was insufficient for a silicon substrate. We had to combine two deposition steps, the first step at –2 V (for 2 min) followed by the second one at –1.3 V for 2 min. High hydrogen gas evolution at –2 V probably disturbed the diffusion layers at local places of the silicon surface and thereby it improved the condition for more homogeneous growth of mercury at –1.3 V. We have to note that any attempt to cover the silicon microdisc surface by a true mercury film at a constant deposition potential was not successful. Only typical mercury hemispheres were observed. A detail of mercury hemispheres grown on the silicon microelectrode array is shown in figure 12.

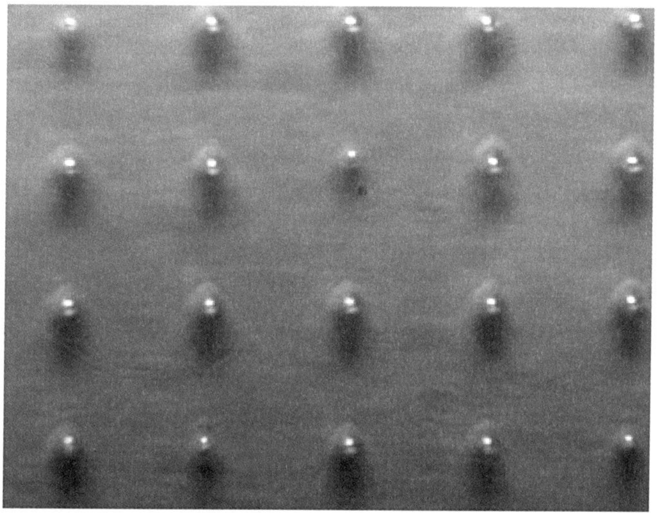

Figure 12. Detail of silicon (p-type) microelectrode array preplated by mercury hemispheres.

When we started the deposition from –1.3 V, only small irregular mercury droplets were formed on silicon discs (figure 13a). It was not possible to cover the silicon surface completely even if the deposition time was prolonged.

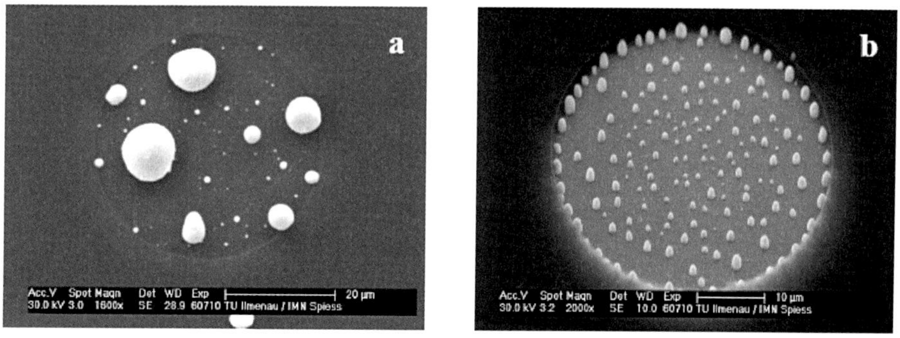

Figure 13. Plating of mercury on silicon surface in 0.1 mol/L acetate buffer (pH 4) without (a) and with 0.2% BHF (b).

We have found a strong influence of the electrolyte on mercury plating. We have investigated different electrolytes such as 0.1 mol/L acetate buffer (pH 4), 0.1 mol/L KNO_3 (acidified by HNO_3 to pH 2.5) and 0.1 mol/L $HClO_4$. We also investigated the effect of a small quantity of buffered hydrofluoric acid (BHF) (0.2%) added to each electrolyte (figure

13b) in order to keep the silicon surface without oxide all the time. The values of the mercury reoxidation current peak in dependence on electrolyte with and without BHF addition are summarized in table 2. One can see a strong effect of BHF in electrolyte on the mercury reoxidation current for p-type Si (Si 9610). We assumed that the presence of the BHF in electrolytes caused that a thin native oxide on silicon was etched but experimental results with n-type Si (Si 9105) are in contradiction with this.

Table 2. Influence of the electrolytes on mercury plating conditions at concentration of Hg(II) 1×10^{-3} mol/L.

Electrolyte (0.1 mol/L)	Hg reoxidation peak [μA]	
	p-Si	n-Si
Acetate	7	12
Acetate + BHF	16	11
KNO_3	75	50
KNO_3 + BHF	220	37
$HClO_4$	40	15
$HClO_4$ + BHF	600	11

We can conclude that the best mercury hemispheres were formed on p-type of silicon in 0.1 mol/L $HClO_4$. Hg reoxidation peak appeared at +380 mV. In the case of n-type of silicon, the deposit was unstable; most of mercury hemispheres or spheres were moved away from the microelectrodes towards the insulator layer. Usually only a few microelectrodes were covered by mercury. Moreover, no effect of BHF addition on mercury plating was observed.

A mercury-plated silicon microelectrode array prepared in a solution of 1×10^{-3} mol/L Hg(II) in 0.1 mol/L $HClO_4$ with BHF (0.2%) was used in *ex-situ* determination of Cu(II), Pb(II) and Cd(II). In figure 14, the current responses of heavy metals with concentration 3.3×10^{-8} mol/L of each are ilustrated using by SWV.

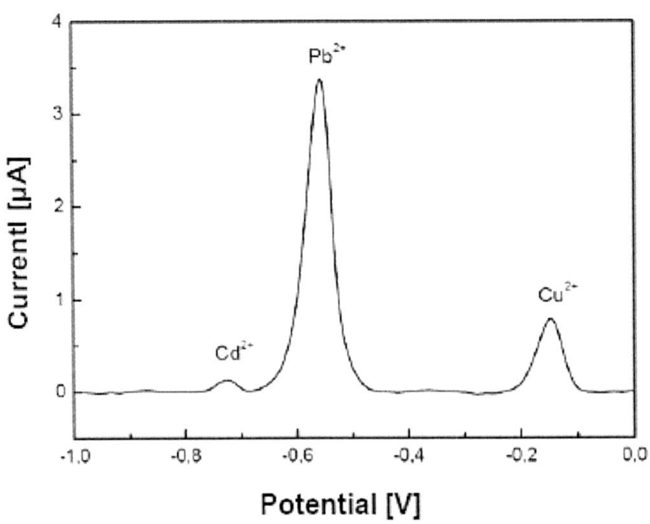

Figure 14. Square-wave stripping voltammogram (E_{step} = 5 mV, E_{pulse} = 25 mV and frequency 25 Hz) obtained with silicon microelectrode array in 0.1 mol/L KNO_3; preconcentrated at -1 V for 120 s.

3.4. Nitrogen-Doped DLC Support for Mercury Microelectrode Arrays

In our preliminary experiments, plating solutions consisting of 0.1 mol/L KNO_3 with different contents of mercuric ions were used. We obtained the highest response in the solution with concentration of Hg(II) 1×10^{-3} mol/L (figure 15).

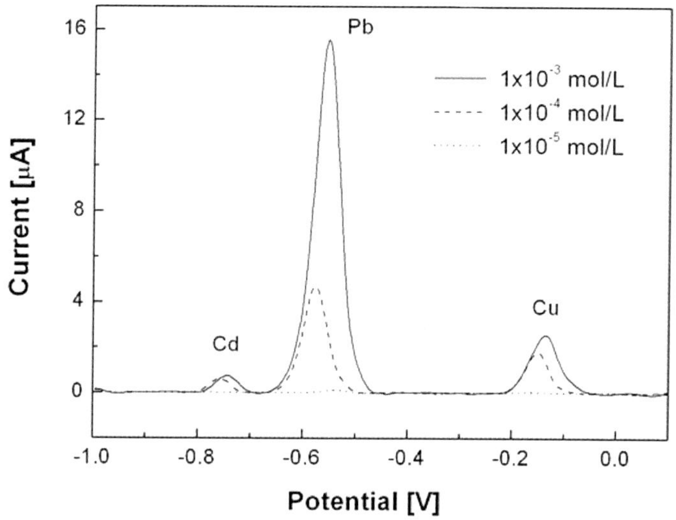

Figure 15. Differential pulse stripping voltammograms (E_{step} = 2 mV, E_{pulse} = 50 mV, 25 ms pulse and 30 mV scan rate). Dependence of current response of nitrogen-doped DLC-Hg microelectrode array on mercuric ion content in plating solution; plating time: 5 min, concentration of metals: 5×10^{-8} mol/L.

In ASV it is very important to choose an optimum preconcentration time. In general, the longer the time, the lower the limit of detection. The dependence of the stripping current on the preconcentration time is given in figure 16.

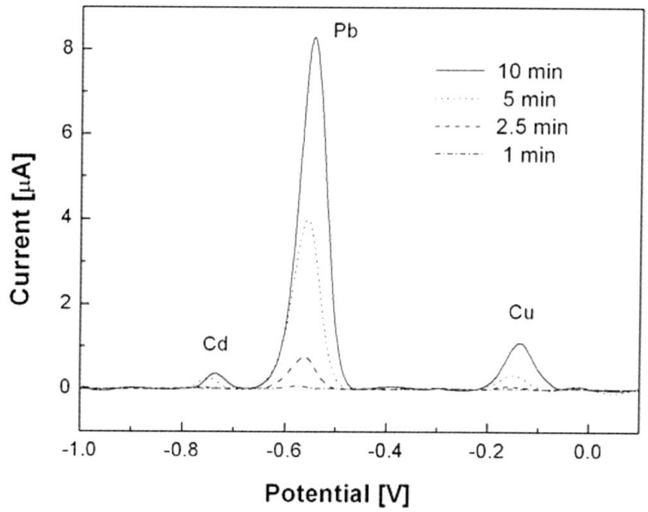

Figure 16. Differential pulse voltammograms (E_{step} = 2 mV, E_{pulse} = 50 mV, 25 ms pulse and 30 mV scan rate). Dependence of current response of nitrogen-doped DLC-Hg microelectrode array on preconcentration time in 0.1 mol/L KNO_3, concentration of metals: 5×10^{-8} mol/L.

A mercury-plated nitrogen-doped DLC microelectrode array for *in-situ* simultaneous determination Cu, Pb and Cd was prepared by co-deposition of Hg(II), Cu(II), Pb(II) and Cd(II) at –0.8 V in a solution of 1×10^{-4} mol/L $Hg(NO_3)_2$ in 0.1 mol/L KNO_3 with different contents of heavy metals from 1.6×10^{-8} to 8.3×10^{-8} mol/L for 5 min preconcentration time. The dependences of current responses on heavy metals concentration, as well as calibration curves inside are shown in figure 17.

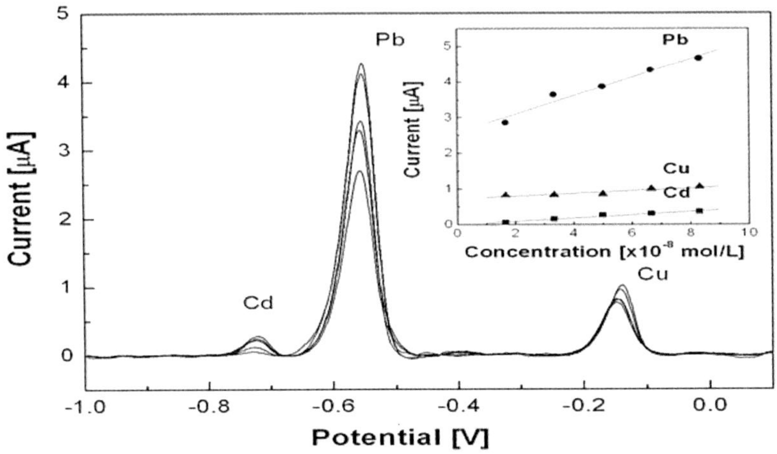

Figure 17. Differential pulse stripping voltammograms (E_{step} = 2 mV, E_{pulse} = 50 mV, 25 ms pulse and 30 mV scan rate). Dependence of current response of nitrogen-doped DLC-Hg microelectrode array on concentration of metals with calibration curve inside; preconcentration time: 5 min.

Different sensitivities are observed of the mercury-plated microelectrode array to each metal. The highest sensitivity is to Pb. Different sensitivities are probably given by unique properties of the electrodes. This is supported by the fact that all variations of sensitivity were reported for this group of metals [2, 51-53].

The influence of strongly complexing agents like thiocyanate (SCN^-) on the stripping current of ASV has already been reported [54]. Addition of a small quantity of SCN^- into the mercury plating solution promoted the formation of the mercury film, improved apparently the sensitivity and repeatability at determination of metals. It was concluded that 5×10^{-3} mol/L SCN^- is the upper concentration limit [54, 55] because for a higher concentration the mercury surface passivates. In figure 18, the dependence of the stripping current on the addition of thiocyanate into the solution is shown. The current peak at +0.1 V comes from oxidation of mercury to complex $Hg(SCN)_2$ [8]. Oxidation of mercury to a mercuric ion occurs at a more positive potential (+0.4 V) than that for the formation of the thiocyanate complex. Under our experimental conditions, the upper concentration limit of (SCN^-) was found to be 2×10^{-3} mol/L. In our case, for higher concentrations the stripping currents decreased. Anyway, the stripping current for Pb was increased 10 times in comparison with that in the absence of thiocyanate.

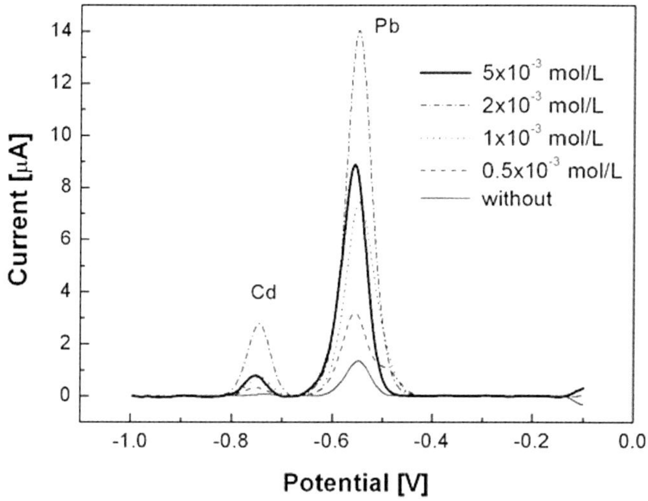

Figure 18. Differential pulse stripping voltammograms (E_{step} = 2 mV, E_{pulse} = 50 mV, 25 ms pulse and 30 mV scan rate). Influence of thiocyanate addition on current response; preconcentration time: 2 min, concentration of metals 5×10^{-8} mol/L.

There are reports on the use of various types of bare carbon (glassy carbon, nitrogen-doped DLC, BDD) for determination of heavy metals by ASV [22, 39, 44]. Deposition of metals ions results in a film of atoms which can occur as a monolayer or multilayer on the carbon surface [39]. Results of the determination of the Pb(II) ions in 0.1 mol/L KNO_3 by DP ASV (figure 19) confirmed our prediction that the sensitivity will be strongly reduced due to the absence of mercury with its accumulating capability. We obtained much lower current responses in comparison with the results obtained on a mercury-plated nitrogen-doped DLC microelectrode array. Almost no current signal was recorded for concentration of Pb(II) 5×10^{-6} mol/L. Similar results, as regards sensitivity in Pb(II) determination on bare carbon, were referred [22, 39, 44].

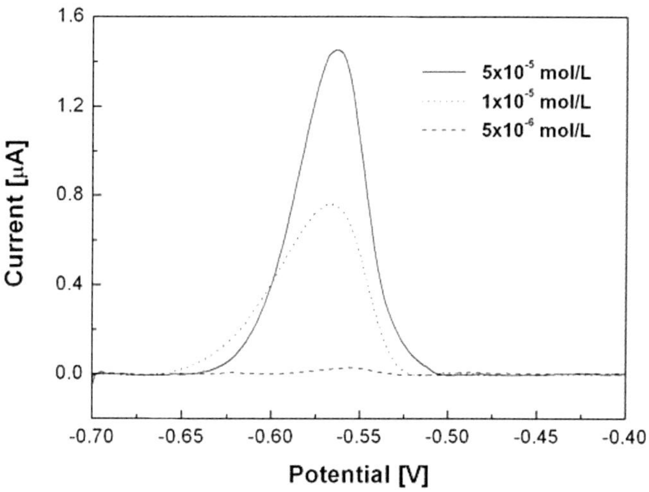

Figure 19. Differential pulse stripping voltammograms (E_{step} = 2 mV, E_{pulse} = 50 mV, 25 ms pulse and 30 mV scan rate). Dependence of current response of bare nitrogen-doped DLC microelectrode array on Pb(II) concentration in 0.1 mol/L KNO_3.

3.5. Nitrogen-Doped DLC Support for Bismuth Film Microelectrode Arrays

Excellent performance of ASV with the bismuth film electrodes in determination of Pb(II), Cd(II) and Zn(II) on various carbon substrates have been already demonstrated [4, 12, 17, 56, 57].

Bismuth-plated nitrogen-doped DLC microelectrode arrays for *in-situ* determination of Pb(II) were prepared by co-deposition of Bi(III) and Pb(II) at –1.3 V vs. Ag/AgCl/Cl^- in a plating solution of $Bi(NO_3)_3$ in a supporting electrolyte with different contents of Pb(II). Stripping currents from reoxidation of Pb were recorded by differential pulse voltammetry. The influence of the concentration of the plating solution on bismuth film formation for *in-situ* technique was investigated in the range from 2.5×10^{-6} to 1×10^{-5} mol/L Bi(III) in 0.1 mol/L acidified KNO_3 (pH 2) (figure 20). For the ASV experiments, concentrations of 2.5×10^{-6} or 5×10^{-6} mol/L were chosen because the stripping current was not increased at higher concentrations than 5×10^{-6} mol/L and, moreover, the bismuth peak shape was deformed.

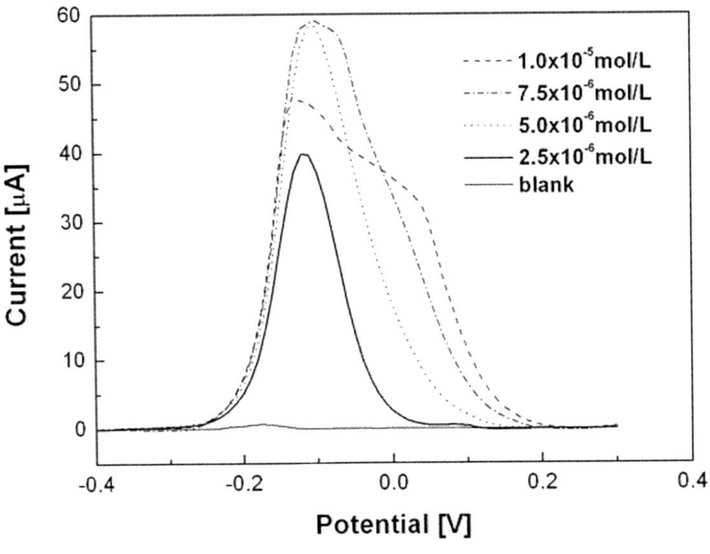

Figure 20. Differential pulse stripping voltammograms (E_{step} = 2 mV, E_{pulse} = 50 mV, 25 ms pulse and 30 mV scan rate). Dependence of current response of nitrogen-doped DLC microelectrode array on Bi(III) concentration in 0.1 mol/L KNO_3 (pH 2); preconcentration time: 120 s.

We have found a minimum effect of different supporting electrolytes (0.1 mol/L acetate (pH 4), nitrate (acidified by HNO_3) or perchlorate solutions) in bismuth plating procedure. No remarkable differences as regards the bismuth film structure and their cyclic voltammograms were observed. The dependence of voltammetric response of bismuth-plated nitrogen-doped DLC array on Pb(II) concentration is shown in figure 21. The stripping current was linear in the range from 2×10^{-8} to 1.2×10^{-7} mol/L concentration of Pb(II) with sensitivity of 2.5µAL/µmol.

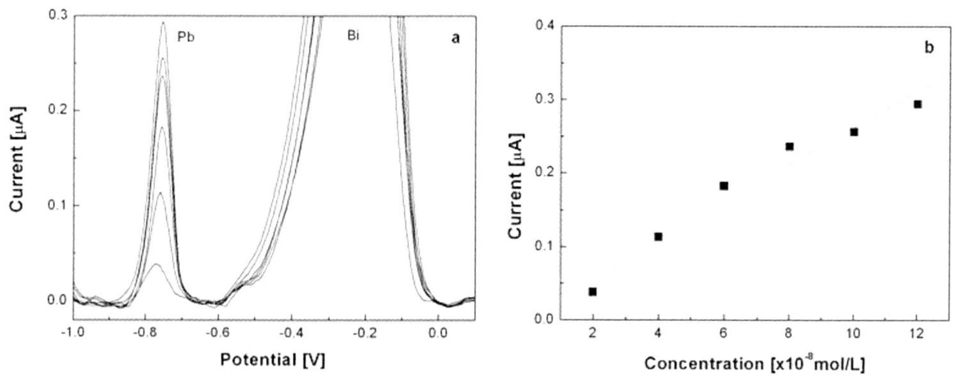

Figure 21. Differential pulse stripping voltammograms (E_{step} = 2 mV, E_{pulse} = 50 mV, 25 ms pulse and 30 mV scan rate). Dependence of current response of nitrogen-doped DLC-Bi microelectrode array on Pb(II) concentration in 0.1 mol/L KNO_3 (pH 2); preconcentration time: 120 s (a) and calibration curve (b).

Important parameters, stability, reproducibility and reliability of nitrogen-doped DLC film supports, multiply used in ASV have been evaluated. From Raman spectra we found no changes in the structure, however, the samples were used 20-times in ASV process (deposition - measuring - cleaning).

3.6. Pencil-Lead Graphite Support for Bismuth Film Electrodes

The behavior of the single disk graphite electrode was characterized by cyclic voltammetry in an 0.1 mol/L acetate buffer (pH 4.5) solution. In a potential range from –1.3 to –0.4 V at a scan rate of 50 mV/s, the effect of the bismuth film was investigated. Negligible difference between voltammograms (with or without bismuth film) was observed despite the fact that various concentration of Bi(III) were used (1×10^{-6} to 1×10^{-5} mol/L).

The disc graphite bismuth film electrode (BiFE) was used for *in-situ* simultaneous determination of Pb(II), Cd(II) and Zn(II) by square wave voltammetry in an aqueous solution.

The SWV parameters which affect the response of the sensor, generally, are SW step increment, SW pulse height and SW frequency [11, 14]. These parameters were investigated using a solution with Pb(II), Cd(II) and Zn(II) concentration of 1×10^{-8} mol/L of each in an 0.1 mol/L acetate buffer.

The effect of the step increment was investigated in the range from 2 to 30 mV (figure. 22). The best stripping response was obtained at 12 mV step increment for Pb(II) and Cd(II), and at 25 mV for Zn(II). For all metals the peak positions as well as the peak widths were independent of the step increment.

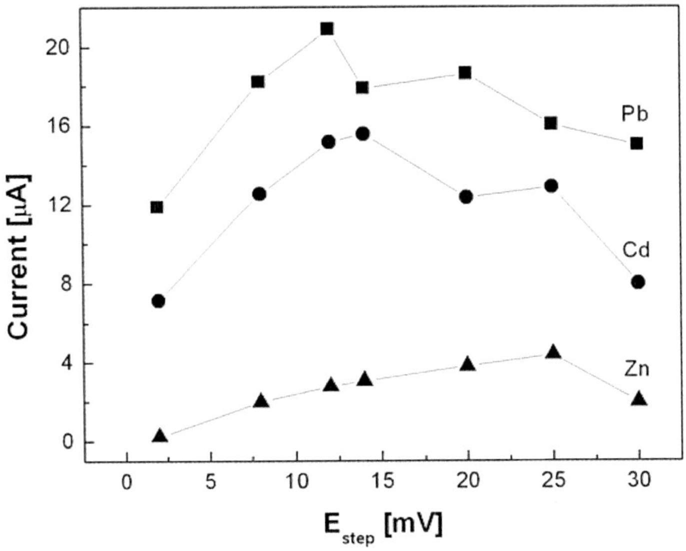

Figure 22. Dependence of current response on step increment of SWV in 0.1 mol/L acetate buffer; concentration of each of metals: 3×10^{-7} mol/L, Bi(III): 1×10^{-6} mol/L, preconcentration time: 60 s.

The SW pulse height was investigated in the range from 20 to 120 mV (figure 23). When the pulse height increased, the stripping response gradually increased. The peak positions were shifted to the cathode direction by about 20 to 30 mV per pulse height increment (20 mV) for all metals. The peak widths were not influenced by the pulse height.

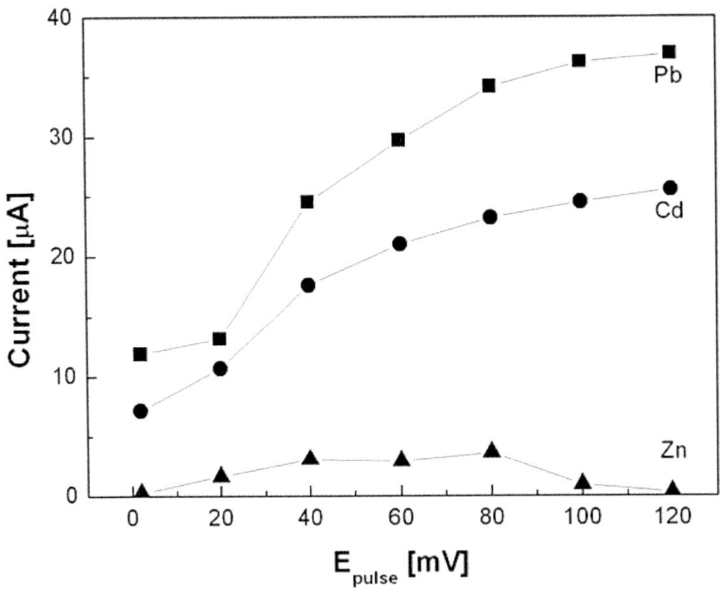

Figure 23. Dependence of current response on pulse height of SWV in 0.1 mol/L acetate buffer; concentration of each of metals: 3×10^{-7} mol/L, Bi(III): 1×10^{-6} mol/L, preconcentration time: 60 s.

The SW frequency had a significant effect on the stripping response (figure 24). The peak heights for Pb(II) and Cd(II) increase up to 120 Hz. Over this frequency, the stripping responses decreased. A negligible effect of frequency on the response was found for Zn(II). The peak positions for all metals were shifted to the cathode direction by about 10 mV per frequency increment (20 Hz). In accordance with the observed results, a step increment of 12 mV, pulse height 80 mV and frequency 120 Hz were chosen for SWV experiments with the pencil-lead graphite electrode.

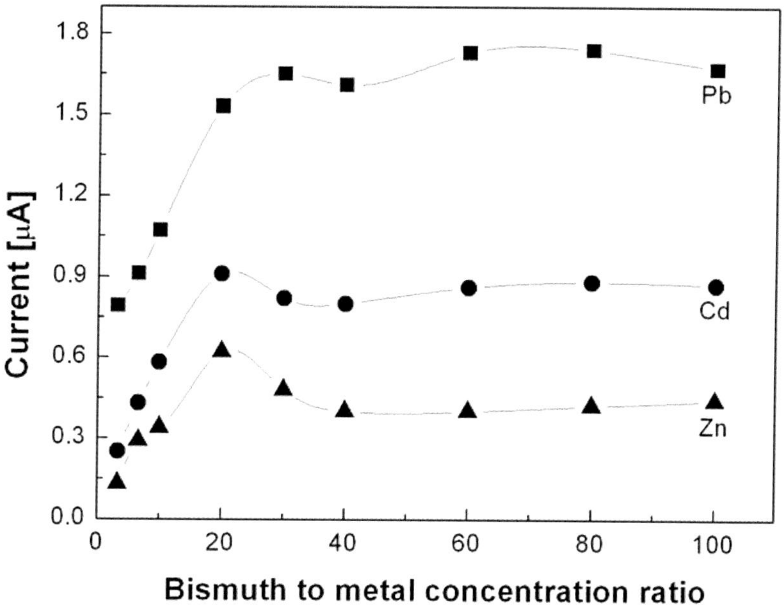

Figure 24. Dependence of current response on frequency of SWV in 0.1 mol/L acetate buffer; concentration of each of metals: 3×10^{-7} mol/L, Bi(III): 1×10^{-6} mol/L, preconcentration time: 60 s.

The concentration of Bi(III) used for *in-situ* formation of the bismuth film on graphite electrodes is known to influence the height of the stripping peaks of the target metals [11, 14, 18]. Stripping voltammograms with the pencil-lead graphite electrode for the solution with 1×10^{-8} mol/L concentration of each of Pb(II), Cd(II) and Zn(II) in 0.1 mol/L acetate buffer are shown in figure 25. They were recorded at different contents of Bi(III) from 3.3×10^{-8} (bismuth-to-metals ratio of 3.3:1) to 1×10^{-6} mol/L (bismuth-to-metals ratio of 100:1). In figure 26 there are shown the dependences of the stripping current on the concentration ratio of Bi(III) to metals in the solution. As the Bi(III) concentration increased, all the peaks increased in height up to a ratio of 20:1. After a slight decrease for Cd(II) and Zn(II) peaks, the current responses of all metals became independent of additional increasing of Bi(III).

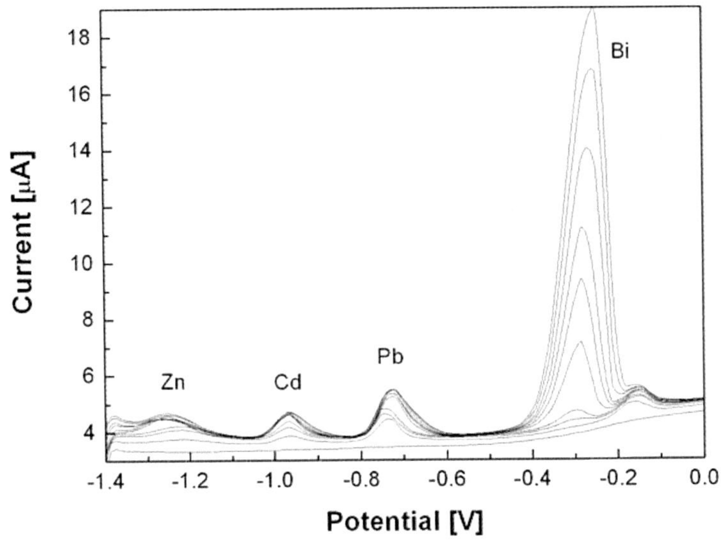

Figure 25. Square-wave stripping voltammograms (E_{step} = 12 mV, E_{pulse} = 80 mV and frequency 120 Hz). Dependence of current response of single graphite BiFE on concentration of Bi(III) (3.3×10^{-8} to 1×10^{-6} mol/L) in 0.1 mol/L acetate buffer (pH 4); concentration of each of metals: 1×10^{-8} mol/L, preconcentration time: 120 s.

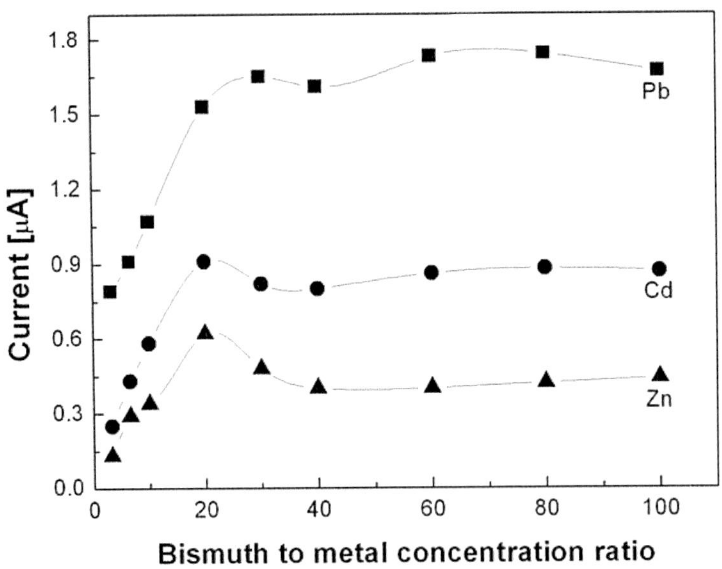

Figure 26. Dependence of current response of single graphite BiFE on concentration ratio of Bi(III) to metals in 0.1 mol/L acetate buffer (pH 4).

Generally, the concentration of Bi(III) in the solution controls the thickness of the bismuth film. Since bismuth-metal alloying occurs mainly on the surface, without any diffusion of the metals into the bismuth film, an increase of the bismuth film thickness should have a little effect on metal stripping. The dependence of the stripping response on Bi(III) concentration at low bismuth-to-metals ratios (< 20) and low target metal concentrations

(1×10^{-8} mol/L) is probably a special feature of *in-situ* prepared electrodes. We suppose that under these conditions the surface of graphite electrode is not covered by bismuth completely.

To improve the electroanalytical performance (especially the sensitivity) of the *in-situ* bismuth electrode, the preconcentration time should be optimized for the stripping responses not to be influenced by saturation. The effect of the preconcentration time was studied with the pencil-lead graphite electrode in the range from 60 to 600 s for a solution with concentration 1×10^{-7} mol/L of each of target metals and 1×10^{-6} mol/L Bi(III) in 0.1 mol/L acetate buffer. The peak heights increased linearly with time up to 360 s (figure 27). At a longer preconcentration time the peak heights deviate from linearity due to surface saturation.

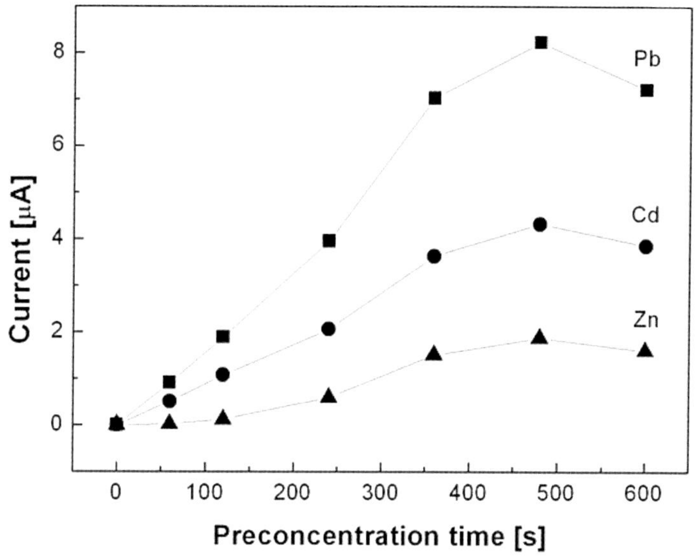

Figure 27. Dependence of current response of single graphite BiFE on preconcentration time in 0.1 mol/L acetate buffer (pH 4); concentration of each of metals: 1×10^{-7} mol/L, Bi(III): 1×10^{-6} mol/L.

The ability of using the BiFE for simultaneous determination of Pb(II), Cd(II) and Zn(II) is illustrated with optimized SWV parameters in figure 28. All the peaks are well resolved and increase linearly with the metal concentration in the range from 1×10^{-8} to 1.2×10^{-7} mol/L. The resulting calibration plots were used for estimating the detection limits.

The calibration data are given in table 3. On the basis of criteria for calculation of the detection limit (Y intercept + 3×SD) the detection limits were estimated as 2.4×10^{-9} mol/L for Pb(II), 2.9×10^{-9} mol/L for Cd(II) and 1.2×10^{-8} mol/L for Zn(II). In addition, one of the features of BiFE is good precision. A series of 10 repetitive measurements of a solution with concentration 1×10^{-8} mol/L of each of Pb(II), Cd(II) and Zn(II) and 2×10^{-7} mol/L Bi(III) at 120 s preconcentration time gave highly reproducible stripping peaks with relative standard deviations 4.5 % for Pb(II), 3.4 % for Cd(II) and 2.7 % for Zn(II). Good precision is attributed to the reproducible bismuth film (2.8 %).

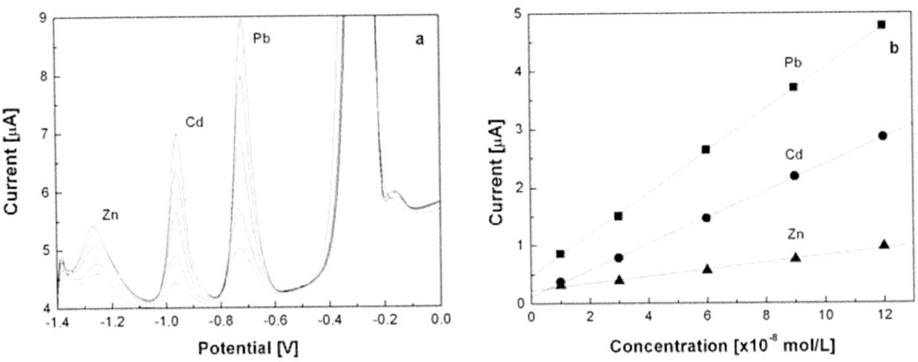

Figure 28. Square-wave stripping voltammograms (E_{step} = 12 mV, E_{pulse} = 80 mV and frequency 120 Hz). Dependence of current response of single graphite BiFE on concentration of metals from 1×10^{-8} to 1.2×10^{-7} mol/L of each; Bi(III): 2×10^{-7} mol/L, preconcentration time: 120 s (a) and calibration curves (b).

Table 3. Calibration data for determination of Pb, Cd and Zn at single graphite BiFE and Bi nitrogen-doped DLC microelectrode array

Ion	Sensitivity [μAL/μmol]	Intercept [μA]	R	SD	Electrode
Pb	35.9	0.465	0.9998	0.029	graphite
Cd	22.7	0.118	0.9998	0.022	graphite
Zn	6.0	0.224	0.9967	0.025	graphite
Pb	2.5	0.005	0.9870	0.019	nitrogen-doped DLC

4. CONCLUSION

In this work, results of heavy metals determination by anodic stripping voltammetry with mercury hemispherical and bismuth film microelectrode arrays are summarized. In this field traditionally used mercury could be replaced by attractive bismuth.

Different supports, high conductive silicon and nitrogen-doped DLC thin films on silicon were used for i*n-situ* or *ex-situ* determination with mercury hemispherical microelectrodes. In the case of bismuth, a single pencil-lead graphite electrode and nitrogen-doped DLC films on silicon were used as a support.

A microelectrode arrays consisting of 50 625 microdiscs with 2 μm in diameter and interelectrode distances of 20 μm were fabricated photolithographically using a positive resist on silicon or nitrogen-doped DLC substrates. It was proved that geometry of this array was sufficient to maintain the advantageous characteristics of microelectrodes.

High conductive (0.008 - 0.024 Ωcm) p- or n-type silicon wafers have been investigated as a support for mercury electroplating. The influence of many parameters, including the concentration of the plating solution, plating potential, kind and concentration of the supporting electrolyte on mercury plating has been studied. We have investigated the influence of different electrolytes (acetate buffer, KNO_3 and $HClO_4$) as well as the effect of a small quantity of BHF (0.2%) added to each electrolyte in order to keep the silicon surface without oxide on mercury plating. The best mercury hemispheres were formed on p-type of silicon in 0.1 mol/L $HClO_4$. Every attempt to cover the silicon microdisc surface by a true mercury film at a constant deposition potential was not successful. The stripping current response of trace heavy metals (Cu(II), Pb(II) and Cd(II) in concentration 3.3×10^{-8} mol/L of each) on *ex-situ* mercury-plated silicon microelectrode array has been demonstrated by SWASV.

The mercury-plated as well as bare nitrogen-doped DLC microelectrode arrays were investigated for application in heavy metals (Cu, Pb and Cd) *in-situ* determination by DPASV. The current responses of all metals were linear with concentration in the range from 1.6×10^{-8} to 8.3×10^{-8} mol/L. The stripping current responses of the bare microelectrode array reached under the same conditions were much lower in comparison with the mercury-plated array. A strong effect was found of adding a small amount of thiocyanate (2×10^{-3} mol/L) into the plating solution on the stripping current responses of metals.

The bismuth-plated nitrogen-doped DLC electrode array was used for *in-situ* determination of Pb(II) by DPV in an aqueous solution. The influence of the plating solution and supporting electrolyte on the stripping responses was investigated. We found that the stripping current for this array was linear in the range of Pb(II) concentration from 2×10^{-8} to 1.2×10^{-7} mol/L. The sensitivity 2.5 μAL/μmol and estimated detection limit 2.2×10^{-8} mol/L were evaluated from the calibration plot. It was found that the analytical performance of the bismuth film microelectrode array was influenced by the properties of nitrogen-doped DLC.

The disc graphite BiFE was used for simultaneous *in-situ* determination of Pb(II), Cd(II) and Zn(II) by SWV in an aqueous solution. The influence of SWV parameters, Bi(III) concentration and preconcentration time was investigated. We found the optimum bismuth-to-metal concentration ratio to be 20. The dependence of the stripping responses on the concentration of target metals was linear in the range from 1×10^{-8} to 1.2×10^{-7} mol/L. Sensitivities of 35.9 μAL/μmol for Pb(II), 22.7 μAL/μmol for Cd(II) and 6.0 μAL/μmol for Zn(II) were achieved by the graphite BiFE. The detection limit was estimated as 2.4×10^{-9} mol/L for Pb(II), 2.9×10^{-9} mol/L for Cd(II) and 1.2×10^{-8} mol/L for Zn(II).

ACKNOWLEDGEMENTS

This work was supported by VEGA grant 1/3095/06 of the Slovak Grand Agency.

REFERENCES

[1] Feeney, R.; Kounaves, S. P. *Electroanalysis* 2000, *12*, 677-684.
[2] Xie, X.; Stuben, D.; Berner, Z.; Albers, J.; Hintsche, R.; Jentzen, E. *Sens. Actuators* 2004, *B 97*, 168-173.
[3] Hutton, E. A.; Hocevar, S. B.; Ogorevc, B. *Anal. Chim. Acta* 2005, *537,* 285-292.
[4] Economou, A. Trends in Analytical Chemistry 2005, 24, 334-340.
[5] Svancara, I.; Vytras, K. *Chem. Listy* 2006, *100,* 90-113.
[6] Tercier, M. L.; Buffle, J. *Electroanalysis* 1993, *5,* 187-200.

[7] Brett, C. M. A.; Lima, J. F. C.; Garcia, B. Q. *Analyst* 1994, *119,* 1229-1233.
[8] Daniele, S.; Bragato, C.; Baldo, M. A. *Anal. Chim. Acta* 1997, *346,* 145- 156.
[9] Nyholm, L.; Bjorefors, F. *Anal. Chim. Acta* 1996, *327,* 211-222.
[10] Kounaves, S. P.; Buffle, J. *J.Electroanal.Chem.* 1987, *216,* 53-69.
[11] Kefala, G.; Economou, A.; Voulgaropoulos, A.; Sofoniou, M. *Talanta* 2003, *61, 603-610.*
[12] Wang, J.; Lu, J.; Kirgoz, U. A.; Hocevar, S. B.; Ogorevc, B. *Anal. Chim. Acta* 2001, *434,* 29-34.
[13] Wang, J.; Kirgoz, U. A.; Lu, J. Electrochem. Comm. 2001, 3, 703-706.
[14] Charalambous, A.; Economou, A. *Anal. Chim. Acta* 2005, *547,* 53-58.
[15] Baldrianova, L.; Svancara, I.; Vlcek, M.; Economou, A.; Sotiropoulos, S. *Electrochim. Acta* 2006, *52,* 481-490.
[16] Baldrianova, L.; Svancara, I.; Economou, A.; Sotiropoulos, S. *Anal. Chim. Acta* 2006, *580,* 24-31.
[17] Wang, J.; Lu, J.; Hocevar, S. B.; Farias, P. A. M. *Anal. Chem.* 2000, *72,* 3218-3222.
[18] Demetriades, D.; Economou, A.; Voulgaropoulos, A. *Anal. Chim. Acta* 2004, *519,* 167-172.
[19] Kounaves, S. P. *Thesis No. 2151,* 1985.
[20] Kounaves, S. P.; Deng, W. *J. Electroanal. Chem.* 1991, *301,* 77-85.
[21] Kounaves, S. P.; Deng, W.; Hallock, P. R.; Kovacs, G. T. A.; Storment, W. *Anal. Chem.* 1994, *66,* 418- 423.
[22] Liu, L.X.; Liu, E. *Surface and Coatings Technology* 2005, *198,* 189-193.
[23] Butsuda, T.; Anita, V.; Saito, N.; Takai, O. *17th International Symposium on Plasma Chemistry,* August 7th – 12th, 2005, Toronto, Canada, 65-66.
[24] Yan, X.; Xu, T.; Chen, G.; Yang, S.; Liu, H. *Applied Surface Science* 2004, *236,* 328-335.
[25] Compton, R. G.; Foord, J. S.; Marken, F. *Electroanalysis* 2003, *15,* 1349-1363.
[26] Rehacek, V.; Hotovy, I.; Spiess, L.; Gubisch, M. *51st Internationales Wissenschaftliches Kolloquium, Technische Universitat Ilmenau,* September 11-15, 2006, Ilmenau, Germany
[27] Rehacek, V.; Hotovy, I.; Vojs, M. *Journal of Physics: Conference Ser*ies, 2007, *61,* 982-986.
[28] Rehacek, V.; Hotovy, I.; Vojs, M.; Mika, F. *Microsyst. Technol.* DOI 10.1007/s00542-007-0432-z
[29] Rehacek, V.; Hotovy, I.; Vojs, M. *Sens. Actuators B* 2007, *127*, 193-197.
[30] Wang, J. *Analytical electrochemistry;* 2nd Edi.; John Wiley and Sons: New York, US, 2000; 75pp.
[31] Aoki, K. *Electroanalysis* 1993, *5,* 627-639.
[32] Heize, J. Angew. Chem. Int. Ed. Engl., 1993, 32, 1268-1288.
[33] Kounaves, S. P.; Deng, W. *Anal. Chem.* 1993, *65,* 375-379.
[34] Stulik, K.; Amatore, C.; Holub, K.; Mareceek, Kutner, W. *Pure Apl. Chem.* 2000, *72,* 1483-1492.
[35] Aoki, K.; Osteryoung, J. *J. Electroanal. Chem.* 1981, *122,* 19-35.
[36] Belmont, C.; Tercier, M. L.; Buffle, J.; Fiaccabrino, G. C.; Koudelka-Hep, M. *Anal. Chim. Acta.* 1996, *329,* 203-214.
[37] Forster, R.J. *Chemical Society Reviews* 1994, 289-297.

[38] Feeney, R.; Kounaves, S.P. *Electroanalysis.* 2000, *12,* 678-684.
[39] Honeychurch, K.; Hart, J. P. *Trends in Analytical Chemistry,* 2003, *22,* 456-469.
[40] Kounaves, S. P.; Deng, W.; Hallock, P.R.; Kovacs, G. T. A.; Storment, C.W. *Anal. Chem.* 1994, *66,* 418-423.
[41] Wu, H. P. *Anal. Chem.* 1993, *65,* 1643-1646.
[42] Amatore, C. *Physical electrochemistry* 1995, Marcel Dekker, New York, 131-208.
[43] Kounaves, S. P.; Deng, W. *J. Electroanal. Chem.* 1991, *301,* 77-85.
[44] Manivannan, A.; Kawasaki, R.; Tryk, D. A.; Fujishima, A. *Electrochimica Acta* 2004, *49,* 3313-3318.
[45] Sonthalia, P.; McGaw, E.; Show, Y.; Swain, G.M. *Anal Chim. Acta.* 2004, *522,* 35-44.
[46] Banks, C. E.; Hyde, M. E.; Tomcik, P.; Jacobs, R.; Compton, R. G. *Talanta,* 2004, *62,* 279-286.
[47] Foord, J. S.; Hao, W.; Eaton, K. *Phys. Stat. Sol.* 2005, *202,* 2116-2121.
[48] Compton, R. G.; Foord, J. S.; Marken, F. *Electroanalysis.* 2003, *15,* 1349-1363.
[49] Yan, X.; Xu, T.; Chen, G.; Yang, S.; Liu, H. *Appl. Surf. Science.* 2004, *236,* 328-335.
[50] Lagrini, A.; Charvet, S.; Benlahsen, M.; Cachet, H.; Deslouis, C. Diamond and Related Materials 2007, 16, 1378- 1382.
[51] Uhlik, A.; Paeschke, M.; Schnakenberg, U.; Hintsche, R.; Diederich, H. J.; Scholz, F. *Sens. Actuators* 1995, *B 24-25*, 899-903.
[52] Silva, P. R. M.; El Khakani, M. A.; Chaker, M.; Dufresne, A.; Courchesne, F. *Sens. Actuators* 2001, *B 76,* 250-257.
[53] Silva, P. R. M.; El Khakani, M. A.; Chaker, M.; Champagne, G. Y.; Chevalet, J.; Gastonguay, L.; Lacasse, R.; Ladouceur, M. *Anal. Chim. Acta* 1999, *385*, 249-255.
[54] Monterroso, S. C. C.; Carapuca, H. M.; Simao, J. E. J.; Duarte, A. C. *Anal. Chim. Acta* 2004, *503*, 203-212.
[55] Javorski, A.; Stojek, Z.; Osteryoung, J. G. *J.Electroanal.Chem.* 2003, *558,* 141-153.
[56] Wang, J. *Electroanalysis* 2005, *17,* 1341-1346.
[57] Svancara, I.; Baldrianova, L.; Tesarova, E.; Hocevar, S. B.; Elsuccary, S. A. A.; Economou, A.; Sotiropoulos, S.; Ogorevc, B.; Vytras, K. *Electroanalysis* 2006, *18,* 177-185.

INDEX

3

A

C

D

E

F

G

H

J

K

L

M

N

O

P

Q

R

S

T

U

V

W

X

Y

Z